A Guide to Plants of Yellowstone
and Grand Teton National Parks

A Guide to Plants of YELLOWSTONE & GRAND TETON *National Parks*

NATURAL HISTORY NOTES AND USES

by Ray S. Vizgirdas

Illustrations by Edna M. Rey-Vizgirdas

THE UNIVERSITY OF UTAH PRESS
Salt Lake City

The Defiance House Man colophon is a registered trademark of the University of Utah Press. It is based upon a four-foot-tall, Ancient Puebloan pictograph (late PIII) near Glen Canyon, Utah.

11 10 09 08 07 1 2 3 4 5

Library of Congress Cataloging-in-Publication Data

Vizgirdas, Ray S., 1960–

A guide to plants of Yellowstone and Grand Teton national parks / by Ray S. Vizgirdas, illustrations by Edna M. Rey-Vizgirdas.

p. cm.

Includes bibliographical references and index.

ISBN-13: 978-0-87480-875-9 (pbk. : alk. paper)

1. Plants—Yellowstone National Park Region—Identification. 2. Plants—Wyoming—Grand Teton National Park Region—Identification. 3. Plants—Yellowstone National Park Region—Pictorial works. 4. Plants—Wyoming—Grand Teton National Park Region—Pictorial works. I. Rey-Vizgirdas, Edna. II. Title.

QK195.V59 2007

581.9786'661—dc22 2006033990

For my son Tomas,
my brother Paulius, and
my good friend Darius.
Enjoy.

CONTENTS

Illustrations

MAP

FIGURES

Tables

Disclaimer

The publisher and the author disclaim any liability for injury that may result from following the instructions in this book for collecting, preparing, or consuming plants. Additionally, it is not our intent to encourage the use of wild plants within the national parks, as it is illegal to collect plants, flowers, rocks, petrified wood, or antlers. The historical and pioneering uses discussed here are presented as a matter of interest because they have helped shape who we are. Enthusiasts are welcome to look at the wild plants, photograph them, enjoy them as they are, and leave them in place to give similar pleasure to those who follow. Efforts have been taken to ensure that the descriptions and drawings are accurate representations of the family, genus, and species noted. It should be understood that growth conditions, improper identification, and varietal differences, as well as an individual's own sensitivity or allergic response, can contribute to a hazard in sampling or using a plant. The reader is encouraged to seek the assistance of experienced botanists in identifying any of the plants discussed in this book.

Preface and Acknowledgments

The Native Americans of the Greater Yellowstone Area (GYA) employed plants in many different ways aside from food and medicine. They used wood to carve implements and containers, to build houses and shelters, and as fuel to heat their homes and to cook and smoke foods. From bark, stems, leaves, and root fibers they made twine, ropes, fishing lines, nets, baskets, bags, mats, and clothing. Plants and plant products were also used as bedding and floor coverings, drying racks, steaming pits, storage vessels for food and water, diapers, abrasives, tinder for fire starting, dressings for wounds, insect repellents, scents, soaps, animal poisons, glues, decorations, and toys. They even used plants as biological indicators for various seasonal and climatological events.

This book provides information on the many plants used by the Native Americans and early settlers of the GYA, as well as insights into their life histories, ecology, and adaptations. Please keep in mind that it is not intended to be a "how-to" book or to advocate or encourage experimentation by the reader: plants have toxic properties and should not be used without thorough knowledge.

Two considerations went into producing this book. First, we wished to address the need for an informative and convenient guide to the useful plants of the Greater Yellowstone Area. The goal was to provide a handy reference primarily for a lay audience but also for those with training in botany. Some of the features of this book will interest travelers and vacationers, weekend botanists, and day hikers as well as more experienced plant enthusiasts.

The second consideration for writing this book revolves around an underlying concern for the environment. Public awareness of environmental issues and interest in our biological heritage are at an unprecedented level, stimulating a tremendous demand for information about plants, animals, and natural communities. We hope this book will help answer that demand, kindling further interest in the biota of the Greater Yellowstone Area and, on a broader level, instilling a greater appreciation and understanding of the natural world.

This book represents the work of many people who have lived, explored, and studied the Greater Yellowstone Area. All we have done is bring this information together to help others enjoy this wonderful and special place. I am particularly indebted to Dr. Don Mansfield of Albertson College of Idaho, Caldwell, Idaho, for his thorough review and thoughtful suggestions. Thank you all.

FIRST PEOPLE

There is little evidence that the early inhabitants of Yellowstone and Grand Teton national parks were truly at home in the high mountains, for the winter season was simply too harsh. Instead, tribes from the surrounding plains established summer camps in the foothills and valleys, making lengthy forays into the mountains to collect food and medicinal plants.

A Guide to Plants of Yellowstone
and Grand Teton National Parks

1

The Greater Yellowstone Area and Its Plants

The moment you step out of your car at a campground or trailhead, you enter into the land of plants. There are the dense forests of pines and spruces, the multitudes of flowers in meadows, and the belly plants of the arctic-alpine zone. No matter what direction you look, you will find some sort of plant growing in what otherwise could be inhospitable conditions.

Life as we know it would end if plants were to disappear. The food we eat, the clothing we wear, the air we breathe, and the houses we live in were derived from the plants that grow around us. Even the heat in our homes, the electric light that turns our nights into day, and the power of our motors are possible only because plants and bacteria of prehistoric days transformed into coal, oil, and gas.

The ability of plants to gather the energy of the sun and store it—not just for their own use but for ours—is an ageless mystery. We continue to study this phenomenon called *photosynthesis*, in which a plant takes up carbon dioxide from the air and water from the ground and, under the power of sunlight and by means of green matter in the leaves, combines them into sugar, releasing oxygen for animals and humans to breathe. In addition to oxygen and sugars, plants build such substances as starches, oils, proteins, and complex vitamins.

The aim of this book is to provide a useful reference to the vascular plants of Yellowstone and Grand Teton national parks and the surrounding forest and wilderness areas. The vascular plants are those plants best known to most people and include the ferns and fern allies, such as horsetails; gymnosperms, such as pines, true firs, and spruces; and plants having true flowers (angiosperms), such as lilies, grasses, and roses. (*Vascular* refers to complex specialized tissue, primarily phloem and xylem, that allows food and water to move through the plant.) In general, the area covered includes the mountains of northwestern Wyoming: the Wind River Range, Snake River Range, Absaroka Range, Gros Ventre Range, Beartooth Mountains, Wyoming Range, and Salt Range, all located within Park, Teton, Lincoln, Sublette, and Fremont counties.

The plants are arranged in a partially botanical and partially alphabetical order. The major plant groups are in their traditional order: ferns and allies; gymnosperms; and flowering plants. The flowering plants are further divided into two subgroups: the dicotyledons (dicots) and monocotyledons (monocots). The ferns, gymnosperms, dicots, and monocots are classed in family groups, which are presented alphabetically by scientific name. Within each

PLANTS AND PEOPLE

It is easy to forget the importance of plants in our lives. Where the land is wild and rugged, plants are still vital for such basic human needs as food, shelter, medicine, and clothing. Humans have been recording the history of plant uses since early times. Expeditions to uncharted lands usually included botanists. Until recently, doctors of Western medicine studied botany because plants were the only source of ingredients for the treatment of disease. Enjoying the beauty of plants is just one of many reasons why protecting native plant species benefits all of us.

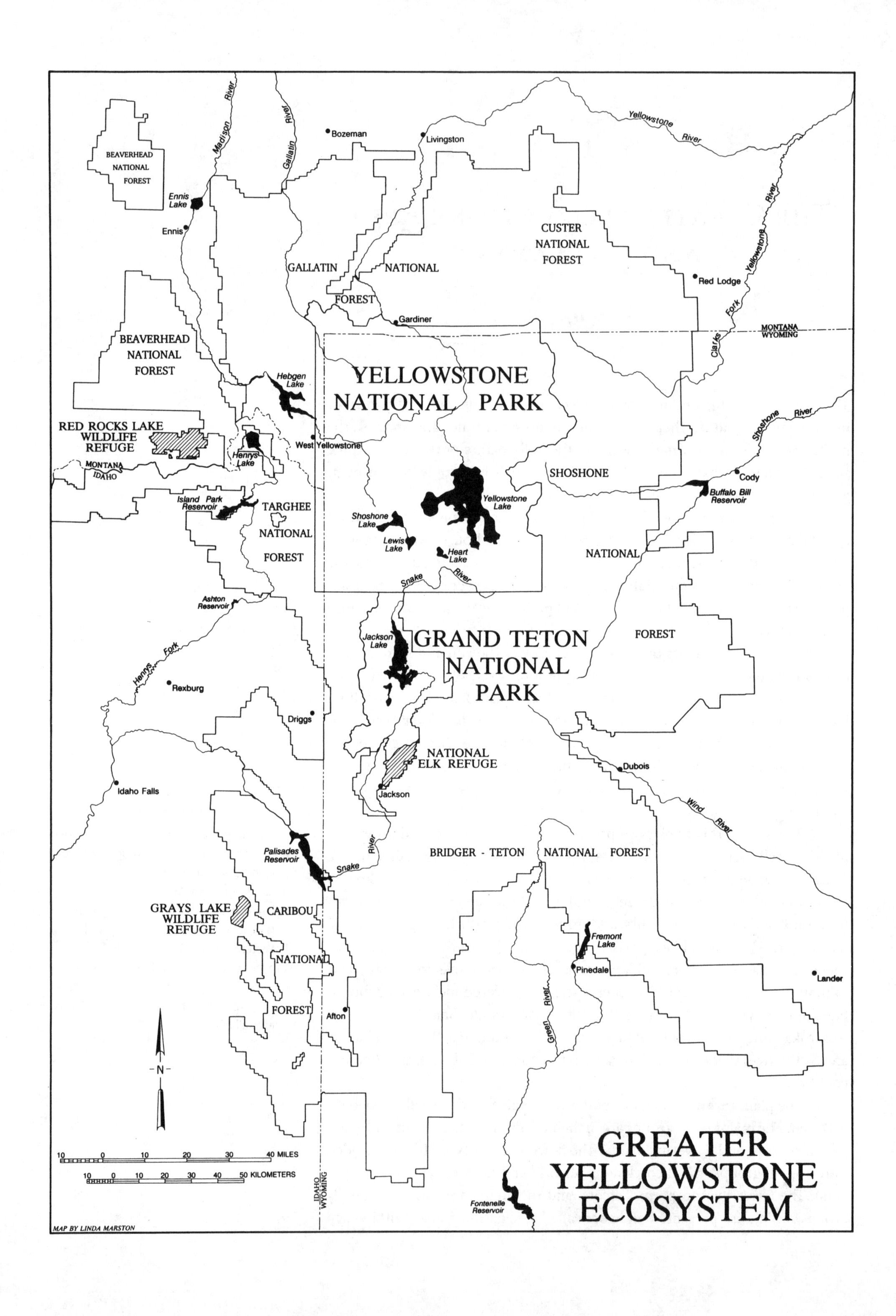
BEAVERHEAD NATIONAL FOREST
Madison River
Gallatin River
Bozeman
Livingston
Yellowstone River
Ennis Lake
Ennis
CUSTER NATIONAL FOREST
GALLATIN
NATIONAL
FOREST
Red Lodge
Yellowstone River
Clarks Fork
Gardiner
MONTANA
WYOMING
BEAVERHEAD NATIONAL FOREST
Hebgen Lake
YELLOWSTONE NATIONAL PARK
RED ROCKS LAKE WILDLIFE REFUGE
Henrys Lake
West Yellowstone
MONTANA
IDAHO
Shoshone River
Cody
SHOSHONE
Island Park Reservoir
TARGHEE
NATIONAL
FOREST
Shoshone Lake
Lewis Lake
Yellowstone Lake
Heart Lake
Buffalo Bill Reservoir
NATIONAL
Snake River
Ashton Reservoir
Jackson Lake
GRAND TETON NATIONAL PARK
FOREST
Henrys Fork
Rexburg
Driggs
NATIONAL ELK REFUGE
Dubois
Idaho Falls
Jackson
Wind River
Palisades Reservoir
Snake River
BRIDGER - TETON NATIONAL FOREST
GRAYS LAKE WILDLIFE REFUGE
CARIBOU
NATIONAL
FOREST
Afton
Fremont Lake
Pinedale
Green River
Lander
N
10 0 10 20 30 40 MILES
10 0 10 20 30 40 50 KILOMETERS
IDAHO
WYOMING
Fontenelle Reservoir
GREATER YELLOWSTONE ECOSYSTEM
MAP BY LINDA MARSTON

family the genera and associated species are also presented in alphabetical order. Common names, though not a reliable identification method, have been included for guidance only and because they are generally of more interest to, and more likely to be known by, the public.

Many published references were consulted, both botanical and ethnographic, in the preparation of this book. The botanical descriptions and the notations on the habitats and distribution of plant species are based in part on personal observation and experience but mainly on various botanical works. *Vascular Plants of Wyoming,* 2nd edition (Dorn 1992), was used a guide, and the dichotomous keys were adapted from many sources, including Dorn 1992, Despain 1975, and Nelson 1992. However, the authority for plant names in this book, at the level of family, genus, and species, is *A Synonymized Checklist of the Vascular Flora of the United States, Canada, and Greenland* (Kartesz 1994). The PLANTS database, which was established by agencies of the U.S. Department of Agriculture and uses Kartesz's 1998 online work as its name authority, was also consulted.

"Of recent years there has come into man's life a new joy. This joy is the acquaintanceship with plants. Nature has long been ready to reveal her secrets, but only to those prepared to hear and see. Gradually a new understanding has arisen between Nature and mankind, and as a result we obtain from such a revelation a joy undreamed of a few years ago.... Plants no longer are lifeless things labeled and grouped under ponderous Latin titles; they are highly developed organisms, which...walk, swim, run, fly, jump, skip, hop, roll, tumble, set traps, and catch fish; decorate themselves that they may attract attention; powder their faces; imitate birds, animals, serpents, stones; play hide and seek; blossom underground; protect their children, and send them forth into the world prepared to care for themselves."

— Royal A. Dixon, *The Human Side of Trees: Wonders of the Tree World,* 1917

THE GREATER YELLOWSTONE AREA

The Greater Yellowstone Area (GYA) extends from its wild core into almost twelve million acres of surrounding federal, state, and private lands. These lands include seven national forests, three national wildlife refuges, Bureau of Land Management land, and about a million acres of private land. The outer boundaries of the GYA spread north to Bozeman and Red Lodge in Montana, east to Cody and Lander in Wyoming, south past Pinedale in Wyoming, and west to Idaho Falls in Idaho. The idea of a "Greater Yellowstone" emerged during the 1960s with the pioneering work of John and Frank Craighead on the population dynamics of grizzly bears in Yellowstone National Park. The concept has been expanded to include buffer zones around what is considered an island of mountains inside a matrix of high, dry plains. The GYA, often referred to as a biogeographic island, currently sustains the largest free-roaming concentration of wildlife in the lower forty-eight states.

The GYA offers a wide variety of activities within some of the most awe-inspiring country found anywhere in North America. From the dramatic peaks of the Grand Tetons, to the amazing colors of the Grand Canyon of the Yellowstone, to the magnificence of the wildlife, ample rewards await those who explore this country. Both Yellowstone and Grand Teton national parks are extremely dynamic geologically, but in very different ways. In Yellowstone, volcanism created the landscape of solidified lava and consolidated ash. In the Tetons, the formation is typical of the Rocky Mountains: uplift along a fault line and subsequent erosion. The Tetons are the youngest mountains in the region, having developed only in the past nine million years.

Yellowstone National Park

Yellowstone, the world's first national park, was established in 1873 and is about 3,574 square miles (55 by 60 miles). Its most prominent geological feature began to develop about two million years ago, when pressure generated by molten rock created an oval-shaped dome that was about 30 by 40 square miles. Eventually, this dome collapsed into a void left by gases, lava, and ash and created a caldera, or depression, in the area of Yellowstone Lake and westward that occupies about one-third of Yellowstone Park. Another caldera formed in the Island Park area. These depressions were filled by lava that continued to flow from cracks in the earth, forming a rolling plateau composed of rhyolite.

The Absaroka Mountains, on Yellowstone's eastern edge, have the highest peaks in the park, with elevations greater than 10,900 feet; the lowest elevation is at Gardiner (5,300 feet). The average elevation in the central plateau area is about 7,800 feet. There are about 150 lakes in the park, making up about 5 percent of the landscape. Most of the lakes are small except for Yellowstone, Shoshone, Lewis, and Heart, which account for about 94 percent of the lake surface area (Gresswell 1984).

Most of the soils in the park developed on glacial till, and there are two kinds of volcanic rock: rhyolite and the nutrient-rich andesite. The highest mean precipitation (approximately 23–28 inches) occurs on the high southwest plateaus and in the Absaroka Mountains to the east. The interior and lower elevations receive less than 11 inches (i.e., about 11 inches at Gardiner). There is great variability from year to year, and most of the precipitation is snow and spring rain.

Grand Teton National Park

Grand Teton National Park, established in 1929, encompasses approximately 485 square miles (310,521 acres). For the mountain men of the past, the Teton Range provided a landmark by which they navigated the wilderness. The Tetons' three most distinctively jagged peaks made a perfect marker, known as Pilot Knobs. The Native Americans of the area referred to them as Tee-win-ot ("three pinnacles"), also a name of one peak in the range. Later on, French voyageurs called them Les Trois Tetons ("the three breasts"), a name that eventually stuck for the peaks (Grand, Middle, and South) and for the entire mountain range.

The Teton Range rises more than 7,200 feet from the basin floor, known as Jackson Hole, to the top of the Grand Teton, an awe-inspiring sight that is not seen anywhere else in the Rocky Mountains. Uplift along the Teton fault, a crack in the crust fifty miles long, created the Teton Range. These mountains are still rising at the rate of about 12 inches every 100 years, while the valley floor east of the range continues to sink.

Early European visitors to the Jackson (Wyoming) area used the word *hole* to describe a broad valley surrounded by mountains. Hence we have Jackson Hole.

Glaciers also contributed to shaping the Tetons and Jackson Hole. The first glacial episode, known as the Buffalo, was the largest and started about 200,000 years ago. Ice covered most of Jackson Hole, from about 98 feet to 2,900 feet. The Bull Lake glacial episode followed the Buffalo and was only about half as large. The Pinedale episode came next, about 18,000 years ago. It was the Pinedale glaciers that created the terminal moraines at the eastern foot of the Tetons, which led to the formation of Jackson, Jenny, Leigh, Bradley, Tagget, and Phelps lakes. The Pinedale also formed the U-shaped valleys and jagged peaks we see today. East of Jenny Lake huge ice blocks were left buried in moraines. The ice eventually melted, and symmetrical depressions called kettles were created.

Linking Yellowstone and Grand Teton is a 7.5-mile corridor known as the John D. Rockefeller Jr. Memorial Parkway. Congress, recognizing Rockefeller's contribution to the creation of Grand Teton National Park, designated this 24,000-acre parkway in his honor in 1972. Managed by the National Park Service, it is a transition between the two parks, combining their characteristics but less spectacular than either. Most of the visitor activities occur around the historic Flagg Ranch, which served as a U.S. cavalry post in 1872.

Wind River Range

The Wind River Range straddles the Continental Divide for about a hundred miles, making it the largest, highest, wildest, and most awesome range in Wyoming. The range officially starts at Togwotee Pass (U.S. Highway 287) at the north and stretches southeasterly to South Pass. The "Winds," as they are often called, are special because of their geology, glaciers, bighorn sheep, and wilderness. Gannett Peak, at 12,804 feet, is the highest point in the state.

The Wind River Range has dramatic beauty and superb alpine scenery, but most impressive is its wild, rugged topography and the strength and endurance needed to get into its heart. The Winds are composed of de facto and designated wilderness. The congressionally designated wilderness areas include the Bridger, Popo Agie (pronounced po-POZ-yuh; in the Crow Indian language *agie* = "river" and *popo* = "head"; hence, "headwaters"), and Fitzpatrick wilderness areas.

WILDERNESS AREAS

Wilderness is a place where the imprint of humans is substantially unnoticed, where natural processes are the primary influences and human activity is limited to primitive recreation and minimal tools. Here we can experience wild places without disturbing or destroying natural processes. Change will occur primarily through natural disturbance rather than human influence. Designated wilderness areas of the GYA include Absaroka-Beartooth (938,750 acres [ac]), North Absaroka (350,488 ac), Washakie (704,529 ac), Teton (585,468 ac), Winegar Hole (10,750 ac), Lee Metcalf (240,000 ac), Jedediah Smith (123,451 ac), Gros Ventre (287,000 ac), Bridger (428,169 ac), Fitzpatrick (198,838 ac), and Popo Agie (101,991 ac).

Gros Ventre Range

The Gros Ventre Range was named for the Gros Ventre (GRO-vont; French for "big belly") Indians, a Blackfeet clan that hunted in the area. The clan's nickname in Plains sign language was a sweeping pass with both hands in front of the stomach, meaning "always hungry," or "beggars," which French trappers misinterpreted as "big belly."

Although not as dramatic as the Tetons, the Gros Ventre scenery is spectacular, with peaks higher than 11,000 feet, many having steep, sheer cliffs with colorful bands; glacier-scoured cirques (basins); and clear mountain streams. The Gros Ventre Range is composed chiefly of folded hard and soft sedimentary rocks. Where the Teton rocks display patterns of black and white, the Gros Ventre have bright reds, pinks, purples, grays, and browns.

On June 23, 1925, the Gros Ventre slide made national news when the north ridge of Sheep Mountain tumbled into the Gros Ventre river valley in less than three minutes, damming the river and creating Lower Slide Lake. The natural dam lasted only two years before saturation and water pressure caused it to collapse on May 18, 1927, releasing a tremendous flood that wiped out the village of Kelly, located about six miles downstream, killing six people. Today the area has been set aside as the Gros Ventre Slide Geological Area.

Red Mountains

The Red Mountains are a small, isolated range topped by Mount Sheridan, which is about 10,308 feet above sea level. It rises above the volcanic plateau of southern Yellowstone and is adjacent to Heart Lake. The area is heavily used by grizzly bears and by a large elk herd during the summer. Seen from Lewis Lake, these mountains look like a high forested range of hills, but the eastern slopes have cirque basins with little lakes and impressive vistas.

Washakie Range

The Washakie Range is located at the southern margin of the Absaroka Range (north of Dubois). It includes the 704,529-acre Washakie Wilderness, which abuts the entire southeast corner of Yellowstone National Park. It is named for Chief Washakie of the Shoshone Indian tribe and contains remnants of petrified forests.

Snake River Range

Wyoming shares the Snake River Range with Idaho. Considered part of the Overthrust Group, it is the northern extension of the larger Salt River Range. The heart of this area is the 135,840-acre Pallisades Wilderness Study Area, where one can explore for days and never see another person. There are fantastic views of the Tetons to the north, lush wildflower displays, twisted and stunted whitebark pines, and incomparable sunrises and sunsets that can be seen in solitude from lofty vantage points.

OVERTHRUST BELT

The sedimentary rocks occurring at the top of the northern Rockies have weathered to create the characteristic shapes that symbolize the highlands of the region. Some of the names applied to these shapes are overthrust mountains, dogtooth mountains, sawtooth mountains, and castellated mountains. Overthrust mountains feature a tilted southwest-facing slope and a steep northeast-facing cliff. This shape illustrates how the blocks of sedimentary rock beds were thrust up in a northeastern direction over other rocks. The Overthrust Belt is an important source of oil and natural gas.

Beartooth Mountains

The Boulder River is the dividing line between the Absaroka Range on the west and the Beartooth Mountains on the east. These two ranges, although adjacent, are distinctly different ecosystems, each with its own geology, topography, and plant species. The Boulder Valley is a beautiful ecosystem edge, adding intrigue and diversity to the area.

The Beartooths are made up of a Precambrian base that, at four billion years old, is among the oldest known rocks on earth. In contrast, the Absarokas, located to the south, are composed of stratified volcanic and softer sedimentary rock of a much younger age. Granite Peak in the Beartooths is about 12,799 feet high.

At least 386 different plant species have been discovered in the Beartooths, giving them one of the richest floras of any mountain range in North America. Dwarf wildflowers and shrubs have adapted to the high mountain climate over millions of years to thrive in microclimates where there are pockets of warmth and moisture in and between the rocks.

Absaroka Range

The Absarokas are probably Wyoming's least known and most misunderstood mountains. The area once was part of the country of the Crow, or Absaroka, Indians, whose name has been pronounced ab-sa-RO-ka, ab-SOR-ka, or ab-SOR-kee, meaning "people of the great winged bird."

The Absarokas are characterized by steep slopes, startling pinnacles, colorful banded cliffs, hidden narrow canyons, swift clear streams, and broad grassy ridge tops above timberline. The acidic, potassium-rich soils have yielded many interesting botanical finds in the form of mustards that are also native to Alaska in permafrost habitats. At least two species of plants here occur nowhere else in the world. Geologically, these mountains are a colossal pile of volcanic lava and breccia laid down some fifty million years ago in "layer cake" fashion and are not extensively folded.

The Absarokas' unique and rare plant species include beargrass, ferns, and the state's only population of high-bush cranberry. Buttercups, shooting stars, and other wildflowers follow the retreating snowbanks each spring, so as you climb higher, you are bound to see wildflowers between June and August. This country is extremely fragile, especially at the higher altitudes, where growth is slower.

Salt River Range

The Salt River Range is another member of the Overthrust Belt, and probably the most impressive. It is bounded by the Grand Canyon of the Snake River to the north and Star Valley and Salt River to the west, and is separated from the Wyoming Range (located to the east) by Greys River. The Salt River Range is about 150 miles long, a north-south-trending range that rises from about 6,600 feet to its highest point at Rock Lake Peak (10,763 feet). The range's special attraction is a geological feature called Periodic Spring, which floods its cave every eighteen minutes. Of three cold-water geysers known to exist, this one is probably the world's largest.

Wyoming Range

The Wyoming Range is the eastern segment and leading edge of the Overthrust Belt. The northern end is quite majestic, but the uplift diminishes as one heads south, where the range becomes more accessible. Named following the creation of Wyoming Territory, this range rises between Greys River to the west and the Green River to the east. The extremely steep west escarpment of the southern part of the range drops impressively from the crest into Greys River, whereas the east slope tends to have a rugged crest ending in long forested benches.

FORESTS AND PLANT COMMUNITIES

Many older field guides used the concept of life zones, developed by C. Hart Merriam, chief of the U.S. Biological Survey in the early 1900s. In Merriam's view, temperature was the controlling factor in the distribution of plants, and it had been observed that as one ascended a mountain, the average temperature fell at a steady rate, in the same fashion that average temperatures fell as one moved north from the equator. Thus the life zones defined by Merriam were mapped against the North American continent and named Transition, Canadian, Hudsonian, and Arctic-Alpine.

Today we know that the patterns of plant distribution are controlled not only by changes in elevation but by many factors, including the availability of moisture during the growing season and the amount of snow accumulation during winter. Moreover, the life-zone scheme has practical difficulties, especially in mountainous areas at elevations near timberline, where the effects of exposure and topography can result in adjacent zones becoming hopelessly intertwined.

For example, on warm ridges facing south or west, trees can advance to an altitude higher than the average timberline on that part of the mountain. On the other hand, in the valley next to this ridge, the effects of shading and cold-air drainage may push the timberline lower. Mix in avalanches, fires, and general soil development, and the situation becomes more complex. With respect to soils, those that have developed on glacial till, for example, often harbor different species than soils formed from decomposed granite.

Therefore, in the Greater Yellowstone Area we can see only a rough correspondence between altitude and Merriam's life zones. What remains valid in the life-zone concept is the idea that predictable associations of plants and animals are to be found in, and indeed characterize, each life zone. Table 1 provides a list of vegetation zones and habitats found in the area. Appendix 2 lists some of the plants and their uses by the habitats in which they occur.

Vegetation of Yellowstone National Park

At the present time, Yellowstone is about 80 percent forested. Douglas-fir (*Pseudotsuga menziesii*) predominates at the lower elevations, from 5,900 to 8,200 feet, and lodgepole pine (*Pinus contorta*) at the middle elevations (7,500

TABLE 1. Vegetation Zones of the Greater Yellowstone Area.

VEGETATION/ HABITAT ZONE	SPECIES
Plains and valleys	Grasslands and shrublands of wheatgrasses (*Agropyron*), fescues (*Festuca*), sagebrush (*Artemisia*), and milkvetch (*Astragalus*)
Sagebrush grasslands	Big sagebrush, bitterbrush (*Purshia*), and fescue
Foothills	Open woodlands and savannahs of limber pine (*Pinus flexilis*), ponderosa pine (*P. ponderosa*), juniper (*Juniperus*), fescue, balsamroot (*Balsamorhiza*), and lupine (*Lupinus*)
Montane	Forests of Douglas-fir (*Pseudotsuga menziesii*), ponderosa pine, aspen (*Populus tremuloides*), snowberry (*Symphoricarpos*), arnica (*Arnica*), aspen, and fairybells (*Disporum*)
Subalpine	Forests of lodgepole pine, subalpine fir (*Abies lasiocarpa*), huckleberry (*Vaccinium*), arnica, and wintergreen (*Pyrola*)
Alpine	Cushions and turfs of moss campion (*Silene acaulis*), alpine forget-me-not (*Eritrichium*), arctic gentian (*Gentianopsis*), and alpine avens (*Dryas*)
Aquatic	Cattails (*Typha latifolia*), duckweed (*Lemna minor*), water buttercup (*Ranunculus aquatilis*), yellow pond-lily (*Nuphar lutea*)

to 8,800 feet). In the higher elevations and on other north-facing slopes there is a mix of spruce (*Picea*) and true firs (*Abies*). Although there is no ponderosa pine (*Pinus ponderosa*) in the park, the species occurs elsewhere in the GYA.

Yellowstone National Park's herbarium has inventoried approximately 1,717 species (counting 190 non-native species), including nonvascular plants such as mosses, liverworts, fungi, and 186 species of lichen. About 7 percent of these species are considered rare.

Aspen (*Populus tremuloides*) is found on only about 10 percent of the landscape, usually occurring in small groves of a few acres at lower elevations to the north and mostly associated with Douglas-fir forests and sagebrush steppe. Interspersed throughout the forests are grasslands, meadows, geyser basins, wetlands, lakes, shrublands, and alpine tundra (above 10,000 feet).

Nonforested vegetation is also common. The foothill grasslands are dominated by bluebunch wheatgrass (*Agropyron*) and Sandberg bluegrass (*Poa*) at the lower elevations. The desert shrublands have fringed sage, Gardiner saltbush (*Atriplex*), prickly pear (*Opuntia*), and winterfat (*Krascheninnikovia*). Within the lower-elevation meadows grow Kentucky bluegrass, sheep sedge (*Carex*), and bearded wheatgrass (*Agropyron*), whereas the higher-elevation meadows have Idaho fescue (*Festuca*), junegrass (*Koeleria*), and thickspike wheatgrass. Meadows in the subalpine forest are dominated by tufted hairgrass (*Deschampsia*), sedges, and a variety of forbs.

The riparian areas in meadows typically have bluegrass (*Poa*), reedgrass (*Calamagrostis*), tufted hairgrass, other grasses, sedges, and forbs. Riparian shrublands have alders (*Alnus*), water birch (*Betula*), shrubby cinquefoil (*Potentilla*), silver sagebrush (*Artemisia*), and willows (*Salix*).

Several unique plants and plant communities can be found around hot springs and geyser basins. Ross bentgrass (*Agrostis rossiae*), for example, is endemic to the park. The mineral-rich water of these thermal features creates a saline environment that is hospitable to halophytes (plants that grow in salty soil) such as seaside arrow-grass (*Triglochin*), alkali cordgrass, meadow barley, and baltic rush.

Vegetation of Grand Teton National Park

The landscape mosaic of Jackson Hole and the Tetons is influenced by its geologic substrate. Glacial moraines and outwash plains, steep mountain slopes, avalanche tracks, the Snake River and its many tributaries, and other geologic features all affect the plant and animal life.

Approximately 58 percent of the Grand Teton National Park is nonforested (alpine tundra, boulder fields, meadows, grasslands, and shrublands). Twenty-eight percent is lodgepole pine forest, 7 percent spruce-fir, 4 percent Douglas-fir, 2 percent whitebark pine (*Pinus albicaulis*), and 1 percent aspen.

Sagebrush steppe dominates the floor of Jackson Hole, except on the moraines, where lodgepole pine and spruce-fir forests are common. Rivers and streams have riparian habitat dominated by alder (*Alnus*), cottonwoods (*Populus*), Engelmann spruce (*Picea engelmannii*), silver buffaloberry, willows (*Salix*), and other species. Aspen groves are found in moist upland areas, and the drier slopes of the Tetons have mixed foothill shrublands or Douglas-fir. At lower elevations Douglas-fir and lodgepole pine are the most common trees. At higher elevations, Douglas-fir is replaced by the subalpine forest.

In the area near Jackson there are at least three distinct sagebrush plant communities. Their distribution appears to be a reflection of different glacial outwash patterns as well as differences in the rock types found in the mountains to the west and east of the valley. Knight (1994:238) provides the following descriptions:

1. The first sagebrush community is located west of the Snake River and north of Jackson Lake Lodge. Here, there is a mosaic of patches of low sagebrush (*Artemisia arbuscula*) interspersed within a matrix of big sagebrush (*A. tridentata*). Low sagebrush is an indicator of shallow soils, and possibly of soils with impeded drainage or lower fertility.

2. The second type of sagebrush community is found east of the Snake River where mountain big sagebrush (*Artemisia tridentata* subsp. *vaseyana*) and Idaho fescue occur. Low sagebrush is rare or absent, and the abundance of mountain big sagebrush suggests a deeper soil with considerable water holding capacity.

3. The third type of sagebrush community is found south of the park near the airport and Blacktail Butte. What is observed here is a sagebrush community characterized with an association of bitterbrush (*Purshia tridentata*) and mountain big sagebrush. The presence of bitterbrush suggests sand or gravel in the soil.

RARE, PROTECTED, AND WEEDY PLANTS

The Native Americans of the area were dependent on nature for all their needs and had an extensive knowledge of which plants (and animals) were edible or useful. Because of this dependence, they shared a strong conservation ethic based on the sanctity of life. Today, with increasing human population and our demands on natural resources, many species are becoming rare because of habitat destruction, competition with non-native species, or other stresses.

Mountain ecosystems evolved in the absence of human activities and have no ready response to some kinds of disturbance. Though appearing

TABLE 2. Some Rare Wyoming Plants in the Greater Yellowstone Area, by Habitat.

HABITAT	SPECIES
Alpine	Rockcress draba (*Draba densifolia* var. *apiculata*), comb-hair whitlow grass (*D. pectinipila*), woolly fleabane (*Erigeron lanatus*), Kirkpatrick's ipomopsis (*Ipomopsis spicata* subsp. *robruthii*), naked-stemmed parrya (*Parrya nudicaulis*), Weber's saw-wort (*Saussurea weberi*)
Aspen woodland	Soft aster (*Aster mollis*)
Barren slopes and ridges	Bastard draba milkvetch (*Astragalus drabelliformis*), Dubois milkvetch (*A. gilviflorus* var. *purpureus*), Owl Creek miner's candle (*Cryptantha subcapitata*), narrowleaf goldenweed (*Haplopappus macronema* var. *linearis*), creeping twinpod (*Physaria integrifolia* var. *monticola*), Rocky Mountain twinpod (*P. saximontana* var. *saximontana*), North Fork Easter daisy (*Townsendia condensata* var. *anomala*)
Bogs	Round-leaved orchid (*Amerorchis rotundifolia*), livid sedge (*Carex livida*), marsh muhly (*Muhlenbergia glomerata*), Greenland primrose (*Primula egaliksensis*), myrtleleaf willow (*Salix myrtillifolia* var. *myrtillifolia*), Rolland's bulrush (*Scirpus rollandii*)
Cushion plant communities	Bastard draba milkvetch (*Astragalus drabelliformis*), Dubois milkvetch (*A. gilviflorus* var. *purpureus*), Owl Creek miner's candle (*Cryptantha subcapitata*), keeled bladderpod (*Lesquerella carinata* var. *carinata*), Payson's bladderpod (*L. paysonii*), North Fork Easter daisy (*Townsendia condensata* var. *anomala*)
Desert grasslands	Contracted Indian ricegrass (*Oryzopsis contracta*)
Disturbed sites	Bastard draba milkvetch (*Astragalus drabelliformis*), Payson's milkvetch (*A. paysonii*), contracted Indian ricegrass (*Oryzopsis contracta*)
Forest meadows and clearings	Hall's fescue (*Festuca hallii*)
Lake and pond shores	Yellowstone sand verbena (*Abronia ammophila*), narrowleaf goldenweed (*Haplopappus macronema* var. *linearis*), marsh muhly (*Muhlenbergia glomerata*)
Lodgepole pine forests	Clustered lady slipper (*Cypripedium fasciculatum*)
Mountain brush	Sweetflowered rockjasmine (*Androsace chamaejasme* var. *carinata*)
Mountain meadows	Soft aster (*Aster mollis*), Hall's fescue (*Festuca hallii*), Absaroka goldenweed (*Pyrrocoma carthamoides* var. *subsquarrosus*)
Riparian areas	Black and purple sedge (*Carex luzulina* var. *atropurpurea*), mountain lady slipper (*Cypripedium montanum*), boreal draba (*Draba borealis*), giant helleborine (*Epipactis gigantea*), marsh muhly (*Muhlenbergia glomerata*), Greenland primrose (*Primula egaliksensis*), persistent sepal yellowcress (*Rorippa calycina*), northern blackberry (*Rubus arcticus* subsp. *acaulis*), myrtleleaf willow (*Salix myrtillifolia* var. *myrtillifolia*)
Rock crevices and ledges	Sweetflowered rockjasmine (*Androsace chamaejasme* var. *carinata*), Wyoming tansy mustard (*Descurainia torulosa*), boreal draba (*Draba borealis*), comb-hair whitlow grass (*D. pectinipila*), Kirkpatrick's ipomopsis (*Ipomopsis spicata* subsp. *robruthii*), shoshonea (*Shoshonea pulvinata*)
Sagebrush grasslands	Soft aster (*Aster mollis*), bastard draba milkvetch (*Astragalus drabelliformis*), contracted Indian ricegrass (*Oryzopsis contracta*)
Subalpine fir/Engelmann spruce forests	Clustered lady slipper (*Cypripedium fasciculatum*)
Talus slopes	Rockcress draba (*Draba densifolia* var. *apiculata*), woolly fleabane (*Erigeron lanatus*), Kirkpatrick's ipomopsis (*Ipomopsis spicata* subsp. *robruthii*), naked-stemmed parrya (*Parrya nudicaulis*), Weber's saw-wort (*Saussurea weberi*), shoshonea (*Shoshonea pulvinata*)
Thermal areas	Ross' bentgrass (*Agrostis rossiae*), giant helleborine (*Epipactis gigantea*)
Wet meadows	Pink agoseris (*Agoseris lackschewitzii*), upward-lobe moonwort (*Botrychium ascendens*), giant helleborine (*Epipactis gigantea*), northern blackberry (*Rubus arcticus* subsp. *acaulis*)
White spruce forest	Round-leaved orchid (*Amerorchis rotundifolia*), red manzanita (*Arctostaphylos rubra*), myrtleleaf willow (*Salix myrtillifolia* var. *myrtillifolia*)
Willow thickets	Upward-lobe moonwort (*Botrychium ascendens*), northern blackberry (*Rubus arcticus* subsp. *acaulis*)

rugged, these environments are fragile and highly susceptible to disturbance, and at times have a low ability to rebound and heal themselves after damage. The degree to which this is true is variable, but the vulnerability of mountain environments to disturbance is well documented (Zwinger and Willard 1972; Price 1981). The flora and fauna of mountain ecosystems are composed of species that are well adapted to cope with environmental extremes, low productivity, and fluctuations within the system. Because of climatic extremes, a brief growing season, lack of nutrients at higher elevations, low biological activity and productivity, their island-like character, the steepness of the slopes, and the basic conservatism of the dominant life forms, mountain environments tend to have a slow rate of restoration to original conditions after disturbance (Price 1981).

Rare and Protected Plants

Federal and state laws list certain plants as rare on the basis of several factors, including the number of individuals in existence, the scope of the geographic area in which the species occurs, and threats to the continued existence of the species. However, some rare plants may frequently be seen in a particular area, probably because of narrow ecological factors. Currently, about 473 plant species of concern are listed in Wyoming's Natural Diversity Database. By avoiding these rare and sensitive species and using common sense, you should be able to enjoy wild plants without appreciably affecting either their population or surroundings (Table 2). You should also check with the local land management agency (e.g., Bureau of Land Management or Forest Service) for its policies regarding native plants.

Weeds

The plants covered in this book include some species that are not native, variously called "weeds," "invasives," and "exotics." A few are descended from garden plants introduced in the early years of settlement in the area, and others arrived with domestic livestock and their feed. Most are casual introductions, having migrated from regions at lower elevations. Many have seeds that are easily carried in on shoes and tires or blown in by the wind. For the most part, these species are restricted in their distribution, growing in places where disturbance of the soil is routine, often following roadsides and trails. The species of greatest concern are those that have the ability to aggressively colonize and push aside native vegetation, and hence they are known as "noxious weeds." Some species that have been identified as noxious weeds in the area are Canada thistle (*Cirsium arvense*), Dalmation toadflax (*Linaria dalmatica*), Dyers woad (*Isatis tinctoria*), leafy spurge (*Euphorbia esula*), musk thistle (*Carduus nutans*), and spotted knapweed (*Centaurea maculosa*).

A number of native species can also behave in a "weedy" manner. Some annuals have the ability to invade and maintain themselves on ground that is subject to disturbance, especially along roadsides and trails. Examples are horsetail (*Equisetum arvense*), bitter cherry (*Prunus emarginata*), and watercress (*Rorippa curvisiliqua*).

Weeds are plants that insist on growing where other plants are desired. It has been said that "weeds are persecuted members of the plant world; and like any persecuted race, they develop unusual powers of adaptation and survival."

— C. J. Hylander, *The World of Plant Life* (1939)

SCIENTIFIC AND COMMON NAMES

Scientific and common names are given for each species. Scientific names are Latin words or words from other languages that have been translated into Latin. They can be intimidating to persons unfamiliar with them, but biologists use them because they have advantages over common names. Rules of nomenclature, established by botanists who name, describe, and study plants, are followed throughout the world. Therefore, scientific names have universal meaning and convey information about the relationships among species.

Each scientific name contains two words. The first is the genus name, and the second is the specific epithet, or species name. Both words are necessary to refer to a particular species (e.g., ponderosa pine, *Pinus ponderosa*). Closely related species (such as lodgepole pine, *Pinus contorta*) have the same genus name but different species names. Many species exhibit obvious patterns of variation throughout their ranges. Distinctive phases may be given subspecies or variety names, usually abbreviated "subsp." and "var.," respectively.

Most plants have common names as well. However, unlike scientific names, common names are not regulated by formal rules and often cause confusion. Many plants have two or more common names, and the same common name can be applied to two entirely unrelated plants. For example, the common dandelion (*Taraxacum officinale*) is known by these other names in different parts of the world: dashelflower, dent-de-lion, dient de leon, gowans, swine snout, lion's tooth, red-seeded dandelion, timetable, white endive, wiggers, puffball, priest's crown, and tarakhshgum. Nevertheless, common names have been retained here since they are generally more interesting to and recognizable by the public. Readers are encouraged to learn to identify plants by both their scientific and common names.

Plant families, in turn, consist of related genera, such as *Arabis*, *Draba*, and *Brassica*, members of the Brassicaceae, or mustard family. Flower features, such as the number of petals, the position of the ovary, number of pistils, and fruits, are important characteristics for identifying families. This book includes representatives of 105 families. Scientific family names are recognized by their *-aceae* ending.

TABLE 3. Common Plant Families of the Greater Yellowstone Area.

COMMON NAME	SCIENTIFIC NAME
Grass family	Poaceae (= Graminae)
Sunflower family	Asteraceae (= Compositae)
Pea family	Fabaceae (= Leguminosae)
Mint family	Lamiaceae (= Labiatae)
Carrot family	Apiaceae (= Umbelliferae)
Sedge family	Cyperaceae
Figwort family	Scrophulariaceae
Mustard family	Brassicaceae (= Cruciferae)
Rose family	Rosaceae

Note: Scientific names in parentheses are former names that have been superseded.

INTRODUCTION TO PLANT FAMILIES

Plants, like all other organisms, are classified in hierarchical categories: kingdom, phylum, class, order, family, genus, and species. This taxonomic hierarchy organizes the great diversity of species encountered in nature, grouping evolutionarily related species in the same categories. Identifying where a plant fits into the taxonomic hierarchy thus serves both the practical function of organizing diversity and the biological function of indicating a species' evolutionary history.

The most useful taxonomic category for the plant world is the family. If you learn the characteristics of the common plant families in the area, you will be able to identify plants much more readily. However, not all plant families are equally represented in all communities. The commonness or rarity of any family can reflect where it first evolved. Alternatively, a plant family may be common in certain communities because its members migrated to an environment that suited them. Nine plant families are very common in the Greater Yellowstone Area (Table 3). If you can recognize these families, it should be easy, by process of elimination, to identify a plant you find in the field.

PLANT CHARACTERISTICS

Before we proceed to the field keys, it may be helpful to review some of the important features of plants. The keys (and many of the plant descriptions) usually refer to specific aspects of a plant. Some of the more common things to look for are a plant's life form, type of leaf, and number of flower parts. The glossary lists the botanic terms that are used to describe plant features.

Life Form

Life form relates to the growth form of a plant. A plant may be a tree, shrub, or herb. Parasitic plants derive their food by attaching themselves to other plants. Some plants are aquatic (growing in water) or grow in mud near water (but not in water). Plants can be annual, biennial, or perennial.

Leaf Type

Leaves may be conspicuous or reduced to spines or scales. Simple leaves are not divided into more than one part, whereas compound leaves are divided into two or more separate parts called leaflets. Lobed leaves are simple leaves that have shallow or deep indentations forming lobes but are not divided into parts. Entire leaves have smooth margins, unbroken by indentations. Toothed leaves have margins indented by fine or coarse teeth or have scalloped edges. Leaves also take on a variety of shapes, including thread-like, grass-like, linear, lance-shaped, heart-shaped, egg-shaped, and arrowhead-shaped. Fleshy plants have succulent leaves. Leaves are attached to the stem in different ways: sessile leaves are directly joined to the stem, without a stalk (petiole), and petioled leaves are connected to the stem by a stalk.

Leaf Arrangement

Leaves have four types based on now they are arranged on plants: basal, opposite, whorled, and alternate. Basal leaves grow directly from the roots of the plant; they do not arise from the stem. Opposite leaves grow in pairs across from each other along a stem. Whorled leaves grow in clusters of three or more around the stem. Alternate leaves grow singly on the stem and vary in position from one side to the other.

Flower Types and Arrangements

At first glance, flowers are either regularly or irregularly shaped. Regular flowers are radially symmetrical with distinct petals or petal-like parts surrounding the center of the flower. Each petal or petal-like part is alike in shape, size, and color. Irregular flowers are not radially symmetrical, and their petals or petal-like parts differ in shape, size, and/or color. Tubular flowers have petals that are fused into lobes that are joined at the base, forming a tube. Some tubular flowers have distinct upper and lower lobes called lips. Inconspicuous flowers do not have petals, or their petals or petal-like parts are so small that they cannot be readily observed.

Flowers are arranged in several ways: many grow singly along the stem; others occur in clusters of various types, such as heads, panicles, racemes, spikes, and umbels. A head is a rounded or flat-topped cluster of sessile flowers, as found in many asters. A panicle is a branched inflorescence with a central axis, as in many grasses. A raceme is an unbranched elongated inflorescence with lateral flowers. A spike is a type of raceme with sessile flowers. Some plants possess a single terminal spike, and other have numerous small spikes called spikelets branching from a central axis or side branches. An umbel is an inflorescence with several branches arising from the end of a peduncle (flowering stalk), as in water parsnip.

Structure of a Flower

A typical flower has *sepals* (usually green but not always) as the outer layer. Collectively, they form the *calyx*. Next is the *corolla*, composed of petals, and in some flowers the petals grow together, forming a tube or a bell. Next inward are the *stamens*, each consisting of a *filament* and the pollen-producing *anther*. One or more *pistils* are at the flower's center, each with an *ovary*, a *style*, and a terminal *stigma* that is rough and sticky to catch pollen. Stamens and pistils are sometimes in separate flowers—sometimes even on the same plant. Once you can identify these four major parts of the flower, how they are arranged and in what numbers, you will begin to see the idea of plant family recognition (e.g., all mustards have 4 separate petals, 4 separate sepals, superior ovary, and 6 stamens, with 4 of them longer than the other 2).

ECOLOGICAL REASONS NOT TO PICK WILDFLOWERS

(1) Flowers are more than beautiful structures that appeal to humans: they exist so the plant can reproduce itself. Many of the most spectacular blossoms are specially designed to attract certain pollinating animals. The number of flowers pollinated combined with their arrangement on the stem can make a difference between reproductive success or failure for the entire year.

(2) Removing flowers from annuals (plants that bloom for only one year and then die) means that the seeds the plant would have made will not be there for next year's wildflower season.

(3) Many species of wildflowers have already suffered great reductions in numbers over the last 100 years because of increasing alterations of their habitat.

(4) It is often difficult to distinguish between common and rare or endangered species of wildflowers. Species that are in danger of extinction may seem abundant to the casual observer, who may be looking at one of only a few remaining areas where the plant is found.

HOW TO USE THE DICHOTOMOUS KEYS

The keys in this book are dichotomous. That is, there are always two alternatives from which to choose an appropriate plant characteristic. Be sure to read both choices before making a decision. To use the keys, you should have a representative portion at hand of the plant you wish to identify, including leaves and flowers, not just the leaves or flowers alone. However, as a general rule, *bring your eye to the flower, don't bring the flower to your eye*. This prevents any unnecessary destruction of plants, which is against the law in some areas this book covers (i.e., national parks and monuments). Appendix 1 provides a key to the plant families.

For those not familiar with the use of keys, they are simply a list of plants arranged by contrasting characters. To find a name of a plant, it is necessary only to determine which of the two contrasting characters best describes the plant. In the sample key below, for example, begin by choosing between the characters in the first couplet (red or blue). If the flower is red, proceed to the next pair and make a choice between green and purple leaves. If the flower is blue, proceed to the third pair and make a choice between tall and short, and so on, until you arrive at the name of the plant (Species A, B, C, D, or E). The description for a particular plant should help ensure that the correct choice was made.

SAMPLE FIELD KEY

1. Flower red 2
1. Flower blue 3

2. Leaves green .... Species A
2. Leaves purple 4

3. Plant tall .... Species B
3. Plant short .... Species C

4. Alpine plant .... Species D
4. Aquatic plant .... Species E

When using the dichotomous keys, consider the following points:

(1) Remember that keys are not perfect and can be difficult to use until you gain experience. Use the glossary if you need help with technical terms.

(2) The keys here are based on an "average" plant. Always examine several specimens before coming to a decision on any point. If there is a difference in the numbers of any part (e.g., some flowers may have four petals and others five), use the number that seems to be the most common.

(3) The plant to be identified may have been left out of the key (intentionally or unintentionally). Also, remember that there are a number of weedy, invasive, and exotic plants that continually "migrate" and may even have become locally common since the last time a concise survey of the area's flora was conducted. For plants that were rare or not well understood when the last major flora was completed, you may want to try other local floras or manuals for help in identification. You may also wish to consult a specialist at a university botany department or natural history museum.

(4) Keep in mind that gymnosperms (conifers) have no ovary and no style or stigma. In most of the genera in the area the naked ovules are borne exposed at the base of a woody scale. The presence of any part of the pistil identifies the plant as belonging to the angiosperms (flowering plants).

(5) Look closely at a flower bud before deciding that the calyx (sepals) are lacking. In some plant species (e.g., *Actea* in Ranunculaceae) the calyx drops off as soon as the flower opens. If only one set of floral parts is present, no matter what the color, it is usually best to consider it the calyx, not the corolla (petals).

(6) The number of locules of the ovary can be determined by cutting open a fruit to see the number of "compartments."

(7) In the sunflower family (Asteraceae) and other families of plants having an inferior ovary, the lower part of the calyx (calyx tube) is often so intimately fused with the ovary that it appears to be part of it. Additionally, in many genera of the Asteraceae the calyx is modified into a pappus.

(8) Sometimes two or three characters are used in each couplet. If one or two seem difficult, use the one that is most obvious.

(9) Because flowering times vary so much according to altitude, latitude, and other ecological conditions, and because the growing season is short in the Rocky Mountains, no attempt has been made to include this information for all species.

"Much of the beauty of Yellowstone National Park is provided by the plant cover draped across its mountains and valleys. Standing on a high point in the park, such as Mount Washburn, one can see below a rich tapestry of colors. Most pieces of the mosaic are subtle shades of green that differ very little from their neighbors. Close observation, however, reveals darker, bluish green Douglas-fir stands; dark green spruce-fir stands; and yellowish green young lodgepole pine stands. These stands form a sharp contrast against the browns and blacks of recently burned forests, the rich green of meadows, or the gray of alpine ridges."

— Don G. Despain, *Yellowstone Vegetation* (1999)

FIELD KEY TO THE MAJOR PLANT GROUPS

1. Plants not producing seeds, reproducing by spores 2
1. Plants producing seeds.. 7

2. Leaves broad, not scale-like; plants usually resembling ferns 3
2. Leaves scale-like, awl-like, or rush-like; plants moss-like 5

3. Leaves 4-foliate, plants growing in mud .... Marsileaceae
3. Leaves entire or dissected, never 4-foliate 4

4. Sporangia large, ringless; fronds not spirally coiled in vernation........................... Ophioglossaceae
4. Sporangia minute, provided with a ring; fronds spirally coiled in vernation .... True Ferns

5. Stems jointed, plants rush-like .... Equisetaceae
5. Stems not jointed... 6

6. Leaves scale-like, plants moss-like .... Lycopodiaceae and Selaginellaceae
6. Leaves awl-like, plants submerged or amphibious .... Isoetaceae

7. Ovules naked, not enclosed in an ovary 8
7. Ovules enclosed in an ovary 9

8. Leaves narrowly linear, leaves and cone scales spirally arranged................................ Pinaceae
8. Leaves mostly small and scale-like, leaves and cone scales opposite or whorled ....Cupressaceae

9. Leaves parallel-veined; flower parts usually in 3's or 6's; embryo one cotyledon...... Monocotyledoneae (Monocots)
9. Leaves usually netted-veined; flowers parts usually in 4's or 5's; embryo normally with 2 opposite cotyledons ... Dicotyledoneae (Dicots)

2

Ferns and Their Allies

The ferns and fern allies, which make up the nonflowering vascular plants, are distinguished from the other vascular plants by reproducing by spores rather than true seeds.

The most conspicuous members of this group are the "true ferns," found in four families in the area. At one time, they were all lumped under one family, Polypodiaceae.

Similar in appearance to the true ferns are the grapeferns, horsetails, quillworts, clubmosses, and lesser clubmosses. Although they are a minor element of the flora in this area, they may be dominant features in some habitats.

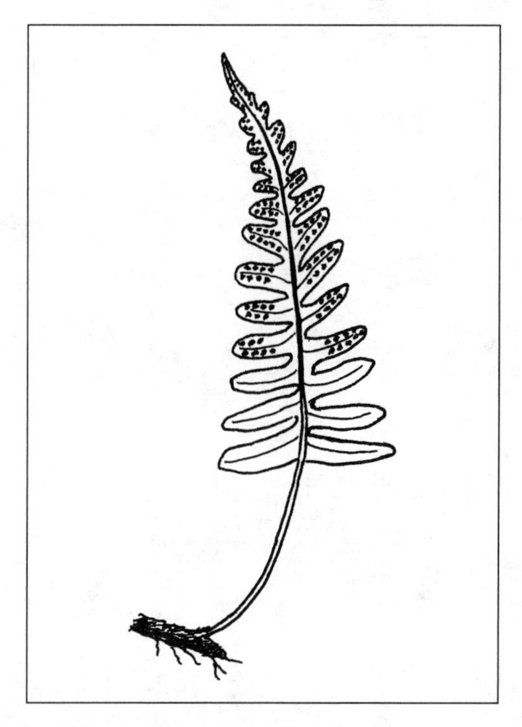

Figure 1. Fern frond with sori.

FIELD KEY TO THE FERNS AND FERN ALLIES

1. Plants with jointed stems and evident leaves (horsetails) ... Equisetaceae
1. Leafy plants; stems are not jointed.................................. 2

2. Plants growing in water or mud 3
2. Plants growing in dry or wet soil but not in water 4

3. Leaves grass-like.. Isoetaceae
3. Leaves compound with 4 leaflets .... Marsileaceae

4. Leaves small and entire; plants moss-like............................5
4. Leaves larger, lobed or compound 6

5. Spore cases in the axils of the leaves of a stalked, terminal spike; spore cases and spores all alike ....Lycopodiaceae
5. Spore cases in the axils of leaves of sessile, somewhat 4-sided spikes, those of the lower part of the spike large and containing a few large (female) spores; those of the upper part smaller and containing many very small (male) spores .... Selaginellaceae

6. Spore cases on a specialized branch or leaf ....Ophioglossaceae
6. Spore cases on the lower surfaces or margins of the leaves...... 7 (Ferns)

7. Fronds of 2 kinds, the sterile broader than the fertile, the fertile taller; leaves under 8 inches long .... Pteridaceae
7. Fronds all alike, at least some usually bearing sori on the underside 8

8. Sori borne along margin of leaflet; indusium absent or sori covered by reflexed leaflet margin 9
8. Sori borne away from margin on underside of fronds or leaflets....... 11

9. Leaflets all entire .... Pteridaceae
9. Leaflets toothed or lobed..10

10. Rhizome and petiole base with linear to ovate scales........ Pteridaceae
10. Rhizome and petiole base covered with felt-like hairs . . Dennstaedtiaceae

LIFE CYCLE OF A FERN

The gametophyte represents the sexual generation of the fern life cycle. It is a tiny body, usually hidden under debris on the forest floor, which bears eggs and sperm. The sporophyte generation is the familiar fern or fern-like plant, which produces spores.

11. Sporangia scattered along veins, not clustered into distinct sori; indusia absent .... Pteridaceae
11. Sporangia clustered in distinct sori; indusia present or not 12

12. Indusia absent .... Dryopteridaceae
12. Indusia present.. 13

13. Sori round, indusium centrally attached............... Dryopteridaceae
13. Sori elongated, indusium laterally attached 14

14. Fronds 1-pinnate or forked .... Aspleniaceae
14. Fronds 2-pinnate or divided .... Dryopteridaceae

---

SPLEENWORT FAMILY (ASPLENIACEAE)

Ferns have a different set of identifying characteristics than flowering plants: fronds (the leaf), pinnae (leaflets of a frond), and sori (reproductive clusters found on the underside of the pinnae). This family contains about 9 genera and 700 species of terrestrial or epiphytic, small to medium-sized ferns. The leaves are simple to several times pinnately divided. The sporangia are gathered into sori on the veins of the leaf.

BRIGHT-GREEN SPLEENWORT (*Asplenium trichomanes-ramosum*)

Description and Habitat. This small, tufted fern grows 2–6 inches tall and has persistent, old stalks from the previous year's growth. The leaves are bright green, soft, and once pinnately divided into 7–20 leaflets. The sori are elongate and occur on the veins. This fern grows in moist crevices of cliffs and ledges or in soil, usually in cool, shaded canyons.

Natural History Notes and Uses. This genus has about 650–750 species of terrestrial and epiphytic ferns with divided leaves. The sori are elongate and occur on the veins. The genus name comes from the Greek *a* ("not"), and *splen* ("spleen"), alluding to the plant's supposed medicinal properties. The Cherokee made an infusion of the plant for "breast diseases" and for coughs. Several species are grown as ornamentals.

Lady Ferns (genus *Athyrium*)

SPLEENWORT (*A. americanum* = *A. distentifolium*)

COMMON LADY FERN (*A. filix-femina*)

Description and Habitat. Lady ferns are medium to large ferns with succulent, straw-colored, tightly tufted, erect to ascending stalks that arise from scaly rootstalks. The leaves are yellow-green in color, 2–4 times pinnately divided, and curve gracefully outward. The sori are oblong to horseshoe-shaped. The indusium is thin and fragile, soon disappearing or sometimes lacking. The name is from the Greek *a*, "without," and *thurium*, referring to a long, oblong shield.

FIELD KEY TO THE LADY FERNS

1. Indusium is crescent-shaped and frayed on the margin; smallest veins are clearly visible from below; leaves are widest at the midpoint of the frond and taper to either end ...... *A. filix-femina*
1. No indusium; veins are generally obscure when viewed from below; leaves are widest at the base, tapering to the tips .... *A. distentifolium*

Natural History Notes and Uses. The name *lady fern* may be derived from a resemblance to the European lady fern, whose spores when placed in the shoes are suppose to render the wearer invisible as well as confer "second sight," useful in finding lost things. These powers vanish when the "seed" is lost.

Lady fern fiddleheads were eaten in the early spring when they were 2–6 inches tall. They were boiled, baked, or eaten raw with grease. The plant's chemical properties led to medicinal uses. The underground stems, pulverized to a powder, were used to drive worms out of the intestinal system, although this use has not been medically recognized (Grillos 1966). A tea made from the root was used as a diuretic, and the powdered root was used externally for sores.

BRITTLE BLADDERFERN (*Cystopteris fragilis*)

Description and Habitat. This 6–16-inch-tall plant is loosely tufted from a short, creeping rhizome. The leaves are thin and delicate in texture. The stipes are brown below, yellowish above, and smooth. The indusia are small, attached at one side and arching back to form a hood. This widely distributed fern is found in the crevices of cliffs and ledges and in soil under rocks, shrubs, or trees.

Natural History Notes and Uses. The genus name comes from the Greek *kystis*, "bladder," and *pteris*, "wing," referring to the shape of the indusium. The Navajo made a cold, compound infusion of the plant to use as a lotion for injuries.

Woodferns (genus *Dryopteris*)

SPREADING WOODFERN (*D. expansa*)

MALE FERN (*D. filix-mas*)

Description and Habitat. Woodferns are medium to tall plants of moist and semi-shaded areas. The leaves are 2–3 times pinnately divided and arise from a short, sturdy rootstalk that is set with chaffy scales and the persistent petiole bases of previous years. The indusium is round or broadly spade-shaped. The sori are borne at the end of the veins on both sides of the ultimate leaflets at midrib.

FIELD KEY TO THE WOODFERNS

1. Leaves triangular in shape, widest at the base, and 3 times compound .... *D. expansa*
1. Leaves elliptic, widest near the middle, and 2 times compound.................................. *D. filix-mas*

Natural History Notes and Uses. The edibility of the various species of woodfern is unknown, but some species are reported to be edible, poisonous, or of medicinal value. Several species contain phloroglucinol derivatives (filicin), which paralyze intestinal parasites. One of the most potent remedies for tapeworm ever recorded in the annals of medicine, male fern has been recommended for this purpose from Greek antiquity to the present day and was listed in the U.S. Pharmacopeia as late as 1965. Tapeworms know no social boundaries—even Louis XVI of France paid a kingly sum for a formula containing this drug. A root tea from a related species, *D. cristata* (crested woodfern), was traditionally used to induce sweating, clear chest congestion, and expel intestinal worms (Foster and Duke 1990).

WESTERN OAKFERN (*Gymnocarpium dryopteris*)

Description and Habitat. Oakfern is a small, delicate fern. The deciduous leaf blades are light green, and the petioles are 4–12 inches long and oriented parallel to the ground. The glabrous to slightly glandular blade is broadly triangular in outline and distinctively divided into 3 triangular leaflets that are 1–2 times pinnately divided into paired asymmetric pinnae. The basal pair are evidently the largest. Oakfern occurs in moist to wet forests, as well as in some riparian areas. This species does well in shade so deep that other undergrowth species cannot exist.

Natural History Notes and Uses. The spore clusters are not covered by an indusium, hence the name *Gymnocarpium*, "naked fruit." Oakfern is an indicator species of moist sites. As moisture levels increase, other species such as horsetails (*Equisetum* spp.), a fern ally (related to ferns), become more abundant and oakferns are displaced. Oakfern is not considered a fire-surviving species but may sprout from rhizomes following light burns. It is slow to return after an intense fire. The spores are stored in soil seedbanks (reserves of viable seeds in the soil), so fires that do not damage upper soil layers may not permanently eliminate oakferns from an area. The spores are well adapted for high wind dispersal. The Cree Indians crushed the leaves to repel mosquitoes and soothe mosquito bites.

Hollyferns (genus *Polystichum*)

NORTHERN HOLLYFERN (*P. lonchitis*)

MOUNTAIN HOLLYFERN (*P. scopulinum*)

Description and Habitat. These ferns have evergreen leaves clustered on a short, vertical rhizome. The leaves have scaly petioles and blades that are 1–2 times pinnately divided into numerous toothed or lobed segments. The name is from the Greek *poly* ("many") and *stichos* ("row") because the sori of some species develop in several rows. The membranous indusium arises from the center of the sori.

FIELD KEY TO THE HOLLYFERNS

1. Leaves only once divided, the leaflets incised less than halfway to midrib .... *P. lonchitis*
1. Leaves twice divided, or if once compound, then incised more than halfway to midrib .... *P. scopulinum*

Natural History Notes and Uses. The leaves of these ferns can be used as a protective layer in pit cooking or as flooring or bedding. *P. scopulinum* is quite aromatic. Although the edibility of these species is unknown, the large rhizomes of a related species, *P. imbricans*, can be roasted over a fire or steamed in a pit oven, then peeled and eaten (Frye 1934; Whittlesey 1985). The cooked rhizomes are also said to be a cure for diarrhea.

Woodsias (genus *Woodsia*)

OREGON WOODSIA (*W. oregana*)

ROCKY MOUNTAIN WOODSIA (*W. scopulina*)

Description and Habitat. These small ferns commonly grow in rocky places. The underground stem is densely tufted and clothed with broad, thin scales. The leaves are clustered, numerous, small, linear to lanceolate-ovate, and 1–2 times pinnate. The sori are round and seated on the back of the free veins, and the indusia are under the sori with star-shaped divisions. The genus name honors English botanist Joseph Woods.

FIELD KEY TO THE WOODSIAS

1. Leaf blades with both glands and separate hairs .... *W. scopulina*
1. Leaf blades glabrous or with only glands.................... *W. oregana*

Natural History Notes and Uses. Native people of Canada used Rocky Mountain woodsia as a sign of water when traveling through the mountains (Turner and Kuhnlein 1991). An infusion was made from a related species (*W. neomexicana*) for use as a douche at childbirth. Some species are cultivated as ornamentals.

BRACKEN FAMILY (DENNSTAEDTIACEAE)

The bracken family has about 18 genera and 400 species of terrestrial, occasionally epiphytic ferns with creeping rhizomes and 2–3-pinnate or rarely simple leaves. This family is very similar to the Pteridaceae (brake fern family), and many older books may still include the western brackenfern as part of that family. Members of the Dennstaedtiaceae are distinguished by scattered (not clustered) fronds and green, grooved petioles (stems). The sori are borne marginally or submarginally.

WESTERN BRACKENFERN (*Pteridium aquilinum*)

Description and Habitat. Western brackenfern is a medium-sized to large plant with decompound, broadly triangular leaves up to 7 feet long including the stipe. The stipes are green or yellowish, and on the surface there are fine white hairs. The sori are marginal and continuous and partially covered by the recurved leaf margins. This is a widely distributed species found in open woods, on rock slides, or on slopes in damp or dry places, up into the high mountains (below 10,000 feet). The generic name comes from the Greek *pteris*, "wing," and is applied to ferns because of their feathery leaves.

Natural History Notes and Uses. The young fronds, or fiddleheads, of brackenfern can be collected, boiled, and dried in the sun. The dried product can then be used as a winter food. Old fronds may be poisonous in large amounts (see caution below). The starchy rhizome (underground stem) is edible after roasting or boiling but is usually tough. The leaves can be used as one of the protective plant layers for pit cooking. Some Native Americans would consume only fiddleheads so that their scent would not scare off deer. A root tea was taken for stomach cramps and diarrhea, and poulticed roots were used for burns and sores. Ashes of the plants have been used as an ingredient in glass and soap.

Caution: Although this plant has traditionally been accepted and harvested as a suitable edible, new evidence indicates that eating sufficient quantities over a period of time may be dangerous (Foster and Duke 1990). The plant is known to contain several poisonous compounds, including a cyanide-producing glycoside (prunasin); an enzyme, thiaminase, that reduces the body's thiamine reserves; and at least two potent carcinogens, quercetin and kaempferol. Another, unidentified toxin is believed to be a naturally occurring, radiation-mimicking substance, also apparently mutagenic and carcinogenic. Bracken has caused many livestock deaths. The risks to humans of eating bracken fiddleheads and rhizomes have not been fully established, but their safety is questionable. Schofield (1989) reports that in Japan bracken is suspected of causing stomach cancer.

COOKING USING A STEAM PIT

The steam-pit cooking method has been used by different peoples around the world for thousands of years. This method of cooking locks in the natural juices and flavor of the food. There are many variations of the pit, but basically all you need to do is dig a hole in the ground about two feet deep and two feet across and line the bottom and sides with flat rocks. One rule of thumb is to have the hole no more than three times the size of the total amount of food. Construct a fire in the pit until the rocks are red-hot. After about an hour or so, carefully remove the coals and place about 6–8 inches of green grass, fern fronds, or other nonpoisonous vegetative matter on top of the hot rocks. Then place your food on top of this layer of plants and cover with an additional 6–8 inches of grass. On top of the grass, place a thin layer of bark slabs or brush to prevent dirt from sifting through to the food. Finally, cover with a mound of dirt and leave it alone for a few hours. When you return, your meal will be cooked and ready to eat.

MAIDENHAIR FAMILY (PTERIDACEAE = ADIANTACEAE)

Pteridaceae produce shiny leaves in clusters, with petioles (leaf stems) brown to black in color. The leaves are at least bipinnately compound (symmetrical around a major axis and sub-axis) and carry the sori (reproductive clusters) along the margins of the leaf or lobes. The sori are not covered by a separate flap (indusia). The plants are rarely over 2 feet tall.

These are small ferns and, except for maidenhair, are mostly found in drier, rocky places where conditions are too harsh for other ferns. An indusium is not present, although in some species the sori are protected by the inrolled margin of the leaflet.

MAIDENHAIR FERN (*Adiantum pedatum* = *A. aleuticum*)

Description and Habitat. About 200 species of *Adiantum* occur worldwide. This is a delicate graceful fern of moist rocky woods and ravines. The leaves vary from about 9 to 24 inches long and have purplish brown, polished stalks. The leaflets are all along the upper sides of the branches of the leaf instead of on both sides, and the circular leaves have no midribs. The sporangia are in sori that are borne on the veins under the frond surface. The genus name is from the Greek word *adianthos*, "unwetted," referring to the ability of the impermeable leaves of some species to shed water.

Natural History Notes and Uses. These black-stemmed ferns can be used in basketry. The herbage is reported to be bitter and causes an increased secretion of mucus. The leaves were used as a tea or syrup to treat colds, coughs, and hoarseness. The rhizomes were used as a stimulant, to soothe the mucous membranes of the throat, and to loosen phlegm. Some of these medicinal uses have been recorded since ancient times.

INDIAN'S DREAM (*Aspidotis densa*)

Description and Habitat. The leaves of this species are long and of two forms; both are glabrous, densely tufted, and originate from a short, much-branched rhizome with persistent, glossy brown petiole bases. This species is found in exposed, rocky habitats.

Natural History Notes and Uses. Related species are said to prevent baldness.

SLENDER LIPFERN (*Cheilanthes feei*)

Description and Habitat. Lipferns are small, densely tufted ferns from thick rhizomes that grow in the crevices of rocks or in the soil under rocks. The genus name is from the Greek, meaning "margin" and "flower," because the sori are clustered on the edge of the frond.

Natural History Notes and Uses. A related species, *C. covillei* (Colville's bead fern), was used medically by Hupa women during childbirth. The Kawaiisu in California drank a tea made from the stems and leaves as a general tonic.

Rock-brakes (genus *Cryptogramma*)

AMERICAN ROCK-BRAKE (*C. acrostichoides*)

FRAGILE ROCK-BRAKE (*C. stelleri*)

Description and Habitat. Rock-brakes are small ferns of rocky places, mostly at high elevations. These plants grow in dense tufts up to 16 inches tall. The sterile leaves are dark green and tripinnate, and the stipes are yellow or straw-colored to the base. The fertile leaves are larger and longer. This species occurs on cliffs, ledges, and talus slopes. The genus name is from the Greek *cryptos*, "hidden," and *gramme*, "line," and refers to the sori being covered by the infolded margins of the pinnules.

FIELD KEY TO THE ROCK-BRAKES

1. Fronds mostly scattered on an elongate slender rhizome; petioles brown to dark purple at the base.................... *C. stelleri*
1. Fronds densely crowded on a short branched rhizome; petioles greenish or straw-colored.................... *C. acrostichoides*

Natural History Notes and Uses. An infusion of the washed, strained fronds of a related species (*C. sitchensis*) was used as an eyewash. The infusion was also taken to treat gallstones.

Cliffbrakes (genus *Pellaea*)

BREWER'S CLIFFBRAKE (*P. breweri*)

(*P. occidentalis* = *P. glabella* subsp. *occidentalis*)

(*P. suksdorfiana* = *P. glabella* subsp. *simplex*)

Description and Habitat. Cliffbrakes are small, tufted ferns growing in the crevices of rocks. The rhizomes are short, thick, creeping, densely brown scaly, and covered with old stipes. The leaves are singly pinnate or bipinnate. The stipes are dark reddish brown. The sori are covered by the recurved margin of the leaf segments. The generic name comes from the Greek *pellos* ("dusky") and refers to the appearance of the leafstalks.

FIELD KEY TO THE CLIFFBRAKES

1. Middle and lower pinnae asymmetrical, mitten-shaped .... *P. breweri*
1. Pinnae seldom mitten-shaped, the lower ones often with 1 or 2 pairs of pinnules................................ *P. glabella*

Natural History Notes and Uses. A refreshing tea was said to be made by the Luiseño Indians of southern California from a related species (*P. mucronata*). They also used the tea medicinally as a decoction to stop hemorrhages, to reduce fevers, as an emetic, and as a wash for skin problems. Other Native Americans used the brown fibers from the rhizome to make basketry patterns.

Caution: The younger leaves and stems are occasionally eaten by sheep and other grazing animals, but they are poisonous and frequently cause death.

HORSETAIL FAMILY (EQUISETACEAE)

Horsetails are an interesting group of plants because they are relicts of a flora that existed during the geological period in which most of the coal deposits were formed. Many different horsetails existed during that time, and some grew to the size of trees. Today the family is represented by only one genus and several species.

Horsetails are related to ferns in that they reproduce by means of spores. The stems are hollow and jointed, and the leaves are reduced to two kinds of structures: toothed sheaths at the joints of the stems, and little shield-shaped scales on which the spore cases are borne. The stems are impregnated with particles of silica, which makes them rough to the touch; they are sometimes called "scouring rushes." They are found in moist to dry areas of sterile or sandy soil in open meadows and woods. Many species of waterfowl, particularly geese, feed on these plants.

Horsetails (genus *Equisetum*)

FIELD HORSETAIL (*E. arvense*)

WATER HORSETAIL (*E. fluviatile*)

SCOURINGRUSH HORSETAIL (*E. hyemale*)

SMOOTH HORSETAIL (*E. laevigatum*)

VARIEGATED SCOURINGRUSH (*E. variegatum*)

Description and Habitat. Horsetails are generally rhizomatous herbs with hollow, grooved, regularly jointed stems that are impregnated with silica. The leaves are reduced in size, appearing as a series of teeth around a joint. The spores are produced in cone-like structures atop the stems. They are found in moist soil along streams and rivers, marshes, and other damp habitats. The name comes from the Latin *equus*, "horse," and *seta*, "bristles."

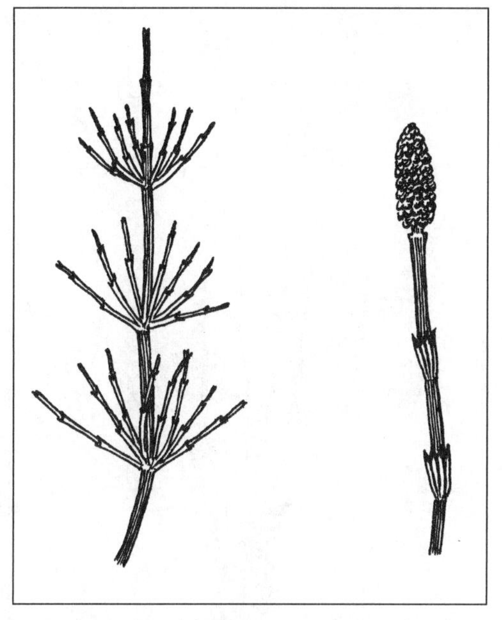

FIGURE 2. Horsetail (*Equisetum* spp.).

FIELD KEY TO THE HORSETAILS

1. Plants with regularly whorled branches on the green stems	2
1. Plants without regularly whorled branches	3
2. Central cavity of the stem large, four-fifths the diameter of the outside of the stem	*E. fluviatile*
2. Central cavity of the stem mostly less than two-thirds the diameter of the outside of the stem	*E. arvense*
3. Teeth of leaf sheaths soon deciduous, sheaths then with a scalloped margin	*E. laevigatum*
3. Teeth of leaf sheaths persistent	4
4. Stems with 14–40 longitudinal ridges	*E. hyemale*
4. Stems with 3–12 longitudinal ridges	*E. variegatum*

Natural History Notes and Uses. All species are useful and identical in their application. The tough outer tissue can be peeled away and the sweet inner pulp eaten in small amounts. In large quantities, defined as greater than 20 percent of body weight by some authorities, they can be toxic. Certain chemicals in this plant are said to destroy specific B vitamins such as thiamine. The enzyme thiaminase is apparently responsible for the poisoning. Cooking destroys this enzyme and renders the plants safe for consumption. The tuberous growth on the roots (actually rhizomes) can be eaten raw in the early spring or boiled later in the season.

Horsetails have an unusual chemistry. Some species contain alkaloids (including nicotine) and various minerals, whereas other species have been known to concentrate gold in their tissues, although not in sufficient amounts to warrant extraction. In the fall the stems become impregnated with silicon dioxide and can be used to scour pots and pans or as a type of sandpaper for wood. Native Americans used horsetails to polish arrow shafts. The silica-rich stems are reputed to be superior to the finest grades of steel wool. Additionally, the high silica content of this herb is said to be effective in strengthening bones and connective tissue (Tilford 1997).

Caution: The waters within which these plants grow may be contaminated.

QUILLWORT FAMILY (ISOETACEAE)

Quillworts are submerged or partially immersed small plants with a cormlike stem crowned by numerous subulate or nearly filiform leaves. There are 2 genera and 75 species worldwide. None are of economic importance.

Quillworts (genus *Isoetes*)

BOLANDER'S QUILLWORT (*I. bolanderi*)

WESTERN QUILLWORT (*I. occidentalis*)

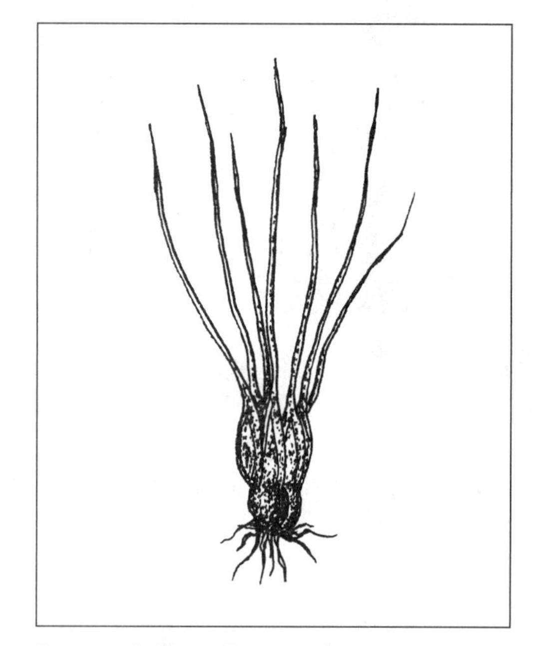

FIGURE 3. Quillwort (*Isoetes* spp.).

Description and Habitat. Quillworts are small, usually erect, grass- or rush-like plants that are submerged or partially immersed in streams, ponds, and lakes. The stem, called a corm, is short, fleshy, and 2–3-lobed and has many roots developing from the base. The sporangia develop on the upper surface of the expanded leaf base. They are solitary, orbicular to ovoid, and usually covered by a thin membranous tissue called *velum*. This velum, or flap, is formed by the infolded leaf margins. The common name is derived from the plants' resemblance to a bunch of quills. The approximately 75 species of quillworts worldwide can be distinguished only by microscopic examination of their spores.

FIELD KEY TO THE QUILLWORTS

1. Leaves acute, not long-pointed . . . . *I. occidentalis*
1. Leaves with a long fine point . . . . *I. bolanderi*

Natural History Notes and Uses. Reportedly, quillworts have been occasionally used as food in Europe (Pfeiffer 1922). We found no record of anything injurious about the plants. The corm serves as a storehouse of food for the plant, as do the subterranean winter organs of many perennials; thus some food value is quite probable (Frye 1934). Other sources indicate that quillworts are rich in starch and oils and could be edible raw or cooked (Weber 1990). The corms are edible but not palatable.

CLUBMOSS FAMILY (LYCOPODIACEAE)

Superficially, these plants look very much like true mosses, but structurally and evolutionarily they are different. Clubmosses are vascular plants with small, evergreen leaves that are usually spirally arranged about the stems. Spores are produced in sporangia that are found in the axils of modified leaves known as *sporophylls*. The sporophylls may be grouped into cones, which may be either stalkless at the tips of branches or on specialized stalks. Early in the year, before the sporangia are produced, clubmosses are difficult to differentiate.

Clubmosses (genus *Lycopodium*)

STIFF CLUBMOSS (*L. annotinum*)

MOUNTAIN CLUBMOSS (*L. selago = Huperzia s.*)

Description and Habitat. There are approximately 450 species of clubmosses worldwide. They have small, narrow, evergreen leaves that are alternate and more or less spirally arranged. The genus *Huperzia* differs from *Lycopodium* in the absence of cones; instead, the sporangia are borne singly in the axils of leaves along the stems and thus are often overlooked. *Huperzia* has a second means of reproduction as well: when the short, bud-like branches called *gemmae* break off, they can form new plants. Mountain clubmoss is found in moist to exposed rocky areas and woods, and stiff clubmoss in moist woods and slopes.

FIELD KEY TO THE CLUBMOSSES

1. Spore-bearing leaves (sporophylls) similar to the vegetative leaves, not in distinct terminal cluster .... *L. selago*
1. Sporophylls clustered into distinct cones at the top of the stems .... *L. annotinum*

Natural History Notes and Uses. The genus name comes from the Greek *lycos* ("wolf") and *pous* ("foot"), from the appearance of the branching shoot tips of several of the species. The fine, yellow spores of clubmoss, which are very light, odorless, and tasteless, are the primary product of importance for this family and were used in Germany as early as 1665. The spores do not contain any medicinal substances but were used by aboriginal peoples in North America as a drying agent for wounds, a treatment for nosebleeds, and a dusting powder to prevent chafing. The spores' fine texture makes them smooth and prevents stickiness. The spores of *L. clavatum* (running clubmoss) were used in decoctions to increase urine flow, to treat severe diarrhea, and to increase the appetite.

To collect spores, people dried the tops of the plants in vessels to prevent the spores from blowing away. After drying, the plants were beaten and rubbed, and the spores were sifted out of the vegetation.

The spores of all clubmosses are also very flammable and were used in fireworks, stage lighting, and early flash photography (Pojar and MacKinnon 1994). When heated slowly, the spores burn quietly, but thrown into a flame, they ignite explosively and produce a great deal of light. The spores also contain a waxy substance that makes them water repellent. It is said that if you coat your hand with clubmoss powder and submerge it in water, it will remain dry.

Caution: Most clubmoss species contain poisonous alkaloids and therefore should not be taken internally (Willard 1992).

PEPPERWORT FAMILY (MARSILEACEAE)

Pepperworts are aquatic or semiaquatic perennial herbs with creeping, hairy rhizomes that root in the mud. The leaves are 4-foliate and clover-like. The sporocarps are borne at the base of the stipe. The family comprises three genera and 70 species distributed worldwide. Two genera are native to the United States and have limited importance as ornamentals.

FIGURE 4. Hairy pepperwort (*Marsilea vestita*).

HAIRY PEPPERWORT (*Marsilea vestita*)

Description and Habitat. This species is found along slow-moving streams, in shallow lakes and ponds, or on their muddy borders. Arising from the rhizome are the distinctive leaves, arranged like those of a four-leaf clover. The leaflets are pale green and hairy. The spores are produced within an elliptical, nut-like body borne on short stalks.

Natural History Notes and Uses. Although there are no documented uses for this species in the area, related species are supposedly edible when dried and ground into flour. For example, the sporocarps of nardoo (*M. drummondii*) found in central Australia are crushed into a powder between stones, made into a dough by adding water, and baked into cakes.

GRAPE FERN FAMILY (OPHIOGLOSSACEAE)

The herbaceous plants of this family have fleshy rhizomes with numerous fibrous, often fleshy roots. The leaves (fronds) consist of two parts, a sterile blade that is simple or compound, and sessile or stalked; and a stalked, spore-bearing spike or panicle. The sterile and fertile parts are borne on an erect common stalk. There are three genera and 70+ species in this family, none of which has economic importance.

FIELD KEY TO THE GRAPE FERN FAMILY

1. Vegetative part of leaf compound or at least not entire . . . . . . *Botrychium*
1. Vegetative part of leaf simple and entire . . . . *Ophioglossum*

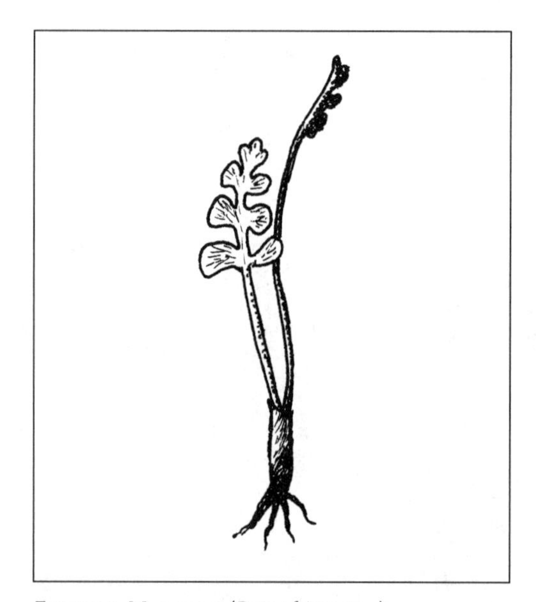

FIGURE 5. Moonwort (*Botrychium* spp.).

Grape Fern, Moonwort (genus *Botrychium*)

COMMON MOONWORT (*B. lunaria*)

LEATHERY GRAPE FERN (*B. multifidum*)

LITTLE GRAPE FERN (*B. simplex*)

RATTLESNAKE FERN (*B. virginianum*)

Description and Habitat. Worldwide there are about 40 species of moonworts, of which six occur in the area. This is a difficult genus to study and requires fully developed specimens for identification, so no field key is provided here. The spore sacs, technically termed the *sporangia*, are borne in grape-like clusters on a naked stalk, not on leaves as in the true ferns. Any *Botrychium* specimens located in the area should be reported to the local land management agency botanist for identification and further study.

Natural History Notes and Uses. Early records appear to indicate that juice extracts from common moonwort were used to stop bleeding and vomiting and to treat bruises. They may have been used as balsams for healing internal wounds (Grillos 1966). Additionally, a poultice or lotion made from the root of rattlesnake fern was used for snakebites, bruises, cuts, and sores. In folk medicine the root tea was used as an emetic.

SOUTHERN ADDER'S-TONGUE (*Ophioglossum vulgatum*)

Description and Habitat. This rhizome-based plant grows to less than one foot high. It has two-part leaves, with a rounded, diamond-shaped sheath and narrow spike. It reproduces by means of spores. The species has been documented growing in meadows, swampy areas, and woods in Yellowstone National Park.

Natural History Notes and Uses. In traditional European folk medicine, leaves and rhizomes were used as a poultice for wounds. This remedy was sometimes called the "green oil of charity." A tea made from the leaves was a traditional European folk remedy for internal bleeding and vomiting.

As in all ferns, the plants seen are the sporophyte stage of the life cycle. The sporophytes can reproduce vegetatively by spreading from the roots, and they also produce spores with a single set of chromosomes, but they do not reproduce sexually. The spores germinate and grow into tiny gametophytes (still with a single set of chromosomes) that accomplish sexual reproduction. Adder's-tongue gametophytes are brown and subterranean and require a cooperative (mycorrhizal) relationship with a fungus to nourish themselves. The fertilization of an egg cell within the gametophyte adds the second set of chromosomes, and the fertilized egg grows into the visible sporophyte. Interestingly, adder's-tongue has more chromosomes than any other vascular plant: up to approximately 1,320 chromosomes (Flora of North American Editorial Committee 1993)! (In comparison, humans have 46.) There has never been an adequate explanation for what, if any, benefit these large numbers of chromosomes might provide.

"Nature is a mutable cloud which is always and never the same."

— Ralph Waldo Emerson

SELAGINELLA FAMILY (SELAGINELLACEAE)

This family of moss-like plants includes many important natural ground covers on rocky or gravelly soils. The sporangia are borne in terminal spikes of sporophylls, the larger ones containing 3 or 4 megaspores, the smaller ones numerous minute microspores.

Members are sometimes called "little clubmoss" because of their resemblance to the clubmoss family (Lycopodiaceae). The points that distinguish these two families are technical and not easy to see without magnification.

Spikemoss (genus *Selaginella*)

LESSER SPIKEMOSS (*S. densa*)

CLUB SPIKEMOSS (*S. selaginoides*)

Description and Habitat. Spikemosses are low, moss-like, mat-forming, leafy, evergreen terrestrial plants, and their leaves are scale-like and less than ⅛ inch long.

FIELD KEY TO THE SPIKEMOSSES

1. Strobili 4-angled (square in cross section); leaves thick and firm, dorsal groove present. . . . . *S. densa*
1. Strobili not 4-angled; leaves soft and thin, without a dorsal groove . . . . *S. selaginoides*

Natural History Notes and Uses. The spores of spikemosses were once used like those of clubmoss (*Lycopodium*) for keeping pills from adhering to one another. Many campers still gather the plants with mosses and other plants to make a soft bed when sleeping on the ground, a rather extravagant practice with such a small plant (Kavash 1979).

Although there are no edible or medicinal uses recorded for these species, some related species, such as *S. concinna* and *S. obtusa*, found on the Reunion Islands in the Indian Ocean, were used medicinally as astringents and blood purifiers and as carminatives in cases of dysentery. In Mexico a variety of *S. rupestris* was used as a home remedy, and in the East Indies *S. convoluta* was considered an aphrodisiac.

3

Gymnosperms

Gymnosperms are a group of plants that do not have flowers and fruits. The seeds are exposed, usually on a cone-like structure. Hence the majority of the gymnosperms are cone-bearing, most commonly known as conifers: spruce, fir, pines, and junipers.

The conifers in this area are all evergreens, the leaves resemble scales or needles, and true flowers are not present. In the genus *Larix* (hemlocks), however, the needles are deciduous. These wind-pollinated plants produce male pollen and female ovules in separate cones—pollen cones and seed- or pinecones, respectively. Junipers differ in having male and female cones on separate plants. For most, the seeds develop on scales in woody cones, whereas in junipers the seed is enclosed in a fleshy, berry-like structure.

FIELD KEY TO THE CONIFERS

1. Leaves scale-like	Cupressaceae
1. Leaves linear or needle-like	Pinaceae

CYPRESS FAMILY (CUPRESSACEAE)

The cypress family is economically important for timber trees and ornamentals. It has 130 species in 19 genera, 5 of which are native to the United States. The heartwood of many species is resistant to termite damage and fungal decay, and therefore it is widely used in contact with soil (e.g., for fenceposts). The premier coffin wood of China, *Cunninghamia lanceolata*, is another member of the family, and *Chamaecyparis* wood is in similar demand in Japan.

Many genera are incorrectly called cedars because their heartwood is as aromatic as that of the true cedars, *Cedrus* (Pinaceae). Wooden pencils are made from incense-cedar (*Calocedrus decurrens*) and eastern redcedar (*Juniperus virginiana*), which is also used for lining "cedar" chests. Wood from species of redcedar (*Thuja*) is used for roof shingles and house siding.

Junipers (genus *Juniperus*)

COMMON JUNIPER (*J. communis*)

CREEPING JUNIPER (*J. horizontalis*)

UTAH JUNIPER (*J. osteosperma*)

ROCKY MOUNTAIN JUNIPER (*J. scopulorum*)

FIGURE 6. Common juniper (*Juniperus communis*).

Description and Habitat. Junipers are evergreen trees and shrubs, with opposite or whorled leaves that are scale-like or linear. The male cones are small, and the female cones are larger and berry-like. There are 1–3 seeds within the berries, which often have a waxy bloom.

FIELD KEY TO THE JUNIPERS

1. Leaves awl-like, in whorls of 3, and whitish on the upper side . . . . . . *J. communis*
1. Leaves scale-like and mostly opposite 2

2. Plants shrubby, usually less than 12 inches tall . . . . . . *J. horizontalis*
2. Plants taller and more tree-like 3

3. Leaves longer than wide, margins entire . . . . . . *J. scopulorum*
3. Leaves about as long as wide, margins with minute teeth . . *J. osteosperma*

Natural History Notes and Uses. Junipers have many things to offer. The berries and twigs can be used to make tea, season game, smoke fish, repel moths, soothe rheumatic pains, and kill germs. The fleshy cones are edible raw but taste better if dried, ground, and used as a flour or flour extender or made into cakes. Cooking the flour with other foods can make it more palatable. The berries can also be roasted and ground and used as a coffee substitute. The Swedes make an extract from the berries which they eat with bread in the same way as butter. The berries of *J. communis* give gin its characteristic flavor.

The boughs can be steeped in hot water for 5–10 minutes to make a beverage. Cooking them in an uncovered pot is recommended to allow the volatile oils to escape. A leaf or berry infusion was used to relieve urinary problems.

Juniper oil extract can be rubbed on skin to repel insects and to relieve pain in muscles and joints. The bark, roots, twigs, and cones furnish red dyes. The white film covering the berries is a type of yeast that can be used to make a primitive sourdough starter (Webster 1980). Indians used the wood for bows.

The shredded bark is excellent tinder for fire starting and can be used as bedding and padding. It is said that some Native Americans ate the inner bark in times of famine. The inner bark was also used for clothes and mattresses and could be worked with the hands until soft enough to use baby diapers or sanitary pads.

ASH CAKES

Make a dough by mixing flour with water. Pat the dough into a patty. The thicker it is, the more doughy it will be, whereas the thinner it is, the crispier it will be. Throw the patty into the ashes (hot coals), let it cook a bit, and you have an ash cake.

PINE FAMILY (PINACEAE)

The pine family comprises trees with needle-like leaves. Except for larches (*Larix*), most species are evergreen. Worldwide there are 250 species in 10 genera, 6 of which are native to the United States. Many are economically important timber trees. Male cones are small, soft, and deciduous after shedding pollen. Females cones (the pinecone) are more robust, with woody, spirally arranged scales bearing seeds on their upper surface. Both winged and nonwinged seeds have evolved, depending on the reproductive strategy of the species.

FIELD KEY TO THE PINE FAMILY

1. Leaves occurring in dense clusters on short spur branches; leaves deciduous .... *Larix*
1. Leaves borne singly or in clusters of 2–5............................ 2

2. Leaves 2–5 per cluster.. *Pinus*
2. Leaves occurring singly 3

3. Branchlets rough where needles have fallen; leaves typically 4-angled and sharply pointed .... *Picea*
3. Branches smooth where needles have fallen; leaves typically flattened and blunt-pointed.......................... 4

4. Cones hanging downward, with 3-lobed bracts protruding from between scales.......................... *Pseudotsuga*
4. Cones erect, without protruding bracts, falling apart at maturity . . . *Abies*

SUBALPINE FIR (*Abies lasiocarpa*)

Description and Habitat. Firs are tall, evergreen trees with whorled branches and flattened, linear, needle-like leaves. The bark of immature stems has "resin blisters." Male cones are small and pendant from the lower side of the branchlets near midcrown. At the top of the crown the female (seed) cones are borne erect at the tips of last season's growth. When these cones ripen in the fall of the first season, they disintegrate, leaving an upright, persistent, bare central axis. Often the lower branches become rooted, forming a circle of smaller trees, or "snow mat." The genus name comes from the Latin *abire*, meaning "to rise," alluding to the great height some species attain. Subalpine fir is a widespread tree at high elevations.

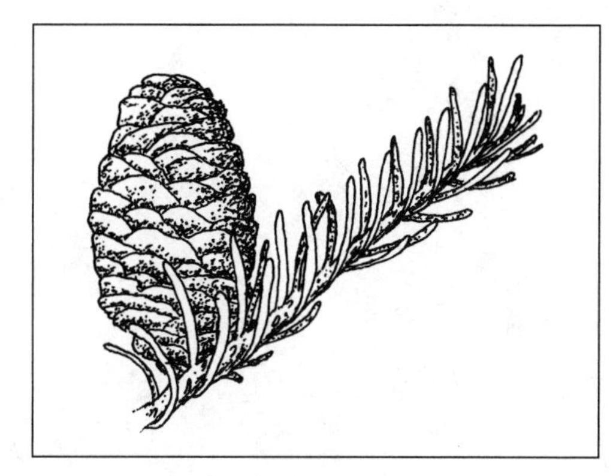

FIGURE 7. True fir (*Abies* spp.).

Natural History Notes and Uses. The inner bark can be eaten raw. Medicinally, the resin, or sap, was brewed as an emetic to "clean out the insides" (Willard 1992). Leaves were used in a poultice for fever and chest colds (Hart 1996). The sap was also chewed for pleasure and as a cure for bad breath. Hart (1996) describes its use by many Native Americans for cuts, wounds, ulcers, sores, skin infections, colds, coughs, and constipation. The needles were pounded into a fine powder, mixed with grease, and then rubbed on the infected skin. The sappy or gummy secretions were also used as an antiseptic for wounds. The boughs can be used for bedding, for covering floors, and as incense.

WESTERN LARCH, TAMARACK (*Larix occidentalis*)

Description and Habitat. Larches are the only northern conifers that drop their needles during autumn. This is a tall tree, growing up to 170 feet, with a narrow crown, long trunk, and thick, deeply furrowed, reddish brown bark. The needles are yellowish green, and the mature seed cones are about 1½ inches long with nearly orbicular scales. Larches usually grow on north-facing slopes and other moist sites at low to middle elevations.

Natural History Notes and Uses. Larch is said to have one of the heaviest, strongest, and toughest softwoods. Because of its durability, it is used for fences, posts, poles, and ties.

The inner bark was reported to be eaten by some Native Americans (Turner and Kuhnlein 1991). The Flathead and Kootenai Indians obtained a sweet syrup, a natural sugar called *galactan*, by hollowing out a cavity in the trunk, allowing about a gallon of the sap to accumulate. Natural evaporation then concentrated the syrup, making it sweeter (Hart 1996), resembling a slightly bitter honey.

The bark was made into poultices and used for chronic eczema, psoriasis, bruises, and wounds (Willard 1992). The strong root fibers were used for sewing and basket weaving. The solidified pitch (the sap hardens when exposed to air) was chewed as gum.

Spruces (genus *Picea*)

ENGELMANN'S SPRUCE (*P. engelmannii*)

WHITE SPRUCE (*P. glauca*)

BLUE SPRUCE (*P. pungens*)

Description and Habitat. Most spruce needles are 4-sided, stiff, and less than one inch long. Woody, peg-like projections help join the needles to the twigs, which feel rough or bumpy to the touch. Spruce trees grow as tall as 150 feet, and most are shaped like pyramids.

FIELD KEY TO THE SPRUCES

1. Cone scales entire on the ends and broadest near the tips ...... *P. glauca*
1. Cone scales ragged on the ends, broadest at the middle or base 2

2. Young twigs with fine hairs; cones less than 2½ inches long .... *P. engelmannii*
2. Young twigs without hair; cones more than 2½ inches long .... *P. pungens*

Natural History Notes and Uses. Spruces have a light, soft, compact, straight-grained wood that is stiff, strong, easy to work, and comparatively free of resin. Spruces are used for pulpwood, light construction, boxes, millwork, and Christmas trees. The resonant wood is also used for sounding boards for pianos and for the bodies of violins and similar instruments. Blue spruce is widely planted in gardens because of its beautiful silver-blue foliage.

The cambium, or inner bark, of Engelmann's spruce can be eaten raw, boiled like noodles, or dried and used as a flour substitute (Pojar and MacKinnon 1994). The inner bark was also used as a laxative. Some aboriginal peoples ate the new growth raw, a good source of vitamin C (Brown 1985). The green spring growth on branches can be steeped in hot water for tea. Spruce beer is made from fermented needles and twigs that have been boiled with honey. The gum can be applied to cuts and wounds as a healing agent and applied to the skin to protect against sunburn (Willard 1992). The crushed needles can be rubbed on traps and skin to camouflage human scent. Although spruce seeds are very small, they are edible raw or cooked.

Pines (genus *Pinus*)

WHITEBARK PINE (*P. albicaulis*)

LODGEPOLE PINE (*P. contorta*)

LIMBER PINE (*P. flexilis*)

PONDEROSA PINE (*P. ponderosa*)

Description and Habitat. There are over 100 species of *Pinus*, making it the largest genus of conifers and the most widespread genus of trees in the Northern Hemisphere. The natural distribution of pines ranges from the arctic and subarctic regions of Eurasia and North America south to subtropical and tropical (usually montane) regions of Central America and Asia. Many pines are fast growing, tolerant of poor soils, and able to thrive in relatively arid regions.

FIELD KEY TO THE PINES

1. Leaves generally 5 in a cluster 2
1. Leaves generally 2 or 3 in a cluster 3

2. Cones mostly over 3 inches long; cone scales thinning near tip .... *P. flexilis*
2. Cones generally less than 3 inches long; cone scales thickened toward tip .... *P. albicaulis*

3. Leaves and cones over 3 inches long; leaves usually in bundles of 3 .... *P. ponderosa*
3. Leaves and cones usually less than 3 inches long; leaves usually in bundles of 2 .... *P. contorta*

Natural History Notes and Uses. One pine in the area, whitebark pine, is rare and rather important. It is most abundant in the Absaroka, Beartooth, and Wind River ranges. Unfortunately, it is declining throughout most of the Rocky Mountains because of fire suppression and disease. The loss of this species would be a severe hardship for wildlife (especially grizzly bears), and the federal government has been petitioned to list the whitebark pine as an endangered or threatened species.

Pines may be divided into two major subgroups: "soft" pines and "hard" pines. In soft pines the needles are usually in bundles of 5, and the cones have no prickles. The wood of the soft pines is straight grained, that is, comparatively free from resin, and easy to work. It is used for rough carpentry, cabinetwork, patterns, toys, crates, and boxes. In contrast, hard pines have 2–3 needles per bundle, and the cones have prickles. The strong, resinous wood of hard pines is used in buildings, bridges, ships, and other types of heavy construction. Because of its durability, the wood of hard pines is valuable for floors, stairs, planks, and beams.

All pines have edible seeds, but they are an erratic food source, abundant in some years and sparse in others. Long poles were used to knock the cones from the branches. One of the easiest ways to gather the nutritious seeds is to heat the green cones until they open. The seeds are best when harvested in fall or early winter, when cones normally release their seeds. They can then be shelled and eaten or ground or roasted and made into a flour. Seeds may contain as much as 15 percent protein, 50–62 percent fat, and 18 percent carbohydrates, with approximately 3,000 calories per pound (Farris 1980; Vizgirdas 1999c).

The inner bark is also edible in an emergency. The tender mucilaginous layer between the bark and the wood was tediously scraped or peeled off and then cooked or ground into meal. The use of the inner bark by Native Americans was so extensive that early explorers reported large areas of trees stripped of bark (Strike 1994).

The firm and unexpanded pollen cones can be boiled and eaten. They have a surprisingly sweet and nonpitchy taste. The edible pollen is usually mixed with flour and used as a soup thickener.

The needles of most pines can be steeped in hot water to make a satisfying tea and are a good source of vitamin C. It takes some practice to steep the right amount of leaves, since too much may be too strong. Additionally, pine cleaning fluid can be extracted from boiling the needles and skimming off the oil-like substance from the surface. It may take a lot of pine needles to get a small cupful.

All pine species have been used medicinally for many centuries. Chewing the sap was said to be soothing for a sore throat. The sap can also be dried, powdered, and applied to the throat with a swab. It was also heated and used as a dressing to draw out embedded splinters or to bring boils to a head. Smeared on a hot cloth, the sap was used much like a mustard plaster in treating pneumonia, sciatic pains, and general muscular soreness. Pine sap can be collected in quantity from cuts, burns, and broken branches. The collected sap can then be heated and formed into balls for future use. Be careful not to expose the sap to flames as it is very flammable.

Pine oil is widely used in massage oil for muscular stiffness, sciatica, and rheumatism and in vapor rubs for bronchial congestion. All pines are rich in resin and camphoraceous volatile oils, including pinene, that are strongly antiseptic and stimulant. The needles and resin produce a brown dye.

Pine roots were valued as twining material for baskets. The roots, about as thick as a pencil, can be several feet long. Roots were collected from opposite sides of a tree each year to prevent destroying the tree. After cleaning, the roots were buried in a heated pit. A fire built over the pit for two days turned the roots a light tan color, and they were then ready to be split into smaller strips.

DOUGLAS-FIR (*Pseudotsuga menziesii*)

Description and Habitat. Douglas-fir is a tall tree, 35–60 feet high, with drooping branches. The leaves are needle-like, blue-green, and spirally arranged on the branches, but they appear to be in a flat spray because the needles are turned at the petiole base. The needles are ¾ to 1½ inches long and pointed at the tip. The cones are cylindrical, 4–6 inches long, with 3-fingered bracts overlapping the scales. These bracts are characteristic of the genus.

Natural History Notes and Uses. Douglas-fir is an important timber species. The wood is resinous with close, even, well-marked grains and is of medium weight, strength, stiffness, and toughness. It is very durable and, when well seasoned, does not warp. It is used in piles, ties, floors, and millwork and for making a variety of items such as handles for tools and spears, spoons, fire tongs, and fishing hooks.

A tea can be made from the needles (Davidson 1919). As in pines, the pitch can be used as a glue for sealing implements and caulking water containers. Medicinally, the sap provided a salve for wounds and skin irritations. The pliable roots have been used in weaving.

FIGURE 8. Douglas-fir (*Pseudotsuga menziesii*).

4

Flowering Plants: Dicots

Here we begin our exploration of the flowering plants, or angiosperms. Angiosperms are the plants that possess true flowers, and they are divided into two major categories: dicots and monocots. The monocots are those flowering plants with one seed leaf (cotyledon), 3 or 6 petals in their flowers, and long, parallel-veined leaves. Examples of monocots are onions, lilies, orchids, grasses and rushes, and cattails. In contrast, most dicots have 4, 5, or more petals in their flowers and net-veined leaves. Appendix 1 provides an additional dichotomous key to the plant families found in the area.

FIELD KEY TO THE DICOT FLOWERING PLANT FAMILIES

1. Flowers without petals (apetalous); calyx often lacking;
   flowers sometimes in catkins. . . . . .Group 1
1. Flowers with petals 2

2. Corolla of separate petals. . . . . Group 2
2. Flowers with petals more or less united into a regular
   or irregular corolla. . . . . Group 3

— *Group 1: Apetalous (without petals) Dicots* —

1. Flowers in catkins; flowers unisexual, plants either male or female. 2
1. Flowers not in catkins. 5

2. Flowers staminate *or* pistillate (never both) in catkins. . . . . Urticaceae
2. Both staminate and pistillate flowers in catkins. 3

3. Ovary becoming a 1-celled, multiseeded pod;
   seeds hairy-tufted . . . . .Salicaceae
3. Ovary becoming a 1-seeded nut, achene, or berry 4

4. Trees or shrubs (not parasitic) . . . . Betulaceae
4. Plants parasitic on trees . . . . Viscaceae

5. Ovary or its cells containing 1–2 ovules (rarely 3–4) 6
5. Ovary or its cells containing many ovules 28

6. Pistils more than 1 (distinct or almost) 7
6. Pistil 1, simple or compound 8

7. Stamens inserted in a ring around the pistil;
   leaves with stipules . . . . *Cercocarpus* in Rosaceae
7. Stamens inserted beneath the ovary; calyx is petal-like. . . Ranunculaceae

8. Ovary superior; calyx sometimes missing 9
8. Ovary inferior, or appearing so. 24

9. Stipules sheathing the stem at the nodes (ochrea). . . . . Polygonaceae
9. Stipules, if present, not sheathing the stem 10

10. Herbaceous plants. 11
10. Trees or shrubs. 22

11. Plants aquatic, usually submerged 12
11. Plants terrestrial. 13

12. Leaves whorled and dissected; style 1 . . . . Ceratophyllaceae
12. Leaves opposite and entire; styles 2 . . . . Callitrichaceae

13. Style 1 (if any); stigma 1. 14
13. Styles 2–3 (or may be branched); 1–4 carpels. 15

14. Flowers unisexual; ovary 1-celled . . . . Urticaceae
14. Flowers bisexual; pod 2-celled and 2-seeded . . . *Lepidium* in Brassicaceae

15. Plant juice milky; ovary and pod tricarpellate . . . . Euphorbiaceae
15. Plant juice not milky; ovary not 3-celled. 16

16. Flowers in involucrate head;
fruit a 3-angled achene . . . . *Eriogonum* in Polygonaceae
16. Flowers not in involucrate head 17

17. Leaves, at least below, covered with stellate hairs . . . . Euphorbiaceae
17. Leaves without stellate hairs 18

18. Leaves opposite 19
18. Leaves alternate 21

19. Plants fleshy . . . . *Salicornia* in Chenopodiaceae
19. Plants not fleshy. 20

20. Flowers in heads or spikes . . . . Amaranthaceae
20. Flowers sessile (in forks of branching
inflorescence) . . . . *Paronychia* in Caryophyllaceae

21. Flowers and bracts scale-like (scarious) . . . . Amaranthaceae
21. Flowers greenish, no scarious bracts . . . . Chenopodiaceae

22. Leaves alternate . . . . Rhamnaceae
22. Leaves opposite 23

23. Fruits tricarpellate, not winged. . . . . Rhamnaceae
23. Fruits winged . . . . Aceraceae

24. Plant parasitic on tree branches . . . . Viscaceae
24. Plants not parasitic on tree branches. 25

25. Aquatic herb. . . . . Haloragidaceae
25. Plants terrestrial. 26

26. Shrubs or trees with leaves having minute
scales or particles. . . . . Elaeagnaceae
26. Herbs with corolla-like calyx. 27

27. Leaves opposite . . . . Nyctaginaceae
27. Leaves alternate . . . . Santalaceae

28. Ovaries 2 or more, separate. . . . . Ranunculaceae
28. Ovary 1. 29

29. Leaves compound . . . . Ranunculaceae
29. Leaves simple 30

30. Sepals distinct. . . . . Caryophyllaceae
30. Sepals more or less united 31

31. Calyx 5-toothed or cleft . . . . *Glaux* in Primulaceae
31. Calyx 4-toothed. . . . . *Synthyris* in Scrophulariaceae

— *Group 2: Free-Petal Dicots* —

1. Stamens more than 10 (more than double the sepals or calyx-lobes) 2
1. Stamens not more than twice as many as the sepals. 16

2. Calyx separate and free from the pistil; ovary superior 3
2. Calyx more or less united to a compound inferior ovary. 13

3. Pistils several to many. 4
3. Pistil 1, styles and stigmas sometimes more numerous 6

4. Aquatic plant with peltate leaves (petiole arises
from under the surface) . . . . Nymphaeaceae
4. Plant terrestrial 5

5. Filaments united into a tube; leaves alternate . . . . Malvaceae
5. Filaments distinct, on the calyx; leaves alternate. . . . . Rosaceae

6. Leaves punctate with translucent dots. Hypericaceae
6. Leaves not as above 7

7. Ovary simple (1 stigma, 1 style). 8
7. Ovary compound (2 or more placentae, styles, or stigmas). 9

8. Ovules 2, fruit a drupe . . . . *Prunus* in Rosaceae
8. Ovules many; leaves 2–3 ternately compound
or dissected. . . . . Ranunculaceae

9. Ovary 1-celled 10
9. Ovary several-celled 12

10. Placentae central; sepals 2 . . . . Portulacaceae
10. Placentae attached to the walls 11

11. Sepals 2, rarely 3. . . . . Papaveraceae
11. Sepals 4 . . . . Capparidaceae

12. Stamens united; land plant . . . . Malvaceae
12. Stamens not united; aquatic plant . . . . Nymphaeaceae

13. Fleshy-stemmed plant, without true leaves;
petals numerous. . . . . Cactaceae
13. Leaves are present 14

14. Sepals or calyx lobes 2 . . . . Portulacaceae
14. Sepals or calyx lobes more than 2. 15

15. Stipules present . . . . Rosaceae
15. Stipules not present; herbage rough to the touch . . . . Loasaceae

16. Stamens equal in number to the petals and opposite them. 17
16. Stamens not equal to the number of petals; or, if equal,
then alternate with the petals 20

17. Ovary 2- to 4-celled. . . . . Rhamnaceae
17. Ovary 1-celled 18

18. Anthers opening by uplifted valves . . . . Berberidaceae
18. Anthers not opening by uplifted valves. 19

19. Style 1, unbranched; stigma 1 . . . . *Lysimachia* in Primulaceae
19. Styles, style branches, or stigmas more than 1;
sepals or calyx lobes 2 . . . . Portulacaceae

20. Calyx free from the ovary (superior) 21
20. Calyx-tube adherent to the ovary, at least to its
lower half (ovary inferior) 49

21. Ovaries 2 or more, separate or united 22
21. Ovary 1 27

22. Stamens inserted at the base of the ovary 23
22. Stamens inserted on the calyx 25

23. Leaves fleshy . . . . Crassulaceae
23. Leaves not fleshy 24

24. Ovary (or ovary lobes) 5, with a common style . . . . Geraniaceae
24. Ovary with separate styles or sessile stigmas . . . . Ranunculaceae

25. Plants fleshy; stamens twice the number of pistils . . . . Crassulaceae
25. Plants not fleshy; stamens not twice as many as the pistils 26

26. Stipules present . . . . Rosaceae
26. Stipules missing . . . . Saxifragaceae

27. Ovary simple, with 1 parietal placenta . . . . Fabaceae
27. Ovary compound, shown by number of its cells,
placenta, styles, or stigmas 28

28. Ovary 1-celled 29
28. Ovary 2- to several-celled (flowers regular) 37

29. Corolla irregular 30
29. Corolla regular 31

30. Petals 4, stamens 6, placentae 2 . . . . Fumariaceae
30. Petals and stamens 5, placentae 3 . . . . Violaceae

31. Ovule 1; trees or shrubs . . . . Anacardiaceae
31. Ovules more than 1 32

32. Ovules at the center of bottom of cell 33
32. Ovules on 2 or more parietal placentae 34

33. Petals not inserted on the calyx . . . . Caryophyllaceae
33. Petals inserted on throat of a bell-shaped
or tubular calyx . . . . Lythraceae

34. Leaves punctate with translucent dots . . . . Hypericaceae
34. Leaves not punctate 35

35. Petals 4–5; stamens as many; fruit a berry . . . . Grossulariaceae
35. Petals 4–5; stamens 6 36

36. Stamens unequal, 2 shorter than the other . . . . Brassicaceae
36. Stamens roughly equal . . . . Capparidaceae

37. Stamens neither just as many nor twice as many as the petals 38
37. Stamens equal to or double the number of petals 40

38. Trees or shrubs . . . . Aceraceae
38. Herbs 39

39. Petals 5 . . . . Hypericaceae
39. Petals 4 . . . . Brassicaceae

40. Ovules and seeds 1–2 in each cell. 41
40. Ovules (and seeds) several to many in each cell 46

41. Herbs 42
41. Shrubs or trees with opposite leaves 45

42. Flowers imperfect . . . . Euphorbiaceae
42. Flowers perfect and symmetrical 43

43. Cells of the ovary as many as the sepals, 5 . . . . Geraniaceae
43. Cells of the ovary twice as many as the sepals 44

44. Leaves abruptly pinnate . . . . Zygophyllaceae
44. Leaves simple . . . . Linaceae

45. Leaves palmately veined. . . . . Aceraceae
45. Leaves pinnately veined . . . . Celastraceae

46. Leaves compound; leaflets 3, obcordate . . . . Oxalidaceae
46. Leaves simple 47

47. Styles 2–5; leaves opposite . . . . Caryophyllaceae
47. Style 1 48

48. Stamens free from the calyx . . . . Ericaceae
48. Stamens inserted on the calyx. . . . . Lythraceae

49. Tendril-bearing herbs . . . . *Echinocystis* in Cucurbitaceae
49. Not tendril-bearing 50

50. Ovules and seeds more than 1 in each cell 51
50. Ovules and seeds 1 in each cell 55

51. Ovary 1-celled 52
51. Ovary 2- to many-celled. 53

52. Sepals or calyx lobes 2; ovules at base of the ovary. . . . . Portulacaceae
52. Sepals of calyx lobes 4–5; placentae 2–3; fruit a berry . . . . Grossulariaceae

53. Stamens inserted on or about a flat disk
that covers the ovary. . . . . Celastraceae
53. Stamens inserted in the calyx 54

54. Style 1; stamens 4–8 . . . . Onagraceae
54. Styles 2–3; stamens 5 or 10. . . . . Saxifragaceae

55. Stamens 5 or 10 56
55. Stamens 4 or 8 57

56. Trees or shrubs. . . . . *Crataegus* in Rosaceae
56. Herbs . . . . Apiaceae

57. Style and stigma 1; fruit a drupe . . . . Cornaceae
57. Fruit not a drupe 58

58. Style 1; stigma 2–4-lobed . . . . Onagraceae
58. Styles or sessile stigma 1 or 4. . . . . Haloragidaceae

— *Group 3: United-Petal Dicots* —

1. Stamens more numerous than the lobes of the corolla 2
1. Stamens not more numerous than the corolla lobes. 7

2. Stamens free from the corolla 3
2. Stamens attached to the base or tube of the corolla 5

3. Styles 5; leaves 3-foliate; may be united at the base .... Oxalidaceae
3. Style 1; leaves simple 4

4. Ovary superior; fruit a capsule or berry .... Ericaceae
4. Ovary inferior; fruit a berry .... *Vaccinium* in Ericaceae

5. Saprophytic herbs (without green foliage) .... Ericaceae
5. Plants green 6

6. Filaments united into a tube .... Malvaceae
6. Filaments free; leaves ternate .... Adoxaceae

7. Stamens equal in number to the corolla lobes
   and opposite them .... Primulaceae
7. Stamens alternate with the corolla lobes or fewer 8

8. Ovary superior 9
8. Ovary inferior 33

9. Corolla regular 10
9. Corolla irregular 28

10. Stamens as many as the corolla lobes 11
10. Stamens fewer than the corolla lobes 25

11. Ovaries 2 (or if 1, 2-horned) 12
11. Ovary 1 13

12. Stamens distinct .... Asclepiadaceae
12. Stamens united .... Apocynaceae

13. Ovary deeply 4-lobed 14
13. Ovary not deeply lobed 15

14. Leaves alternate .... Boraginaceae
14. Leaves opposite .... Lamiaceae

15. Ovary 1-celled; seeds several to many 16
15. Ovary 2–10-celled 18

16. Leaves entire, opposite .... Gentianaceae
16. Leaves toothed, lobed, or compound 17

17. Whole upper surface of the corolla white-bearded;
    leaflets 3, entire .... Menyanthaceae
17. Corolla not conspicuously bearded; leaves,
    if compound, with toothed leaflets .... Hydrophyllaceae

18. Stamens free or almost free from the corolla .... Ericaceae
18. Stamens on the tube of the corolla 19

19. Stamens 4 20
19. Stamens 5 21

20. Stems with opposite leaves; corolla petaloid .... Verbenaceae
20. Stem wanting; corolla scarious, at least on margins .... Plantaginaceae

21. Fruit 2–4 nutlets .... Boraginaceae
21. Fruit a many-seeded capsule or berry 22

22. Styles 2 23
22. Style 1, often branched 24

23. Capsule mostly 4-seeded .... Convolvulaceae
23. Capsule many-seeded .... Hydrophyllaceae

24. Branches of the style (or the lobes of the stigma) 3 . . . . Polemoniaceae
24. Branches of the style or lobes of the stigma 2,
or wholly united . . . . Solanaceae

25. Stamens with 4 anthers, in pairs . . . . Verbenaceae
25. Stamens with 2 anthers, rarely 3 26

26. Ovary 4-lobed . . . . *Lycopus* in Lamiaceae
26. Ovary 2-celled, not 4-lobed 27

27. Stemless; corolla scarious . . . . Plantaginaceae
27. Leafy stemmed; corolla not scarious . . . . *Veronica* in Scrophulariaceae

28. Stamens with 5 anthers . . . . *Verbascum* in Scrophulariaceae
28. Stamens with 2 or 4 anthers 29

29. Ovules solitary in the 1–4 cells 30
29. Ovules 2 to many in each cell; ovary 1–2-celled 31

30. Ovary 4-lobed; style arising from between the lobes . . . . Lamiaceae
30. Ovary not lobed; style arising from its apex . . . . Verbenaceae

31. Parasitic plants, without green foliage . . . . Orobanchaceae
31. Plants not parasitic 32

32. Ovary 1-celled; stamens 2; aquatic plant . . . . Lentibulariaceae
32. Ovary 2-celled; placentae in the axis,
usually many-seeded . . . . Scrophulariaceae

33. Tendril-bearing herbs; anthers often united . . . . Cucurbitaceae
33. Not tendril-bearing 34

34. Stamens separate 35
34. Stamens united by their anthers; these joined in a ring or a tube 38

35. Stamens free from the corolla; equal in number
to its lobes . . . . Campanulaceae
35. Stamens inserted on the corolla 36

36. Stamens 1–3, always fewer than the corolla lobes . . . . Valerianaceae
36. Stamens 4–5; leaves opposite or whorled 37

37. Leaves opposite or a leaf with the stem apparently passing through
it (perfoliate), but neither whorled nor with stipules . . . . . . Caprifoliaceae
37. Leaves either opposite and stipulate, or whorled
and without stipules . . . . Rubiaceae

38. Flowers separate, not involucrate;
corolla irregular . . . . *Lobelia* in Campanulaceae
38. Inflorescence a compact head of small
individual flowers surrounded by an involucre . . . . Asteraceae

MAPLE FAMILY (ACERACEAE)

The maple family consists of two genera and about 200 species distributed worldwide. In general, they are shrubs or trees with opposite leaves that may be simple or compound. The flowers are small, usually appearing before the leaves. The family is of economic importance as a source of timber, ornamentals, and sugar. Maple wood is heavy, tough, compact, and very hard. Its pale brown color, dense even grain, and fine texture make it one of the best woods for furniture, veneers, and flooring. It is also used in making violins, pianos, and tool handles.

Maple (genus *Acer*)

ROCKY MOUNTAIN MAPLE (*A. glabrum*)

BIG TOOTH MAPLE (*A. grandidentatum*)

BOXELDER (*A. negundo*)

Description and Habitat. Maples are deciduous trees or shrubs with male and female flowers on the same or separate plants. The flowers are arranged in racemes, corymbs, or umbels. The fruits are winged schizocarps that resemble tiny helicopters when blown by the wind. Maples are usually found in moist places in canyons, on hills, and along streams from low to high elevations. Of the approximately 15 species of *Acer* native to the United States, 3 can be found in the area.

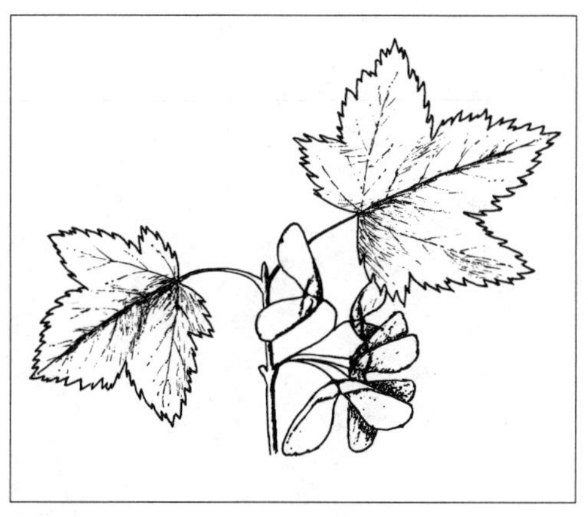

FIGURE 9. Maple (*Acer* spp.) leaves and fruit.

FIELD KEY TO THE MAPLES

1. Leaves compound with 3–5 leaflets . . . . *A. negundo*
1. Leaves simple 2

2. Seed area of samara strongly wrinkled; fruit wings broad; petals commonly present . . . . *A. glabrum*
2. Seed area of samara not strongly wrinkled; fruit wings narrow; petals lacking . . . . *A. grandidentatum*

Natural History Notes and Uses. Sap can be harvested in much the same way as in the eastern sugar maple. To obtain sap, simply bore a small hole into the tree a couple feet above the ground in the springtime. The sunny side of the tree is usually the ideal spot to bore. Insert a small, grooved, wooden peg into the hole. This peg will be the spigot. If the tree is ready, sap will immediately begin to flow after drilling. Hang or place a container under the spigot to collect sap. After collecting some sap, seal the hole to protect it from infection and further sap loss while it heals.

The sap must next be boiled down because most of it is water. Only a small fraction of the original volume collected will be left. Boil the sap to the desired taste. As an alternative to boiling, the collected syrup can be allowed to freeze overnight, which allows the water to separate from the sap. The frozen water can be easily discarded (Hart 1996).

The inner bark of all maples can be eaten in emergencies. A tea made from the inner bark of boxelder is used to induce vomiting. The young shoots of mountain maple can be eaten like asparagus. The winged seeds of boxelder can be roasted and eaten (Harrington 1967).

Native Americans used the young saplings for basketry. The saplings were split into quarters and used as a white wrapping or sewing strand in coiled baskets. In some places, maple thickets were intentionally manipulated by burning the old growth to promote new growth. These straight, uniform shoots were highly valued as basketry material. Maple wood has been used to make snowshoe framing, mush paddles, and other household utensils. Knots and burls on tree trunks can be used for making bowls, dishes, and other items. Gum from the buds was mixed with animal fat and used as a hair tonic. The inner bark can be shredded and twisted into a coarse rope.

MOSCHATEL FAMILY (ADOXACEAE)

This family of dwarf herbs has 1–3 times compound leaves with 3 leaflets on each final branch. It is characterized by opposite leaves and small 5-petaled flowers in dense umbels or panicles; the fruit is a berry. In older classifications the family comprised a single species, *Adoxa moschatellina*. The genera *Sambucus* (elderberries) and *Viburnum* (viburnums) were previously included in the honeysuckle family (Caprifoliaceae), but genetic tests showed they are better placed here. However, in this book we have retained *Sambucus* and *Viburnum* within the Caprifoliaceae.

MUSKROOT (*Adoxa moschatellina*)

Description and Habitat. Muskroot is a musky-scented perennial plant with delicate stems. The basal leaves are long-stalked and divided into 3 compound leaflets. The leaflet blades are thin and round-toothed. There is one pair of smaller, 3-parted leaves attached just above the middle of the stem. The compact terminal inflorescence comprises 5 greenish flowers. The central flower in the inflorescence has 2 sepals and a 4-lobed corolla, and the lateral flowers have 3 sepals and a 5-lobed corolla. The fruit is a drupe.

Natural History Notes and Uses. This perennial herb is native to northern regions of Eurasia and North America and has greenish white, musk-scented flowers, hence the common name muskroot. It is found in small numbers on shady, moist, moss-rich cliffs and ledges in coniferous forest (i.e., Engelmann spruce and Douglas-fir). The genus name is Greek, *a* ("without") and *doxa* ("repute" or "glory"), alluding to the flowers.

PIGWEED FAMILY (AMARANTHACEAE)

The pigweed, or amaranth, family contains more than 60 genera and 900 species distributed worldwide. Many are weedy species of little economic importance. Some species of *Amaranthus* are cultivated for their red pigmentation, and some are grown as cereal crops. The family name originates from the Greek *amarantos*, which means "unfading," possibly alluding to the "everlasting" quality of the papery perianth parts.

Pigweed, Amaranth (genus *Amaranthus*)

PROSTRATE PIGWEED (*A. albus*)

MAT AMARANTH (*A. blitoides*)

POWELL'S AMARANTH (*A. powellii*)

REDROOT AMARANTH (*A. retroflexus*)

Description and Habitat. Amaranths are herbaceous annuals with small greenish flowers and alternate entire or wavy-margined leaves. They occur in many different habitats and often hybridize, making identification difficult.

FIELD KEY TO THE AMARANTHS

1. Flowers occurring in small clusters in the axils of leaves; leaf blades usually less than 1½ inches long 2
1. Flowers occurring are both axillary and terminal; leaf blades generally greater than 1½ inches long 3

2. Sepals 3 .... *A. albus*
2. Sepals 4 or 5 .... *A. blitoides*

3. Stamens 3 .... *A. powellii*
3. Stamens 5 .... *A. retroflexus*

Natural History Notes and Uses. Used by Native Americans, the dried small black seeds of amaranth have been found in many archaeological remains. Seeds of all species can be eaten whole as a cereal or ground into meal and made into cakes. The seeds are best collected in summer when the plants are fully mature. To free the seeds from their husks, rub the seed clusters between your hands. You can then winnow the seeds if there is a breeze or, if the air is calm, slowly pour the seeds out of your hands and blow the chaff away. The seeds contain approximately 15 grams of protein per 100 grams, more than is found in rice and corn and equal to if not surpassing that found in wheat. When ground into a flour, amaranth has a distinctive flavor that is a bit strong used alone. We find it is better when mixed with other flours for breads and pancakes.

The highly nutritious amaranth contains more fiber and calcium than any other "cereals," in addition to a wide spectrum of vitamins and minerals, including vitamins A and C, calcium, magnesium, and iron. Amaranth is rich in the amino acid lysine, a product scarce in true cereal grains, thereby providing a more balanced source of protein.

The edible young shoots and leaves have a pleasant taste if eaten as a potherb soon after collection. Since the plants can accumulate nitrates, it is wise not to consume large quantities where nitrate fertilizers are used. Livestock losses have occurred as a result of excessive amaranth consumption. The leaves of amaranth contain oxalic acid, which binds with calcium, restricting its absorption by the body. As long as your diet contains plenty of calcium from other sources, eating amaranth and other vegetables that contain oxalic acid (e.g., spinach, wood sorrel) should not cause any health problems.

Amaranth also has astringent properties and can be used for treating diarrhea, dysentery, excessive menstrual flow, hemorrhaging, and hoarseness. Amaranth is also helpful in treating mouth and throat inflammations and sores. Steeping dried leaves in boiling water (the more leaves steeped, the stronger the infusion) was considered a valuable remedy.

In the Midwest several species of amaranth are being grown as agricultural crops for use in cereals and bread. Amaranth is photosynthetically efficient and produces a high yield of both greens and seeds. It was an important food in the past and may become an important one in the future (National Research Council 1985).

AN ESSENTIAL AMINO ACID

Lysine is one of numerous amino acids that is needed for growth and repair of tissue. It is considered one of the nine "essential" amino acids because it comes from outside sources such as foods or supplements. Like all amino acids, lysine functions as a building block for proteins and has a key role in the production of various enzymes, hormones, and disease-fighting antibodies. Though many foods supply lysine, the richest sources are red meats, fish, and dairy products. Vegetables, on the other hand, are generally a poor source of lysine, with the exception of legumes (beans, peas, lentils).

SUMAC FAMILY (ANACARDIACEAE)

There are approximately 79 genera and 600 species in this family. Products originating from the sumac family include resins, oils, lacquers, edible fruits, ornamentals, and tannic acid. Although many members of this family produce edible fruits, the resinous oils can produce extreme dermatitis in sensitive individuals. The family contains the infamous poison ivy (*Toxicodendron rydbergii*), poison oak (*T. diversilobum*), and poison sumac (*T. vernix*). Two species are in the area: *T. rydbergii* and *Rhus trilobata*.

FIGURE 10. Skunkbush (*Rhus trilobata*).

FIGURE 11. Poison ivy (*Toxicodendron rydbergii*).

FIELD KEY TO THE SUMAC FAMILY

1. Mature fruit is red or orange and hairy.................. *Rhus trilobata*
1. Mature fruit white........................... *Toxicodendron rydbergii*

SKUNKBUSH SUMAC (*Rhus trilobata*)

Description and Habitat. Approximately 60 species of this genus occur worldwide. Skunkbush is a shrub that grows on dry sunny slopes and has compound leaves with 3 leaflets, of which the middle leaflet is largest. Skunkbush has small yellow-green flowers that bloom before the leaves come out. The red-orange fruits are sour-tasting. The genus name comes from the Greek *rhous*, which is the name of a bushy sumac.

Natural History Notes and Uses. The berries of skunkbush can be eaten raw or soaked in cold water to make a refreshing drink. Malic acid (the cause of tartness in apples) flavors the sour fruits of the sumacs.

Tannic acid, which is present in all parts of the plant, can be used in tanning leather. The leaves, branches, and fruits provide colorfast dyes for wool. The stem produces a yellow dye and the berries a tan or beige dye. Since tannic acid acts as a natural mordant (dye fixer), the fiber does not need to be treated with other chemicals (Bryan and Young 1940).

The slender, flexible branches of skunkbush can be used for weaving baskets, as they are somewhat vine-like (Barrett 1908). The branches can also be used as chewing sticks to clean teeth and massage gums. Take a small stem several inches long, remove the outer bark, and chew on the tip to soften fibers. Since some people may have an allergic reaction to the oils of sumac, this should be done sparingly.

WESTERN POISON IVY (*Toxicodendron rydbergii*)

Description and Habitat. Poison ivy is an erect trailing shrub that spreads from a creeping rhizome and forms open colonies. The leaves are shiny, deep green, and divided into 3 egg-shaped leaflets. The flowers are small and greenish white, and the fruits (drupes) are white, fleshy, and 1-seeded.

Natural History Notes and Uses. The foliage of poison ivy is ***poisonous***, causing contact dermatitis. The skin irritant is found in the sap. The danger of developing an itchy or painful rash from contact with the sap is greatest in spring and summer when the sap is abundant and the plant is easily bruised. Shortly after contact, the symptoms include itching, burning, and redness. Small blisters may appear after a few to several hours. Severe dermatitis, with large blisters and local swelling, can remain for several days and may require hospitalization. Secondary infections may occur when the blisters are broken. To help alleviate itching immediately, wash thoroughly with soap and water as soon as possible after contact.

Since droplets of the irritating chemical can be carried in smoke on dust particles and ash, do not burn poison ivy. Smoke carries the oil and can produce a rash over the whole body. If the oil is inhaled, a rash can develop in the throat, bronchial tubes, and lungs. The oil can even spread through the body in the bloodstream. Although poison ivy can cause havoc for humans, the berries are relished by birds such as robins, cedar waxwings, flickers, and woodpeckers.

URUSHIOL

The heavy oil found in all parts of poison ivy is called urushiol. It is incredibly potent, and an amount the size of a pinhead can cause rashes and blisters in sensitive people. It is also long lasting, so much so that botanists handling hundred-year-old specimens have developed rashes. Urushiol is confined to the canals inside the plant's leaves, stems, and roots, and it oozes out readily when bruised or chewed on. It takes the oil only ten minutes to penetrate the skin, so washing quickly with soap and water is the simplest field remedy.

CARROT FAMILY (APIACEAE = UMBELLIFERAE)

Approximately 300 genera and 3,000 species are found in this family. About 25 percent of the genera and 10 percent of the species are native to the United States. The carrot family is of considerable economic importance because of numerous food plants, condiments, ornamentals, and poisonous species. Familiar members include carrot (*Daucus*), parsnip (*Pastinaca*), celery (*Apium*), anise (*Pimpinella*), and parsley (*Petroselinum*).

Members of this family are short to tall perennials (or biennials), generally with hollow stems, pinnately divided leaves, and small flowers in flat-topped umbels. The bases of stem leaves are usually expanded and sheathe the stalk. The umbels may be themselves composed of smaller umbels, called *umbelets*, each of which is on a special stem called a *ray*. Each flower has 5 sepals and 5 petals, most commonly white or yellow in color. The inferior ovary develops into a dry fruit called a *schizocarp*, which at maturity splits into a pair of seeds held by a short, stem-like structure. The fruits are variously ribbed or winged.

The essential oil extracted from the fruits of several members of the Apiaceae can cause nervous disorders in high doses. The excessive use of these fruits as a condiment should be avoided. The fruits of this family are generally referred to as "seeds." However, although small and dry, they are botanically fruits, not seeds.

Note: No wild members of this family should be eaten until they have been accurately identified. Correct identification usually requires the schizocarp.

FIELD KEY TO THE CARROT FAMILY

1. Leaves all simple and entire . . . . *Bupleurum*
1. Leaves mostly compound or deeply cleft 2

2. Fruits and ovaries armed with prickles . . . . *Sanicula*
2. Fruits and ovaries glabrous or hairy but without prickles 3

3. Lower leaves highly dissected, appearing lacy or fern-like with the ultimate segments or lobes less than ⅛ inch wide and ½ inch long 4
3. Lower leaves mostly 1–2 times divided into definite leaflets greater than ½ inch wide or ½ inch long or both 9

4. Plants with purple spots on stems; found in moist disturbed areas . . . . *Conium*
4. Stems not distinctly purple-spotted; usually native habitats 5

5. Fruits with wavy-margined wings on the outer face; ultimate umbels subtended by bracts on only one side ...... *Cymopterus*
5. Wings of fruits, if present, not wavy-margined; bracts subtending ultimate umbels, if present, not 1-sided 6

6. Compound umbels subtended by narrow bracts............... *Carum*
6. Compound umbels without bracts or bracts minute 7

7. Fruit with more or less corky, winged margins and a flattened outer face, barely convex .... *Lomatium*
7. Corky wings, if present, on the face as well as the margins; outer face convex, not flattened 8

8. Petioles with winged margins nearly to the first division of the leaf blade; base of stem and petioles purple .... *Musineon*
8. Petioles winged only near the base; base of stem and petioles not purple..................................... *Ligusticum*

9. Leaves only once divided into 3 large, maple leaf–shaped leaflets; plants very large .... *Heracleum*
9. Leaves more than once divided or leaflets not maple leaf–shaped... 10

10. Fruits spindle- or club-shaped, at least 4 times as long as wide... *Osmorhiza*
10. Fruits more elliptic, less than 3 times as long as wide................ 11

11. Plants at least 12 inches tall; leaves mostly once divided into long, linear, entire-margined leaflets.................. *Perideridia*
11. Plants smaller or with wider or toothed leaflets 12

12. Compound umbels subtended by green, sometimes leaf-like bracts 13
12. Compound umbels without bracts................................ 15

13. Lower leaves twice divided; veins of the leaves ending in the sinuses between the marginal teeth .... *Cicuta*
13. Lower leaves only once pinnately divided; veins not ending in the sinuses between the marginal teeth 14

14. Some leaflets of upper leaves lobed halfway to the midrib; fruit round in outline .... *Berula*
14. Leaflets of upper leaves merely toothed; fruit elliptic in outline..... *Sium*

15. Flowering stems less than 3 inches long; root is a nearly globose tuber.. *Orogenia*
15. Flowering stems mostly greater than 4 inches long; if shorter, then root not a globose tuber 16

16. Flowers yellow 17
16. Flowers white or purple 18

17. Basal leaves with leaflets greater than 6 inches wide; plants of disturbed areas.................................. *Pastinaca*
17. Basal leaves with leaflets less than 6 inches wide; plants of native habitats .... *Lomatium*

18. Fruits flattened on the outer face with marginal wings wider than those of the face................................ *Angelica*
18. Fruits convex on the outer face with marginal wings as wide as those of the face .... *Ligusticum*

Angelica (genus *Angelica*)

LYALL'S ANGELICA (*A. arguta*)

SMALL-LEAF ANGELICA (*A. pinnata*)

ROSE ANGELICA (*A. roseana*)

Description and Habitat. The angelicas are herbaceous perennials from stout taproots. The leaves are compound; the leaflets are broad and distinct, and dentate or lobed. The flowers are white, pink, or purplish, and the fruit is strongly flattened dorsally.

FIELD KEY TO THE ANGELICAS

1. Ovary and fruit glabrous	*A. arguta*
1. Ovary and fruit hairy	2
2. Leaves pinnate, smooth beneath	*A. pinnata*
2. Leaves mostly ternate pinnate, usually rough to the touch beneath	*A. roseana*

Natural History Notes and Uses. Although there are a number of edible species of angelica, we have not found any information regarding the edibility of these species. Therefore, the internal use of any angelica species in the area is ***not recommended*** until studies have been conducted concerning their toxicity and because they superficially resemble the poisonous water hemlock (*Cicuta maculata*). All species of angelica contain furanocoumarins, which increase skin photosensitivity and may cause dermatitis (Kingsbury 1964; Muenscher 1962).

Little is known about the medicinal aspects of these species. Angelica has been described as an antispasmodic, expectorant, diaphoretic, diuretic, an effective astringent for the stomach lining, and a menstrual stimulant that helps reduce cramps. Poultices from the mashed roots of angelica were applied for arthritis, chest discomfort, and pneumonia.

Warning: Angelica closely resembles the highly poisonous water hemlock (*Cicuta maculata*). Positive identification of the plants is paramount. The identification of young plants of angelica and water hemlock can be made by examining the leaf venation. The leaf edges of angelica are serrate and pinnately divided into opposing pairs, as in water hemlock, but the leaf veins extend from the midribs to the outer tips of the serrations. Water hemlock has leaf veins terminating within the notches of the serrations. Keep in mind, however, that the plant you are observing may actually be poison hemlock (*Conium maculatum*) (Bomhard 1936).

CUT-LEAF WATERPARSNIP (*Berula erecta*)

Description and Habitat. Cut-leaf waterparsnip superficially resembles a fern but is a perennial aquatic herb that grows up to about 3 feet tall. The leaves are pinnate with toothed segments. The inflorescences are erect and leafy and bear small umbels of white flowers.

COUMARIN

Coumarin occurs in many plants and can be found in the seed coats, roots, leaves, flowers, fruits, and stems. It seems to play an important role in a plant's defense against herbivory and has some antimicrobial properties. Coumarin has a sweet smell, familiar to anyone acquainted with new-mown sweetclover hay. Insects reject the bitter taste of coumarin.

Dicoumarol is produced from coumarin and interferes with the body's ability to produce vitamin K (essential to blood clotting). Therefore, dicoumarol acts as an anticoagulant. When livestock eat sweetclover hay that has spoiled, the dicoumarol thins their blood, leading to internal (or external) bleeding. Derivatives of dicoumarol (i.e., warfarin) have been used in rat poisons.

Natural History Notes and Uses. An infusion of the whole plant was used as a wash for rashes and athlete's foot. The leaves and blossoms were used as food by the Apache, although it is unknown how they were prepared. The plant is known to produce coumarins and therefore should be considered toxic.

AMERICAN THOROW WAX (*Bupleurum americanum*)

Description and Habitat. This yellow-flowered plant has undivided leaves (a rather unusual trait for a member of the carrot family), but they are linear and toothless. There is a leaf-like bract at the base of each small flower head and again at the base of the main cluster.

Natural History Notes and Uses. *Bupleurum* has 70–100 species of annual and perennial herbs and shrubs with simple leaves and umbels of typically yellow flowers. Several species are grown as ornamentals. In China, thorow wax plants (often in combination with other plants) have been used to relieve the side effects of steroids.

CARAWAY (*Carum carvi*)

Description and Habitat. This glabrous biennial has a taproot and slender, leafy stems and grows 1–3 feet tall. The basal leaves have petioles and blades 3–7 inches long and are pinnately divided into highly dissected and feather-like segments. The stem leaves are smaller with wider sheaths. The white flowers occur in numerous compound umbels of 7–14 rays. The fruits are elliptical to almost round in shape and have low ribs on the face.

Natural History Notes and Uses. Caraway seed has been a popular and widely used spice for several thousand years. The plant is native to the Mediterranean regions of Europe, Asia, and North Africa. It is cultivated in northern Europe, Russia, and the United States, though for many years the Netherlands has been the main supplier in world trade, with up to 10,000 hectares under cultivation. The seeds are used for spicing a wide variety of baked products, cooked meats, salads, pickles, and drinks: the characteristic flavor of caraway derives from the essential oil, which is extracted from small channels within the seeds. This oil is valued in pharmacology as an antispasmodic and carminative; it is used in perfumery and also in preparations such as gargles and mouthwashes, having antibacterial properties. More recently, it has been discovered that the largest constituent of the oil, carvone, has potential uses as an insect repellent, a suppressant of sprouting in stored potatoes, and an inhibitor of the growth of some fungi.

SPOTTED WATER HEMLOCK (*Cicuta maculata*)

Description and Habitat. This species is found in marshes and along the edges of streams and ponds from low to middle elevations. Water hemlock is a stout perennial from fleshy, fascicled roots. Leaves are 1–2 times pinnately divided into narrowly lance-shaped, sharply toothed leaflets. The veins of the leaflets terminate at the notches between the teeth. Numerous white to greenish flowers are arranged in compound umbels. The fruits are slightly flattened with thickened ribs on the faces. The bruised foliage produces a musky odor. These plants flower from June to September.

FIGURE 12. Spotted water hemlock (*Cicuta maculata*).

Natural History Notes and Uses. This is an extremely **poisonous** plant if ingested. In fact, water hemlock has been described as **the most violently poisonous vascular plant in North America**. The entire plant contains cicutoxin, a resin-like substance that depresses the respiratory system, with the root being particularly dangerous. A single mouthful of the plant is capable of killing an adult. This species, as well as poison hemlock, has been used throughout the ages to execute criminals and kings. Many children have been fatally poisoned by blowing into whistles made from hollow stems of water hemlock. In Oregon, Native Americans soaked arrows in *Cicuta* juice, rattlesnake venom, and decayed deer liver to poison arrow tips for hunting (Pojar and MacKinnon 1994). Water hemlock roots were mashed and smeared on a hot stone and then pressed against sore arms or legs to relieve the pain.

The following graphic description of poisoning due to the ingestion of water hemlock in Europe should instill into the minds of wild food gatherers the need to positively identify a plant species before eating it.

> When about the end of March 1670, the cattle were being led from the village to water at the spring, in treading the river banks they exposed the roots of this *Cicuta* (water hemlock) whose stems and leaf buds were now coming forth. At that time two boys and six girls, a little before noon, ran out to the spring and the meadow through which the river flows, and seeing a root and thinking that was a golden parsnip, not through the bidding of any evil appetite, but at the behest of wayward frolicsomeness, ate greedily of it, and certain of the girls among them commended the root to others for its sweetness and pleasantness, wherefore the boys, especially, ate quite abundantly of it and joyfully hastened home; and one of the girls tearfully complained to her mother she had been supplied meagerly by her comrades, with the root.
>
> Jacob Maeder, a boy of six years, possessed of white locks, and delicate though active, returned home happy and smiling, as if things had gone well. A little while afterwards he complained of pain in his abdomen, and, scarcely uttering a word, fell prostrate to the ground, and urinated with great violence to the height of a man. Presently he was a terrible sight to see, being seized with convulsions, with the loss of all his senses. His mouth was shut most tightly so that it could not be opened by any means. He grated his teeth; he twisted his eyes about strangely and

blood flowed from his ears. In the region of his abdomen a certain swollen body of the size of a man's fist struck the hand of the afflicted father with the greatest force, particularly in the neighborhood of the ensiform cartilage. He frequently hiccupped; at times he seemed to be about to vomit, but he could force nothing from his mouth, which was most tightly closed. He tossed his limbs about marvelously and twisted them; frequently his head was drawn backward and his whole back was curved in the form of a bow, so that a small child could have crept beneath him in the space between his back and the bed without touching him. When the convulsions ceased momentarily, he implored the assistance of his mother. Presently, when they returned with equal violence, he could not be aroused by no pinching, by no talking, or by no other means, until his strength failed and he grew pale; and when a hand was placed on his breast he breathed his last. These symptoms continued scarcely beyond a half hour. After his death, his abdomen and face swelled without lividness except that a little was noticeable about the eyes. From the mouth of the corpse even to the hour of his burial green froth flowed very abundantly, and although it was wiped away frequently by his grieving father, nevertheless new froth soon took its place. (Jacobson 1915:14)

POISON HEMLOCK (*Conium maculatum*)

Description and Habitat. Poison hemlock is a biennial with a stout taproot and a disagreeable odor when crushed. The stems are purple-blotched and hollow, and the leaves are pinnately dissected and have a lacy appearance. The flowers are white in compound umbels, and the fruits are egg-shaped, flattened, with prominent, wavy ribs. The plant is usually found in disturbed sites and waste places at low elevations. It blooms from April to September.

Natural History Notes and Uses. This is another **extremely poisonous** plant. Death reportedly results from the ingestion of the leaves, roots, or seeds. The ancient Greeks used poison hemlock as a humane method of capital punishment, most famously in the case of Socrates, executed in 399 B.C. It is said to be quite painless, and the mind remains clear to the end. Introduced from Europe, poison hemlock has established itself as a common weed (Hardin and Arena 1974; Turner and Szczawinski 1991).

Spring Parsley, Cymopterus (genus *Cymopterus*)

PLAINS SPRING PARSLEY (*C. acaulis*)

EVERT'S SPRING PARSLEY (*C. evertii*)

SPRING PARSLEY (*C. longilobus = Pteryxia hendersonii*)

LONGSTALK SPRING PARSLEY (*C. longipes*)

SNOWLINE SPRING PARSLEY (*C. nivalis*)

Description and Habitat. The species of *Cymopteris* are low perennial herbs with long, thick taproots. The leaves are 2–4 times pinnately divided into small ultimate segments, and the flowers are yellow or white in terminal, compound umbels. The round fruits have winged ribs on the outer faces. Most of the species occur in dry soils or on gravelly slopes.

FIELD KEY TO THE SPRING PARSLEYS

1. Leaves clustered at the top of a naked stem 2
1. Leaves clustered at the base of the plant............................. 3

2. Dorsal wings of fruit equal to lateral ones; ultimate leaf segments strongly overlap .... *C. longipes*
2. Dorsal wings of fruit smaller than lateral ones; little or no overlapping of ultimate leaf segments .... *C. acaulis*

3. Leaf blades over 2 inches long; petals yellow; ovary and fruit usually glabrous .... *C. longilobus*
3. Leaf blades mostly 2 inches or less long; petals white; ovary and fruit rough to the touch................................. 4

4. Fruit with narrow wings dorsally; ovary and fruit glabrous..... *C. nivalis*
4. Fruit with stout ribs dorsally; ovary and fruit rough to the touch... *C. evertii*

Natural History Notes and Uses. All species produce edible roots. We found the older roots more fibrous than the younger ones. The root can be used in stews or it can be boiled or roasted in a pit, mashed, and dried into cakes. When dried, it will keep indefinitely. During the Lewis and Clark expedition this food was known as *kouse* (from the French, meaning "bread of cows") (Harrington 1967; Hart 1996). The old roots can also be used as an effective insect repellent when boiled. Just sprinkle the tea around camp and in sleeping areas.

The upper parts of the plants have been used raw or as potherbs. If cooked, they require several changes of water. The seeds of some species are edible when ground and used as flour.

COMMON COWPARSNIP (*Heracleum maximum*)

Description and Habitat. Cowparsnip is a stout perennial that can grow more than 7 feet tall. The lower leaves are 3-lobed, resembling a maple leaf, and up to 14 inches long. The white flowers are in compound umbels, and the fruits are egg-shaped with only the marginal ribs winged. It is usually found in moist soils around streams, seeps, and avalanche chutes up to subalpine environments. The genus is named for Hercules (from the Greek *Heracles*), who is reputed to have used it as a medicine. It blooms from April to June.

Natural History Notes and Uses. The young stems of cowparsnip can be peeled and eaten raw but are best when cooked. The hollow base of the plant can be cut into short lengths and used as a substitute for salt when eaten or cooked with other foods. The young leaves are also edible after cooking, but we find them not very tasty. The leaves can also be dried and burned and the ashes used as a salt substitute. Strong and bitter, the cooked roots are said to be good for digestion and for relieving gas and cramps.

The seeds can be sparingly added to salads for seasoning. The mature, green seeds have a mild anesthetic action on tissues in the mouth: when gently chewed and sucked, they will numb the tongue and gums in a manner similar to clove oil.

Medicinally, root pieces were placed in tooth cavities to stop pain. An infusion for a sore throat can be made by soaking the mashed root in water. Root tea was used for colic, cramps, headaches, sore throats, and flu.

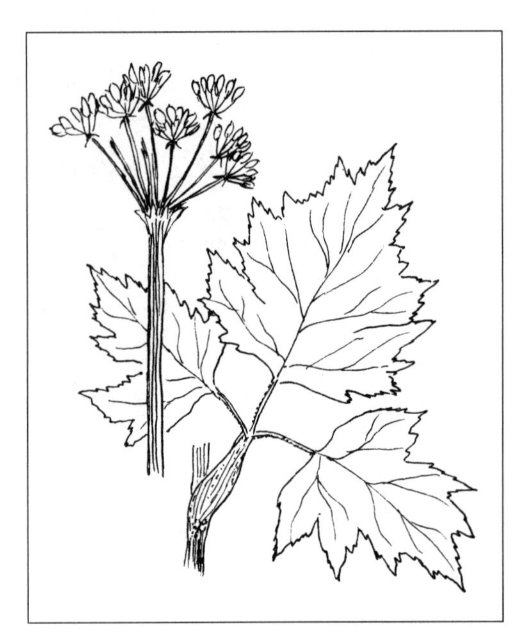

FIGURE 13. Cowparsnip (*Heracleum lanatum*).

The leaves of cowparsnip are large enough to be used as a toilet paper substitute and as a mild insect repellent. However, since furanocoumarin is present in the sap and the outer hairs, this use may be a problem for people with sensitive skin. When the sap makes contact with the skin, it can cause a sunburn effect (redness, blistering, and running sores) when exposed to light. As a poultice, the leaves were used externally for sores, bruises, and swellings (Muenscher 1962; Pojar and MacKinnon 1994).

The older stems, before the flower cluster unfolds, can be peeled and the inner tissue eaten raw or cooked, although it has an unpleasant taste. Cooking it in a couple of changes of water usually improves the taste and digestibility. In any case, cowparsnip is considered an excellent survival plant in the mountains. It was probably the most intensively used springtime green among Native Americans (Strike 1994).

Warning: Do not confuse this plant with other species in the same family that are highly toxic (i.e., *Cicuta* spp. and *Conium maculatum*).

Licorice-root (genus *Ligusticum*)

CANBY'S LICORICE-ROOT (*L. canbyi*)

FERNLEAF LICORICE-ROOT (*L. filicinum*)

Description and Habitat. Licorice-roots are native perennials with taproots that are sheathed by old leaf bases at the crown. The leaves have winged petioles, and the leaf blades are 1–3 times pinnately dissected. The flowers occur in compound umbels that are subtended by bracts, which are inconspicuous or even lacking. The petals are white to pink, and the fruits are oblong to elliptical with mostly winged ribs.

FIELD KEY TO THE LICORICE-ROOTS

1. Leaf segments are all linear, less than ⅛ inch wide .... *L. filicinum*
1. Leaf segments are broader than linear, usually more than ⅛ inch wide .... *L. canbyi*

Natural History Notes and Uses. Some of the species contain alkaloids, but the green stems and roots of *L. filicinum* may be eaten raw or cooked. Several species have been used medicinally. They contain volatile and fixed oils and a very bitter alkaloid that has been shown to increase blood flow to coronary arteries and the brain.

L. canbyi has been described as an antiviral, expectorant, anesthetic to the throat, and diaphoretic and as useful in the treatment of upper respiratory infections (Tilford 1993). Native Americans commonly chewed the dried roots for relief from sore throats, colds, toothache, headache, stomachache, fever, and heart problems (Hart 1996).

Biscuitroot, Desert Parsley (genus *Lomatium*)

WYETH BISCUITROOT (*L. ambiguum*)

TAPERTIP DESERT PARSLEY (*L. attenuatum*)

COUS BISCUITROOT (*L. cous*)

FERNLEAF BISCUITROOT (*L. dissectum*)

DESERT BISCUITROOT (*L. foeniculaceum*)

KING DESERT PARSLEY (*L. graveolens*)

GRAY'S BISCUITROOT (*L. grayi*)

BIGSEED BISCUITROOT (*L. macrocarpum*)

NUTTALL'S BISCUITROOT (*L. nuttallii*)

NORTHERN IDAHO BISCUITROOT (*L. orientale*)

NINELEAF BISCUITROOT (*L. triternatum*)

Description and Habitat. The biscuitroots are perennial plants with thick roots and leaves that are divided several times from the base. The white, yellow, pink, or purplish flowers are in compound umbels. The fruits are flattened and elliptical to oval, and the margins may or may not be winged. Most species are found in dry ground or rocky situations. The genus name means "small border," alluding to the wings of the fruit.

FIGURE 14. Biscuitroot (*Lomatium* spp.).

FIELD KEY TO THE BISCUITROOTS

1. Leaves appearing lace-like 2
1. Leaves not appearing lace-like 8

2. Plants more than 10 inches tall; leaf blade 5–13 inches long . . *L. dissectum*
2. Plants generally less than 10 inches tall; leaf blade less than 5 inches long 3

3. Plant with conspicuous, minute hairs 4
3. Plants without hairs or slightly rough to the touch 6

4. Ovaries and fruits hairy all over . . . . *L. foeniculaceum*
4. Ovaries and fruits without hairs 5

5. Involucel bractlets with dense, short, wool-like, tangled hairs . . . . *L. macrocarpum*
5. Involucel bractlets not as above . . . . *L. orientale*

6. Bractlets of involucel lance- to egg-shaped or elliptic . . . . *L. cous*
6. Bractlets of involucel linear to lance-shaped 7

7. Ultimate leaf segments conspicuously broadened near middle . . . . *L. attenuatum*
7. Ultimate leaf segments linear . . . . *L. grayi*

8. Involucel lacking . . . . *L. ambiguum*
8. Involucel present 9

9. Plants with minute hairs, at least on the flowering stalk . . . *L. triternatum*
9. Plants lacking hairs 10

10. Leaves somewhat fleshy . . . . *L. nuttallii*
10. Leaves not fleshy . . . . *L. graveolens*

Natural History Notes and Uses. All *Lomatium* species have edible roots and were an important staple among many Native Americans. They can be eaten raw or cooked or can be dried and ground into flour. The flour can then be kneaded into dough, flattened into cakes, and dried in the sun or baked. Some of the species we tried were too resinous to enjoy. With trial and error you will be able to find the more palatable species.

The green stems can be eaten after boiling in the springtime, but as summer progresses they become tough and fibrous. A tea can be brewed from the leaves, stems, and flowers. The tiny seeds are nutritious raw or roasted and can be dried and ground into meal.

The plants are rich in vitamin C. The seeds were chewed for colds and sore throats, and sap from the roots was used to treat cuts and sores. A poultice of pulverized roots was applied to a newborn baby's umbilical cord to facilitate healing. The roots were also chewed to relieve sore throats. Studies have shown that fernleaf biscuitroot has an ability to kill certain forms of influenza virus, especially those that infect the respiratory tract. It also has other antimicrobial and immuno-stimulating qualities (Whittlesey 1985).

Caution: As with any member of the carrot family, positive identification is important before consumption. Strike (1994) indicates that some indigenous peoples considered the mature stalks, leaves, roots, and flowers of *L. dissectum* poisonous. In fact, the roots were used as a fish poison and insecticide by some native people in the West. The plant contains phototoxic compounds of the furanocoumarin group, and one or more of these compounds is responsible for the poisonous properties found in the chocolate-tipped roots.

LEAFY WILD PARSLEY (*Musineon divaricatum*)

Description and Habitat. This perennial grows from a purple taproot. The plants are about 4–8 inches tall. The stems spread out from the root crown and then curve upward. The stalked leaves are thick and glossy and are divided 2 or 3 times into leaflets and lobes. The tiny yellow flowers are borne in small stalked clusters atop a main stalk. A whole flat-topped group of flower clusters is about 1 or 2 inches wide. The fruits are egg-shaped and evidently ribbed but without wings.

Natural History Notes and Uses. The genus is similar to *Lomatium* and *Cymopterus* but is distinguished by the completely wingless fruit. The generic name *Musineon* is surely a corruption, possibly of the Greek *mouseion*, "museum" or "shrine of Muses." Leafy wild parsley was first described for science by the eminent

German botanist Frederick Pursh (1774–1820). Although there are no known economic uses for this species, the fleshy root was used for food, raw or cooked, by the Blackfoot and Crow Indians.

GREAT BASIN INDIAN POTATO (*Orogenia linearifolia*)

Description and Habitat. This insignificant little plant, also known as snowpeas, has umbels that are only about ¼ inch across. What is remarkable, however, are the blooms that develop when thousands of plants emerge and flower simultaneously, shortly after snow melt in April in soggy areas of the higher foothills.

Natural History Notes and Uses. The common name refers to its edible roots. The roots can be boiled, steamed, roasted, or baked in any way used for preparing potatoes. When small, they can be eaten raw. The roots can also be cooked and mashed into cakes for drying, and when protected from moisture, they will keep a long time. The hard cakes can be soaked and cooked in stews.

Plant hunter Sereno Watson (1826–92) collected Indian potato in Utah's Wasatch Mountains while a member of the King expedition. *Orogenia* is derived from two Greek words, *oros* ("mountain") and *genea* ("race"), referring to the fact that it grows in the foothills and lower mountains. The species name refers to its narrow leaves.

FIGURE 15. Indian potato or snowpeas (*Orogenia linearifolia*).

Sweet Cicely, Sweetroot (genus *Osmorhiza*)

SWEET CICELY (*O. chilensis* = *O. berteroi*)

BLUNTSEED SWEETROOT (*O. depauperata*)

WESTERN SWEETROOT (*O. occidentalis*)

Description and Habitat. Sweet cicelys are herbaceous perennials from stout roots, with leaves twice divided into 3's. The flowers are borne in open, compound umbels that arise from leaf axils. The fruit is spindle-shaped and compressed along the sides. An examination of the fruits is important for differentiating the species, since the shape of the stylopodium (base of the style), a small structure at the tip of the fruit, is critical. Sweet cicelys occur on moist slopes, in open areas, and in forests.

FIELD KEY TO THE SWEET CICELYS

1. Flowers yellow and fruit glabrous; plants smell like anise . . . . *O. occidentalis*
1. Flowers are white to pink or purple; fruits hairy 2

2. Fruits concavely narrowed to a beak-like tip; flowers white . . . *O. chilensis*
2. Fruits abruptly rounded at the tip, no beak; flowers sometimes purplish or pink . . . . *O. depauperata*

Natural History Notes and Uses. The leaves of sweet cicely, which is also known as "dryland parsnip," can be boiled and eaten. The roots were dug in the spring and either pit-cooked or boiled as a vegetable. To us the taste is reminiscent of baby carrots.

The roots of *O. occidentalis* taste and smell like licorice or anise and can overwhelm the taste buds if eaten in large amounts. In small quantities this species can liven up the taste of teas or meals that are otherwise bland or unpleasant. Uses of other related species include a poultice from the roots for boils, cuts, sores, and wounds, and root tea for sore throat and upset stomach.

Caution: Western sweetroot resembles the very poisonous water hemlock (*Cicuta maculata*), but the strong smell of anise gives it away as sweetroot. Also, the venation of water hemlock is unique among the Apiaceae (carrot family). Sweet cicely can be confused with baneberry (*Actea rubra*), but baneberry is easy to identify by the cluster of red or white berries.

Yampah (genus *Perideridia*)

BOLANDER'S YAMPAH (*P. bolanderi*)

COMMON YAMPAH (*P. montana* = *P. gardneri*)

FIGURE 16. Yampah (*Perideridia* spp.).

Description and Habitat. The yampahs are biennial or perennial herbs with fascicled tuberous roots and pinnate leaves. The calyx teeth are well developed. The petals are white or pinkish, and the stylopodium is conic. The fruit is nearly terete or somewhat flattened laterally.

FIELD KEY TO THE YAMPAHS

1. Main leaves somewhat dissected; petioles dilated; fruit oblong	*P. bolanderi*
1. Main leaves only once or maybe twice pinnate or ternate; petioles not dilated; fruit oval	*P. montana*

Natural History Notes and Uses. All the species within this genus are edible. They were an important food of many indigenous peoples from British Columbia to California and the Great Basin region (Coffey 1993; Elias and Dykeman 1982). The finger-like roots have a pleasant, nutty flavor when eaten raw and resemble carrots when cooked. They are best when dug up before the flowers appear. The roots should be washed and peeled before cooking. They can be easily dried and will keep well for future use. When dried, the roots can be pounded and ground into flour or mashed into cakes. The seeds may be parched and ground or eaten whole.

WESTERN SNAKEROOT (*Sanicula graveolens*)

Description and Habitat. Western snakeroot is a perennial with a single stem arising from a taproot. The stem is often branched near its base, and these branches may be further branched above. The leaves are alternate and are all found on the stems; the lower leaves are well developed and petiolate. These often appear to come separately from the ground but are indeed attached to the main stem. The blades of the lower leaves are ternately to pinnately compound with 3–7 or more leaflets. Each leaflet is further pinnatifid with the lowest pair of leaflets separate from the rest by an entire leaf rachis or vein. The upper leaves are sessile and reduced in size and tend to have fewer leaflets.

Natural History Notes and Uses. Western snakeroot may be found from southern British Columbia south to California and east to western Montana and northwestern Wyoming. It is also found in Chile and Argentina. The herbage contains various alkaloids and should therefore be regarded as inedible.

SHOSHONE CARROT (*Shoshonea pulvinata*)

Description and Habitat. This mat-forming perennial herb grows ½ to 3 inches tall, with the mats up to 18 inches across. The leaves are once pinnately compound with 2–5 pairs of leaflets. The inflorescence is a compact, flat umbel less than ½ inch wide and scabrous, and the minute flowers are yellow. The fruits are sessile, oblong in shape, and prominently ribbed.

Natural History Notes and Uses. This species is endemic to southern Montana and the Absaroka and Owl Creek mountains of Wyoming (Fremont, Hot Springs, and Park counties). It is found growing in shallow, stony calcareous soils of exposed limestone outcrops, ridgetops, and talus slopes between elevations of 5,900 and 9,200 feet.

WATERPARSNIP (*Sium suave*)

Description and Habitat. Waterparsnip is a stout plant up to 5 feet tall. The leaves are pinnately divided, the flowers are white, and the fruit is oval. It is usually found in water or swampy areas in the mountains.

Natural History Notes and Uses. This highly variable species can be found throughout the West in wet habitats. The plant is very variable and can be hard to identify in the field. The characteristics to look for are the bracts and bracteoles subtending the rays and raylets of the inflorescence and the serrulate to serrate margins of the leaflets.

The long, fleshy root of waterparsnip, which is edible raw or cooked, has a sweet, carrot-like taste. The leaves and younger stems are also edible after cooking, but we found them better when boiled until tender. The older plants and flowers should be avoided because they are toxic and have been suspected of poisoning a wide range of livestock.

Warning: The plant is very similar in form and habitat to *Cicuta maculata* (water hemlock), which is the most poisonous vascular plant in North America. Both plants produce white flowers in umbrella-like clusters, and both grow in swampy ground at lake or pond edges. Waterparsnip has leaves that are once compound, whereas the leaves of water hemlock are 3 times compound. Water hemlock also has a distinctive turnip-like swelling at the base of the stem, which is usually chambered when cut open vertically and exudes a yellowish liquid along the cut surface. If in doubt, leave it alone!

MEADOW ZIZIA (*Zizia aptera*)

Description and Habitat. The plant has several 1½-inch clusters of small, bright yellow flowers growing in an open mounded form with lush, heart-shaped foliage. This plant has several common names including Golden Alexander, Heartleaf Alexander, Meadow Zizia, and Heart-leaved Meadow Parsnip. Meadow zizia ranges from Connecticut in the Northeast, south to subtropical Florida, and west to the Rocky Mountains and the Pacific Northwest of Canada.

Natural History Notes and Uses. Nonreproductive plants form compact rosettes up to 6 inches in diameter. Reproductive plants may attain heights of 3 feet and produce compound umbels of small, bright yellow flowers from May to July. These flowers are pollinated by a variety of bees and flies, some species of which focus exclusively on *Zizia* species. Although the plant produces defensive compounds, including the unique furanocoumarin apterin (named for the species), a number of insect species are herbivores on leaves, stems, and seeds of *Zizia*.

The presence of secondary compounds such as apterin may contribute to the potential medicinal value of *Zizia* species. For example, *Zizia* roots have been used by Native Americans as a tea to cure fevers, and the plant has been referred to as a vulnerary (wound-healing) agent (Foster and Duke 1990). This species has also been used to induce sleep and to alleviate syphilis (Foster and Duke 1990). However, the specific medicinal properties of *Zizia* have not been documented. If the species possesses medicinal value and has been used by Native Americans and/or early colonial settlers, this feature could in part explain its curious distribution in many parts of the United States.

DOGBANE FAMILY (APOCYNACEAE)

The dogbane family has about 200 genera and 2,000 species, mostly found in the tropics. Nearly all the members of this family are ***poisonous***, usually have milky juice, and have opposite leaves. The flowers have 5 petals that are fused into a tube, at the base of which the 5 stamens are borne. The 2 pistils are united only at the single, enlarged stigma and develop into 2 elongated, pod-like follicles. Some of the well-known genera are ornamentals such as *Vinca minor* (periwinkle) and *Nerium oleander* (oleander). *Rauwolfia serpentina* (Indian snakeroot), a tropical tree, yields a drug used in the treatment of high blood pressure, and periwinkle is a source of an anti-tumor drug.

Dogbane (genus *Apocynum*)

SPREADING DOGBANE (*A. androsaemifolium*)

INDIAN HEMP (*A. cannabinum*)

Description and Habitat. Dogbanes are perennial herbs with milky juice, leaves that are opposite, and pink, bell-shaped flowers that are borne in cymes. There is considerable hybridization between species. *Apocynum* is Greek meaning "noxious to dogs."

FIELD KEY TO THE DOGBANES

1. Flowers pinkish . . . . *A. androsaemifolium*
1. Flowers greenish . . . . *A. cannabinum*

Natural History Notes and Uses. The primary use of dogbanes is for fiber. The stem fibers are strong and can be used for rope, mats, baskets, bowstrings, fishing lines and nets, sewing, animal trap triggers, snares, cordage for bow-and-drill fire making, and weaving. One of the easiest ways to isolate the fibers is to soak the stems in water. Archaeologists working in Utah have discovered dogbane fiber nets dating back to about 5000 B.C. Many Native American tribes used dogbane to make rabbit-catching nets. Measuring about 200 feet long, 3–4 feet high, and with a 3-inch opening, the nets were propped on sticks across level ground. The men formed a line some distance away and advanced toward the nets, beating the brush with sticks and driving the rabbits into the nets (Ebeling 1986).

Dogbanes should be considered ***poisonous*** to humans if ingested. However, some authorities indicate that the small seeds can be parched and ground into meal to make fried cakes. Strike (1994) believes the seeds eaten were those of *A. pumilum* (mountain dogbane), considered a subspecies of *A. androsaemifolium.*

Dogbanes were extensively used as medicine by aboriginal peoples. They contain highly toxic glycosides and resins, and their primary medicinal constituents are cymarin and apocannocide. The cardiac glycosides may be useful in treating malignant tumors. Millspaugh (1974) describes spreading dogbane as an emetic that does not cause nausea, a cathartic, and a powerful diuretic and sudorific; it is also an expectorant and antisyphilitic.

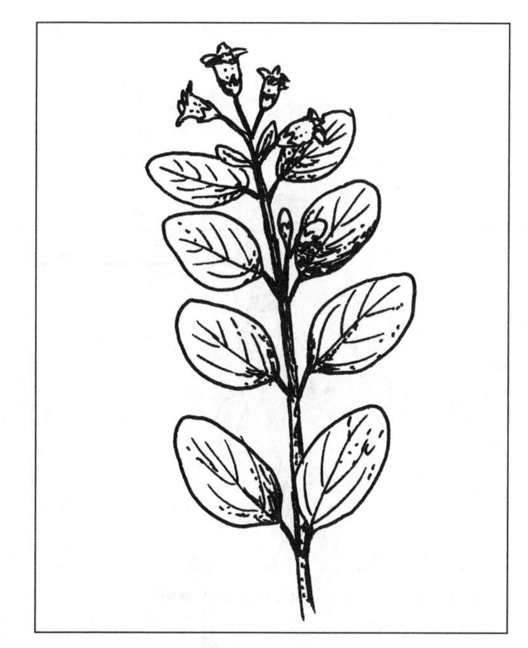

FIGURE 17. Dogbane (*Apocynum* spp.).

"Fiber plants are second only to food plants in terms of their usefulness to humans and their influence on the advancement of civilization. Tropical people use plant fibers for housing, clothing, hammocks, nets, baskets, fishing lines, and bow strings. Even in our industrial society, we use a variety of natural plant fibers.... In fact, the so-called synthetic fibers now providing much of our clothing are only reconstituted cellulose of plant origin."

— Mark Plotkin, *Biodiversity* (1988)

MILKWEED FAMILY (ASCLEPIADACEAE)

About 250 genera and 2,000 species are found in the milkweed family worldwide. Milky sap in the stems, leaves, and flowers inspired its common name. The family is of moderate economic importance as a source of ornamentals, latex, fibers, medicinal plants, and a few food plants. The sap contains latex, and in a few species it may yield industrially important hydrocarbons. The flowers are 5-parted.

The flowers are highly specialized to bring about cross-pollination through the aid of insects. From the base of the stamens grows a corolla-like structure called the crown. The 5 lobes of the crown are united at the base to form a column, and each bears a spur or horn on its inner surface. The structure surrounds the stamens and forces the anthers to touch around the pistil. The filaments are united into a column, and the sepals are turned back when the flower is in bloom. If you insert a pin into one of the slits of the crown of the flower and pull upward, you will usually find 2 pollen masses hanging on the pin. When a bee alights on the crown to obtain nectar, its foot is likely to slip into one of the slits, and when the foot is withdrawn, the pollen masses cling to it just as they did to the pin. When the bee flies to another flower, the pollen masses may come into contact with the stigmas, and pollination is thereby accomplished. The fruits of milkweed are pods containing many seeds that have long, silky hairs.

Milkweed (genus *Asclepias*)

PALLID MILKWEED (*A. cryptoceras*)

SHOWY MILKWEED (*A. speciosa*)

GREEN MILKWEED (*A. viridiflora*)

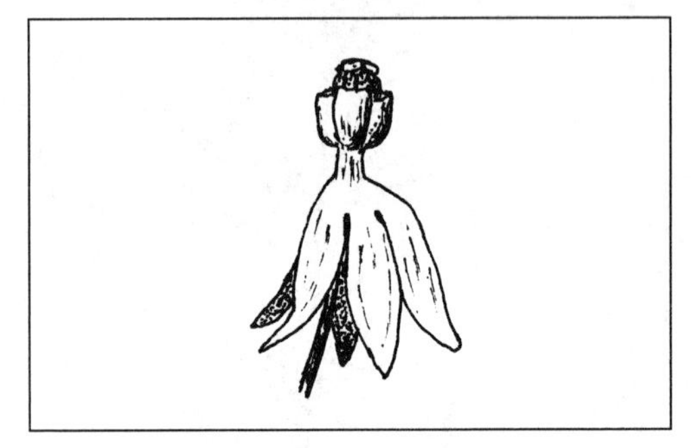

FIGURE 18. Milkweed (*Asclepias*) flower.

Description and Habitat. The milkweeds are erect or decumbent herbs from deep perennial roots. The leaves are opposite or whorled, and the corolla is deeply 5-parted with the segments reflexed. The corona hoods each have an incurved horn within. The name *Asclepias* refers to Asklepios, the Greek god of medicine. The larvae of the monarch butterfly (*Danaus plexippus*) feeds on the leaves of milkweeds.

FIELD KEY TO THE MILKWEEDS

1. Leaves oval-shaped, barely longer than wide; lower half of the stem is more or less glabrous .... *A. cryptoceras*
1. Leaves usually twice or more as long as wide; lower half of stem usually hairy 2

2. Corolla mostly greenish .... *A. viridiflora*
2. Corolla usually pink, rose, green, or yellow .... *A. speciosa*

Natural History Notes and Uses. Almost every book on edible plants in the United States lists milkweeds as being edible. It should be noted that in most cases they are referring to the eastern species of *Asclepias syriaca*, which does not occur in the area, and the western species of *A. speciosa* (showy milkweed) and *A. asperula* (spider milkweed). These latter two species contain cardiac glycosides, which can cause severe poisoning if not properly prepared or cooked.

Cardiac glycosides are an important class of naturally occurring drugs whose actions include both beneficial and toxic effects on the heart. Plants containing cardiac steroids have been used as poisons and heart drugs since at least 1500 B.C. Throughout history these plants or their extracts have been variously used as arrow poisons, emetics, diuretics, and heart tonics. Cardiac steroids are widely used in the modern treatment of congestive heart failure and for treatment of atrial fibrillation and flutter, yet their toxicity remains a serious problem.

With respect to the two western species mentioned above, Harrington (1967) suggests gathering plants when they are 6 inches tall and then boiling for 15–20 minutes in at least two or three changes of water. We tried five to seven changes of water, and the plants were still bitter! The unopened flower buds can be served like broccoli by boiling in at least three changes of water.

A strong fiber can be obtained from the inner bark to make rope, fishing line, clothing, and nets. Archaeologists have discovered clothing that was made from the fibers more than 10,000 years ago. The silky floss found in mature milkweed seed pods were used in making candle wicks, and the fiber can be spun like cotton. The floss is buoyant and water resistant and makes a good insulator. During World War II schoolchildren in Canada harvested milkweed floss from the wild for the U.S. Navy to use as a substitute for kapok in life vests. The dried pods were used as utensils. The sap was used as an adhesive.

Milkweeds contain asclepain in their plant parts and sap. Asclepain is a proteolytic enzyme that gives credence to the old pioneer remedy of applying the white juice daily to get rid of warts. Some Native American tribes used to collect the milk of *A. speciosa* and roll it in hand until it became firm enough to chew as gum, but it was not swallowed.

Milkweeds have been used in folk medicine for hundreds of years. The powdered roots of several species were reportedly used to treat wounds, pulmonary diseases, rheumatism, and gastrointestinal problems, among other ailments (Coffey 1993). Many modern medicines were originally derived from poisonous plants. Perhaps research will validate some of the medicinal uses of milkweeds and provide us with new medicines from the old (Lewis and Elvin-Lewis 1977).

Additionally, milkweeds have the potential to furnish an exciting array of products for industry and the home. In the future, as petroleum products dwindle, perhaps we will find ourselves taking a closer look at the possibilities of cultivating milkweeds for fiber, hydrocarbons, and medicines.

Warning: Milkweeds can be confused with other plants that produce milky juice such as dogbane (*Apocynum*), which is considered poisonous. Some species of milkweed at certain stages are poisonous to animals and could affect humans when eaten raw.

SUNFLOWER FAMILY (ASTERACEAE = COMPOSITAE)

A very diverse family with more than 20,000 species, the sunflower, or composite, family is the second largest plant family in the world. It contains many economically important crop plants such as sunflowers, lettuce, and artichokes. Numerous edible and useful composites are found in the area. The pollen of many composites is allergenic. The colorful flowers of many species produce yellow and orange dyes.

Although the family is considered by many botanists a "difficult" group, composites in general are relatively easy to recognize. The small flowers are arranged in heads that at first appear to be an individual flower but may actually consist of several to hundreds of florets (little flowers). Each flower has an inferior ovary, 5 stamens fused at the anthers, and 5 fused petals. The flowers at the center of the head are disk flowers; the peripheral ones are called ray flowers. Surrounding the head is a series of bracts called the involucre. The calyx (called the pappus and usually modified into thin hairs for dispersal), if present, crowns the summit of the ovary in the form of awns, capillary bristles, scales, or teeth. Nearly all composites are herbs or shrubs. Table 4 summarizes the various tribes of the family. Identification of composites is easier if you use 10- or 14-power magnification to discern the fine details of the phyllaries, pappus, and achenes. Careful dissection of the heads may also be required.

FIELD KEY TO THE SUNFLOWER FAMILY

1. Flowering heads composed of ray flowers, disk flowers absent; sap milky . . . . Group I
1. Flowering heads composed partly or entirely of disk flowers; sap not milky 2

2. White-woolly herbs with discoid flowering heads; involucral bracts papery, at least the upper portion . . . . Group II
2. Plants without the above combination of characters; involucral bracts usually greenish 3

WILDERNESS CORDAGE

The survival and continued existence of early humans was as much dependent on fiber as on food. The cordage produced from the fibers of wild plants can be used to make blankets, sandals, baskets, clothing, nets for fishing, and snares for capturing small game animals. In a wilderness situation, you might be surprised how important a piece of string or cordage can be. There are many species of plants in the northern Rockies that have fiber in the stem, leaves, or bark which can in one way or another be used as cordage. Some of the species discussed in this handbook are milkweed (*Asclepias* spp.), dogbane (*Apocynum* spp.), sagebrush (*Artemisia* spp.), cottonwood (*Populus* spp.), willow (*Salix* spp.), juniper (*Juniperus* spp.), thistle (*Cirsium* spp.), sunflower (*Helianthus* spp.), yucca (*Yucca* spp.), and nettle (*Urtica* spp.). There are probably other species that can be used, and finding out which ones is a matter of experimentation.

One of our favorite activities when teaching wilderness survival courses is to have participants twist about 50 feet of cordage. After accomplishing that task, students appreciate how much easier it is to go to the hardware store and buy it.

Twisting cordage is relatively easy once the fiber has been extracted from the plant. In most cases, the fiber is located in the outer part of the plant stem. The fibers can be removed by rubbing the stem between the hands or by carefully pounding it with a rounded rock or mallet. It is important not to break the length of the fiber while doing this. This process should result in soft, thread-like fibers.

To twist the fiber into a short piece of cordage, simply roll the length of fiber down your leg with an open palm until it is rounded and reasonably uniform in diameter. However, if longer cordage is required for fishing, sewing, nets, or clothing, it will be necessary for you to twist and splice. The following directions are for right-handed people. If you are left-handed, simply reverse the process.

With the strand of fiber in your left hand, bend it about in half so that you have two uneven lengths. Pinch the loop you've created with the left thumb and forefinger. With your right hand, grab the strand on the outside, twist it in the outward direction about halfway, and then lay it over the inside strand. Move your left thumb and forefinger down to hold it together. Again, with your right hand, grab the new outside strand, twist it out, and again lay it over the inside strand and reposition your left thumb and forefinger down to hold it together. Repeat this a few more times. When you are about 2–3 inches from the end of the shorter strand, take another length of fiber and lay it on the shorter piece and twist it as though it were part of the shorter strand. Continue twisting as before until you reach the 2- or 3-inch mark with the other strand. Again, attach a new strand and twist it as part of the new one. This is called splicing. If you are doing this for the first time, you'll soon find muscles in your fingers you never knew you had.

TABLE 4. Sunflower Family Tribes.

TRIBE	DESCRIPTION	REPRESENTATIVE GENERA
Anthemideae	Aromatic plants with lacy leaves. Disk flowers only or disk and ray flowers. The flower head bracts have membranous edges at tips.	*Achillea, Anthemis, Artemisia, Matricaria, Tanacetum*
Astereae	Disk flowers only or disk and ray flowers. If both flowers are present in the same head, disk flowers are usually yellow and ray flowers a contrasting color. Flower head bracts are greenish and shingled in graduated lengths, making several rows.	*Aster, Chrysopsis, Conyza, Erigeron, Grindelia, Gutierrezia, Haplopappus, Solidago, Townsendia*
Cichorieae	Ray flowers only. Stamens and pistils are present in each flower. Milky sap.	*Agoseris, Cichorium, Hieracium, Lactuca, Lygodesmia, Malocothrix, Microseris, Prenanthes, Sonchus, Stephanomeria, Taraxacum, Tragopogon,*
Cynareae	Disk flowers. Spiny plants that are thistle-like.	*Arctium, Carduus, Cirsium, Saussurea, Silybum*
Eupatorieae	Disk flowers white, sometimes tinged with pink.	*Eupatorium, Brickellia*
Helenieae	Disk flowers only, or ray and disk flowers. No papery scales in any flowers.	*Helenium, Chaenactis, Eriophyllum*
Heliantheae	Disk flowers only, or ray and disk flowers. One papery scale below all disk flowers. Pappus not hairy.	*Ambrosia, Balsamorhiza, Bidens, Gaillardia, Helianthella, Helianthus, Hulsea, Hymenopappus, Hymenoxys, Iva, Madia, Ratibida, Rudbeckia, Viguiera, Wyethia, Xanthium*
Inuleae	Head of disk flowers; anthers are tailed.	*Antennaria, Anaphalis, Gnaphalium*
Senecioneae	Ray flowers none or vestigial. Pappus composed of capillary bristles.	*Arnica, Senecio, Petasites, Tetradymia*

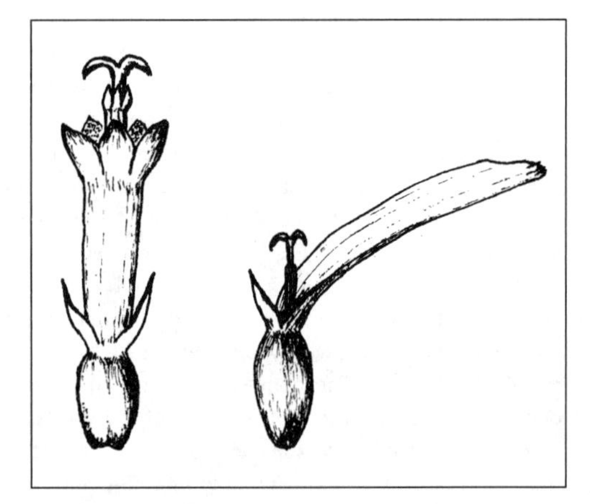

FIGURE 19. Sunflower family flower types: disk (tubular) flower and ray (ligulate) flower.

3. Flowering heads composed entirely of disk flowers, ray flowers absent . . 4
3. Flowering heads with both ray and disk flowers 5

4. Corolla of disk flowers yellow or orange . . . . Group III
4. Corolla of disk flowers purple, blue, pink, white, or absent. . . . . Group IV

5. Ray flowers yellow or orange . . . . Group V
5. Ray flowers purple, blue, pink, or white . . . . Group VI

— *Group I* —

1. Flowers yellow or orange 2
1. Flowers purple, blue, pink, or white. 13

2. Stems entirely leafless or with a few leaf-like bracts in the inflorescence; leaves all basal 3
2. Stems with leaves at least on the lower portion. 9

3. Flowering heads solitary on the naked stem. 4
3. Flowering heads 2 to many in a branched inflorescence 6

4. Involucral bracts in 2 distinct series, the outer often reflexed; achenes with spines or points near the summit of the body . . . *Taraxacum*
4. Involucral bracts in 1 or more overlapping series, never reflexed; achenes without spines or points 5

5. Pappus bristles broad and flattened toward the base; achenes without a narrow beak between the body and the pappus. . . . . *Microseris*
5. Pappus bristles not broad and flattened toward the base; achenes with a narrow beak between the body and the pappus . . *Agoseris*

6. Involucral bracts glabrous . . . . *Microseris*
6. Involucral bracts hairy 7

7. Involucre ¼ to 1 inch high; some leaves usually with backward-pointing lobes .... *Crepis*
7. Involucre less than ¼ inch high; leaves entire or toothed, without backward-pointing lobes .... *Hieracium*

8. Leaves with entire or shallowly toothed but not prickly margins 9
8. Leaves usually deeply lobed; if not, then margins prickly 11

9. Involucral bracts hairy .... *Hieracium*
9. Involucral bracts glabrous 10

10. Stem leafy on the lower half only .... *Microseris*
10. Stem leafy to the top .... *Tragopogon*

11. Achenes not flattened, plants not prickly .... *Crepis*
11. Achenes usually flattened, plants sometimes prickly on the leaf margins 12

12. Mature achenes without a slender beak between the body and pappus; heads with 85–200 flowers .... *Sonchus*
12. Mature achenes with or without a beak; heads with 11–50 flowers .... *Lactuca*

13. Flowers white; leaves not deeply lobed .... *Hieracium*
13. Flowers usually purple, blue, or pink; if white, then leaves deeply lobed 14

14. Pappus of minute scales .... *Cichorium*
14. Pappus of thin, hair-like bristles 15

15. Flowers pink; pappus bristles branched and feather-like .... *Stephanomeria*
15. Flowers blue or white; pappus bristles unbranched .... *Lactuca*

— *Group II* —

1. Involucral bracts almost entirely white and papery, without hairs .... *Anaphalis*
1. Involucral bracts with a portion that is colored or hairy 2

2. Plants annual .... *Gnaphalium*
2. Plants perennial 3

3. Plants with a taproot, without tufted basal leaves .... *Gnaphalium*
3. Plants with fibrous roots, stolons often present .... *Antennaria*

— *Group III* —

1. Shrubs with woody stems 2
1. Plants herbaceous 4

2. Involucre of 4–5 bracts .... *Tetradymia*
2. Involucre of 6 to many bracts 3

3. Leaves deeply divided or lobed at the tip .... *Artemisia*
3. Leaves entire .... *Chrysothamnus*

4. Leaves opposite .... *Arnica*
4. Leaves alternate 5

5. Pappus of slender bristles 6
5. Pappus lacking or consisting of a low ridge at the top of the achene 7

6. Leaf blade deeply lobed into linear segments .... *Erigeron*
6. Leaf blade shallowly lobed or entire .... *Senecio*

7. Flowering heads solitary at the ends of branches; plants annual . . . . . *Matricaria*
7. Heads many in inflorescence; plants biennial or perennial 8

8. Inflorescence narrow or spreading, not flat-topped . . . . . *Artemisia*
8. Inflorescence flat-topped . . . . . *Tanacetum*

— *Group IV* —

1. Leaves with spines at least ⅛ inch long on the margins 2
1. Leaves without spines on the margins 3

2. Flowers stalks straight, heads held erect . . . . . *Cirsium*
2. Flowerstalks bent, heads nodding . . . . . *Carduus*

3. Involucral bracts spiny, fruit of some species a bur 4
3. Involucral bracts not spiny 6

4. Spines of involucral bracts straight . . . . . *Centaurea*
4. Spines of involucre hooked 5

5. Involucral bracts united to form a solid bur; flowers 2, inconspicuous . . . . . *Xanthium*
5. Involucral bracts separate; flowers many, apparent . . . . . *Arctium*

6. Pappus of hair-like bristles, these sometimes branched and feathery 7
6. Pappus of scales (broader than bristles) or lacking) 11

7. Pappus bristles branched and feathery 8
7. Pappus bristles unbranched 9

8. Leaves lance-shaped, many greater than ¼ inch wide . . . . . *Saussurea*
8. Leaves linear . . . . . *Liatris*

9. Basal leaves large and conspicuous at flowering time . . . . . *Petasites*
9. Basal leaves mostly deciduous by flowering time 10

10. Involucral bracts mostly of equal length . . . . . *Eupatorium*
10. Involucral bracts not equal in length, in 2 to several series . . . . . *Brickellia*

11. Leaves pinnately divided or deeply lobed 12
11. Leaves toothed or entire, not deeply lobed or divided . . . . . *Iva*

12. Flowers without corollas, plants annual . . . . . *Ambrosia*
12. Corolla present, plants perennial . . . . . *Chaenactis*

— *Group V* —

1. Pappus composed of simple or branched hair-like bristles 2
1. Pappus a low crown or lacking or composed of scales or awns, wider than hair-like bristles 9

2. Leaves opposite . . . . . *Arnica*
2. Leaves alternate or basal 3

3. Involucral bracts of equal length . . . . . *Senecio*
3. Involucral bracts in 2 to several series, often overlapping 4

4. Leaves mainly basal, stem leaves reduced 5
4. Stem leaves conspicuous 6

5. Involucral bracts glabrous, black at tip . . . . . *Haplopappus*
5. Involucral bracts hairy, green throughout . . . . . *Erigeron*

6. Involucre less than ¼ inch high, rays less than ¼ inch long 7
6. Involucre greater than ¼ inch high, rays greater than ¼ inch long 8

7. Flowering heads solitary. . . . . *Haplopappus*
7. Flowering heads several to many . . . . *Solidago*

8. Plants herbaceous with basal leaves deciduous by flowering time . . . . *Heterotheca*
8. Plants shrubs or herbaceous with persistent basal leaves . . . *Haplopappus*

9. Lower stem leaves pinnately divided 10
9. Lower stem leaves with entire, toothed, or lobed margins, not deeply divided 12

10. Plants covered with dense, white hair . . . . *Eriophyllum*
10. Plants not densely white-hairy 11

11. Flowering heads columnar in shape, more than ¼ inch tall. . . . . . *Ratibida*
11. Flowering heads more disk-shaped, less than ¼ inch tall. . . . . . . *Anthemis*

12. Rays 3-lobed at the tip 13
12. Rays entire-margined at the tip. 14

13. Rays reflexed downward, disk flowers yellow. . . . . *Helenium*
13. Rays not reflexed, disk flowers purple . . . . *Gaillardia*

14. Involucral bracts hairy 15
14. Involucral bracts glabrous 18

15. Stem leaves lacking, only 1–2 narrow bracts present, basal leaves large . . . . *Balsamorhiza*
15. Stem leaves not greatly reduced 16

16. Plants occurring at timberline or above . . . . *Hulsea*
16. Plants of lower elevations (montane). 17

17. Flowering heads large, rays more than ¼ inch long. . . . . *Helianthus*
17. Flowering heads small, rays less than ¼ inch long. . . . . *Madia*

18. Lower stem leaves opposite. . . . . *Bidens*
18. Lower stem leaves alternate. 19

19. Rays large, more than ¾ inch long . . . . *Wyethia*
19. Rays small, less than ¾ inch long 20

20. Involucral bracts resinous . . . . *Grindelia*
20. Involucral bracts not resinous. . . . . *Gutierrezia*

— *Group VI* —

1. Flowering heads without true rays, but marginal disk flowers enlarged and resembling ray flowers . . . . *Centaurea*
1. True ray flowers present. 2

2. Pappus of simple or branched hair-like bristles. 3
2. Pappus of scales or awns or lacking 8

3. Plants annual . . . . *Conyza*
3. Plants biennial or perennial. 4

4. Teeth of leaf margins spine-toothed . . . . *Machaeranthera*
4. Leaves without spines on the margins. 5

5. Basal leaves triangular in outline, white-woolly beneath. . . . . *Petasites*
5. Basal leaves narrower in outline, usually not white-woolly. 6

6. Involucre usually more than ¼ inch high, heads (not including rays) generally more than ¾ inch wide, basal leaves numerous and linear lance-shaped. . . . . *Townsendia*
6. Flowering heads usually smaller, or if as large, then leaves different 7

FIGURE 20. Yarrow (*Achillea millefolium*).

7. Involucral bracts subequal (not quite equal) and not much overlapping. . . . . *Erigeron*
7. Involucral bracts in 2 to several series and overlapping. . . . . *Aster*

8. Rays less than ¼ inch long . . . . *Achillea*
8. Rays more than ¼ inch long. 9

9. Papery, scale-like bracts between the disk flowers, plants ill-smelling . . . . *Anthemis*
9. Bracts between the disk flowers lacking, plants not ill-smelling. . . . . *Matricaria*

COMMON YARROW (*Achillea millefolium*)

Description and Habitat. Yarrow is a strongly scented perennial herb with alternate leaves that are finely dissected and appear feathery. The white or sometimes yellow flowers are borne in a flat-topped corymb. Yarrow is widespread and can be found in a variety of habitats from low elevations to above timberline. The generic name honors Achilles, who as a child was dipped into a yarrow bath by his mother to make him invincible. Since she held him by his heels, he was made vulnerable through his "Achilles heel."

Natural History Notes and Uses. Yarrow is often referred to as "poor man's pepper." The leaves can be dried, ground, and used as seasoning. The young leaves can be added to salads. The aromatic leaves were also placed in freshly split fish to expedite drying.

Medicinally, the leaves and stems can be dried, boiled in water, strained, and drunk to remedy a run-down condition or to help with an upset stomach. Taken as a hot infusion, yarrow will increase body temperature, open skin pores, and stimulate perspiration, making it a valuable herb for colds and fevers. The juice can be used as an eyewash to reduce redness. Leaves can be used to stop bleeding in small wounds and to heal rashes when applied directly to the skin. Leaves were also chewed to relieve toothaches. A poultice of mashed leaves can be applied to swellings or sores. More than 200 biologically active compounds have been identified from the species; some are known to be quite toxic (Foster and Duke 1990). Prolonged use of yarrow may cause allergic rashes and make the skin more sensitive to sunlight.

Rubbing the plant on one's clothing and skin was an ancient prescription for repelling biting insects. The stalks burned on coals were said to deter mosquitoes. The leaves were used in herbal snuffs and smoking tobaccos. Yarrow has also been used as a hops substitute for brewing "yarrow beer."

False Dandelion, Agoseris (genus *Agoseris*)

ORANGE AGOSERIS (*A. aurantiaca*)

PALE AGOSERIS (*A. glauca*)

Description and Habitat. False dandelions are annual or perennial, tap-rooted herbs with milky juice that resemble common dandelions (*Taraxacum*). The flowers are all ray, yellow or occasionally orange. The pappus is white with hair-like bristles. The fruit (achene) is conspicuously 10-nerved. False dandelions occur on moist to dry ground, in meadows and open areas at various elevations. The genus name is from the Greek, meaning "goat chicory."

FIELD KEY TO THE FALSE DANDELIONS

1. Flowers orange when fresh; achene beak ½ to 2 times as long as body . . . . *A. aurantiaca*
1. Flowers yellow when fresh; achene beak less than half as long as body . . . . *A. glauca*

Natural History Notes and Uses. The leaves and roots of some species are edible when cooked but are bitter, especially late in the season. Strike (1994) indicates that the seeds were eaten by the Chumash Indians in southern California. The sap from the leaves of some species, when hardened, can be used as chewing gum. Since the sap from some species is very thick and insoluble, it is useful for waterproofing containers (e.g., coiled baskets) and footwear.

FIGURE 21. Mountain dandelion (*Agoseris* spp.).

FLATSPINE BURR RAGWEED (*Ambrosia acanthicarpa*)

Description and Habitat. Flatspine burr ragweed, or sand-bur, as it is sometimes called, is an erect, branched annual that grows to about 30 inches tall from a slender taproot with sparingly hispid stems, winged petioles, and bipinnatifid leaves almost ovate in outline. The numerous staminate flowers heads are about ⅛ inch in diameter and hang down from higher up on the stem, and the pistillate heads are 1-flowered in the leaf axils. The fruit is an ovoid bur with 9–18 straight, flattened spines that are pale golden in color. Flatspine burr ragweed is a common weedy plant that prefers sandy soils, creek bottoms, and roadsides. It blooms from August to November.

Natural History Notes and Uses. Since the wind-blown pollen is highly allergenic, this ragweed, like others, is a notorious cause of hay fever where the plants are common. The genus name is from the Greek and Latin and refers to an early name for aromatic plants. *Ambrosia* is from the Greek, meaning "food of the gods," and is the classical name for various plants. Its application to these weedy specimens is obscure. The Greek words *akantha* (a thorn or prickle) and *karpos* (fruit) refer to the spines on the fruit.

Other species of *Ambrosia* were used by Native Americans. For example, *Ambrosia trifida* was cultivated in prehistoric times for its edible seeds in the midwestern United States. A tea from the leaves of *A. trifida* was formerly used for fevers, diarrhea, dysentery, and nose-bleeds and was gargled for sore throats. Other species were used in teas for various medicinal purposes. The heated leaves of *A. psilostachia* were used as a poultice to ease aching joints, and a decoction was used to bathe sores and burns. Native Americans rubbed the leaves of *A. artemisiifolia* on insect bites, infected toes, minor skin eruptions, and hives. A tea was used for fevers, nausea, mucous discharges, and intestinal cramping.

WESTERN PEARLY-EVERLASTING (*Anaphalis margaritacea*)

Description and Habitat. This perennial grows up to 36 inches tall and has white woolly herbage. The leaves are alternate, entire, and lanceolate to linear or oblong. The sessile leaves are also woolly beneath but soon becoming green and glabrous above. The flowering heads form a flat-topped, terminal cluster. The involucral bracts are papery, pearly white, and imbricated in several series. Pearly-everlasting is found in openings along trails and on talus slopes. It blooms from June to August.

Natural History Notes and Uses. The herbage of western pearly-everlasting has been used as a tobacco substitute to relieve headaches. As a tea, the plant has been used for colds, bronchial coughs, and throat infections. The whole plant can be used as a wash or poultice for external wounds. It has also been used for rheumatism, burns, sores, bruises, and swellings.

FIGURE 22. Pussytoes (*Antennaria rosea*).

Pussytoes (genus *Antennaria*)

PEARLY PUSSYTOES (*A. anaphaloides*)

FLATTOP PUSSYTOES (*A. corymbosa*)

WHIP PUSSYTOES (*A. flagellaris*)

WOOLLY PUSSYTOES (*A. lanata*)

RUSH PUSSYTOES (*A. luzuloides*)

ROCKY MOUNTAIN PUSSYTOES (*A. media*)

LITTLE-LEAF PUSSYTOES (*A. microphylla*)

SMALL-LEAF PUSSYTOES (*A. parvifolia*)

SHOWY PUSSYTOES (*A. pulcherrima*)

RACEME PUSSYTOES (*A. racemosa*)

UMBER PUSSYTOES (*A. umbrinella*)

Description and Habitat. Pussytoes are herbaceous perennials, often mat-forming. The heads are discoid with small white flowers surrounded by bracts that are typically hairy below with a smooth and membranous portion varying in color from white to pink to dark brown or black. The pappus is composed of numerous hairy bristles. The various species can be found in dry, open habitats or in moist or seasonally wet places from the foothills to alpine areas.

FIELD KEY TO THE PUSSYTOES

1. Plants are mat-forming, having numerous stolons, and leafy 2
1. Plants not mat-forming; without stolons 9

2. Plants generally less than 2 inches tall; heads found among the leaves, and solitary 3
2. Plants more than 2 inches tall; heads well above the basal leaves. 4

3. Heads solitary; stolons naked .... *A. flagellaris*
3. Heads 3–8; stolons leafy .... *A. parvifolia*

4. Basal leaves definitely less hairy above than below, becoming green above and white tomentose beneath .... *A. racemosa*
4. Basal leaves similar on both sides, pubescent, silver or gray 5

5. Terminal, scarious part of outer involucral bracts brown to black-green 6
5. Terminal, scarious part of involucral bracts white or pink, maybe with a dark spot at the base.................................. 7

6. Terminal part of involucral bract entirely black-green and often sharply pointed .... *A. media*
6. Terminal part of involucral bract becoming white, or the whole scarious bract remaining brown; tip blunt ... *A. umbrinella*

7. Involucral bracts with conspicuous dark spot at base of scarious part...................................... *A. corymbosa*
7. Involucral bracts without dark spot at base of scarious part............ 8

8. Flowering heads relatively large, the involucre usually ¼ to ½ inch high....................................... *A. parviflora*
8. Flowering heads smaller, the involucre less than ¼ inch high .... *A. microphylla*

9. Involucre scarious throughout, glabrous or almost .... *A. luzuloides*
9. Involucre hairy, less scarious on lower parts of the bracts............. 10

10. Plants less than 8 inches tall; involucre blackish, dark brown or green, and may be white at the tip............. *A. lanata*
10. Plants more than 8 inches tall; involucre brownish to white........... 11

11. Involucre mostly brownish green...................... *A. pulcherrima*
11. Involucre mostly white.............................. *A. anaphaloides*

Natural History Notes and Uses. The sap from the stems of most species can be chewed like gum and has some nutritive value. Moore (1979) indicates that a tablespoon of the chopped plant steeped in hot water is an excellent remedy for liver inflammation. It has also been used as an astringent for the intestinal tract. Leaves can be poulticed for use on bruises, sprains, and swelling. The blossoms can be boiled and used to bathe sore or ulcerated feet, or mashed and applied to sores. *A. microphylla* was chewed as a cough remedy by the Thompson Indians in British Columbia. The tiny leaves were also dried, stripped, and used as one of the ingredients in Indian tobacco (Teit 1930).

GOLDEN CHAMOMILE, DOGFENNEL (*Anthemis tinctoria*)

Description and Habitat. Golden chamomile is an annual or short-lived perennial herbaceous plant with radiate flowering heads composed of white or yellow ray flowers. An introduced European weed, it is usually found at the lower elevations as an escapee from gardens and cultivation.

Natural History Notes and Uses. A tea can be made from related species to induce sweating and vomiting. An astringent and diuretic, it has been used for ailments such as fevers, colds, diarrhea, dropsy, rheumatism, and headaches (Foster and Duke

1990). The leaves were rubbed on insect bites and stings. Golden chamomile was originally considered a noxious weed of clover fields but has since been brought into cultivation for horticultural purposes.

LESSER BURDOCK (*Arctium minus*)

Description and Habitat. Lesser burdock is a coarse biennial with large rounded to ovate leaves. There are several flowering heads composed of pink or purplish tubular flowers. This naturalized species from Europe is an occasional weed at the lower elevations. It flowers from June to August. *Arctium lappa* (greater burdock) also occurs in the area and can be used in much the same way as lesser burdock.

Natural History Notes and Uses. Rich in vitamins and iron, the young leaves and shoots can be gathered for use as a potherb or eaten raw in salad. The plant has a strong rank taste and an objectionable odor. The inner pith-like material of the young stems can be eaten raw, but we find it better when boiled in one or two changes of water. The roots of young plants can be sliced and cooked, then eaten. The older roots can be roasted and ground for use as a tea or coffee substitute. The seeds can be dampened and grown as sprouts.

The medicinal uses of the plant predate its use as a food plant. The Chinese are said to have used the plant as a blood purifier for thousands of years. Research has confirmed the usefulness of common burdock in the treatment of rheumatism, water retention, and high blood pressure. As a wash, it was used externally for hives, eczema, and other skin problems. The crushed seeds were used as a poultice. The tall, rigid stems were used as drills for primitive fire-starting techniques. The burs can be used as a sort of survival Velcro for holding clothes together.

It is unclear exactly when our ancestors started to use fire. Among the earliest signs are layers of ash found in a cave (Zhoukoudian) in China. These date back about half a million years, when the cave was inhabited by *Homo erectus*, the species that preceded humans. Making fire, as opposed to starting it from a natural blaze, is a much more recent skill. It is believed that this ability probably started less than 25,000 years ago.

Arnica (genus *Arnica*)

CHAMISSO ARNICA (*A. chamissonis*)

HEARTLEAF ARNICA (*A. cordifolia*)

FOOTHILL ARNICA (*A. fulgens*)

BROADLEAF ARNICA (*A. latifolia*)

SPEARLEAF ARNICA (*A. longifolia*)

HAIRY ARNICA (*A. mollis*)

PARRY'S ARNICA (*A. parryi*)

RYDBERG'S ARNICA (*A. rydbergii*)

Description and Habitat. These perennials arise from a rhizome or caudex, and the rootstalk produces a cluster of leaves the first year and a flowering stem the next. The leaves are simple and opposite. Flowering heads, 1 to several per stem, are radiate or discoid. The pappus consists of many capillary bristles. The name translates as "lamb's skin" and refers to the modified leaves (bracts), which are usually woolly.

FIELD KEY TO THE ARNICAS

1. Stem leaves mostly 5–12 pairs 2
1. Stem leaves fewer than 4 pairs 3

2. Bracts with a tuft of long hairs at the tip; flower heads hemispheric in shape .... *A. chamissonis*
2. Bracts without a distinct tuft of hair at tip; flower heads bell-shaped .... *A. longifolia*

3. Flower heads without rays and nodding in the bud .... *A. parryi*
3. Flower heads radiate 4

4. Pappus straw-colored or tawny .... *A. mollis*
4. Pappus white 5

5. Basal leaves broadly arrow- or heart-shaped, less than 3 times as long as wide 6
5. Basal leaves narrow to lance-shaped, more than 3 times as long as wide 7

6. Leaves mostly heart-shaped .... *A. cordifolia*
6. Leaves mostly ovate .... *A. latifolia*

7. Flower heads with 7–10 flowers .... *A. rydbergii*
7. Flower heads with 10–25 flowers .... *A. fulgens*

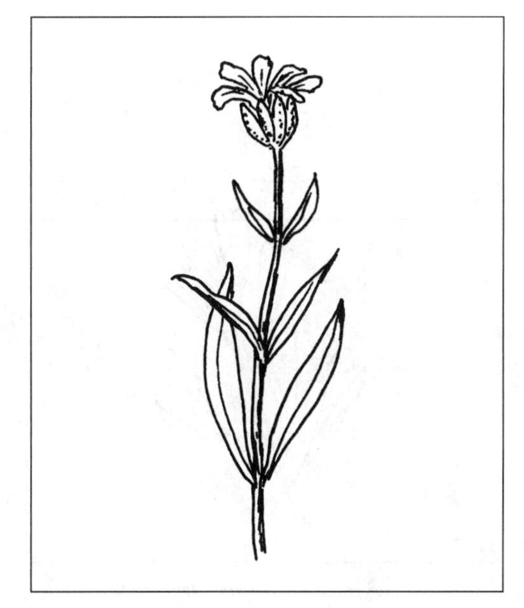

FIGURE 23. Arnica (*Arnica* spp.).

Natural History Notes and Uses. All the species are reported to be ***poisonous*** if taken internally. Arnica contains arnicin, choline, a volatile oil called arnidendiol, angelic and formic acid, and other unidentified substances that can alter cardiovascular activity. The Food and Drug Administration lists arnica as "unsafe" and bans its use for human consumption. Moore (1979) states that arnica is an *external* remedy only. The chopped plant is steeped in rubbing alcohol for about a week and squeezed through a cloth. The liniment is then used for joint inflammations, sprains, and sore muscles. It should not be used if the skin is broken since it is toxic if it enters the bloodstream. Arnica is useful as a topical preparation for bruises, sprains, and other closed injuries. When gathering, grasp the plant at the base of the stem just below the ground to leave the rhizome for continued growth. Wear gloves because the volatile oils can be absorbed.

Arnica flowers and roots have also been used extensively in European folk medicine as pain relievers, expectorants, and stimulants. Johann Wolfgang von Goethe (1749–1832), the German philosopher and poet, drank arnica tea to ease his angina in old age (Coffey 1993).

Warning: All species of *Arnica* are reported to be poisonous if taken internally.

Sagebrush, Sagewort, Wormwood (genus *Artemisia*)

LOW SAGEBRUSH (*A. arbuscula*)

FIELD SAGEWORT (*A. campestris*)

SILVER SAGEBRUSH (*A. cana*)

WORMWOOD, TARRAGON (*A. dracunculus*)

FRINGED SAGEWORT (*A. frigida*)

LONGLEAF SAGEBRUSH (*A. longifolia*)

LOUISIANA SAGEWORT (*A. ludoviciana*)

MICHAUX'S SAGEBRUSH (*A. michauxiana*)

NORWAY SAGEWORT (*A. norvegica = A. arctica*)

BLACK SAGEBRUSH (*A. nova*)

BIRDFOOT SAGEBRUSH (*A. pedatifida*)

ALPINE SAGEBRUSH (*A. scopulorum*)

BUD SAGEBRUSH (*A. spinescens = Picrothamnus desertorum*)

BIG SAGEBRUSH (*A. tridentata*)

THREE-TIP SAGEBRUSH (*A. tripartita*)

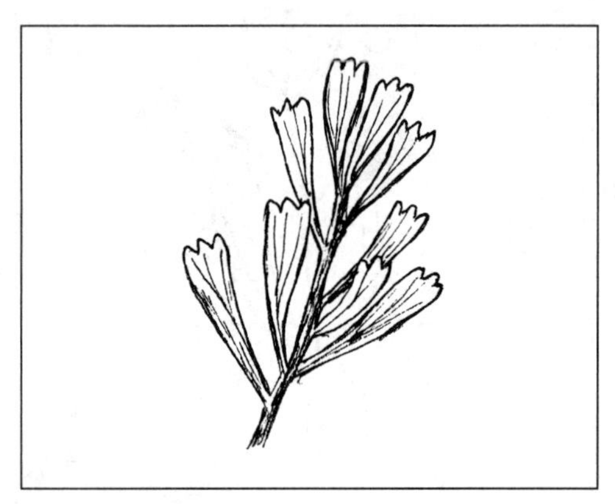

FIGURE 24. Sagebrush (*Artemisia tridentata*).

Description and Habitat. *Artemisia* sagebrushes include annual, biennial, and perennial herbs and shrubs. They are mostly aromatic with entire or dissected leaves. The flower heads are small, inconspicuous, and composed of disk flowers. The genus name honors Artemisia, wife of Mausolus, who was the king of Caria, a province in Asia Minor. After the king's death in 350 B.C., Artemisia built the renowned Mausoleum, one of the Seven Wonders of the World.

FIELD KEY TO THE SAGEBRUSHES

1. Flowers all alike, perfect; plants are shrubs, not spiny 2
1. Flowers of 2 kinds, marginal flowers pistillate,
   the inner flowers perfect but sometimes sterile; herbs or subshrubs. . . . 6

2. Leaves 3-toothed or 3–6-parted at the apex. 3
2. Leaves all linear, entire or almost . . . . . *A. cana*

3. Leaf segments long-linear or filiform. . . . . *A. tripartita*
3. Leaf segments toothed or lobed at apex, or entire 4

4. Flowering heads few, rather large,
   in a narrow panicle; leaves lobed . . . . . *A. arbuscula*
4. Flowering heads many, small; leaves 3-toothed. 5

5. Shrubs less than 15 inches tall; involucres glabrate . . . . . *A. nova*
5. Shrubs 15–80 inches tall; involucres tomentose . . . . . *A. tridentata*

6. Plants spiny, subshrub; white tomentose;
   spring-blooming . . . . . *A. spinescens*
6. Plants not spiny 7

7. Inner flowers sterile, the style entire, the ovary aborting;
   plants herbaceous 8
7. Inner flowers also fertile, the style 2-cleft 11

8. Herbage green, glabrous; leaves entire . . . . . *A. dracunculus*
8. Herbage grayish white; leaves pinnatifid or dissected. 9

9. Flowers brown . . . . . *A. campestris*
9. Flowers yellow 10

10. Plants 11–27 inches tall. . . . . *A. campestris*
10. Plants less than 6 inches tall . . . . . *A. pedatifida*

11. Receptacle with long woolly hairs 12
11. Receptacle without long woolly hairs among the flowers 13

12. Heads small, less than ¼ inch wide, and numerous. . . . . *A. frigida*
12. Heads larger, ¼ to ½ inch wide, few or solitary . . . . *A. scopulorum*

13. Plants silky-haired or glabrate, never tomentose; leaves twice 3–7-parted, primarily basal . . . . *A. norvegica*
13. Plants tomentose, at least on the lower surface of the leaves; leaves present on stem 14

14. Involucres densely tomentose 15
14. Involucres glabrous or glabrate, at least in age. . . . . *A. michauxiana*

15. Leaves usually glabrate on upper surface in age. . . . . *A. ludoviciana*
15. Leaves tomentose on both surfaces 16

16. Leaf margins rolled toward the lower side . . . . *A. longifolia*
16. Leaf margins not as above . . . . *A. ludoviciana*

Natural History Notes and Uses. The seeds of many species are edible raw or as flour. The seeds and peeled shoots of *A. douglasiana* and *A. ludoviciana* were eaten raw by Native Americans.

Herbage of various *Artemisia* species may be toxic if eaten in large amounts but may be used in small quantities to flavor stews, soups, and other foods. A tea from leaves was a cure for colds and sore eyes and was used as a hair tonic. Some of the "softer" species can be used as toilet paper and foot deodorant. Crushed leaves can be mixed with stored meat to maintain a good odor. Since many species are aromatic, they can be added to buried food caches to mask the odor and rubbed on the body to mask human scent while hunting. The wood of *A. tridentata* is a good material for fire drills. Although cordage can be made from the bark, it is not very strong.

Many species of *Artemisia* have been used as medicine by Native Americans and were used in sweathouses to relieve various ailments. A strong tea of *A. ludoviciana* was used as an astringent for eczema and as a deodorant and antiperspirant for underarms and feet. A weak tea was used for stomachaches. For sinus ailments, headaches, and nose-bleeds, a leaf snuff was used. A leaf or root tea of *A. dracunculus* was used for colds, dysentery, and headaches and to promote an appetite. The leaves were poulticed and used for wounds and bruises. Moore (1979) says that *A. tridentata* is strongly antimicrobial and was used as a disinfectant and cleansing wash. Volatile oils in *A. tridentata* are responsible for its pungent aroma and are so flammable that they can cause even green plants to burn. It should also be noted that the Food and Drug Administration classifies *Artemisia* as an unsafe herb containing "a volatile oil which is an active narcotic poison" (Duke 1992:222).

WILDERNESS FOOD STORAGE PIT

To store foods for extended periods of time, Native Americans used storage pits. After a hole was dug, moisture was removed from the soil by lining the pit with hot rocks and allowing it to steam. With the rocks left in place, the pit was then lined with dried grasses, and the food was placed inside. On top of the food, dried bark from junipers or other plants high in tannic acid was placed to repel insects. On top of this, dried, aromatic, nonpoisonous leaves such as sagebrush were placed to disguise the smell of the food. Finally, the pit was covered with a thick layer of dirt and heavy rocks to prevent animals from getting to the stored food.

Aster (genus *Aster*)

Description and Habitat. The many aster species are perennial herbs with alternate leaves. The ray flowers are pistillate, containing female parts that range in color from blue, purple, and white to pink (never yellow). The central disk flowers are usually yellow. The involucral bracts are in many overlapping series, like shingles on a roof. The pappus is composed of capillary bristles. Several genera (e.g., *Eucephalus*, *Eurybia*, *Ionactis*, and *Symphyotrichum*) have been split from the genus *Aster* on the basis of characteristics of the leaves, phyllaries, and overall habit. As used here, *Aster* encompasses its traditional definition. Included here are plants with leafy or leafless, branched or unbranched stems and toothed or nontoothed leaves. The phyllaries overlap in 3–5 rows, may be sparsely hairy, and may be glandular. The ray flowers are blue to purple, and the pappus consists of numerous slender hairs.

Natural History Notes and Uses. Asters usually flower from late summer into fall. A related genus, *Erigeron* (fleabane), is often confused with *Aster*, but fleabanes usually flower from late spring to midsummer and the bracts of the involucre are in 1–2 series. Asters are found in various habitats from low elevations into the alpine zone. The genus name is from Greek and Latin, meaning "star," referring to the radiating ray flowers.

The leaves of several species were boiled and eaten by some Native American tribes. The Cheyenne Indians used a tea from *A. foliaceus* in the form of eardrops to relieve earaches. *Aster laevis* (smooth aster) was burned to create smoke in a sweatbath, and the crushed foliage of *A. hesperius* (= *A. lanceolatus* subsp. *hesperius*) was sprinkled on live coals and inhaled to treat nosebleeds (Chamberlain 1901; Fernald and Kinsey 1958; Hart 1981).

FIGURE 25. Arrowleaf balsamroot (*Balsamorhiza sagittata*).

Balsamroot (genus *Balsamorhiza*)

HOARY BALSAMROOT (*B. incana*)

CUTLEAF BALSAMROOT (*B. macrophylla*)

ARROWLEAF BALSAMROOT (*B. sagittata*)

Description and Habitat. Balsamroots are low perennial herbs with thick rootstalks, and the leaves are mostly basal, large, and long-petioled. The yellow flowering heads are large and showy, mostly on long peduncles. Balsamroot is often confused with *Wyethia* (mule's ears), which can be found in similar habitats. However, *Wyethia* leaves lack the fuzzy gray appearance of some balsamroots.

FIELD KEY TO THE BALSAMROOTS

1. Leaves entire . . . . *B. sagittata*
1. Leaves deeply cleft or pinnatifid 2

2. Plants with long, soft, tangled, or cottony hairs on the leaves and stem . . . . *B. incana*
2. Plants with short stiff hairs . . . . *B. microphylla*

Natural History Notes and Uses. Although *B. sagittata* is considered one of the most versatile sources of food, it is not necessarily palatable. The plants contain a strong, bitter, pine-scented sap. The large taproot, root crowns, young shoots, young leafstalks and leaves, flower budstalks, and seeds were all eaten by various Native Americans. The larger mature leaves were often used in food preparation (i.e., wrapping).

The woody taproot of perhaps all species is edible raw or cooked. The polysaccharide inulin is its major carbohydrate. The roots can be collected throughout the year but are very difficult to dig out. In some species the taproot may be as large as one's forearm. Cooking the roots is yet another challenge. One method we have used involves peeling the roots by pounding them to remove the bark. They are then pit-cooked for 24 or more hours. When properly cooked, the roots turn brownish and sweet-tasting. Another way to prepare the roots is to pit-steam large quantities for a day and then mash and shape them into cakes for storage. Cooked this way, the roots were called *pash* or *kayoum* (Hart 1996).

The young shoots are edible raw or pit-cooked before they emerge in early spring. The young stems and leaves can also be eaten raw or boiled as greens. The older stems are fibrous and tough and require additional boiling.

The flower budstalks are collected while the buds are still tightly closed, then peeled and eaten raw or cooked as a green vegetable. They have a slightly nutty taste.

When harvested from dried heads, the seeds can be roasted and eaten or ground into flour. The chaff is usually removed by winnowing.

The roots are said to be antimicrobial, expectorant, disinfectant, and immuno-stimulant. They can be mashed and applied to swellings and insect bites. Native Americans considered a boiled solution from the root of *B. hirsuta* (= *B. hookeri* var. *neglecta*) (neglected balsamroot) to be an excellent medicine for stomachaches and bladder troubles. The mashed roots of arrowleaf balsamroot were also used by Native Americans to treat swellings or insect bites.

NODDING BEGGARTICKS (*Bidens cernua*)

Description and Habitat. This annual plant is usually found in moist areas. The stems are erect and have opposite, simple leaves. There are few or no yellow ray flowers but many yellow disk flowers.

Natural History Notes and Uses. The genus name is Latin meaning "with two prongs or points," in reference to the prominent prongs on the seeds. Once these seeds penetrate the clothing, they are very difficult to remove. The species name is from the Latin *cernuus*, meaning "inclining the head or stooping," hence "nodding." The small seeds are edible (Weber 1990). The young leaves of related species are cooked and eaten in tropical Asia.

Brickellbush (genus *Brickellia*)

TASSEL-FLOWER BRICKELLBUSH (*B. grandiflora*)

LITTLE-LEAF BRICKELLBUSH (*B. microphylla*)

Description and Habitat. The two species of *Brickellia* that occur in the area are perennial herbs with fibrous roots. The disk flowers are all tubular, and white or creamy to pink-purple. They can be found in a variety of habitats at the lower elevations. This is a large and complex genus consisting mostly of shrubs.

FIELD KEY TO THE BRICKELLBUSHES

1. Plant shrub; leaf blade less than 1 inch long . . . . *B. microphylla*
1. Plant perennial herb; leaf blade greater than 1 inch long . . . . *B. grandiflora*

Natural History Notes and Uses. Moore (1989) states that a tea or tincture from *B. grandiflora* has three distinct uses: lowering blood sugar in certain types of diabetes, stimulating hydrochloric acid secretions by the stomach, and stimulating bile synthesis and gallbladder evacuation. Others species were also probably used medicinally by Native Americans.

TINCTURES

Here is one general procedure for making tinctures. First, place the fresh plant in a glass jar and cover with 80 proof brandy or vodka. Keep the jar in a warm dark place for about two weeks, shaking it daily. After two weeks, strain the herbs through muslin or two layers of cheesecloth. Squeeze well to extract as much fluid as possible. Discard the herbs and bottle the tincture. The dosage varies depending on the purpose of the herb. For sundew (*Drosera*), 3–6 drops in a cup of water is recommended. Vinegar can be used in place of alcohol, but vinegar tinctures have a shorter shelf life (1–2 years) than alcohol tinctures (30–40 years). Vinegar tinctures are often used for babies, alcoholics, and persons with liver problems.

NODDING PLUMELESS THISTLE (*Carduus nutans*)

Description and Habitat. This biennial has a strong, simple or sparingly branched stem that grows up to 7 feet tall. The stems are winged, and the wings beset the spines. The leaves are deeply lobed to pinnately divided, and the lobes are tipped with a spine. Smaller spines are distributed along the margins. The purple flower heads are solitary, large, and nodding. *Carduus* is distinguished from the similar-looking *Cirsium* in that the pappus of *Carduus* is simple and smooth, not a plume. These are weedy species that may occasionally be found along roadsides and other waste places at lower elevations.

Natural History Notes and Uses. Kirk (1975) indicates that the pith of four species (unspecified), without the easily removed rind, may be boiled in salted water and seasoned in various ways. The dried flowers may be used as a rennet to curdle milk. Strike (1994) notes that the raw or cooked leaves and stems and the raw buds were also eaten.

Star-thistle, Knapweed (genus *Centaurea*)

BACHELOR'S BUTTON (*C. cyanus*)

SPOTTED KNAPWEED (*C. maculosa = C. biebersteinii*)

STAR-THISTLE (*C. repens = Acrotilea r.*)

Description and Habitat. *Centaurea* species are annual, biennial, or perennial herbs. The herbage is often densely white-hairy when young, becoming glabrous with age. The flowering heads terminate the branches of open, leafy inflorescences. The flower heads are composed of disk flowers, but the marginal flowers are often sterile and have an enlarged corolla, making them appear to be ray flowers. The flowers are white to various shades of purple and blue.

FIELD KEY TO THE STAR-THISTLES

1. Marginal flowers similar to center ones . . . .*C. repens*
1. Marginal flowers enlarged and resemble ray flowers 2

2. Lower leaves deeply pinnately divided into linear segments . . *C. maculosa*
2. Lower leaves lobed or toothed and not divided
into many segments .... *C. cyanus*

Natural History Notes and Uses. Historically, knapweed was used as a topical vulnerary, sore throat remedy, and appetite stimulant. *C. cyanus* is considered a powerful nervine, and Native Americans used it for venomous bites, indigestion, jaundice, and eye disorders. Culpeper (1972:87) writes, "Knapweed gently heals up running sores, both cancerous and fistulous, and will do the same for scabs of the head." Though the formulations and preparations used might be considered questionable, the plants are abundantly available and probably warrant further investigation.

C. maculosa has become a serious weed in many areas. Sometimes it grows so profusely that it crowds out other species of plants, making the area uninhabitable for native plants and animals. This genus apparently causes an inability to swallow if ingested by horses, resulting in death.

DOUGLAS' DUSTY-MAIDEN (*Chaenactis douglasii*)

Description and Habitat. This short-lived perennial is usually about 8 inches tall and sparingly branched from the base. The leaves are crinkly and highly dissected, resembling those of a carrot except for a covering of cobwebby hairs that gives the plant its "dusty" appearance. The basal leaves are up to 3 inches long, but those on the upper plant are smaller. About 10–20 creamy white to pale pink flowers are clustered in heads up to an inch wide. The fruits are club-shaped achenes bearing papery scales. The species can be found in open, dry, and rocky habitats from the lower elevations into the alpine. The genus is endemic to the western United States.

FIGURE 26. Dusty-maiden (*Chaenactis douglasii*).

Natural History Notes and Uses. The generic name stems from a botanical combination of the Greek words *chainein*, "yawn" or "gape," and *actis*, "rays," undoubtedly in reference to the 2-lobed petals found on this genus. The plant was first described for science in 1840 by Sir William J. Hooker (1785–1865), who dedicated the specific epithet to Scottish botanist David Douglas (1798–1834). Douglas discovered hundreds of new plants during his explorations of the American West.

An infusion of the plant was used as a wash for chapped hands, insect bites, boils, tumors, and swellings by the Okanagon and Thompson. A decoction of the plant was used for indigestion, coughs, and colds. A strong decoction of the plants was applied to snakebites by the Thompson, Okanagon, and Paiute.

Rabbitbrush (genus *Chrysothamnus*)

SPEARLEAF RABBITBRUSH (*C. linifolius*)

RUBBER RABBITBRUSH (*C. nauseosus = Ericameria n.*)

PARRY'S RABBITBRUSH (*C. parryi = Ericameria p.*)

GREEN RABBITBRUSH (*C. viscidiflorus*)

FIGURE 27. Rabbitbrush (*Chrysothamnus nauseosus*).

Description and Habitat. The four species of *Chrysothamnus* in the area are shrubs with alternate, sessile, entire, and linear leaves. The flowering heads are composed of 5–30 yellow disk flowers and typically bloom in late summer and fall. Rabbitbrush is found in dry, open places at low to middle elevations. *Chrysos*, "gold," and *thamnos*, "shrub," refer to shrubby plants with golden yellow flowers. *Ericameria* is from the Greek *erica* (*ereika*), "heath," and *meris* or *meros*, "division or part," referring to the heath-like leaves.

FIELD KEY TO THE RABBITBRUSHES

1. The branches of the plants are covered with close, felt-like hairs........2
1. The branches are less hairy or glabrous..............................3

2. Flower head in a cyme or corymb; involucral bracts obtuse or acute at tip.................................. *C. nauseosus*
2. Flower heads in racemes or maybe a panicle; some involucral bracts with needle-like tips ....*C. parryi*

3. Plants are shrubs over 16 inches tall, growing in alkaline areas; leaves are elliptic .... *C. linifolius*
3. Plants shrubs, usually less than 16 inches tall, growing in non-alkaline areas; leaves variable and often twisted..... *C. viscidiflorus*

Natural History Notes and Uses. A tea was reported to be made from the twigs of *C. nauseosus* for relief from chest pains, coughs, and toothaches. The leaves and stems were also boiled and the liquid used to wash itchy areas.

Great Basin Indians chewed the stems of rabbitbrush to extract the latex. They believed that chewing rabbitbrush relieved both hunger and thirst. The secretion obtained from the top of the roots can also be chewed as gum.

The rubber shortage of World War II stimulated research on rabbitbrush and other rubber-producing plants. Rabbitbrush produces a high-quality rubber called chrysil that vulcanizes easily. Extraction of this rubber for economic reasons at this point is not feasible. Because of its rubber-based compound, rabbitbrush will burn even if wet or green. Navajo Indians derived a yellow dye from the flowers and a green dye from the inner bark.

CHICORY COFFEE ADDITIVE

To make a coffee additive, dig up chicory roots in the fall through spring, scrub them, and slice in half. Roast them in an oven at a low temperature (250 degrees) for 2–4 hours, or until they become dark brown and brittle. Break up and grind as you would coffee. One part chicory to four parts coffee is a common ratio for brewing.

CHICORY (*Cichorium intybus*)

Description and Habitat. Chicory is a perennial herb that grows up to 3 feet tall and has dandelion-like leaves. The blue flower heads, which can be seen from spring to fall, are composed of 15–20 or more ray flowers. The sap is milky. Chicory is a plant of waste places and is found at the lower elevations. Introduced from Europe, it now grows throughout the United States.

Natural History Notes and Uses. Although the roasted root was used for coffee, it is not considered a very satisfactory substitute by itself. Many coffee producers use chicory as a coffee additive.

The young basal leaves and flowers buds hidden at the base of the leaves are edible and best if collected from fall to spring. Because they are bitter, we found it necessary to boil them in at least one to three changes of water. When collected very young, the plants are milder when eaten raw. In some European countries the buds are pickled and canned (Tull 1987).

Thistle (genus *Cirsium*)

CANADIAN THISTLE (*C. arvense*)

GRAYGREEN THISTLE (*C. canovirens*)

EATON'S THISTLE (*C. eatonii*)

ELK (EVERTS') THISTLE (*C. foliosum*)

JACKSON HOLE THISTLE (*C. subniveum*)

WAVYLEAF THISTLE (*C. undulatum*)

BULL THISTLE (*C. vulgare*)

Description and Habitat. Most native thistle species go unnoticed. Only a few introduced thistles have become weedy pests. There are approximately 160 native thistle species in North America, with at least 110 species north of Mexico and 50 south of the border.

Thistles are characterized as biennial or perennial herbs with alternate leaves that are lobed or cleft with spines. The red, yellow, or white heads are showy, and the involucral bracts are overlapping. The native and introduced species can be found in a wide variety of habitats from the foothills to the higher elevations. *Cirsium* comes from the Greek *kirsos,* meaning "swollen vein," for which thistles (*kirsios*) were a reputed remedy.

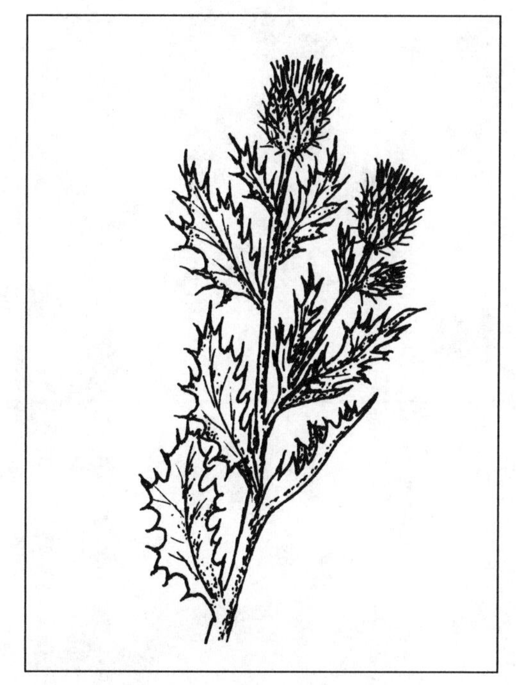

FIGURE 28. Thistle (*Cirsium* spp.).

FIELD KEY TO THE THISTLES

1. Upper surface of leaves with short, appressed, yellowish prickles . . . . *C. vulgare*
1. Upper surface of leaves glabrous to tomentose, without the prickles 2

2. Plants perennial, with creeping roots; involucre ½ to ¾ inch tall . . . . *C. arvense*
2. Plants taprooted; involucre more than ¾ inch tall 3

3. Stem leaves extending downward along the stem 4
3. Stem leaves clasping the stems 5

4. Wings on the lower leaves longer than wings at the top . . . . *C. subniveum*
4. Wings on the middle and upper leaves longer than wings on the lower leaves . . . . *C. eatonii*

5. Flowering heads in terminal clusters . . . . *C. foliosum*
5. Flowering heads solitary or in loose clusters 6

6. Leaves densely white-woolly beneath . . . . *C. undulatum*
6. Leaves sparsely white-woolly, gray-green . . . . *C. canovirens*

Natural History Notes and Uses. Thistles are known for their spiny leaves and stems, although the amount of armature varies. Many native birds and insects use the various thistle species as a source of food and nest material. Goldfinches feed on several thistle species, and bees collect nectar from the plants to make an especially sweet honey.

Everts Ridge, just northwest of Mammoth in Yellowstone National Park, is named after Truman Everts, a member of the 1870 Washburn-Langford-Doane expedition. In September 1870 he became separated from his group near Yellowstone Lake, losing his way and breaking his eyeglasses. His horse bolted, taking everything with it except a penknife and a pair of opera glasses. He kept warm at night by sleeping next to hot springs near Heart Lake until he ended up badly burning himself. After about 37 days of being lost, Everts was finally discovered—shoeless, frostbitten, emaciated, and raving like a madman...but alive.

Many butterfly larvae, especially in the metalmark group, use thistle as their main or only food source. The pollen of thistle flowers is fed on by wasps, syrphid flies, and beetles, which in turn provide a food source for other wildlife. Several species of weevils feed on the developing seeds.

Thistles have also been used as a deterrent to invasion. When Scotland was threatened by Vikings, the Scots piled the beaches with thistles and waited. Arriving in the night, the Norsemen, who were wearing sandals, leaped onto the thistle-strewn beach and let out cries of pain and curses. Thus warned, the Scots were able to drive the Norsemen back to their ships. That is how the thistle became Scotland's heraldic emblem and the source of its motto, "Touch Me Who Dares." Since 1687 induction into Scotland's Order of the Thistle has been a great honor (Haughton 1978).

Although thistles were not a major food source in the past, they were used when needed. Here is our favorite story about how useful thistles can be in emergency situations. In 1870 Truman Everts, a participant in the early explorations of the Yellowstone Park region, became lost for more than a month and subsisted on thistles. He apparently had lost his glasses and was able to identify thistles by touch.

Thistles are difficult to collect, but they are well worth the pain. All species have roots that can be eaten raw, boiled, or roasted. Some have roots that turn sweet when roasted. The immature flower buds (asparagus-like) can be eaten raw or cooked. Young leaves, de-thorned, are edible raw, and a tea can be brewed from all leaves. The peeled young stems may be cooked as greens and resemble celery in taste. The older stalks are also edible but are somewhat more fibrous and bitter. The seeds can be boiled and eaten in the same manner as sunflower seeds, or they can be ground into flour for baking.

Medicinally, thistle stalks were chewed to ease stomach pains. Pounded stalks were used as a salve for facial sores or on infected wounds. A decoction made from thistle roots was used to relieve asthma.

When well dried and de-thorned, the stems can be used as hand drills for starting fires. The stem fibers of any thistle species can be used as thread or crude cordage. To obtain the fiber, simply soak the stalks in water for a day or more to loosen them from the outer layer. The downy part of the seed heads makes good insulating material and a good tinder additive.

CANADIAN HORSEWEED (*Conyza canadensis*)

Description and Habitat. Canadian horseweed is also known as horseweed, Canadian fleabane, and fleabane. It is an annual weed similar to *Erigeron* that grows to about 2 feet tall and has numerous, narrow leaves (the species was once called *Erigeron canadensis* and *Leptilon canadense*). There are numerous white flower heads. Canadian horseweed is usually found growing in waste places at elevations below 6,000 feet.

Natural History Notes and Uses. A native of North America, Canadian horseweed was introduced around the mid-seventeenth century into Europe, where it became widely known for its tonic and astringent properties. A tea was made from the entire dried plant and used for gravel dropsy, diarrhea, and scalding urine. Native Americans used the plant in the form of a tea for leucorrhea and applied the solution to external sores in cases of gonorrhea (Callegari and Durand 1977). Foster and Duke (1990) indicate that *E. canadensis* (= *C. canadensis*) was used as a folk diuretic and an astringent, and to treat diarrhea, kidney stones, nose-bleeds, fevers, and cough. The leaves and tops of horseweed can be pounded and eaten uncooked (Strike 1994). The name *fleabane* suggests the plant's value as an insecticide or repellent, and there may be scientific data to support this use.

Hawksbeard (genus *Crepis*)

TAPERTIP HAWKSBEARD (*C. acuminata*)

ELEGANT HAWKSBEARD (*C. elegans*)

DWARF ALPINE HAWKSBEARD (*C. nana*)

WESTERN HAWKSBEARD (*C. occidentalis*)

FIDDLELEAF HAWKSBEARD (*C. runcinata*)

Description and Habitat. Hawksbeards are perennial, taprooted herbs with milky juice. The leaves are alternate or all basal, and the yellow flowers are all ray. The various species can be found in dry, open places at lower elevations to gravelly or rocky places in alpine or subalpine areas. This is a rather difficult genus to work with in the field as hybridization, apomixis (asexual seed production), and polyploidy (multiplication of entire chromosome complements) are common.

FIELD KEY TO THE HAWKSBEARDS

1. Dwarf plants of alpine areas, less than 3 inches tall . . . . *C. nana*
1. Plants usually taller than 8 inches, below alpine 2

2. Plants glabrous or glaucous but never woolly 3
2. Plants woolly or covered with fine down. 4

3. Heads with 20 or more flowers . . . . *C. runcinata*
3. Heads with 6–12 flowers . . . . *C. elegans*

4. Involucres glabrous, 5–7-flowered . . . . *C. acuminata*
4. Involucres obviously hairy . . . . *C. occidentalis*

Natural History Notes and Uses. The stems and leaves of *Crepis* were eaten by Native Americans. The Karok Indians of northern California peeled the stems of *C. acuminata* before eating (Baker 1981). The seeds or whole plant of *C. acuminata* were thoroughly crushed and applied as a poultice to breasts after childbirth to induce milk flow. The root of the plant was used to remove a foreign object from the eye. For other eye problems, the root was also ground into a smooth powder and sprinkled in the eye. Several applications were necessary.

FETID MARIGOLD, DOGWEED (*Dyssodia papposa*)

Description and Habitat. Fetid marigold is an ill-scented annual up to 20 inches tall, but most specimens are less than a foot tall. The leaves are 1–2 inches long, mostly opposite, and pinnately divided into narrow lobes that are sometimes again divided. The leaves are mostly hairless but are irregularly dotted with orange, sunken oil glands. The flower heads are bell-shaped and less than ½ inch wide. The few ray flowers are inconspicuous, but the 12–50 disk flowers are yellow. The mature achenes are about ⅛ inch long and bear a tuft of bristles.

Natural History Notes and Uses. The generic name *Dyssodia* was taken from the Greek for "disagreeable odor." The specific epithet *papposa* was used because of the conspicuous pappus, or tuft of bristles, on the achene. Fetid marigold was first described for science under a different genus by the French botanist Etienne Pierre Ventenat (1757–1808); it was placed in its currently accepted genus in 1891 by the eminent American botanist Albert Spear Hitchcock (1865–1935).

The leaves and heads are dotted with orange oil glands. The oils and resins in the plant can irritate nose membranes, so it is rarely consumed by livestock. The Dakota gave the plant to horses to treat coughs. The Lakota breathed a preparation of the plant to treat headaches. The Omaha induced nose-bleeds by snuffing powdered leaves of fetid marigold, believing this would relieve headaches. The Navajo used a poultice of chewed leaves to treat ant bites. Pioneers used the leaves to make a tea to settle the stomach. The Dakota observed that prairie dogs had a taste for fetid marigold and that it often grew in prairie dog towns.

ASTER VERSUS FLEABANE

One of the best ways to distinguish an *Aster* from an *Erigeron* is to note the phyllaries (bracts under the flowering head). In *Erigeron* the phyllaries are in 1 or 2 series, whereas 3 to 5 series is typical for *Aster*.

Daisy, Fleabane (genus *Erigeron*)

Description and Habitat. Many species of fleabane occur in the West. They are characterized as annual, biennial, or perennial herbs with alternate or basal leaves. The flowering heads are radiate with narrow ray flowers that may be white, pink, blue, purple, or occasionally yellow. The numerous disk flowers are yellow, and the pappus is composed of capillary bristles. The various species bloom mostly in the spring and early summer, except at the higher elevations (see *Aster*), and can be found in a variety of habitats. Fleabanes resemble asters but are distinguished by fewer rows of involucral bracts. Fleabanes also bloom earlier than asters. The genus name comes from the Greek *eri* ("early") and *geron* ("old man"). The common name comes from the belief that these plants repelled fleas. This genus is rather difficult to work with in the field.

Natural History Notes and Uses. Fleabanes are listed as astringent and diuretic. The disk flowers of a related species, *E. philadelphicus* (Philadelphia fleabane), were powdered to make a snuff to cause one to sneeze and break up a cold or catarrh. A tea from the entire plant of *E. annuus* (eastern daisy fleabane) was used to treat a sore mouth. The dried roots, stems, and flowers of *E. peregrinus* (wandering fleabane) were steeped in hot water, and the patient would breathe the vapors to relieve chest congestion. Fleabanes may cause dermatitis in some people.

WOOLLY ERIOPHYLLUM (*Eriophyllum lanatum*)

Description and Habitat. Eriophyllum is a densely woolly-hairy perennial. The leaves are highly variable, ½ to 3½ inches long, and entire to pinnately lobed (depending on variety). Both the ray and disk flowers are bright yellow-orange in color.

Natural History Notes and Uses. This is an exceedingly variable species with many different varieties. The plants may be erect or sprawling and vary from 4 to 18 inches in height. The genus name is from the Greek, *erion*, meaning "wool," and *phyllon*, "leaf." Eriophyllum seeds were parched and ground into a flour by Cahuilla and Luiseño Indians in California.

SPOTTED JOE PYE WEED (*Eupatorium maculatum*)

Description and Habitat. The plant is 2–6 feet in height, and the flowers are pink to purple-pink and arranged in a flat-topped cluster. The leaves are elongate, with dentate outer margins, and arranged in whorls of 4–5. The stem is purple or green spotted with purple.

Natural History Notes and Uses. When in full bloom, the flowering tops can be gathered and stripped from the stalk, then dried and used to make a tea or tonic that causes vomiting (Sweet 1976). The tea has been used as a cold tonic by Native Americans, and a hot infusion is used for malarial fever (Sweet 1976). *Eupatorium maculatum* is thought to have beneficial uses as either a kidney tonic or a urinary tonic. Herbal remedies are prepared only from the root.

COMMON GAILLARDIA (*Gaillardia aristata*)

Description and Habitat. Gaillardia is a perennial herb with a slender taproot. The leaves are linear to lance-shaped. The flowering heads are solitary or few-flowered with yellow ray flowers and purplish disk flowers. It is found in open places at low and middle elevations.

Natural History Notes and Uses. The Blackfoot Indians drank a tea made from the root for gastroenteritis, and the chewed, powdered root was applied to skin disorders. They also bathed sore nipples of nursing mothers in a tea made from the plant and used the liquid as an eyewash or nose drops.

FIGURE 29. Common gaillardia (*Gaillardia aristata*).

Cudweed (genus *Gnaphalium*)

CUDWEED (*G. microcephalum = Pseudognaphalium canescens*)

WESTERN MARSH CUDWEED (*G. palustre*)

CUDWEED (*G. viscosum = Pseudognaphalium v.*)

Description and Habitat. Cudweeds are woolly annual, biennial, or short-lived perennial herbs with alternate leaves. The plants are often confused with *Antennaria*, but cudweeds have both male and female flowers on the same plant and are taprooted. The disk flowers are yellow or whitish. The species are found from low to middle elevations in moist, open areas with well-drained soils.

PYRROLIZIDINE ALKALOIDS

Most alkaloids are amino acid derivatives and have no certain role in plant metabolism except to repel insects and herbivore predators with their bitter taste and often neurotoxic properties. Pyrrolizidine alkaloids (PAs) are of special interest because some of them have been shown to cause toxic reactions in humans, primarily veno-occlusive liver disease, when ingested with foods or herbal medicines. Comfrey, a well-known medicinal herb and popular herb tea around the world, contains PAs that are capable of causing liver damage. The PAs are primarily found in members of three plant families: Asteraceae, Boraginaceae, and Fabaceae.

FIELD KEY TO THE CUDWEEDS

1. Involucre is less than ¼ inch high; plants are less than 8 inches tall; many-branched . . . . . *G. palustre*
1. Involucre more than ¼ inch high; plants more than 8 inches tall; not many-branched 2

2. Plants glandular-hairy; plants sticky . . . . . *G. viscosum*
2. Plants not glandular-hairy . . . . . *G. microcephalum*

Natural History Notes and Uses. The bruised plant assists in healing wounds, and steeping the leaves in cold water is used for increasing perspiration. Some species contain pyrrolizidine alkaloids and should be regarded as potentially toxic.

A related species, *Gnaphalium leucocephalum,* found in the Southwest, appears to lend itself well to relieving initiatory stages of bronchitis when there is a dry and painful hacky cough. The tea is useful in dislodging impacted mucus that is difficult to expectorate. Its mild diaphoretic properties are excellent for dry feverish states in which the body is hot and flushed. Lending credence to cudweed's beneficial effect on the pneumic environment are its genus-wide, mild antimicrobial properties; a number of cudweed species have been found to inhibit *Staphylococcus* and *Streptococcus* varieties, making the application to bacterial-associated bronchitis more potent.

Most other species of *Gnaphalium,* particularly the aromatic varieties, can probably be used medicinally as well. The great array of species in the West has led to a wide use of these plants. Both *Antennaria* and *Anaphalis* are closely related to *Gnaphalium* and are used similarly, although they are less expectorating and stimulating in nature.

FIGURE 30. Gumweed (*Grindelia squarrosa*).

CURLYCUP GUMWEED (*Grindelia squarrosa*)

Description and Habitat. This species is a biennial or short-lived perennial of waste places at low elevations. The leaves are alternate and have toothed margins. The middle and upper leaves clasp the stems. The regularly overlapping involucre bracts glisten with a gummy resin, the source of the plant's strong, pungent odor. The outer, green bracts are evidently reflexed, and the ray and disk flowers are bright yellow.

Natural History Notes and Uses. In general, gumweeds are considered toxic, and the toxicity appears to depend on the soil in which they grow. Cattlemen consider curlycup gumweed noxious because it increases greatly on native pastures that are heavily grazed; the plant also invades seeded pastures. Resin from the plant has a long history of use in home remedies as a sedative and expectorant and for treatment of burns, poison ivy, and whooping cough.

The sticky flower heads of *G. squarrosa* were used as a chewing gum. The young leaves make an aromatic, bitter tea. The flower heads can be boiled in water and used as an external remedy for skin diseases, scabs, and sores. A hot poultice of the plant was used for swellings. A tea made from the plant was used for coughs, pneumonia, bronchitis, asthma, and colds.

Caution: *G. squarrosa* tends to concentrate selenium and is therefore considered toxic.

BROOM SNAKEWEED (*Gutierrezia sarothrae*)

Description and Habitat. This sticky, glandular perennial appears almost shrubby and grows 6–24 inches high. The leaves are entire, linear-filiform, and less than ⅛ inch wide. The flowering heads are small, numerous, and with whitish, leathery, involucral bracts. The ray flowers are yellow, number 3–8 per head, and are ⅛ inch long. There are 3–8 disk flowers, and the pappus consists of 2–8 stiff awns. Broom snakeweed grows on dry slopes up to 8,000 feet and blooms from May to October. Another common name, matchweed, refers to the match-like appearance of the flower heads.

Natural History Notes and Uses. As with many aromatic plants, this species was used medicinally. The plant was boiled to make a tea for colds, coughs, and dizziness. The tops of fresh, mature snakeweed were boiled until strong and dark. The liquid could be drunk for lung trouble and colds or applied externally for skin ailments such as heat rash, poison ivy, and athlete's foot. For respiratory ailments, the root was boiled in water and the steam inhaled (Moerman 1986).

Goldenweed, Haplopappus (genus *Haplopappus*)

STEMLESS GOLDENWEED (*H. acaulis* = *Stenotus a.*)

THRIFT GOLDENWEED (*H. armerioides* = *Stenotus a.*)

ALPINE TONESTUS (*H. lyallii* = *Tonestus l.*)

NUTTALL'S GOLDENWEED (*H. nuttallii* = *Machaeranthera grindelioides*)

SHRUBBY GOLDENWEED (*H. suffruticosus* = *Ericameria suffruticosa*)

Description and Habitat. *Haplopappus* is a genus of North and South American perennial herbs or shrubs with yellow flowers. Some classifications (as noted above) include species placed in other genera. The generic name for goldenweed was derived from the Greek *haplous*, "single," and *pappos*, "pappus." A pappus is a ring of bristles or scales on the seeds (achenes) of members of the aster family.

FIELD KEY TO THE GOLDENWEEDS

1. Leaves with small spines over the surface.................. *H. nuttallii*
1. Leaves not as above.. 2

2. Involucral bracts leaf-like in appearance............................. 3
2. Involucral bracts not leaf-like...................................... 4

3. Plants more than 3 feet tall .... *H. suffruticosus*
3. Plants less than 2 inches tall .... *H. lyallii*

4. Involucral bracts green throughout......................... *H. acaulis*
4. Involucral bracts green only at tips .... *H. armerioides*

Natural History Notes and Uses. The seeds and stems were eaten by many Native Americans.

Sneezeweed (genus *Helenium*)

SNEEZEWEED (*H. autumnale*)

SNEEZEWEED (*H. hoopesii* = *Hymenoxys h.*)

Description and Habitat. The sneezeweeds are tall-growing autumnal plants that are closely related to the sunflowers (*Helianthus*). The blooming period is from June to the end of September, and during this time the plants are covered with flowers of mahogany crimson, coppery bronze, lemon yellow, and light and deep rich yellows. The flowers are flat and are borne in large heads or clusters. The plants grow 1–6 feet high. The plant is so named because the appearance of the rays suggests that it has just sneezed.

FIELD KEY TO THE SNEEZEWEEDS

1. Stem winged by the decurrent leaf bases .... *H. autumnale*
1. Stem not winged, leaves not decurrent..................... *H. hoopesii*

Natural History Notes and Uses. Powdered sneezeweed was used as snuff to induce sweating, which in turn relieved the congestion of head colds.

Animals that feed on *H. hoopesii* may become affected with "spewing sickness." The disease gets its name from its most characteristic sign: chronic vomiting or spewing. In the western states sneezeweed is common to many mountain ranges. Sheep are frequently poisoned by sneezeweed, but cattle are rarely poisoned. Animals eat sneezeweed during the summer and fall when other forage is scarce or has become less palatable. All the plant parts are **poisonous** to humans.

False Sunflower (genus *Helianthella*)

FIVE-NERVE HELIANTHELLA (*H. quinquenervis*)

ONE-FLOWER HELIANTHELLA (*H. uniflora*)

Description and Habitat. False sunflowers are taprooted perennials with alternate or opposite leaves. The leaves are simple and often triple-nerved. The ray flowers are large, yellow, and neutral, whereas the disk flowers are numerous, fertile, and yellow to brownish purple. The involucral bracts are more or less foliaceous. Achenes are strongly compressed at right angles to the involucral bracts, and the pappus often has 2 slender awns.

FIELD KEY TO THE FALSE SUNFLOWERS

1. Leaves with two prominent pairs of lateral veins; involucral bracts ovate or lanceolate in shape; receptacular bracts soft and scarious; rays pale yellow .... *H. quinquenervis*
1. Leaves more or less triple-veined; involucral bracts mostly lance-linear; receptacular bracts firm; rays bright yellow ..... *H. uniflora*

Natural History Notes and Uses. *Helios* is from the Greek for "sun," and *anthos* means "flower" (see *Helianthus*). *Helianthella*, therefore, is the diminutive of *Helianthus* and thus means "little sunflower." The species name is Latin for "five-nerved" (probably referring to the leaf veins).

Five-nerve helianthella secretes nectar from special glands called extrafloral nectaries. Although extrafloral nectar is not attractive to pollinators, it is attractive to ants. Ants collect this resource and in return protect the plant from herbivores and seed predators, an interaction that is mutually beneficial. This type of mutualism is very common, and up to one-third of all plant species in tropical regions possess extrafloral nectaries.

One-flower helianthella was used medicinally by the Paiute Indians as a dermatological aid. A hot poultice of mashed roots was applied to swellings and sprains. The Shoshone made an infusion of the root as a wash or compress for headaches. They also made a poultice of mashed roots for rheumatism of the shoulder or knee.

Sunflower (genus *Helianthus*)

COMMON SUNFLOWER (*H. annuus*)

NUTTALL'S SUNFLOWER (*H. nuttallii*)

PRAIRIE SUNFLOWER (*H. petiolaris*)

Description and Habitat. Sunflowers are coarse herbs, annual and perennial, often with tall stems. The leaves are simple, the lower ones opposite, the others sometimes alternate. The flower heads are showy with bright yellow ray flowers. Involucral bracts are green and herbaceous. Other genera, including *Wyethia, Balsamorhiza,* and *Arnica,* are often mistaken for *Helianthus.* The genus name comes from the Greek *helios anthes*, "sun flower."

FIGURE 31. Sunflower (*Helianthus nuttallii*).

FIELD KEY TO THE SUNFLOWERS

1. Rhizomatous perennial.................................. *H. nuttallii*
1. Taprooted annuals.. 2

2. Leaves with heart-shaped bases; involucral bracts with long tapered tips................................... *H. annuus*
2. Leaves not heart-shaped at the base; tips of involucral bracts not long and tapered .... *H. petiolaris*

Natural History Notes and Uses. The largest member of this genus is *H. annuus*, a valuable and useful plant. It has been cultivated in the United States since before Columbus. Other species of *Helianthus* may be used similarly.

The seeds may be eaten raw or roasted, then ground into meal and made into bread. The roasted shells can be used as a coffee substitute. To separate large amounts of seeds from shells, first grind them coarsely, then stir vigorously in water: the shells will float, and the seeds will sink to the bottom. The tiny unopened flower buds are also edible and taste something like artichokes. To reduce their bitterness, boil in two to three changes of water. Serve with lemon and melted butter.

Sunflower oil can be extracted from the seeds for cooking and can also be used in making soap, paints, varnishes, and candles. It is extracted by simply boiling the crushed seeds and then skimming the oil from the surface of the water. The pulp remaining after the oil is extracted provides food for livestock.

Medicinally, the crushed roots can be applied to bruises. Other uses of sunflower include fiber obtained from the stalks for cordage, weaving, and sewing. The Chinese reportedly use the stalk fibers in fabrics and the pulp for paper production; the Russians used the stalks as buoyant material for life preservers. Purple and black dyes can be obtained from the seeds and a yellow dye from the flowers.

BUTTERFLIES AND EDIBLE PLANTS

From an ecological perspective, all life ultimately depends on other forms of life to survive and reproduce. As John Muir (1838–1914) once said, "When we try to pick out anything by itself, we find it hitched to everything else in the Universe" (Muir 1988:110).

One way to locate edible and useful plants is to watch butterflies (order Lepidoptera). Most butterflies are closely tied to certain species or groups of plants to complete their life cycle. The larvae (caterpillars) feed on those host plants. Adult female butterflies select the proper larval food plant by "smelling" with their with antennae and "tasting" with sensory receptors on their feet. Eggs are laid on a specific host plant so that hatching caterpillars can begin to feed right away.

Here are some butterflies and their host plants and uses:

- Rocky Mountain parnassian (*Parnassius smintheus*) feeds on stonecrop (*Sedum*). The leaves, stems, and rhizomes of these plants are edible raw or cooked and are high in vitamins A and C.
- Western white (*Pontia occidentalis*) is usually associated with mustards (Brassicaceae). Mustards such as *Lepidium*, *Brassica*, *Arabis*, and *Sisymbrium* are edible as greens.
- Milbert's tortoiseshell (*Nymphalis milberti*) is associated with nettle (*Urtica*). These plants provide useful fibers for cordage, and the plants are edible when prepared as potherbs.
- Fritillary (*Speyeria* spp.) is found on violets (*Viola*), which are edible raw or mixed with other greens in a salad.

HAIRY GOLDEN ASTER (*Heterotheca villosa*)

Description and Habitat. This hairy perennial with yellow flower heads in branched clusters is found almost everywhere in dry places from Canada to the west-central and western United States. This species has also been known as *Chrysopsis villosa*. It is a complex species with many varieties that are distinguished by their hairiness and other subtle characteristics. These plants have endured dozens of other name changes, but they have retained their eye-arresting beauty and pungent-sweet aroma no matter what they have been called.

Natural History Notes and Uses. The genus *Heterotheca* is complex and not completely understood. According to Weber (2001:83), this group of plants "exhibits an enormous range of variability. Many species and varieties have been proposed for what seem to be nodes of stability within a mass of variable characters."

The genus name comes from the Greek *hetero*, meaning "different," and *theca*, meaning "case," referring to the differing seeds of the ray and disk flowers. This is an unusual characteristic, for other sunflowers produce identical seeds from both the outer ray flowers and the inner disk flowers. This species was first collected by Thomas Nuttall "on the Missouri" and was named *Amellus villosus* by Frederick Pursh in 1814.

The Navajo used this species to induce vomiting as part of a treatment for sexually transmitted diseases and indigestion. Additionally, the roots were heated and applied to cavities to relieve toothaches.

Hawkweed (genus *Hieracium*)

WHITE HAWKWEED (*H. albiflorum*)

ORANGE HAWKWEED (*H. aurantiacum*)

HOUNDSTONGUE HAWKWEED (*H. cynoglossoides*)

SLENDER HAWKWEED (*H. gracile*)

WOOLLYWEED (*H. scouleri*)

Description and Habitat. Hawkweeds are fibrous-rooted perennial herbs with milky juice. With one exception, the flowering heads have bright yellow to orange ray flowers, and there are no disk flowers. The phyllaries are in 1–3 series and are worth a look with a hand lens, being ornamented variously with hairs and glands. The pappus is composed of stiff, brown bristles. Hawkweeds are found in a variety of habitats up to the subalpine.

FIELD KEY TO THE HAWKWEEDS

1. Flowers white	*H. albiflorum*
1. Flowers yellow or orange	2
2. Flowers red-orange	*H. aurantiacum*
2. Flowers yellow	3
3. Leaves mostly basal; leaves may be glabrous or with short hairs	*H. gracile*
3. Stem leaves present; leaves with some long hairs	4
4. Upper stem leaves almost glabrous	*H. scouleri*
4. Upper stem leaves densely long-hairy	*H. cynoglossoides*

Natural History Notes and Uses. The common name comes from the ancient Greek belief that hawks would tear apart a plant called hieracion (from the Greek *hierax,* meaning "hawk") and wet their eyes with the juice to clear their vision. The green plant and juices of white-flowered hawkweed may be used as chewing gum, although they are best when dried first. The plant was also used to ease toothaches, to cure warts, as an astringent in treating hemorrhages, and as a general tonic (Strike 1994).

PAIN HULSEA (*Hulsea algida*)

Description and Habitat. This attractive yellow composite grows in crevices on talus slopes at treeline and flowers toward the end of August. It has serrated leaves and a thick stem, and the yellow flowering head is reminiscent at first of a dandelion, but on closer examination one sees that the leaves are thick and sticky and the flower head has both ray and disk flowers. Like many of the composites, the plant gives off a pronounced aromatic odor.

Natural History Notes and Uses. The plant is named for U.S. Army physician and botanist Dr. Gilbert White Hulse (1807–1883); *algida* means "cold" in Latin, a reflection of the plant's environment at high elevations.

FINELEAF HYMENOPAPPUS (*Hymenopappus filifolius*)

Description and Habitat. Very fine thread-like mounds of leaves are evident for weeks before tall slender stalks arise and are then topped by several small yellow flowers. This plant, like many members of the Asteraceae family composed of only disk flowers, is easily overlooked.

Natural History Notes and Uses. *Hymenopappus* refers to the membranous pappus (small scales, bristles, or hairs at the apex of the seed), and *filifolius* is from the Latin *fili* ("thread") and *folius* ("leaf"). A poultice of the chewed root was applied to swellings. A decoction of the plants was taken by the Navajo for coughs. The leaves were boiled, rubbed with cornmeal, and baked into bread. The roots were chewed as chewing gum.

Hymenoxys (genus *Hymenoxys*)

HYMENOXYS (*H. acaulis = Tetraneuris a.*)

HYMENOXYS (*H. grandiflora = Tetraneuris g.*)

PINGUE RUBBERWEED (*H. richardsonii*)

HYMENOXYS (*H. torreyana = Tetraneuris t.*)

Description and Habitat. Hymenoxys are perennial plants. The yellow ray flowers are 3-lobed at the tip, and the pappus is composed of scales. The generic name was compounded from the Greek *hymen* ("membrane") and *oxys* ("sour"), likely in allusion to the translucent scales at the base of the flowers and the sour or bitter taste of several of the species.

FIELD KEY TO THE HYMENOXYS

1. Leaves entire and all basal 2
1. Leaves, or some of them, ternately or pinnately divided 3

2. Leaves densely and conspicuously glandular-punctate ..... *H. torreyana*
2. Leaves not as above...................................... *H. acaulis*

3. Bracts of the involucre similar.......................... *H. grandiflora*
3. Bracts of the involucre in 2 dissimilar series............ *H. richardsonii*

Natural History Notes and Uses. The Hopi Indians used *H. acaulis* as an antirheumatic drug for severe pains in hips and back. A poultice of the plant was applied to those areas, especially during pregnancy. The plant was also used to make a stimulating drink. Native Americans in New Mexico made a chewing gum from the bark and roots from *H. richardsonii*, which was often called "rubber plant."

Povertyweed (genus *Iva*)

IVA POVERTYWEED (*I. axillaris*)

GIANT SUMPWEED (*I. xanthifolia*)

Description and Habitat. Povertyweeds are plants with creeping roots. The greenish, inconspicuous flower heads lie close to the stem in the axils of the small upper leaves. Four or 5 bracts enclose the heads in a cup. Although only about ⅛ inch wide, each head has 5–8 female flowers (peripheral) and 8–20 male flowers (central). There are no ray flowers. The fruits are tiny achenes about ⅛ inch long. These plants are widespread natives of salt marshes and alkali plains.

FIELD KEY TO THE POVERTYWEEDS

1. Plants annual; leaves spade-shaped to triangular .... *I. xanthifolia*
1. Plants perennial; leaves broadly linear to oblong............ *I. axillaris*

Natural History Notes and Uses. There are about 15 species in the genus, and all are found in the Western Hemisphere. Iva povertyweed was described for science in 1814 by the eminent German-American botanist Frederick Traugott Pursh (1774–1820). He was the first to publish on the many new plants collected by the Lewis and Clark expedition (1804–6). The pollen of this plant is highly allergenic, and the plants may cause contact dermatitis in sensitive individuals. The derivation of the generic name is unknown.

Giant sumpweed is a widespread native of western North America and a desirable component of salt marsh and alkali plains communities. However, it sometimes forms large clonal colonies on sites subjected to disturbances such as cultivation or overgrazing. Once established, colonies are difficult to eradicate. Studies suggest that the plants have allelopathic properties, inhibiting the growth of other plants by releasing chemical substances. Colonies in cultivated fields can significantly reduce crop yields. Handling the foliage may cause contact dermatitis, and inhaling the pollen can trigger allergic responses in sensitive individuals. Elsewhere the plant has been associated with selenium poisoning of livestock grazing on high-selenium soils, but animals seldom consume the unpalatable foliage.

Wild Lettuce (genus *Lactuca*)

BLUE LETTUCE (*L. oblongifolia*)

PRICKLY LETTUCE (*L. serriola*)

Description and Habitat. The wild lettuces are annual to perennial herbs with milky sap. The leaves are alternate, often lobed. The flowering heads are composed of ray flowers, and the pappus consists of numerous long, soft bristles.

FIELD KEY TO THE WILD LETTUCES

1. Flowers blue .... *L. oblongifolia*
1. Flowers yellow .... *L. serriola*

Natural History Notes and Uses. Collected in the late fall to early spring, the plants should be boiled in a couple of changes of water to reduce the bitterness. The earlier or younger the plant is collected, the better the flavor. Because of the latex sap, raw greens can cause upset stomach if eaten in quantity. In sensitive people the latex can cause dermatitis. These wild plants contain more vitamin A than spinach and a good quantity of vitamin C. An extract of the white sap from two species of *Lactuca* in Europe has been used to replace opium in cough remedies. The extract, lactucarium, is reported to be a mild sedative. The plants also contain a mildly narcotic compound in the latex. The active constituents increase during flowering and are relatively low in young plants.

DOTTED GAYFEATHER (*Liatris punctata*)

Description and Habitat. This perennial grows from a heavy, taprooted root crown to about 10–18 inches tall. The narrow, sandpapery leaves point strongly upward and are covered with tiny dots of resin. At the bottom the leaves are about 4 inches long, but they get progressively smaller toward the top of the plant. One to two dozen light purple (rarely white) flower heads about ½ inch wide form spikes on the upper third of the plant. The tiny achenes (seeds) bear a plume of bristles.

Natural History Notes and Uses. The root crowns of this plant were used as food by Native Americans in New Mexico. Roots of some other members of the genus *Liatris* were used as tonics and stimulants.

There are about 30 species of *Liatris*; all are found in temperate North America from the Atlantic Ocean to the Rocky Mountains. The derivation of the name *Liatris* is unknown. *Punctata* means "with colored or translucent dots or pits" in Latin. Dotted gayfeather was described for science in 1834 by the famous British botanist Sir William Jackson Hooker in his monumental *Flora Boreali-Americana*.

RUSH SKELETONPLANT (*Lygodesmia juncea*)

Description and Habitat. This plant looks like a slender bare branch about 6–12 inches tall, but close inspection reveals a few tiny, needle-like leaves on the upper branches. Single pink flower heads about ½ inch wide appear at the tips of the branches. Each head contains 5 flowers. Skeletonplant is a perennial from long taproots. The fruits are achenes bearing bristles to aid in transport by wind. The genus name is from the Greek *lygos*, meaning "pliant twig," and *desme*, meaning "bundle," all referring to the flower stems—a bundle of pliant twigs. *Juncea* means "stiff" in botanical Latin. Skeletonplant was first described for science by German botanist Frederick Pursh in 1829.

Natural History Notes and Uses. The stems of rush skeletonplant often have bladder-like insect galls, which are caused by a parasitic insect, the gall wasp *Anistrophus pisum*. The milky juice of the plant provides a chewing gum when coagulated. Mexican children have collected the small yellow balls that form on the plant and chewed them as gum, which turns bright blue on chewing. American Indians used extracts from this plant for eyewash. The plant has been reported poisonous by stockmen in the West.

Many Native American tribes used this plant for medicinal purposes. After childbirth the Cheyenne, Omaha, and Sioux drank an infusion of the plant to increase milk flow. The Lakota and Ponca used an infusion to treat diarrhea. The Blackfoot took an infusion for coughs, heartburn, and kidney ailments and used an infusion to treat saddle sores on their horses.

Spiny Aster (genus *Machaeranthera*)

HOARY ASTER (*M. canescens*)

TANSEYLEAF ASTER (*M. tanacetifolia*)

Description and Habitat. There are about 30 species of *Machaeranthera* in temperate North America. The generic name was derived from the Greek *machaira*, "sword," and *anthera*, "anther," in reference to the pointed pollen-bearing organs.

FIELD KEY TO THE SPINY ASTERS

1. At least the lower leaves are 1–2 times pinnatifid .... *M. tanacetifolia*
1. Leaves are entire or merely shallowly lobed or toothed..... *M. canescens*

Natural History Notes and Uses. The Hopi made a beverage from the taproot of a related species, *M. grindelioides*, which was used for coughs.

MOUNTAIN TARWEED (*Madia glomerata*)

Description and Habitat. Tarweeds are annuals with a tar scent of varying intensity. The leaves are narrow, usually opposite below and alternate above. The flower heads are composed of inconspicuous yellow ray flowers. Tarweed can be found at moderate elevations in open, grassy, or vernally moist areas.

Natural History Notes and Uses. The seeds of *M. glomerata* may be eaten raw or cooked, or dried and ground into meal. The scalded seeds also yield a nutritious oil. All tarweeds were used medicinally by Spanish settlers. An oil of excellent quality was made from their seeds in this country before olives were readily available.

Strike (1994) indicates that tarweed seeds were collected and stored until needed. Native Americans often used the seeds for making pinole. In California tarweed seeds were pulverized and eaten dry (Barrett and Gifford 1933). When tarweed seeds had matured but the plants were still green, Hupa Indians burned the areas where the plants grew. The seeds were then gathered from the scorched plants, and because they needed no further parching, they were crushed into flour. The roots of some species were also eaten.

DESERT DANDELION (*Malacothrix sonchoides*)

Description and Habitat. The short, lobed, basal leaves are diagnostic even before the flower emerges. The plant is typically 2–14 inches tall, often with outward-leaning flower stems, and the flower heads are composed only of ray flowers.

Natural History Notes and Uses. The Greek *malacos* and *trichos*, for "soft hairs," refer to the downy hairs of the pappus. The flower of desert dandelion looks like the flower of sow thistle (*Sonchus oleraceus*), hence the species name, *sonchoides*. Several Native American tribes ate the seeds of a related species (*M. californica*).

PINEAPPLE WEED, DISC MAYWEED (*Matricaria discoidea* = *M. matricarioides*)

Description and Habitat. These annual or biennial herbs have a branched habit. The leaves are alternate and pinnately lobed or divided. The small, terminally arranged flower heads are composed of disk or ray flowers. *Matricaria* are introduced plants that are circumboreal in distribution. The scientific name is from the Latin *mater* ("mother") or *matrix*, referring to the plants' reputed medicinal value.

Natural History Notes and Uses. A delicious tea can be made from the dried flowers of the plant. The leaves are edible but bitter. The medicinal uses of pineapple weed are identical to those of chamomile (*Anthemis* sp.). Used as a tea, it is a carminative, antispasmodic, and mild sedative.

NODDING MICROSERIS (*Microseris nutans*)

Description and Habitat. This annual grows 4–24 inches tall and has mostly basal leaves and a leafless stem. Leaves are 1½ to 10 inches long, linear to narrow elliptic in shape, and entire or with a few teeth or pinnatifid. The flowering heads are terminal on the stems. Involucres are ⅝ to 1½ inches long. The heads are composed of all ray flowers, and the ligules are yellow. Achenes are not beaked, and the pappus has 5 silvery, papery scales that are tipped with an awn or bristle. *Microseris* grows in grassy places or in open woods below 6,000 feet. It flowers from April to June.

Natural History Notes and Uses. The slender roots of this species are apparently edible raw.

Microseris (genus *Nothocalais*)

MEADOW MICROSERIS (*N. nigrescens*)

WEEVIL MICROSERIS (*N. troximoides*)

Description and Habitat. These taprooted perennials have basal leaves that are entire or ciliate-margined. The flowers are yellow, drying to a purplish color. The pappus consists of capillary bristles or slender scales.

FIELD KEY TO THE MICROSERIS

1. Leaves are narrow, about 20–50 times as long as wide, with wavy margins. . . . . *N. troximoides*
1. Leaves about 5–20 times as long as wide, with margins scarcely wavy . . . . *N. nigrescens*

Natural History Notes and Uses. The genus name is from the Greek *nothos* ("false") and *calais* (a mythological figure with scales on his back). The genus is sometimes placed in *Microseris* and is similar to *Agoseris*.

ARROWLEAF SWEET COLTSFOOT (*Petasites sagittatus*)

Description and Habitat. This perennial has creeping rhizomes and thick, succulent, densely hairy stems. Its arrowhead-shaped leaves are large, basal, and toothed. The flowering heads are composed of white or purplish ray and disk flowers. The plant can be found in wet habitats at low elevations.

Natural History Notes and Uses. The young leaves and flowering stems may be eaten raw as a salad or cooked as a potherb (Porsild 1957). Schofield (1989) and Hellar (1958) mention that the leaves and flowering stems were used as a food in Alaska and Siberia by Eskimo groups. The young leaves are edible, but because of their felt-like texture, they are not very pleasant. The aboveground parts have a mild, distinctive, salty taste and could be used as a salt substitute. A salt substitute was also made by burning the leaves of a related species, *P. speciosa*, and using the ashes.

The species is famed for its cough-relieving abilities. Syrups have been recommended for coughs, bronchial congestion, and shortness of breath. The plant contains *petasin*, a highly effective antispasmodic. As a decoction, it can be used to wash skin eruptions. A poultice made from the leaves was used for sores, insect bites, swelling, and pain. However, the plant should be used in moderation as it contains alkaloids that can irritate the liver in large quantities.

The leaf or flower infusion or tincture is used for diarrhea, probably due to its astringent action. Crushed coltsfoot leaves or a decoction can be applied externally for insect bites, inflammations, general swellings, burns, erysipelas, leg ulcers, and phlebitis.

OPPOSITE-LEAF BAHIA (*Picradeniopsis oppositifolia*)

Description and Habitat. The plant was formerly placed in the genus *Bahia*. It is a perennial herb or subshrub about 6 inches tall with 1–8 stems arising from a thin taproot. Fine hairs and glands on the leaves and stems make the plants appear gray-olive green. Leaves are opposite and parted 3 times into narrow segments. Each much-branched stem bears 5–20 flower heads about ⅜ of an inch wide. Each head bears 4–7 ray flowers that are paler yellow than the 30–40 disk flowers. Seeds (achenes) are only about ⅛ inch long and have no long bristles to carry them in the wind.

Natural History Notes and Uses. The generic name *Picradeniopsis* means "looks like Picradenia" (another plant in the family) and is derived from the Greek *picros*, "sharp," and *adenia*, "gland," likely in reference to the achenes. The specific name means "opposite-leaved" in Latin. Opposite-leaf bahia was first described for science in 1818 by the eminent English botanist-naturalist Thomas Nuttall (1786–1859). The Navajo used a poultice of chewed leaves to treat red ant bites. The plant was also used as a panacea, or "life medicine."

UPRIGHT PRAIRIE CONEFLOWER (*Ratibida columnifera*)

Description and Habitat. This coneflower is a perennial about 18 inches tall. Several stems usually grow from the crown of a taproot. Leaves about 3 inches long are divided into 5–9 narrow leaflets. The upper third of the stem is bare except for the flower heads. The flower heads consist of several hundred tiny, purplish brown flowers that form a cylinder about an inch long. At the bottom of the cylinder appear about a half-dozen bright yellow rays about an inch long. Fruits are tiny, winged achenes about 1⁄16 inch long. The plant is sometimes grown as an ornamental.

Natural History Notes and Uses. The genus *Ratibida* was named by wanderer-botanist Constantine Rafinesque-Schmaltz (1773–1840), who often assigned unexplained names to plants. The specific epithet *columnifera* is Latin, meaning "bearing columns," in reference to the long, cylindrical flower heads.

Prairie coneflower is consumed by elk, mule deer, white-tailed deer, and pronghorn antelope. Seeds of prairie coneflower provide fair feed for upland game birds, small non-game birds, and small mammals. Native Americans made tea from the leaves and dye from the flowers. Cheyenne Indians boiled prairie coneflower leaves and stems to make a solution to apply externally to draw poison from rattlesnake bites and to provide relief from poison ivy.

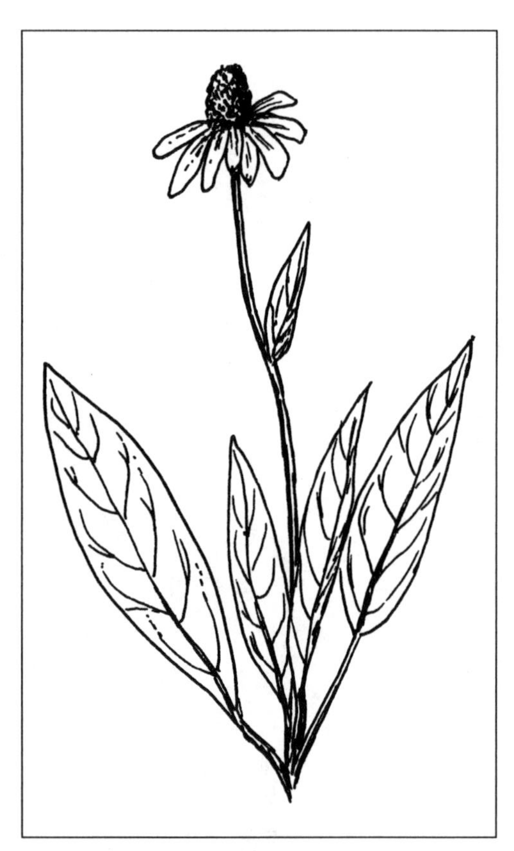

FIGURE 32. Coneflower (*Rudbeckia* spp.).

WESTERN CONEFLOWER (*Rudbeckia occidentalis*)

Description and Habitat. The flowering heads are solitary and terminal, large and showy. Both ray and disk flowers are present. The disk flowers are conical, greenish yellow, and elevated above ray flowers. The ray flowers are yellow, and there are 8–21 per head.

Natural History Notes and Uses. The genus name honors the Swedish father and son who were professors of botany and predecessors of Linnaeus: O. J. Rudbeck (1630–1702) and O. O. Rudbeck (1660–1740).

Several species are suspected of poisoning livestock when eaten in quantity. A root tea from *R. hirta* (black-eyed susan) was used for worms and colds and as an external wash for sores, snakebites, and swelling. The root juice can also be used for earaches. A root tea from *R. lacinata* (cutleaf coneflower) was drunk for indigestion, a poultice made from the flowers was applied to burns, and the cooked spring greens were eaten for "good health." Some Native Americans cooked the entire flower bud of coneflower for food and collected the ripe seeds in fall to grind into meal.

WEBER'S SAW-WORT (*Saussurea weberi*)

Description and Habitat. Weber's saw-wort is a perennial that arises from a woody rootstalk. The leaves are lance-shaped and coarsely toothed and have a broad petiole. The foliage is covered with long, tangled hairs when young but becomes glabrous with age. The purple flowering heads occur in a dense cluster subtended by leaves at the top of the stem. The broad involucral bracts of the heads have rounded tips, and all the flowers are disk flowers (no ray flowers). The achene has a bristly pappus on top, the longest bristles united at the base.

Natural History Notes and Uses. The genus was named for Theodore de Saussure (the son; 1767–1845) and Horace Benedict de Saussure (the father; 1740–1799), two eminent Swiss naturalists. Horace was a geologist, botanist, and early mountain climber. The genus contains about 300 species of perennial herbs with simple to pinnately lobed leaves and flowers in daisy-like heads that lack distinct rays. Some species are grown as ornamentals.

Groundsel (genus *Senecio*)

Description and Habitat. Members of this genus are sometimes quite different from one another. They are annual, biennial, or perennial herbs with alternate or basal leaves. The flower heads are yellow, and the pappus is made up of hair-like bristles. Groundsels can be found in various habitats and elevations. *Senecio* is one of the largest genera of plants, with nearly 2,000 to 3,000 species distributed worldwide. Approximately 100 species are found in the western United States. Some species have been placed in other genera such as *Packera*.

Natural History Notes and Uses. Many species contain highly toxic alkaloids and should therefore be avoided. *S. douglasii* (Douglas' groundsel) was used medicinally by southwestern Native Americans as a laxative, although misuse could result in death. Strike (1994) indicates that young *Senecio* leaves were eaten by Native Americans as cooking herbs. The seeds may also have been eaten by the Chumash Indians. *Senecio* leaves were apparently used to line earth ovens.

Goldenrod (genus *Solidago*)

CANADA GOLDENROD (*S. canadensis*)

GIANT GOLDENROD (*S. gigantea*)

MISSOURI GOLDENROD (*S. missouriensis*)

VELVETY GOLDENROD (*S. mollis*)

MOUNTAIN GOLDENROD (*S. multiradiata*)

BABY GOLDENROD (*S. nana*)

MT. ALBERT GOLDENROD (*S. simplex*)

GOLDENROD (*S. sparsiflora = S. velutin*)

Description and Habitat. The various species of goldenrod are perennial herbs with fibrous roots. The leaves are alternate, simple, and either toothed or entire. The heads are made up of yellow ray flowers. Goldenrods may be found in dry to moist habitats from the foothills to timberline, often in dense patches. *Solidago* means "to make whole."

FIELD KEY TO THE GOLDENRODS

1. Basal leaves not well developed	2
1. Basal leaves usually well developed	7
2. Leaf surfaces usually not hairy	3
2. Leaf surfaces hairy	5
3. Largest leaves at middle of stem; lower leaves deciduous	*S. gigantea*
3. Largest leaves toward the base; lower leaves persistent	4
4. Stems hairy below inflorescence	*S. multiradiata*
4. Stems rarely hairy below inflorescence	*S. missouriensis*
5. Ray flowers about 13 per head; largest leaves generally over 4 times as long as wide	*S. canadensis*
5. Ray flowers about 8 or fewer per head; largest leaves about 4 times or less as long as wide	6
6. Leaves often moderately hairy; involucral bracts broadest near middle	*S. mollis*
6. Leaves only sparsely hairy; involucral bracts mostly broadest at base and acute at tip	*S. sparsiflora*
7. Leaves hairy	*S. nana*
7. Leaves not hairy but may have ciliated margins	8
8. Lower leaves with strongly ciliate-margined petiole; ray flower about 13 per head	*S. multiradiata*
8. Lower leaves without ciliate-margined petioles; ray flowers about 8 per head	*S. simplex*

Natural History Notes and Uses. Young leaves can be prepared as potherbs or added to soups. Depending on habitat, age, and personal preference, their palatability is quite variable. The dried leaves and dried, fully expanded flowers can be used to make a tea. The seeds can be used to thicken stews. Large amounts of the raw herbage should be avoided as it may be toxic.

Medicinally, *Solidago* was employed for stopping internal and external bleeding. An antiseptic lotion may be made by boiling the stems and leaves or by using dry, powdered leaves. The powdered dry leaves were also sprinkled on cuts as a styptic. For insect bites and minor scrapes, apply fresh, crushed, or chewed leaves. A tea wash is said to be good for rheumatism, neuralgia, and headaches.

The fluffy down from the flower heads is a good additive for tinder bundles. All goldenrods contain small quantities of natural rubber and were once cultivated as a domestic source.

Sowthistle (genus *Sonchus*)

FIELD SOWTHISTLE (*S. arvensis*)

SPINY SOWTHISTLE (*S. asper*)

COMMON SOWTHISTLE (*S. oleraceus*)

Description and Habitat. Introduced from Europe, sowthistles are weedy perennials and annuals with alternate leaves that are entire to pinnately divided. The leaf bases have eared-shaped lobes, and the margins are prickly. The flower heads are composed of entirely yellow ray flowers. The pappus is bristly. The common name is said to be derived from the observation that pigs eagerly consumed the plants. In general, they occur at the lower elevations in gardens and waste places.

FIELD KEY TO THE SOWTHISTLES

1. Plants perennial with rhizome-like roots .... *S. arvensis*
1. Plants annual with a taproot 2

2. Lobes at the base of leaves are sharp-pointed .... *S. oleraceus*
2. Lobes at the base of leaves are rounded .... *S. asper*

Natural History Notes and Uses. The young plants of all three species can be prepared as a potherb. As they get older they become increasingly bitter. We found that boiling them in at least two changes of water makes them a little more palatable. Since the plants have an abundance of soluble vitamins and minerals, use only a minimum amount of water and boil briefly. The milky gum obtained from *S. oleraceus* was once used in treating opium addiction, and Native Americans used a tea made from the leaves of *S. arvensis* to calm nerves (Foster and Duke 1990). In Europe a poultice from the leaves was used as an anti-inflammatory.

NARROW-LEAF WIRELETTUCE (*Stephanomeria tenuifolia* = *S. minor*)

Description and Habitat. These annual or perennial herbs have milky juice and are more or less branched. The leaves are small and often scale-like. The flowers are pink and composed of ray flowers. Narrow-leaf wirelettuce is found in dry, open places at low and middle elevations.

Natural History Notes and Uses. Related species have been used medicinally. *Stephanomeria virgata* exudes a milky sap and was used as an eye medication by the Kawaiisu Indians in California. The sap of another species, *S. pauciflora,* was used as a chewing gum.

COMMON TANSY (*Tanacetum vulgare*)

Description and Habitat. This plant grows up to 60 inches tall and emanates a peculiar spicy odor. The leaves are alternate, deep green in color, deeply cleft, and deeply cut again to give a fine-toothed appearance. The flower heads are small, bright yellow, and composed of disk flowers only.

Natural History Notes and Uses. Common tansy, also known as golden buttons and garden tansy, is a perennial herb native to Europe and has a long history of medicinal use. It was first introduced to North America for use in folk remedies and as an ornamental plant. It is considered a noxious weed in many areas.

The first historical records of common tansy cultivation are from the ancient Greeks, who used it for a variety of ailments. It was grown in the garden of Charlemagne in the eighth century and in the herb gardens of Swiss Benedictine monks as a treatment for intestinal worms, rheumatism, fevers, and digestive problems. In medieval times common tansy in large doses was used to induce abortions. Cases of cattle abortions in Pennsylvania have been circumstantially linked to common tansy, but these reports are not well substantiated. Ironically, smaller doses of common tansy were thought to prevent miscarriage and enhance fertility. Common tansy is still used in some medicines and is listed in the *United States Pharmacopoeia* as a treatment for colds and fever.

Warning: The plants contain alkaloids that are toxic to both humans and livestock if consumed in large quantities. Cases of livestock poisoning are rare, however, because tansy is unpalatable to grazing animals. Human consumption of common tansy has been practiced for centuries with few ill effects, yet the toxic properties of the plants are cumulative and long-term consumption of large quantities has caused convulsions and even death. In addition, hand pulling of common tansy has been reported to cause illness, suggesting that toxins may be absorbed through unprotected skin.

FIGURE 33. Common dandelion (*Taraxacum officinale*).

Dandelion (genus *Taraxacum*)

WOOLBEARING DANDELION (*T. eriophorum*)

ROCK DANDELION (*T. laevigatum*)

COMMON DANDELION (*T. officinale*)

DANDELION (*T. scopulorum* = *T. lyratum*)

Description and Habitat. Dandelions need very little introduction. All species are taprooted perennials with milky juice and leaves that form a dense, basal rosette. The solitary flower head is composed of bright yellow ray flowers. They are found in a variety of habitats up to the alpine zone. *Taraxos* means "disorder," and *akos* means "remedy." The plant was an ancient remedy for ailments ranging from spring doldrums to mononucleosis. The latex-like sap was a folk medicine for warts. Red-seeded dandelion (*T. laevigatum*), formerly considered a separate species, is now referred to as *T. officinale*. The native species in the area are *T. eriophorum* and *T. scopulorum*.

FIELD KEY TO THE DANDELIONS

1. Leaves mostly lobed more than halfway to midrib; outer involucral bracts normally reflexed 2
1. Leaves mostly lobed less than halfway to midrib; outer involucral bracts often erect to spreading 3

2. Achenes red, purple, or reddish brown at maturity .... *T. laevigatum*
2. Achenes olive to brown at maturity .... *T. officinale*

3. Achenes red or reddish brown at maturity; leaves entire or slightly toothed .... *T. eriophorum*
3. Achenes straw-colored or olive to brown or black at maturity; leaves often lobed or coarsely toothed.......... *T. scopulorum*

Natural History Notes and Uses. Common dandelion was introduced into North America by European settlers as a food crop and a medicinal cure-all. Every part of common dandelion is edible. The young leaves may be eaten raw or cooked like spinach. The older leaves are also edible, but we find it is better to boil them in one or two changes of water to eliminate the bitterness that comes with age. The plants are high in vitamins A and C and a good source of B complex, iron, calcium, phosphorous, and potassium. The roots can be eaten raw or boiled as a vegetable, baked as potatoes, or added to soups and stews. The roasted root can be used as a substitute for coffee, but it lacks the caffeine buzz. The flower buds can be pickled and added to meals such as omelets. In general, common dandelion is good for blood circulation.

The less common native species of *Taraxacum* may also be edible, but no authenticated information has been located. Kirk (1970) believes that the other native species have similar qualities. During World War II a species of dandelion was cultivated by the Russians as a commercial source of rubber. In spite of its numerous uses, many Americans consider the common dandelion a pest and spend a great deal of time and money trying to eradicate it.

Horsebrush (genus *Tetradymia*)

SPINELESS HORSEBRUSH (*T. canescens*)

NUTTALL'S HORSEBRUSH (*T. nuttallii*)

SHORTSPINE HORSEBRUSH (*T. spinosa*)

Description and Habitat. Horsebrushes are rather low-growing, multibranched, and unarmed or spiny shrubs, found either as well-scattered individuals or as small colonies mixed in with other vegetation. Some species may reach heights of 6–8 feet but are more commonly 3 feet or less. Reproduction is from wind-dispersed seeds and from sprouting of root crowns and rhizomes in longspine horsebrush, hairy horsebrush, spiny horsebrush, and cotton horsebrush. Horsebrush is commonly associated with the sagebrush vegetation type, but the genus has widespread occurrence from barren slopes and alkaline plains upward into the pinyon-juniper and yellow pine types. The genus name comes from the Latin *tetradymos*, meaning "four-sided," referring to the shape of the flower bracts. The plants are easily mistaken for rabbitbrush (*Chrysothamnus*).

FIELD KEY TO THE HORSEBRUSHES

1. Plants not spiny .... *T. canescens*
1. Plants spiny.. 2

2. Flowers 5–9 per head; heads solitary in upper axils . . . . *T. spinosa*
2. Flowers 4 per head; heads clustered at ends of branches. . . . . . *T. nuttallii*

Natural History Notes and Uses. Horsebrush provides ground cover and soil stability. It is generally considered to have low forage value, although buds and new leaders are consumed by cattle, sheep, goats, antelope (*Antilocapra americana*), and mule deer (*Odocoileus hemionus*) (McArthur et al. 1988). Most species are poisonous to livestock (Johnson 1974; Kingsbury 1964). The flowers are used by small moths, bees, flies, and beetles (McArthur et al. 1988), and gelechiid moths form galls in leaves and stems (Hartman 1984). Late-season flowering species of horsebrush provide an attractive contrast to the vegetation types of dry areas.

Sheep that feed on horsebrush just following or in conjunction with black sage (*Artemesia nova*) and then are exposed to bright sunlight may develop a characteristic swelling of the head called bighead. Sheep grazing horsebrush may die without developing bighead. The toxins that have been identified in horsebrush are resins and furanoeremophilanes (furanosesquiterpenes) such as tetradymol. Sheep vary considerably in their susceptibility to horsebrush, but larger amounts often result in death. The horsebrush toxins synergize with the toxins of other sagebrushes, especially black sagebrush.

NORTHERN GREENTHREAD (*Thelesperma marginatum* = *T. subnudum*)

Description and Habitat. Northern greenthread is a perennial about 8 inches tall. Stems usually are single or rarely are clustered from creeping rootstalks. The 3-inch-long leaves are mostly basal and once or twice divided into 3–9 thread-like sections. The ½-inch-wide, yellow flower heads have no ray flowers, only disk flowers. Heads have bracts with membranous margins. Fruits are tiny achenes about ¼ inch long.

Natural History Notes and Uses. There are only about 10 species in the strictly American genus *Thelesperma*. The generic name was compounded from the Greek *thele*, "nipple," and *sperma*, "seed," because the achenes have tiny papillae. The specific epithet means "margined" in Latin. Greenthread was described for science by the Swedish-born American botanist Peter Axel Rydberg (1860–1931), curator of the New York Botanical Garden and author of several major floras of the American West. Some *Thelesperma* species were used to make beverage plants, dyes, and medicines for treating tuberculosis.

Townsendia (genus *Townsendia*)

HAIRY TOWNSENDIA (*T. condensata*)

HOOKER'S TOWNSENDIA (*T. hookeri*)

COMMON TOWNSENDIA (*T. leptotes*)

PARRY'S TOWNSEND DAISY (*T. parryi*)

SWORD TOWNSENDIA (*T. spathulata*)

Description and Habitat. These taprooted annual, biennial, or perennial herbs have alternate or all basal leaves. The flowering heads are solitary, few, or many-flowered. The ray flowers are blue or purple to pinkish or white. The achenes are flattened, 2-nerved, and usually pubescent.

FIELD KEY TO THE TOWNSENDIAS

1. Involucral bracts with long-tapering points .... *T. parryi*
1. Involucral bracts not long-tapering 2

2. Plants with long, tangled, woolly hairs 3
2. Plants hairy but not as above 4

3. Involucre ¼ to ½ inch wide, about ¼ inch long .... *T. spathulata*
3. Involucre ½ to 1½ inches wide, about ½ inch long .... *T. condensata*

4. Involucral bracts with tuft of tangled hairs at tip; leaves densely hairy .... *T. hookeri*
4. Involucral bracts not as above; leaves more or less glabrous .... *T. leptotes*

Natural History Notes and Uses. There are about 20 species in this western North American genus. The genus is distinguished from asters by the pappus, which consists of several or many awns or scales, or both, instead of hair-like bristles. The genus name honors David Townsend (1787–1858), an amateur botanist of Pennsylvania.

Salsify, Goatsbeard (genus *Tragopogon*)

YELLOW SALSIFY (*T. dubius*)

MEADOW SALSIFY (*T. pratensis*)

Description and Habitat. The salsifies are introduced, taprooted, biennial herbs with milky juice. The leaves are alternate, entire, sessile, and clasping at the base, and they taper to a long point. The flower heads are solitary and composed of pale yellow or purple ray flowers. The heads open early in the day, close about noon, and remain closed on cloudy, rainy days. They are found in many habitats at lower elevations. The genus name is Greek, for "goat's beard," probably referring to the thin, tapering, tufted, grass-like leaves.

FIGURE 34. Salsify (*Tragopogon* spp.).

FIELD KEY TO THE SALSIFIES

1. The stem below the flower head is swollen and hollow .... *T. dubius*
1. The stem not swollen as above .... *T. pratensis*

Natural History Notes and Uses. The fleshy roots of these salsifies can be eaten raw or after cooking but are somewhat small, fibrous, and tough. The flavor resembles that of an oyster, an acquired taste! Salsify root has been cultivated for more than two thousand years in the Mediterranean. The young leaves and stems of all species can be eaten after boiling until tender. The coagulated sap can be used as chewing gum and as a remedy for indigestion.

PERENNIAL GOLDENEYE (*Viguiera multiflora = Heliomeris m.*)

Description and Habitat. Perennial goldeneye is a slender, much-branched perennial plant that grows up to 3 feet tall. From May to October it produces golden yellow blooms at the ends of wiry stems. This plant is found on dry slopes.

Natural History Notes and Uses. The genus was named for D. A. Viguier, librarian and botanist of Montpellier, France. The seeds of this species have been used as food by Native Americans.

FIGURE 35. Mule's ears (*Wyethia* spp.).

Mule's Ears (genus *Wyethia*)

MULE'S EARS WYETHIA (*W. amplexicaulis*)

SUNFLOWER WYETHIA (*W. helianthoides*)

Description and Habitat. Mule's ears are stout perennial, simple-stemmed herbs with large, erect, alternate leaves. The ray flowers are long and yellow or white. The heads are usually solitary. All species have leaves on the stems, distinguishing them from *Balsamorhiza*, which has leaves only at the base.

FIELD KEY TO THE MULE'S EARS

1. Ray flowers yellow	*W. amplexicaulis*
1. Ray flowers white or pale cream	*W. helianthoides*

Natural History Notes and Uses. The plant's genus was named for Nathaniel Wyeth, an early fur trader. He established Fort Hall, a trading post near present-day Pocatello, Idaho. In 1834 Wyeth accompanied a botanist on an expedition across the country, and this plant was collected and described. *W. helianthoides* is similar in appearance to *W. amplexicaulis*, which has yellow ray flowers. Both seem to hybridize where they overlap.

The seeds are edible and somewhat resemble sunflower seeds in taste. The roots of *W. helianthoides* can be eaten after they have been cooked for a day or two in a steam pit. Regardless of how long the roots are cooked, however, the smell is almost intolerable. The flowerstalks of *Wyethia* were also eaten as a vegetable by the Shoshone. A decoction of leaves was used as a bath, producing profuse sweating. The leaves are considered **poisonous** and should not be taken internally.

ROUGH COCKLEBUR (*Xanthium strumarium*)

Description and Habitat. Cocklebur is a coarse annual weed of uncertain origin that has a cosmopolitan distribution. The stems of the plants are simple, and the leaves are alternate. The flower heads are solitary or clustered in the leaf axils. The bur (seed) has conspicuous, slender, hooked prickles. Cockleburs can be found in the lower elevations.

Natural History Notes and Uses. Cockleburs have been used medicinally by many aboriginal people throughout North and South America, as well as by the Chinese. At Zuni Pueblo the seeds were ground, mixed with cornmeal, made into cakes or balls, and steamed.

Historically, the roots of *X. strumarium* were used for scrofulous tumors (related to the lymph glands in the neck). The species was also used for rabies, fevers, and malaria. It possesses diuretic, fever-reducing, and sedative properties. Native Americans used a leaf tea for kidney disease, rheumatism, tuberculosis, and diarrhea and as a blood tonic. The seeds have germicidal qualities and were ground and applied to wounds. The seeds also contain an oil that can be used as lamp fuel.

BARBERRY FAMILY (BERBERIDACEAE)

There are nine genera and 590 species in this family, which is distributed throughout the Northern Hemisphere and South America. Two genera are native to the United States. This is a diverse family of perennial herbs and shrubs. The flowers have six or more stamens that split open by two hinged valves to splatter pollen over insects as they crawl by. Several species are cultivated as ornamentals.

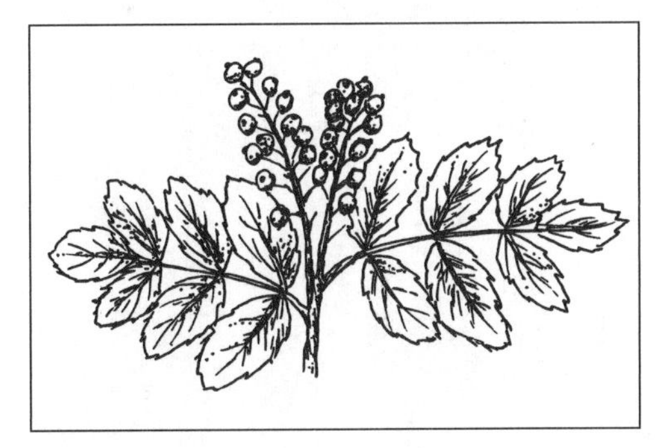

FIGURE 36. Barberry (*Berberis* spp.).

OREGON-GRAPE (*Berberis repens*)

Description and Habitat. The genus *Berberis* are rhizomatous, viny, or upright shrubs with pinnately compound, evergreen leaves. The leaflets have spiny margins, and the yellow flowers are in 3 whorls that are interpreted as bracts, sepals, and petals. The fruits are blue to purple and have a waxy covering.

Natural History Notes and Uses. In some plant guides barberries are those plants with simple leaves and spiny stems whereas Oregon-grapes have compound leaves and spiny blades. In these cases, they have been treated as separate genera: *Berberis* and *Mahonia*, respectively.

The blue berries are edible raw or can be dried for winter use or added to soups to improve flavor. We like them best when picked right off the plant. They also make good jellies.

Medicinally, the bark may be boiled and the infusion used to wash sores on the skin and in the mouth. The plants contain berberine, a bitter alkaloid that gives roots their distinctive yellow color and usefulness as a digestive tonic. Berberine stimulates the involuntary muscles and possesses antipyretic, laxative, and antibacterial qualities. The liquid obtained from the root by chewing was placed on injuries and wounds.

Berberine-containing plants are used medicinally in virtually all traditional medical systems and have a history of usage in Ayurvedic and Chinese medicine dating back at least 3,000 years. Berberine has demonstrated significant antimicrobial activity against bacteria, fungi, protozoans, viruses, helminths, and chlamydia.

A yellow dye can be obtained by boiling bark and roots. The whitish film on the berries is a yeast that can be used in making a primitive sourdough starter (Webster 1980).

Caution: The roots should be considered toxic, and the spines on the leaves may inject fungal spores into the skin.

BIRCH FAMILY (BETULACEAE)

Of the six genera and 150 species in the birch family, five genera are native to the United States. Members of this family are trees and shrubs with deciduous, simple, alternately arranged leaves that have toothed margins. The unisexual minute flowers are arranged in catkins, and male and female flowers are borne on the same plant. The male catkins are soft and pendant. After releasing abundant pale yellow pollen in early spring, they drop off. The hard, cone-like female catkin is either erect or pendant. Most members of this family grow in moist soil, particularly along streams. Economic products include lumber, edible seeds, and oil of wintergreen.

Oil of wintergreen (methyl salicylate) is a folk remedy for body aches and pains and is known for its astringent, diuretic, and stimulant properties. **Caution:** Oil of wintergreen, used externally for pain (i.e., muscular, joint, arthritic, and rheumatic), may cause irritation to the skin. As a precaution, test by using a few drops of wintergreen oil in a carrier oil or liniment.

FIELD KEY TO THE BIRCH FAMILY

1. The female cones are clustered and each one falls intact; the bark is not aromatic . . . . *Alnus*
1. The female cone is solitary and disintegrates when mature; bark is aromatic . . . . *Betula*

Alder (genus *Alnus*)

MOUNTAIN ALDER (*A. incana*)

SITKA ALDER (*A. viridis*)

Description and Habitat. The two species of *Alnus* in the area are small trees or shrubs with smooth bark that is reddish or gray-brown. The leaves are egg-shaped and have serrate edges. The male catkins are grouped near the ends of branches and drop off after pollen is shed. The female catkin is cone-like and persistent. The fruits are flattened achenes with lateral wings or just a membranous border. These plants are usually associated with riparian and wetland sites at low to middle elevations.

FIELD KEY TO THE ALDERS

1. Catkins appearing on previous year's growth before leaves unfurling; female catkins ("cones") on peduncles that are much shorter than the body . . . . *A. incana*
1. Catkins and leaves appearing at the same time on twigs of the current season; female catkins ("cones") on peduncles that are longer than the catkin body . . . . *A. viridis*

Have you ever tried to carve a depression in wood to make a spoon, cup, or bowl with only a knife? It can be frustrating. To simplify the task, try using hot coals. Place a coal where you want the wood to be hollowed out, then blow on the embers with a steady, thin stream of air to keep the coal glowing. A straw is often helpful in directing the air stream. After the coals have burned down a bit, scrape out the charcoal with a stone or knife. Repeat the process with fresh coals until the depression is formed.

Natural History Notes and Uses. The edible catkins are high in protein but don't taste very good. They are more tolerable if they are nibbled raw, added to soups, or dried and powdered and used as a spice. The inner bark is palatable for only a short time in the spring when it is less bitter. A patch of bark is removed from the tree, and the tissue scraped off and eaten fresh or dried in cakes.

The bitter leaves and inner bark act on the mucous membranes of the mouth and stomach to stimulate digestion. A tea made from the leaves was used as a wash or a soothing remedy for poison oak or poison ivy, insect bites, and other skin irritations. Used fresh, the inner bark is emetic, taken to induce vomiting if poisonous substances are ingested. A decoction from alder bark was used to treat colds and stomach trouble and to reduce pain from burns and scalds. Chewing alder bark is said to turn one's saliva red, which was used to dye basketry material (Reagan 1934).

Alder is valued for its hardwood and is useful for open fires as it does not readily spark. Native peoples used it widely for woodworking, making it into objects such as dishes, spoons, and platters. The wood was also used for making fire-drill sets. The astringent bark and woody cones were used for tanning leather. A black-brown dye from the bark was used for coloring fishing nets to make them less visible. Since alders usually grow in the vicinity of free-flowing water, they are considered a botanical indicator of water. The roots have small nitrogen nodules that improve the soil for other plants. Alders are good for controlling erosion and floods and for stabilizing streambanks.

Birch (genus *Betula*)

BOG BIRCH (*B. nana* = *B. glandulosa*)

WATER BIRCH (*B. occidentalis*)

Description and Habitat. Birches are deciduous shrubs or trees with simple, alternate, and sharply toothed leaves. The branchlets are often set with resinous, aromatic, wart-like glands, and the winter buds have several scales. The catkins have unisexual flowers and expand in early spring. The male and female flowers are borne on the same plant. The pendulous, cylindrical-shaped male catkins are borne in clusters of 2 or more; the cone-shaped female catkins are borne singly and erect. Fruits are small, dry, and 1-seeded and have wing-like margins. Birches can be found along streams and in wet meadows and bogs from the foothills to upper montane zone.

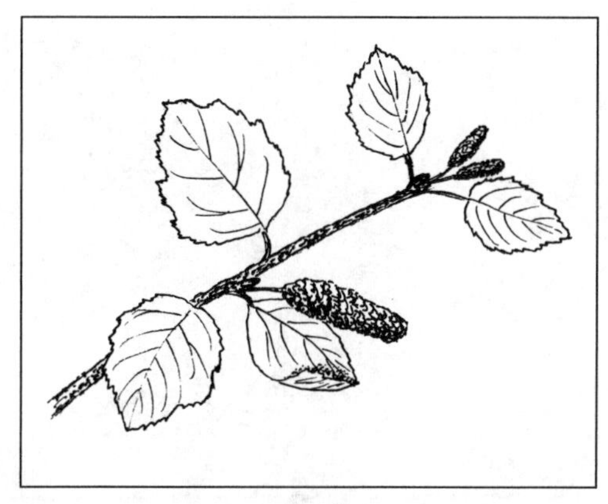

FIGURE 37. Water birch (*Betula occidentalis*).

FIELD KEY TO THE BIRCHES

1. Leaves about ½ to 1 inch long with rounded teeth and oval to obovate . . . . *B. nana*
1. Leaves longer than 1 inch and with pointed teeth . . . . *B. occidentalis*

Natural History Notes and Uses. Young birch leaves can be added to salads. The inner bark can be dried and ground into flour, and the twigs can be steeped in hot water for a tea. The juice of birch leaves makes a good mouthwash. A tea was made from the leaves of a related species, *B. papyrifera* (paper birch), and a poultice of the boiled bark was used to treat bruises, wounds, and burns. Birch contains a significant amount of methyl salicylate and is often used in teas for headaches and rheumatic pain. Birch is highly regarded as a medicinal plant in Russia and Siberia for treating arthritis (Tilford 1997).

Since it burns even when wet, birch bark makes a good tinder. Some Native Americans chopped the new soft wood very fine and mixed it with tobacco. The sap, collected in much the same manner as maple, was sometimes made into vinegar. The best time for tapping is early spring, before the leaves unfurl. The hardwood of some species is used for veneers, cabinet work, and interior finish. It is also used for paper pulp, wooden wares, and novelties. The twigs of some species are distilled to produce oil of wintergreen.

Caution: Oil of wintergreen, used externally for pain (i.e., muscular, joint, arthritic, and rheumatic), may cause irritation to the skin. As a precaution, test by using a few drops of wintergreen oil in a carrier oil or liniment.

BORAGE FAMILY (BORAGINACEAE)

The borage family has approximately 100 genera and 2,000 species, with 22 genera native to the United States. Members of the Boraginaceae can be identified by their alternate leaves, round stems, coiled racemes, and 5-parted radially symmetrical flowers. The corolla has a narrow tube that is abruptly flared at the top. The fruit, composed of 4 nutlets, is helpful for correctly identifying the species of this family. The name *borage* comes from the Middle Latin *burra*, meaning "rough hair" or "short wool," as many of the plants in the family are covered with stiff hairs.

Note: The classification of this family is based primarily on the structure of the fruit. In many cases it is difficult to recognize the genus and almost impossible to obtain a precise identification of the species if the specimens lack mature fruits.

FIELD KEY TO THE BORAGE FAMILY

1. Ovary merely 4-grooved; style arising from apex of ovary; flowers in dense 1-sided spikes .... *Heliotropium*
1. Ovary deeply 4-parted; style arising between the lobes of the fruit...... 2

2. Fruit with hooked or barbed prickle, at least on the margins 3
2. Fruit unarmed, or if prickly, the prickles not hooked or barbed......... 5

3. Corolla reddish purple; prickles covering nutlets completely.................................... *Cynoglossum*
3. Corolla blue or white; prickles mostly along the margins of nutlets 4

4. Plants annual; pedicels erect; style surpassing the nutlets....... *Lappula*
4. Plants perennials or biennials; pedicels reflexed; style often surpassed by the nutlets .... *Hackelia*

5. Corolla blue or occasionally pinkish, tubular to funnel-form . . . *Mertensia*
5. Corolla white, yellow, orange, or if blue, flowers are salverform or rotate... 6

6. Flowers yellow or orange 7
6. Flowers white, blue, pink-purple 9

7. Plants annual .... *Amsinckia*
7. Plants perennial... 8

8. Stigmas 2...*Lithospermum*
8. Stigmas solitary, capitate .... *Cryptantha*

9. Plants less than 4 inches tall, densely caespitose, forming cushion-like mats; flowers bright blue............. *Eritrichium*
9. Plants usually taller 10

10. Flowers blue or pink-purple 11
10. Flowers white or cream-colored.................................. 12

11. Calyx in fruit much enlarged, flaring, irregularly cut, not deeply parted; plant creeping or lying on the ground .... *Aspergo*
11. Fruit not as above .... *Myosotis*

12. Nutlets with a groove-scar or slit running most of their length on the ventral side .... *Cryptantha*
12. Nutlets smooth or keeled 13

13. Nutlets smooth, winged, or keeled around tip and lateral edges .... *Myosotis*
13. Nutlets keeled on ventral side and wrinkled, not winged or keeled around tip and lateral edges .... *Plagiobothrys*

MENZIES' FIDDLENECK (*Amsinckia menziesii*)

Description and Habitat. Fiddlenecks are coarse annual herbs with stiff hairs. The flowers are in a scorpion tail–like spike. This species is usually found below 6,000 feet in dry areas. The genus is named for W. Amsinck, an early-nineteenth-century patron of the botanic garden in Hamburg, Germany.

Natural History Notes and Uses. The seeds of this species were pounded into flour, made into cakes, and eaten without cooking. Coastal western tribes also ate the young fiddleneck leaves. A related species, *A. douglasiana,* was apparently used medicinally (unspecified) (Strike 1994).

FIGURE 38. Fiddleneck (*Amsinckia menziesii*).

Miner's Candle, Catseye (genus *Cryptantha*)

Description and Habitat. The many species of *Cryptantha* are annual or perennial herbs that are rough to the touch. They have linear or spatulate leaves, and the inflorescence is scorpion tail–like with small white or yellow flowers. The corollas are small and white with well-developed appendages inside below the lobes (fornices). *Cryptantha* are usually found in dry, open areas at various elevations. Small-scale characteristics separate *Cryptantha* from other borages.

Natural History Notes and Uses. The genus is native to western North America and South America. The seeds of some *Cryptantha* may have been eaten by the Chumash Indians in California (Sweet 1976).

HOUND'S-TONGUE, GYPSYFLOWER (*Cynoglossum officinale*)

Description and Habitat. Hound's-tongue is a taprooted perennial herb. The stem and leaves are covered with long, spreading hairs. The flowers are dull reddish purple, and the nutlets are densely covered with short, hooked prickles. The plant is an introduced Eurasian weed and is common in disturbed areas, especially along logging roads and heavily used pastures.

Natural History Notes and Uses. These plants contain cynoglossine, consolin, and allantoin, a highly effective healing substance that was sometimes used to relieve pain. The use and effects of hound's-tongue are similar to those of comfrey in treating skin and intestinal ulcerations (Moore 1979). The root and leaves of the plant make an effective tea for a sore throat accompanied by a dry hot cough. A leaf poultice was used for insect bites and piles and other minor injuries such as bruises and burns.

Caution: Though the young leaves of this plant are edible in small amounts after boiling, internal use of hound's-tongue is not recommended. The plant contains potentially carcinogenic alkaloids that may be harmful to the liver if ingested in large quantities.

Alpine Forget-me-not (genus *Eritrichium*)

HOWARD'S ALPINE FORGET-ME-NOT (*E. howardii*)

ARCTIC ALPINE FORGET-ME-NOT (*E. nanum*)

FIGURE 39. Alpine forget-me-not (*Eritrichium nanum*).

Description and Habitat. Alpine forget-me-nots are dwarf, cushion-like perennial herbs. The leaves are densely crowded on numerous short shoots and more or less hairy. The blue flowers with a yellow center occur in clusters. There are 1–4 nutlets.

FIELD KEY TO THE ALPINE FORGET-ME-NOTS

1. Mature leaves densely hairy; hairs not forming tuft at apex of leaf . . . . *E. howardii*
1. Mature leaves loosely long-hairy; hairs sometimes forming tuft at apex of leaf. . . . . *E. nanum*

Natural History Notes and Uses. The showy blue flowers of arctic alpine forget-me-not are pollinated by a wide array of insects ranging from syrphids to bees. Seeing these beautiful, blue alpine plants is a special treat for hikers and climbers, the just reward for reaching the "top of the mountain."

Stickseed (genus *Hackelia*)

MANYFLOWER STICKSEED (*H. floribunda*)

JESSICA STICKTIGHT (*H. micrantha*)

COMMON STICKSEED (*H. patens*)

Description and Habitat. Stickseeds are mostly tall, taprooted perennials with numerous blue or white flowers with yellow centers. The following species grow in moist to medium-dry soils in the foothills to above 8,000 feet.

FIELD KEY TO THE STICKSEEDS

1. Flowers white . . . . *H. patens*
1. Flowers dark blue. 2

2. Prickles on nutlets confined to the margins; basal leaves soon withering; flowers about ½ inch across the top . . . . *H. floribunda*
2. Prickles present on both the face and margins of the nutlet; basal leaves persistent; flowers about ¼ inch across the top . . . . . . *H. micrantha*

Natural History Notes and Uses. Each of the flowers gives rise to 4 small nutlets that possess rows of barbed prickles down their edges, hence the name *stickseed.* It is these prickles that readily enter clothing or the fur of animals but keep the seeds from being pulled out again. The seeds are then transported, and the plants become established a long distance from where they originated. The genus is named for Joseph Hackel, a Czech botanist who lived from 1783 to 1869.

There are no reports of any of the species being eaten by humans. The majority of the species are considered noxious, in that their fruits tend to cling in great numbers to wool shirts, socks, and trousers. The prickles on the fruit of *H. floribunda* can cause skin irritation and swelling, but this species and a related one, *H. hispida* (showy stickseed), were reported to be used medicinally by several Native American tribes.

SALT HELIOTROPE (*Heliotropium curassavicum*)

Description and Habitat. The small flowers are blue or white, and the 5 petals are fused to form a tubular corolla with free lobes at the apex. The flowers are arranged in strongly curled sprays, and the stem and leaves are smooth and fleshy. The leaves are elongate, alternate, pointed at the tip, and with smooth outer margins. The plant grows 6–18 inches tall and is found in open mats on the ground.

Natural History Notes and Uses. The genus name is from the Greek *helios*, meaning "sun," and *tropos*, "turning." In some parts of the West this species is called quail plant, after the birds that feed on its fruit; the Spanish name, *cola de mico*, meaning "monkey tail," describes the coiled flower cluster.

The ashes of the burned plant are used for salt. Consumption of a tea made from the leaves of this plant can cause liver disease. Sensitivity to a toxin varies with a person's age, weight, physical condition, and susceptibility. Children are most vulnerable because of their curiosity and small size. Toxicity can vary in a plant according to season, the plant's different parts, and its stage of growth; and plants can absorb toxic substances, such as herbicides, pesticides, and pollutants, from the water, air, and soil.

Stickseed (genus *Lappula*)

STICKSEED (*L. redowskii* = *L. occidentale*)

EUROPEAN STICKSEED (*L. squarrosa*)

Description and Habitat. Stickseed is a taprooted annual herb with linear, alternate leaves. The plants are densely hairy throughout. The flowers are blue (rarely white), and the fruits have prickles about the edges. Stickseeds are usually found at the lower elevations in dry, disturbed habitats up to 8,000 feet. *Lappula* is Latin, the diminutive of *lappa*, meaning "bur." Members of this genus resemble *Myosotis* and *Cryptantha*.

FIELD KEY TO THE STICKSEEDS

1. Marginal prickles on nutlets in at least 2 rows. . . . . *L. squarrosa*
1. Marginal prickles on nutlets in 1 row. . . . . *L. redowskii*

Natural History Notes and Uses. Medicinally, a poultice from stickseed was applied to sores caused by biting insects; a cold infusion was used as a lotion for sores and swellings. Additionally, the roots of European stickseed placed on a hot stone and allowed to smoke were used as an inhalant, and a snuff was made from the root for headaches (Moerman 1977).

Gromwell, Stoneseed, Puccoon (genus *Lithospermum*)

NARROWLEAF GROMWELL (*L. incisum*)

WESTERN GROMWELL (*L. ruderale*)

Description and Habitat. Gromwells are annual or perennial plants that are sparsely to densely hairy. The flowers occur in terminal, narrow, leafy branches or in the leaf bases. The petals are yellow to white, and the nutlets are mostly oval and smooth, pitted, or wrinkled. The genus name means "stone seed," referring to the hard nutlets.

FIELD KEY TO THE GROMWELLS

1. Flowers greenish yellow; tube of flowers formed by united petals that are about ¼ inch long; plants usually over 12 inches tall. . . . . *L. ruderale*
1. Flowers bright yellow; tube of flowers ½ to 1½ inches long; plants less than 12 inches tall. . . . . *L. incisum*

Natural History Notes and Uses. Lithospermum was used by Native Americans throughout the West as a medicine and food. Although little is known about its chemistry, the effectiveness of several species as contraceptives and as depuratives for skin conditions warrants further investigation. Extracts of *L. ruderale* appear to contain a natural estrogen that interferes with hormonal balances in the female reproductive system. The Shoshone Indians used a tea made from the plant to treat diarrhea and as a female contraceptive, which caused permanent sterility after six months of continued use (Craighead et al. 1963). A salve from powdered and moistened leaves and stems was used for rheumatic and other pains where the skin was not broken.

FIGURE 40. Bluebells (*Mertensia* sp.).

Bluebell (genus *Mertensia*)

ALPINE BLUEBELLS (*M. alpina*)

MOUNTAIN BLUEBELLS (*M. ciliata*)

OBLONGLEAF BLUEBELLS (*M. oblongifolia*)

Description and Habitat. All mertensias have 5 sepals enclosing 5 petals. Together these form a tube that flares more or less abruptly into the free part of the petals. German botanist Karl Heinrich Mertens (1796–1830) collected plants while on a Russian scientific expedition to Alaska in 1827. He applied the name *Mertensia* to honor his father, Franz Karl Mertens (1764–1831), also a botanist. The various mertensias are found primarily in northwestern North America. The common name sometimes used, lungwort, comes from a European species with spotted leaves, believed to be a remedy for lung disease.

FIELD KEY TO THE BLUEBELL

1. Plants usually over 15 inches tall and growing in moist or shady areas; stem leaves with distinct lateral veins, with the middle leaves usually over 2 inches long . . . . *M. ciliata*
1. Plants generally less than 15 inches tall and often growing in dry or open areas; stem leaves typically without distinct lateral veins, with middle leaves rarely over 2 inches long 2

2. Filaments attached in the corolla tube, the anthers not projecting beyond the throat . . . . *M. alpina*
2. Filaments attached near the throat of the corolla tube, the anthers projecting beyond the throat . . . . *M. oblongifolia*

Natural History Notes and Uses. Bluebells are often overlooked in guides to edible plants. The flowers can be nibbled on raw or added to salads. Since the leaves are a bit hairy, we found them better when chopped up and added to soups. Bluebells may contain alkaloids and other constituents that can be toxic if consumed in large quantities (Schofield 1989).

M. ciliata is a favorite browse plant for elk, deer, and other animals, and it is common to find matted areas where large animals have bedded down in the thick plant growth.

Forget-me-not (genus *Myosotis*)

FORGET-ME-NOT (*M. alpestris*)

FIELD FORGET-ME-NOT (*M. arvensis*)

FORGET-ME-NOT (*M. micrantha = M. stricta*)

TRUE FORGET-ME-NOT (*M. scorpioides*)

Description and Habitat. Members of this genus are annual to perennial plants with glabrous or hairy, but not bristly, foliage. The flowers occur in narrow, terminal, spirally coiled branches or in the axils of the upper leaves. The corollas are blue or white and divided into a distinct tube and limb. The nutlets are smooth and shiny.

FIELD KEY TO THE FORGET-ME-NOTS

1. Plants with weak stems, often curved at the base; occurring in wet soils or standing water . . . . *M. scorpioides*
1. Plants growing erect, occurring in moist to dry habitats 2

2. Plants perennial; corolla limb about ¼ inch wide . . . . *M. alpestris*
2. Plants annual or biennial and often occurring in disturbed habitats; corolla limb much less than ¼ inch wide 3

3. When in fruit, the flowerstalks are about as long or longer than the calyx . . . . *M. arvensis*
3. When in fruit, the flowerstalks are shorter than the calyx . . . . *M. micrantha*

Natural History Notes and Uses. These plants are pollinated by insects even though the pollen is very small, which is usually indicative of wind-pollinated plants. There are a number of stories associated with the common name of forget-me-not. The best-known tells of a couple strolling along the Danube. The woman noticed a clump of pretty blue flowers on the streambank and her lover tried to pick some, but as he reached for them, he slipped and fell into the water. As he fell, he was able to toss the flowers to the woman and call out, "Forget me not!"

Popcorn Flower (genus *Plagiobothrys*)

FINE-BRANCHED POPCORN FLOWER (*P. leptocladus*)

SCOULER'S POPCORN FLOWER (*P. scouleri*)

Description and Habitat. Popcorn flowers are annuals with alternate or opposite leaves. The white, salverform flowers are in scorpioid racemes. There are many species, which occur in moist soil.

Natural History Notes and Uses. A rouge was obtained from the stem base and roots of some species of *Plagiobothrys*. Many species have purple dye in the stems and roots. When these parts are pressed and dried in a folded sheet of clean paper, a mirror-image pattern results. The shoots and flowers of related species (*P. fulvus*) were eaten, and the leaves were eaten as greens. *P. nothofulvus* was rubbed between the hands by the Kawaiisu in California to dye their hands purple or to dye their faces (Zigmond 1981).

MUSTARD FAMILY (BRASSICACEAE = CRUCIFERAE)

There are approximately 375 genera and 2,000 species in this family worldwide. Mustard flowers are easy to recognize: they have 4 petals in the form of a cross, hence the name *cruciform*. There are usually 6 stamens, four of which are longer than the remaining two. The fruit is a pod, either long and thin (silique) or short and wide (silicle), with a partition down the middle dividing the seeds into two chambers.

Mustards are a friendly family with a characteristic peppery taste, and it is safe to experiment with them. They are associated with the Roman practice of soaking seeds in newly fermented grape juice (called *must*), drunk as a stimulant to prepare armies for battle. Cauliflower, turnip, radish, cabbage, rutabaga, and watercress are among the economically important plants in this family. Despite the high nutritional value of mustards and the tenacity of the plants, they are largely neglected (Vizgirdas 2000a). In America mustards are often regarded as pests, but in Europe and Asia various species are widely cultivated.

FIELD KEY TO THE MUSTARD FAMILY

1. Pods with a long stipe (½ to 1 inch long) . . . . *Stanleya*
1. Pods sessile or only a short stipe 2

2. Fruit not splitting open . . . . *Chorispora*
2. Fruit dehiscent by valves 3

3.Mature pods (silicles) less than 4 times as long as broad 4
3.Mature pods (silique) more than 4 times as long as broad 22

4. Silicles strongly flattened, not swollen 5
4. Silicles less strongly flattened, distinctly swollen 12

5. Fruit flattened parallel to the septum; pods circular in outline 6
5. Fruit flattened at right angles to the septum 8

6. Seeds 1 to each chamber of fruit . . . . *Alyssum*
6. Seeds more than 1 to each chamber 7

7. Style 1 mm or more long; seeds winged . . . . *Berteroa*
7. Style lacking; seeds not winged . . . . *Draba*

8. Seeds 1 to each chamber of the fruit 9
8. Seeds more than 1 in each chamber 10

9. Pods ovate-cordate, acute at apex, not winged . . . . *Cardaria*
9. Pods circular, elliptic, or rarely ovate in shape,
notched at the apex, usually winged above . . . . *Lepidium*

10. Leaves (some of them) pinnatifid or lobed . . . . *Capsella*
10. Leaves not pinnatifid or lobed 11

11. Plants glabrous . . . . *Thlaspi*
11. Plants densely stellate pubescent . . . . *Lesquerella*

12. Pods compressed at right angles to the septum 13
12. Pods little or not at all compressed at right angles to the septum 15

13. Petals white; herbage glabrous and glaucous . . . . *Lepidium*
13. Petals yellow; herbage stellate pubescent 14

14. Pods in a twin-like pair or with an apical sinus . . . . *Physaria*
14. Pods not in a twin-like pair, nor with an apical sinus . . . . *Lesquerella*

15. Plants acaulescent and glabrous, submerged in water,
with awl-shaped leaves . . . . *Subularia*
15. Plants leafy-stemmed, pubescent, not usually submerged
in water and leaves not awl-shaped 16

16. Petals usually yellow 17
16. Petals white 18

17. Plants glabrous or nearly so and succulent . . . . *Rorippa*
17. Plants pubescent with usually forked hairs . . . . *Descurainia*

18. Stem leaves with ear-shaped lobes, sagittate clasping 19
18. Stem leaves not with ear-shaped lobes or sagittate clasping 20

19. Herbage and fruit with gray pubescence;
racemes short and dense . . . . *Cardaria*
19. Herbage and fruit not with gray pubescence;
racemes long and loose . . . . *Camelina*

20. Pubescence silvery-stellate . . . . *Lesquerella*
20. Pubescence not silvery-stellate 21

21. Style very short; seeds winged . . . . *Berteroa*
21. Style lacking; seeds not winged . . . . *Draba*

22. Siliques with a long seedless beak . . . . *Brassica*
22. Siliques merely tipped with a short style or sessile stigma 23

23. Pods 4-sided, not flattened with the partition; seeds not winged 24
23. Pods flattened parallel to the partition; seeds often winged 26

24. Anthers curved or twisted . . . . *Thelypodium*
24. Anthers not curved or twisted 25

25. Pubescence of simple hairs or none . . . . *Sisymbrium*
25. Pubescence of stem of 2-branched closely appressed hairs . . . . . *Erysimum*

26. Pubescence, when present, of simple, unbranched hairs 27
26. Pubescence usually of at least some branched or stellate hairs 29

27. Flowers white . . . . *Cardamine*
27. Flowers yellow or yellowish 28

28. Seeds in 1 row in each cell; pods somewhat 4-angled . . . . *Barbarea*
28. Seeds in 2 rows in each cell; pods terete . . . . *Rorippa*

29. Flowers yellow or yellowish 30
29. Flowers not yellow 31

30. Pubescence of the stem of 2-branched closely appressed hairs; petals ⅛ inch or usually much more long . . . . *Erysimum*
30. Pubescence of the stem not of 2-branched closely appressed hairs; petals less than ⅛ inch long . . . . *Descurainia*

31. Pods strongly flattened parallel to the partition; leaves simplex . . . . *Arabis*
31. Pods terete or almost or 4-sided; leaves compound or simple 32

32. Leaves compound . . . . *Smelowskia*
32. Leaves simple though deeply cleft in some cases 33

33. Flowers more than ½ inch long; plants perennial . . . . *Hesperis*
33. Flowers less than ½ inch; plants annual, biennial, or perennial . . . . *Halimolobos*

Alyssum, Madwort (genus *Alyssum*)

PALE MADWORT (*A. alyssoides*)

DESERT MADWORT (*A. desertorum*)

Description and Habitat. These small annual plants have foliage that appears dull gray due to the presence of dense, star-shaped hairs. The leaves are alternate and simple, and the flowers are short-stalked on the terminal portion of the stems. The flowers have light yellow petals that quickly fade to white. Fruits are egg-shaped or round in outline and have winged margins and a short style.

FIELD KEY TO THE MADWORTS

1. Fruits are hairy; sepals are persistent at base of mature fruit . . . . *A. alyssoides*
1. Fruits not hairy; sepals fall off as the fruit matures . . . . *A. desertorum*

Natural History Notes and Uses. Both species are Eurasian weeds and are widespread throughout most of the United States. The genus name is from the Greek *a* ("without") and *lyssa* ("rabies"), because the plants supposedly provided a cure for rabies. The small leaves of pale madwort are mild tasting and can be eaten raw.

Rockcress (genus *Arabis*)

DRUMMOND'S ROCKCRESS (*A. drummondii*)

TOWER ROCKCRESS (*A. glabra*)

HAIRY ROCKCRESS (*A. hirsuta*)

LEMMON'S ROCKCRESS (*A. lemmonii*)

LYALL'S ROCKCRESS (*A. lyallii*)

LITTLELEAF ROCKCRESS (*A. microphylla*)

NUTTALL'S ROCKCRESS (*A. nuttallii*)

HOLBOELL'S ROCKCRESS (*A. pendulocarpa = A. holboellii*)

SICKLEPOD ROCKCRESS (*A. sparsiflora*)

Description and Habitat. The rockcresses are biennial or perennial herbs with stellate hairs. The flowers are in racemes, usually white to purple in color. The fruits are linear siliques, usually flattened parallel to the partition. The many species of *Arabis* are found in a variety of habitats at various elevations. This is a difficult group to work with in the field and requires the mature fruit for identification.

FIGURE 41. Rockcress (*Arabis drummondii*).

FIELD KEY TO THE ROCKCRESSES

1. Mature fruits held erect, ascending or at right angles to the stem....... 2
1. Mature fruit drooping or reflexed downward 9

2. Basal leaves broadly lance-shaped or elliptical in outline and often held flat on the ground.................................. 3
2. Basal leaves narrowly lance-shaped and more erect.................. 5

3. Seeds in 2 rows (at least below) in each chamber of the fruit ... *A. glabra*
3. Seeds in a single row in each chamber 4

4. Stem leaves with small lobes at the base; fruits erect .... *A. hirsuta*
4. Stem leaves without small lobes at the base; fruits ascending outward from the stem.................... *A. nuttallii*

5. Seeds in 2 rows in each chamber of the fruit .... *A. drummondii*
5. Seeds in a single row in each fruit chamber.......................... 6

6. Stems with unbranched, spreading hairs at the base ..... *A. microphylla*
6. Stems glabrous or with branched hairs at the base.................... 7

7. Normally a single stem from a simple root crown; plant usually more than 12 inches tall .... *A. sparsiflora*
7. Several stems from a branched root crown, generally less than 12 inches tall.................................... 8

8. Basal leaves densely hairy; fruits occurring on one side of the stem.................................... *A. lemmonii*
8. Basal leaves usually glabrous or almost; fruits not as above ..... *A. lyallii*

9. Stems usually more than 8 inches tall .... *A. pendulocarpa*
9. Stems usually less than 8 inches tall...................... *A. lemmonii*

Natural History Notes and Uses. The crushed plant of *A. puberula* served as a liniment or mustard plaster. At least *A. glabra* and *A. hirsuta* have been eaten raw or cooked. They have a pleasant and pungent taste. The leaves were sometimes preserved in salt. Some species of rockcress were eaten by Native Americans, and an infusion was used to cure colds (Strike 1994).

Yellowrocket (genus *Barbarea*)

AMERICAN YELLOWROCKET (*B. orthoceras*)

GARDEN YELLOWROCKET (*B. vulgaris*)

Description and Habitat. The yellowrockets are biennial or perennial plants with more or less angled, erect stems and glabrous leaves that are pinnately divided with a large terminal lobe.

The yellow flowers occur on short stalks on the upper portion of the stems. The long, linear fruits are 4-angled and sometimes have a small beak at the tip. The seeds occur in a single row in each of the two chambers of the fruit. The genus is named for Saint Barbara and is derived from the fact that the young leaves can be eaten on Saint Barbara's Day in early December.

FIELD KEY TO THE YELLOWROCKETS

1. Fruit conspicuously beaked; basal leaves with more than 1 pair of lateral lobes .... *B. orthoceras*
1. Fruit with a beak at the top; basal leaves with 1 pair of lateral lobes .... *B. vulgaris*

Natural History Notes and Uses. The young stems and leaves of both species can be eaten raw in salads or prepared as a potherb. We like to boil the plants in at least two changes of water.

HOARY FALSE MADWORT (*Berteroa incana*)

Description and Habitat. This annual plant grows up to about 36 inches tall. The lower leaves are broadly linear, 1–2 inches long, and entire-margined with short petioles. These leaves eventually wither as the plant matures. The upper leaves are smaller and sessile. The foliage is grayish in color and has a dense covering of branched hairs. The flowers occur at the terminal portion of the stem, and the notched petals are white and about ½ inch long. The fruits are slightly inflated and elliptical in shape and have a prominent, persistent style at the tip.

Natural History Notes and Uses. This introduced species from Europe is usually seen in old fields and along roadsides. The genus was named for Carlo Guiseppe Bertero, a Piedmontese botanist. The species name means "gray." The notched petals of the white flowers bear a resemblance to those of species of the genus *Draba*. However, *B. incana* is a taller plant with leaves on the stem and an early spring flowering period. The round, ovate seed pods distinguish this plant from species of *Capsella*, *Lepidium*, and *Thlaspi*, which have flattened seed pods.

The plant produces many seeds important to wildlife such as songbirds and small mammals. Horses become intoxicated after eating green or dried plants. When mixed with alfalfa hay, *B. incana* can remain toxic for up to nine months. The toxic dose has not been determined.

Mustard (genus *Brassica*)

INDIA MUSTARD (*B. juncea*)

MUSTARD (*B. kaber = Sinapsis arvensis*)

Description and Habitat. Brassicas are large annuals with showy yellow flowers. The pods are round or 4-sided in cross section, with a conspicuous beak. The various species can be found in waste places and fields at lower elevations. *Brassica* is the Latin name for cabbage.

FIELD KEY TO THE MUSTARDS

1. Beak of the fruit usually with a single seed at base, the beak and valves with 3 raised nerves....................... *B. kaber*
1. Beak of the fruit usually lacking a seed at base, usually 1-nerved, and the valves with 1 raised nerve........... *B. juncea*

Natural History Notes and Uses. All species of *Brassica* have leaves that are edible as greens and are an excellent source of vitamins A, B, and C, and calcium and potassium. The older leaves should be boiled in at least one change of water. The seeds contain thiocyanate and may cause goiter if consumed in large amounts. The seeds can be ground or crushed to a flour and applied as a mustard plaster. The plaster is a long-used remedy for aches and pains. Mustard oil is also a caustic irritant and can discolor and blister the skin if left on too long. The flower buds are rich in protein. Table mustard comes from *B. nigra* (black mustard). In China mustard oil from *B. rapa* (rape mustard) was used for illumination until the introduction of kerosene.

False Flax (genus *Camelina*)

LITTLEPOD FALSE FLAX (*C. microcarpa*)

GOLD-OF-PLEASURE (*C. sativa*)

Description and Habitat. *Camelina* is a European and Asian genus of annual or biennial herbs with all but one species, *C. sativa*, occurring wild or as a weed. The leaves are entire or toothed, those on the stem sessile and clasping at the base. The yellow flowers are small and occur in rather long racemes.

FIELD KEY TO THE FALSE FLAXES

1. Stems hairy.. *C. microcarpa*
1. Stems smooth or nearly so.................................. *C. sativa*

Natural History Notes and Uses. *C. sativa* has been cultivated since the Neolithic Age for the fibers of its stem and the edible oil contained in the seeds. It is said that its oil is similar to linseed oil.

SHEPHERD'S PURSE (*Capsella bursa-pastoris*)

Description and Habitat. Shepherd's purse is a pubescent annual with leaves mostly in a basal rosette. The petals are white and the pods obcordate (heart-shaped), strongly flattened contrary to the narrow septum. *Capsella* means "little box," referring to the fruit, as does *bursa-pastoris*, which means "purse of the shepherd." This is a common weed on dry or disturbed soil. It blooms most of the year.

Natural History Notes and Uses. Shepherd's purse has been used as food for thousands of years. The seeds were found in the stomach of Tollund man in Denmark (approximately 500 B.C. to A.D. 400) and during excavations of the Catal Huyuk site in south-central Turkey, dated to approximately 5950 B.C. (Tull

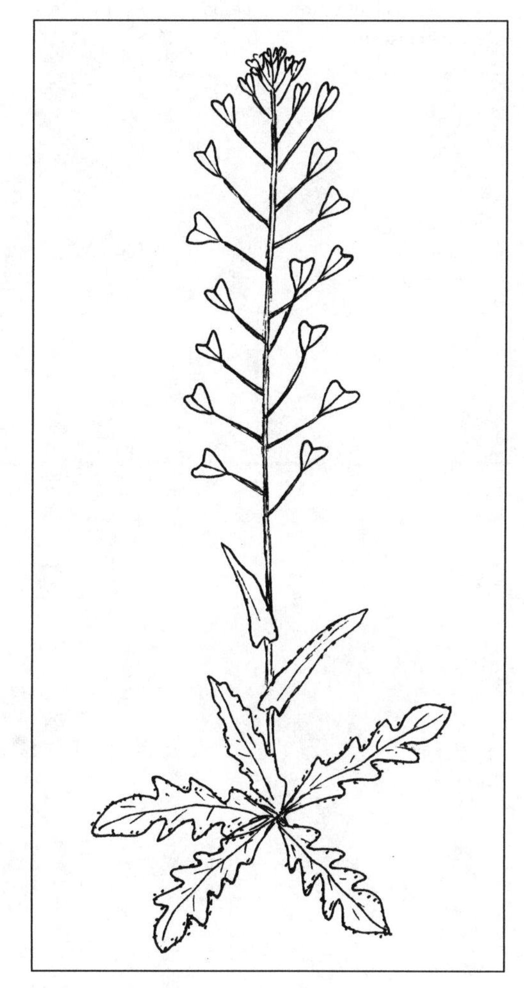

FIGURE 42. Shepherd's purse (*Capsella bursa-pastoris*).

1987). The seeds of shepherd's purse may be parched and eaten or ground into flour. The whole pod, with the seeds beaten out, can be added to salads or soups or dried for winter use. The young leaves can be prepared as a potherb and are a good source of vitamin C. With age, they develop a peppery taste. The entire herb (leaves, stems, green seed pods) can be chopped and added to soups. The roots may be ground or chopped and used as a ginger substitute. The seeds are known to cause blistering of the skin.

The plant is extremely high in vitamin K, the blood-clotting vitamin. Mash or chew the leaves and hold them on a cut. The juice of the plant on a ball of cotton was used to stop a nose-bleed. Shepherd's purse also contains significant amounts of calcium, potassium, sulfur, and ascorbic acid. As a decoction, shepherd's purse has been used to treat hemorrhoids, diarrhea, and bloody urine. The decoction has a gentle detergent action and is very cleansing to the skin.

BIOLOGICAL CONTROL OF MOSQUITOES

The seeds of shepherd's purse have been used against mosquitoes. When the seeds are sprinkled on water where mosquitoes breed, the mucilage on the seeds kills the larvae. One pound of seeds is said to destroy ten million larvae. In water the seeds release a mosquito attractant and gummy substance that binds the mouth of the larva to the seed; in addition, it releases a substance that destroys the larvae. Despite their pesky nature to humans, however, mosquitoes have an important ecological role, providing food for many fish, birds, and other insects.

Bittercress (genus *Cardamine*)

IDAHO BITTERCRESS (*C. oligosperma*)

PENNSYLVANIA BITTERCRESS (*C. pennsylvanica*)

Description and Habitat. The species of *Cardamine* are annuals or perennials with entire or pinnate leaves. The flowers are white or purple. The pods are elongate and flattened. The genus name comes from the Greek, indicating that some bittercresses were thought to have heart-strengthening qualities.

FIELD KEY TO THE BITTERCRESSES

1. Lateral leaflets of stem leaves are lance-shaped to almost round . . . . *C. oligosperma*
1. Lateral leaflets of stem leaves almost linear . . . . *C. pennsylvanica*

Natural History Notes and Uses. The plants can be eaten raw in salads, but we suggest cooking them in at least one change of water to improve the taste. Some plants in this genus were reputed to have medicinal qualities that were used in the treatment of heart ailments.

Whitetop (genus *Cardaria*)

WHITETOP (*C. chalapensis*)

HAIRY WHITETOP (*C. pubescens*)

Description and Habitat. Whitetops are rhizomatous perennials with erect or ascending stems and simple leaves. The herbage is pubescent with simple hairs, and the stalked flowers occur on the upper portion of the stems in a crowded, flat-topped inflorescence. The fruits are egg-shaped to almost round and have a persistent style at the top.

FIELD KEY TO THE WHITETOPS

1. Sepals, fruits, and pedicels hairy . . . . *C. pubescens*
1. Sepals, fruits, and pedicels not hairy . . . . *C. chalapensis*

Natural History Notes and Uses. These species are aggressive perennials native to southwest Asia. They were likely introduced in shipments of contaminated alfalfa seed from Turkestan into North America over a period of 40–50 years. All species readily establish in disturbed areas in range and wildlands and are favored during years of above-average precipitation. Once established, colonies are difficult to eliminate because of deep, persistent roots. Cultivation can facilitate the spread of the plants by dispersing root fragments. However, repeated cultivation (bimonthly to monthly) can destroy colonies in 2–4 years. Flooding an area with 6–10 inches of water for two months can eliminate troublesome infestations.

The leaves of a related species, *C. draba* (= *Lepidium d.*), can be eaten raw or cooked. The young inflorescences resemble small broccoli and are eaten in much the same way.

CROSSFLOWER, BLUE MUSTARD (*Chorispora tenella*)

Description and Habitat. Crossflower is a leafy, taprooted winter annual. The plants branch primarily from the base, with some branching of the flowering stems as well. The rosette leaves of seedling plants are deeply lobed or toothed, and the mature plants are 6–18 inches tall, with nearly unlobed leaves that are toothed or wavy. Flowering stems are leafy on the lower portions. Both leaves and stems are covered with gland-tipped hairs that make the plants feel sticky. The 4-petaled flowers appear in early spring and are purple to somewhat pinkish. Seed pods are thin, slightly curved, elongate, and seedless (sterile) for about a third of their length. Unlike the seed pods of many other mustards, which split open along the length of the pod at maturity, the pods of this species break apart crosswise into two-seeded sections.

Natural History Notes and Uses. This plant is a native of Russia or southwest Asia. It was first documented in this country in Lewiston, Idaho, in 1929 and has spread throughout the western plains states, the western portion of the United States, and southern Canada. Crossflower likely was introduced into the United States by accident in imported seed, as is true with many members of the mustard family.

The genus name is from the Greek words *chori*, meaning "separate," and *spora*, "seed," and refers to the constricted seed shape. The species name means "slender" and probably refers to the overall plant or flower shape.

Tansy Mustard (genus *Descurainia*)

SIERRAN TANSY MUSTARD (*D. californica*)

MOUNTAIN TANSY MUSTARD (*D. incana*)

WESTERN TANSY MUSTARD (*D. pinnata*)

TANSY MUSTARD (*D. sophia*)

WYOMING TANSY MUSTARD (*D. torulosa*)

Description and Habitat. Tansy mustards are annual or biennial herbs with leaves that are 1–3 times pinnately divided. The foliage is covered with simple, branched, or short gland-tipped hairs. The flowers are cream-colored or light yellow, and the pods are long, narrow, and 3-sided to nearly round in cross section. These are weedy species occurring in disturbed soils at the lower elevations.

FIELD KEY TO THE TANSY MUSTARDS

1. Lower leaves 2 or 3 times compound . . . . *D. sophia*
1. Lower leaves mostly once compound 2

2. Fruits widest near the middle . . . . *D. californica*
2. Fruits linear or widest at the top 3

3. Fruits club-shaped, usually not torulose . . . . *D. pinnata*
3. Fruits linear and often alternately swollen and constricted (torulose) . . . 4

4. Fruits hairy, long-tapering to tip, and strongly torulose. . . . . . . *D. torulosa*
4. Fruits not hairy, often abruptly round at tip . . . . *D. incana*

Natural History Notes and Uses. All species of tansy mustard are edible as greens but are bitter. The seeds can be parched, ground, and prepared as mush. The seeds were parched by tossing in a basket with hot stones or live coals, then ground into a fine flour and made into mush. Because of its peppery taste, the mush was often mixed with the flour of other seeds to make it more palatable. Young leaves can be boiled or roasted between hot stones and eaten as green vegetables. The seeds were also used in poultices for wounds. However, one species, *D. pinnata*, is reported to be ***poisonous*** in large quantities, causing blindness and paralysis of the tongue.

Wyoming tansy mustard is known only from the high volcanic mountains in northwest Wyoming and Pine Butte in southwest Wyoming. The species was removed from the state's sensitive list primarily because of taxonomic questions and the apparently large amount of suitable habitat that has not been searched for the plant. Wyoming tansy mustard may in fact be a minor variant of the common *D. incana*. Successful searches for additional populations were conducted in the 1980s, suggesting that additional searches in suitable habitat will yield more populations. The remaining known populations are scattered and extremely small, making this species quite vulnerable (Dorn 1988).

Whitlow Grass (genus *Draba*)

Description and Habitat. These low, pubescent perennials have basal leaves in a tuft or cushiony rosette. The leaves are lanceolate, and the flowers occur in a terminal raceme. The small flowers are white or yellow that fades to white with age. The pods are egg-shaped, elliptical, or club-shaped and sometimes twisted. The various *Draba* species occur on shaded slopes and rocky areas. *Draba* is the Greek word for "acrid" and was applied to similar mustards known to the Greeks thousands of years ago.

The genus is large, and identifying species is difficult. Close inspection of the hairs of the leaves is important for distinguishing the species and worth the attention for the intricate forms the hairs take.

Natural History Notes and Uses. The various species are edible but generally unpalatable. Whitlow grasses were formerly used for treating "whitlows," inflammations of the fingertip. Whitlow is a name applied loosely to any inflammation involving the pulp of the finger and attended by swelling and throbbing pain.

Wallflower (genus *Erysimum*)

WALLFLOWER (*E. asperum* = *E. capitatum*)

SHY WALLFLOWER (*E. inconspicuum*)

Description and Habitat. Wallflowers are annual, biennial, or perennial herbs that are often taprooted. The leaves are usually narrow, and the flowers occur in dense racemes. The sepals are erect and narrow with the two outer usually saccate at the base.

FIELD KEY TO THE WALLFLOWERS

1. Petals greater than ½ inch long	*E. asperum*
1. Petals less than ½ inch long	*E. inconspicuum*

Natural History Notes and Uses. Wallflowers include both popular garden species and many wild forms. There are about 80 species native to southwest Asia, the Mediterranean region, and North America. They are small, short-lived perennial herbs or subshrubs, reaching 4–50 inches tall, with bright yellow to red or pink flowers produced throughout the spring and summer. *Erysio* means "to draw out," as in drawing out pain or causing blisters. The generic name also stems from the Greek *eryomai*, meaning "help" or "save," from the medicinal properties of some species claimed by the early European herbalist-physicians.

Wallflowers were once used as a poultice. Additionally, an infusion of dried, pulverized *E. capitatum* was rubbed on the head and face to prevent sunburn or to alleviate heat exposure. Chewed wallflower root was used to massage the backs of pneumonia patients (Strike 1994). *Erysimum* species are used as food plants by the larvae of some lepidopterans, including garden carpet.

SLENDER MOUSE-EAR-CRESS, ROD HALIMOLOBOS (*Halimolobos virgata*)

Description and Habitat. Slender mouse-ear-cress is a biennial species with one to several stems. The stems often branch and grow 4–16 inches tall. The plant has a grayish appearance due to a covering of simple and branched hairs. The leaves taper to a slender basal stalk and may be pointed or blunt at the tip. The leaf edges may be smooth or wavy-toothed. The several leaves on the stem decrease in size going up the stem but are widest at the base, where they clasp the stem. The flowers occur on slender stalks and open as they develop at the tip of the stem. The sepals are purplish, and the petals are white with pink or lavender veins. The fruits are cylindrical and upright at the ends of their stalks, and the numerous seeds are crowded in two rows on the pod, which fits closely enough to make the seeds visible.

Natural History Notes and Uses. Slender mouse-ear-cress is found only in western North America. In the United States it occurs in Montana, eastern Idaho, Wyoming, Colorado, Utah, Nevada, and southeastern California.

Slender mouse-ear-cress can be distinguished from other, similar members of the mustard family (*Arabis* species) by its white flowers opening while still on the stem tip, by the presence of the many-branched hairs, and by the fruit, which is round in cross section. Some *Arabis* may have white flowers, but only flower buds are found at the stem tips, and the flowers do not open until the stem has grown longer. In *Arabis* species, hairs are absent or unbranched on the stems and leaves. The fruits of *Arabis* are usually somewhat flattened with smooth surfaces, which do not show the seeds inside.

Interestingly, the Committee on the Status of Endangered Wildlife in Canada listed this species as endangered in 1992 and down-listed it to threatened in 2000 when a population was found in Alberta. It is listed as a threatened species in Saskatchewan in *The Wild Species at Risk Regulations* and is protected on private, provincial, and federal lands under part V of Canada's Wildlife Act.

SWEET ROCKET, DAME'S ROCKET (*Hesperis matronalis*)

Description and Habitat. Sweet rocket is a tall herbaceous biennial or a short-lived perennial that is widely planted and escapes readily. The flowers are variable in color, ranging through many shades from white to pink to purple. It is frequently mistaken for "wild phlox," but the 4 petals and conspicuous fruits 2–5 inches long clearly identify it as a mustard family plant rather than a phlox, which would have 5 petals and much smaller, inconspicuous capsules. It does best in moist or wet woods but can tolerate a wide variety of other habitats.

Natural History Notes and Uses. A native European, sweet rocket has naturalized through most of the United States. It can be long established in some of these "wild" situations, freely propagating by seed.

The leaves, flower buds, and flowers are edible raw and are excellent in salads. However, the plant has diaphoretic and diuretic properties.

Note: The sale of sweet rocket to addresses in the state of Colorado is prohibited.

Pepperweed (genus *Lepidium*)

COMMON PEPPERWEED (*L. densiflorum*)

MOUNTAIN PEPPERWEED (*L. montanum*)

CASPING PEPPERWEED (*L. perfoliatum*)

VIRGINIA PEPPERWEED (*L. virginianum*)

Description and Habitat. The four species of *Lepidium* are annual or perennial plants that are widely distributed. One species is grown for salad. The genus name is Greek and refers to a little scale, describing the shape of the pods.

FIELD KEY TO THE PEPPERWEEDS

1. Base of upper stem leaves clasping or wrapping around the stem . . . . . *L. perfoliatum*
1. Base of stem leaves not clasping or wrapping around the stem 2

2. Style as long as or longer than notch of the fruit . . . . . *L. montanum*
2. Style lacking or at least shorter than notch of the fruit 3

3. Petals none or vestigial . . . . . *L. densiflorum*
3. Petals conspicuous . . . . . *L. virginianum*

Natural History Notes and Uses. The young stems and leaves may be eaten raw or dried for future use. The plants contain vitamins A and C, iron, and protein. The seed pods and seeds can be used as a flavoring. For poison ivy rash or scurvy, fresh *L. virginianum* plants were bruised or a tea was made from the leaves.

Bladderpod (genus *Lesquerella*)

ALPINE BLADDERPOD (*L. alpina*)

GREAT PLAINS BLADDERPOD (*L. arenosa*)

IDAHO BLADDERPOD (*L. carinata*)

FOOTHILL BLADDERPOD (*L. ludoviciana*)

PAYSON'S BLADDERPOD (*L. paysonii*)

Description and Habitat. There are about 90 species of bladderpods, which are annual to perennial, densely hairy herbs with small flowers. They generally occur on dry slopes and flower from May to June. The genus is named after Leo Lesquereux, a late-nineteenth-century American paleobotanist. *Lesquerella* and *Physaria* are very similar, and species of one genus can easily be mistaken for members of the other.

FIELD KEY TO THE BLADDERPODS

1. Fruits are flattened at a right angle to the partition 2
1. Fruits are inflated or sometimes flattened parallel to the partition, but only along the margins 3

2. Fruits with strongly flattened margins and strong keels on flattened side . . . . . *L. carinata*
2. Fruits not flattened on the margins nor with strong keels on the sides . . . . . *L. paysonii*

3. Pedicels in fruits S-shaped or uniformly curved upward, or rarely straight; fruit not spherical . . . . . *L. alpina*
3. Pedicels in fruit uniformly recurved, not S-shaped; fruits typically spherical 4

4. Flowers yellow; leaves narrow and linear . . . . *L. ludoviciana*
4. Flowers tinged reddish or purplish; leaves broader than linear and sometimes toothed . . . . *L. arenosa*

Natural History Notes and Uses. Bladderpods have been developed into hundreds of ornamentals as well as food plants such as cauliflower, cress, radish, kohlrabi, turnip, and rutabaga. There are about 90 species worldwide, and the center of their range seems to be the southwestern United States and Mexico. The casual observer may mistake bladderpods for one of the more common mustards, but where most mustards have elongated pods, the pods of bladderpods are rounded.

The seeds of bladderpods contain an oil that is similar to castor oil, which could make the plants valuable as a commercial crop. Castor oil is used in many products, including fibers, paints, resins, lubricants, cosmetics, and hydraulic fluids. Since the 1950s the U.S. Department of Agriculture has been researching the domestication of *Lesquerella* and in the 1980s began breeding programs with various eastern U.S. species.

A related species, Fendler's bladderpod, was prepared as a tea by the Navajo as a remedy for spider bites (Willard 1992).

NAKED-STEMMED WALLFLOWER (*Parrya nudicaulis*)

Description and Habitat. Naked-stemmed wallflower, also called naked-stemmed parrya, is a glandular, perennial herb with stems up to 8 inches high. The stout, woody rootstalk is branched and covered by old leaf bases. Leaves are mostly basal and have oblanceolate, entire to coarsely toothed, stalked blades 5–25 mm wide. The flowers have 4 pink to lavender (sometimes white) petals and 4 purple glandular or glabrous sepals and are arranged in a raceme. The oblong, flattened fruits are usually over ¾ inch long and constricted between the seeds (torulose).

Natural History Notes and Uses. This plant is typically found on steep, unconsolidated talus slopes of gray limestone or pinkish sandstone in the alpine or upper subalpine zones. These sites usually have very low vegetative cover (less than 25%) and are inhabited mostly by low cushion plants and alpine willow species. Occasionally, colonies can be found on moist grassy hummocks on low saddles.

The fruits superficially resemble those of many members of the pea family (Fabaceae) but differ in having an internal dividing membrane (replum). The flowers are very fragrant. The range of variation in this species is unusually great, and the same variants occur in both the North American and Eurasian populations.

Twinpod (genus *Physaria*)

COMMON TWINPOD (*P. didymocarpa*)

SNAKE RIVER TWINPOD (*P. integrifolia*)

Description and Habitat. Twinpods are tuft-forming perennials that have prostrate stems and clustered basal leaves with entire or toothed margins. The plants are covered with star-shaped (stellate) hairs. The flowers are yellow and occur on the terminal portion of the stem. The fruits are 2-lobed and inflated and have a notch at the top between the lobes.

FIELD KEY TO THE TWINPODS

1. Mature fruits usually ½ inch or less wide . . . . *P. didymocarpa*
1. Mature fruits generally over ½ inch wide . . . . *P. integrifolia*

Natural History Notes and Uses. The twinpods are often difficult to distinguish from the bladderpods (*Lesquerella*), but the inflated fruits of twinpods are divided by a notch into 2 balloon-like lobes. The fruits of bladderpods are nearly round with a slender style at the tip. The genus name is from the Greek *physa* ("bladder"), alluding to the inflated fruit. This is a genus of about 14 species of perennial herbs. Some species are cultivated as ornamentals.

Yellowcress, Watercress (genus *Rorippa*)

PERSISTENT SEPAL YELLOWCRESS (*R. calycina*)

BLUNT-LEAF YELLOWCRESS (*R. curvipes*)

CURVE-POD YELLOWCRESS (*R. curvisiliqua*)

WATERCRESS (*R. nasturtium-aquaticum*)

BOG YELLOWCRESS (*R. palustris*)

Description and Habitat. Yellowcresses are taprooted annuals or rhizomatous perennials with simple or pinnately divided leaves. The flowers are yellow or white; the pods are elliptical to linear and 3-sided to slightly compressed. The species occur in moist, wet, or aquatic habitats up into the middle elevations.

FIELD KEY TO THE YELLOWCRESSES

1. Flowers white; most of the leaves pinnately compound . . . . *R. nasturtium-aquaticum*
1. Flowers yellow; leaves usually not compound 2

2. Plants perennial, with rhizomes . . . . *R. calycina*
2. Plants annual or biennial, without rhizomes 3

3. Pedicels usually as long as or longer than fruits; plants of moist areas . . . . *R. palustris*
3. Pedicels usually shorter than fruits 4

4. Fruits narrowly elliptical . . . . *R. curvipes*
4. Fruits more linear . . . . *R. curvisiliqua*

FIGURE 43. Watercress (*Rorippa* spp.).

Natural History Notes and Uses. The herbage of watercress is edible if the waters in which they grow are not polluted. However, finding unpolluted water may be difficult. One suggestion is to soak the fresh greens in a disinfectant or treat the water with water purification tablets or a tablespoon of bleach in a quart of water. Then rinse the greens well in potable water to remove the chemicals. The peppery-tasting plants were eaten raw or cooked as a potherb. A good source of vitamins, watercress is listed as efficient in preventing scurvy. Medicinally, the plant was used for freckles, pimples, and liver and kidney troubles.

Note: Bittercress (*Cardamine brewerii*) looks similar to watercress. To quickly differentiate the two species, look at the fruits. Bittercress fruits are linear and narrow; watercress fruits are round or 4-angled in cross section.

TALL TUMBLE MUSTARD (*Sisymbrium altissimum*)

Description and Habitat. This annual herb has small flowers that are yellow or white; the fruits are linear. This species was introduced from Europe and is widespread throughout the United States. It is usually found in waste places and disturbed habitats at low elevations.

Natural History Notes and Uses. The seeds of a related species, *S. officinale* (hedge mustard), were parched and then ground into a flour. Tall tumble mustard makes a fine potherb. As with other mustards, it is best to cook the plants in a couple of changes of water.

ALPINE SMELOWSKIA (*Smelowskia calycina*)

Description and Habitat. This gray matted plant has white or purplish flowers in dense racemes held well above thick clusters of basal leaves. The flower petals are each about ¼ inch long, and the leaves are ½ inch long, oblong, pinnately divided into narrow segments, and on stiffly hairy stalks as long as the blades. The fruit is a lanceolate-shaped pod about ¼ to ½ inch long and held erect.

Natural History Notes and Uses. The genus name honors T. Smelowski, an eighteenth-century Russian botanist. This arctic-alpine plant is relatively common in stony, open soil areas of windswept ridges and summits near or above timberline. The plants generally bloom in May or June, but those that occur in shady sites where snow may remain late into the summer season may produce lax stems and elongated leaves and bloom later or not at all.

These plants are often infected by rust fungi, which disfigures the leaves, causing the flowers to abort. At first the rust appears as many pimple-like bumps on the leaves, which soon become powdery with spores.

Prince's Plume (genus *Stanleya*)

DESERT PRINCE'S PLUME (*S. pinnata*)

HAIRY PRINCE'S PLUME (*S. tomentosa*)

Description and Habitat. Prince's plumes have yellow flowers that occur in elongated racemes. The fruits (siliques) are borne on a long stipe. The plants are usually found in sagebrush habitats at low elevations. The genus is named for Lord Edward Stanley, a British ornithologist who lived from 1775 to 1851. Prince's plume was described for science by the German botanist Frederick Pursh (1774–1829), who first published on the many plants collected by Lewis and Clark.

FIELD KEY TO THE PRINCE'S PLUMES

1. Lower leaves and stem densely long hairy; claws or petals glabrous . . . . *S. tomentosa*
1. Lower leaves and stem not hairy or only sparsely short hairy; claws of petals hairy . . . . *S. pinnata*

Natural History Notes and Uses. These plants host a wide variety of native insects, including the butterflies that are its pollinators, and is always interesting to examine. The tall stalks make good cut flowers, and the plants may flower a second time after cutting. The foliage may have an odd and somewhat unpleasant odor. This is the smell of organic selenium-bearing chemicals. The plant is an accumulator of this element and is usually found on high-selenium soils derived from fine-textured sedimentary rocks. Selenium is essential for animals, but too much of it can be toxic. The unpleasant smell is not likely to be a problem in the home garden. The tiny orange seeds of prince's plumes are borne in long, pendant pods that are easily stripped from the plants when dry.

The tender leaves and stems of all species can be prepared in much the same way as cabbage. The Navajo, Paiute, and Hopi people are all known to have eaten its leaves, but because of their high selenium content, they were twice boiled, which is said to remove the toxic mineral. The seeds can be collected, parched, and then ground into a flour. They can be eaten as a mush or used in making breads. Prince's plume is a Navajo "life medicine" and has been used as an emetic and to treat glandular swellings.

WATER AWLWORT (*Subularia aquatica*)

Description and Habitat. Awlwort is a small, rather uncommon, underwater herb that looks like a tuft of quill-like leaves with a leafless flowerstalk growing 1–4 inches tall. The small, white flowers appear in a loose cluster along the ends of the stalks.

Natural History Notes and Uses. The plant is found in shallow water in clear, cold lakes and slow streams, often on sandy or gravelly sediments. It is usually an annual plant, although it may overwinter if completely submersed. It is an inconspicuous plant that is overlooked by most people and is probably more common than collecting records indicate. In Michigan the plant is considered an endangered species.

The genus name is from the Latin *subula,* meaning "awl"; *aquatica* is from the Latin *aquaticus,* meaning, as one might expect, "living, growing, or found in or by the water; aquatic." The common name comes from the long, pointed leaves and the Anglo-Saxon *wort,* meaning "plant."

Thelypody (genus *Thelypodium*)

ENTIRE-LEAVED THELYPODY (*T. integrifolium*)

NORTHWESTERN THELYPODY (*T. paniculatum*)

Description and Habitat. These herbs are biennial or perennial, with or without hairs. The stem leaves are often arrowhead-shaped. The flowers are purple to white, and the pods are rounded to flattened parallel to the partition.

FIELD KEY TO THE THELYPODY

1. Stem leaves sagittate or clasping at the base . . . . *T. paniculatum*
1. Stem leaves not sagittate or clasping at the base . . . . *T. integrifolium*

Natural History Notes and Uses. The genus was named by Austrian botanist Stephan Ladislaus Endlicher (1804–49). The genus name is from the Greek *thely* ("woman") and *pod* ("foot" or "tender"), but the significance of these words as applied to these two plants is lost in history. Perhaps the long, slender pods are supposed to be the shape of a woman's foot. A few sources indicate that the name refers to the stipe (the stalk) of the ovary. Native people ate the leaves and seeds of a related species (*T. flavescens*).

Thomas Nuttall collected the first specimen of entire-leaved thelypody on his 1834–37 trip across the Louisiana Territory with the Wyeth expedition. He named the plant *Pachypodium integrifolium* in 1838. It was renamed *Thelypodium integrifolium* in 1842 by Endlicher.

Pennycress (genus *Thlaspi*)

FIELD PENNYCRESS (*T. arvense*)

ALPINE PENNYCRESS (*T. montanum*)

MEADOW PENNYCRESS (*T. parviflorum*)

Description and Habitat. Pennycresses are glabrous annuals or perennials with leaves that have entire or toothed margins. The stem leaves have small lobes at the base that clasp the stem. The flowers occur on the bractless, terminal portion of the stems. The petals are white, and the flattened fruits are egg-shaped to narrowly elliptical with a notch at the top and broad or narrow winged margins. *Thlaspi*, a name used by the ancient Greek physician Dioscorides, is from the Greek *thlao*, "to flatten," and *aspis*, "shield," alluding to the shape of the fruits. Dioscorides, who lived in Rome during the time of Nero, is famous for his book *De Materia Medica*, the precursor to all modern pharmacopeias.

FIELD KEY TO THE PENNYCRESSES

1. Plants annual . . . . *T. arvense*
1. Plants perennial 2

2. Petals ⅛ inch long; styles very short . . . . *T. parviflorum*
2. Petals more than ⅛ inch long; styles more than 1⁄16 inch long . . . . *T. montanum*

Natural History Notes and Uses. Field pennycress was introduced to North America from Eurasia at a very early date and is conceded to be a significant agricultural weed that competes keenly with crops for moisture and space, causing profound reductions in yield. Adding to the problem is seed distribution by spring floods. Once the species is established, the soil becomes contaminated with its seeds. The plant bears an unpleasant odor, making it easy to identify. Because of its abominable smell, it became widely known as stinkweed. The name is easily understood by anyone who has ever handled the weed or tasted milk or butter from a cow that has eaten it.

Field pennycress is a popular food plant in various parts of the world, often being cultivated in Europe. It is used when the shoots are young and tender, eaten raw as a salad or cooked as a potherb like spinach. It is high in vitamin C and contains a relatively large amount of sulfur; it may have the health effects of sulfur and molasses. To prepare, boil the shoots for 15–25 minutes and change the water once or twice. Even then a slight bitterness is present, so mix the greens with those from some blander-tasting plants like pigweed (*Amaranthus*). The young tender leaves, when used in a salad, are also rather bitter-tasting, so either mix them with those from other plants or use a strong-flavored salad dressing. The seeds and fruits have been used to flavor other food. But be cautious, as field pennycress has caused illness when fed to cattle in hay.

Caution: The entire plant of field pennycress is considered to be poisonous, but the seeds especially contain isoallyl thiocyanates and irritant oils. Symptoms include oral and gastrointestinal irritation, which leads to head shaking, salivating, colic, abdominal pain, vomiting, and possibly diarrhea. Generally, symptoms do not occur unless large quantities are consumed over a period of time.

CACTUS FAMILY (CACTACEAE)

There are over 140 genera and 2,000 species within the cactus family worldwide. Approximately 16 genera are native to the United States. Native to the Western Hemisphere, cacti have been spread all over the world, frequently carried by explorers and other travelers. Cacti are typically succulent spiny herbs of diverse form. One distinctive feature is the presence of *areoles*, which are round or elongated spots or openings that may be raised or pitted and usually are arranged in rows or spirals over the surface of the plant. The spines grow from the areoles. The flowers have many sepals, petals, and stamens and an inferior ovary. Economic products from this family include ornamentals, edible fruits, nopalitas, and the hallucinogenic peyote. The spines of some cacti were once used as phonograph needles.

Cacti are as evocative of the West as sagebrush. They have a number of distinctive shapes, their flowers are often large and attractive, and they have evolved over aeons of time to be perfectly at home in what we humans usually call "a hostile environment."

FIELD KEY TO THE CACTUS FAMILY

1. Stems flat or cylindrical and jointed. . . . . *Opuntia*
1. Stems oval or globose 2

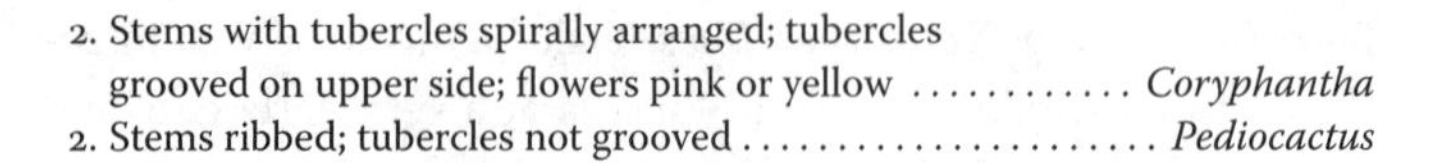

2. Stems with tubercles spirally arranged; tubercles grooved on upper side; flowers pink or yellow .... *Coryphantha*
2. Stems ribbed; tubercles not grooved .... *Pediocactus*

PINCUSHION CACTUS (*Coryphantha missouriensis = Escobaria m.*)

Description and Habitat. This cactus is so small that it easily escapes our eyes. The plant generally grows to only 2–3 inches tall, including the half of the plant body that is below ground level. The flowers are greenish yellow and are followed by seeds that mature into bright red berries the year after the flower blossoms.

Natural History Notes and Uses. The generic name means "top-flowered" in botanical Latin, in reference to the placement of the flower atop the stem. The species name means "of Missouri," in reference to the river, not the state. This species was described for science in 1827 by English horticulturist and ornithologist Robert Sweet (1783–1835). Sweet was the monographer of the geranium family. The 60 or so cacti from this genus come from the southwestern United States and northwestern Mexico. It is closely related and similar to the genus *Mammillaria.*

The rounded stems or "balls" of this perennial plant have to carry on the photosynthesis because there are no true leaves. These stems, which protrude part way through the ground, may be found singly or branched into groups of up to a dozen. Numerous spine-bearing tubercles dot the plant.

Several species of this genus contain alkaloids that are chemically similar to epinephrine. The Comanche are said to have used the spines of this cactus to punish unfaithful wives. Members of the Crow tribe ate the ripe fruit.

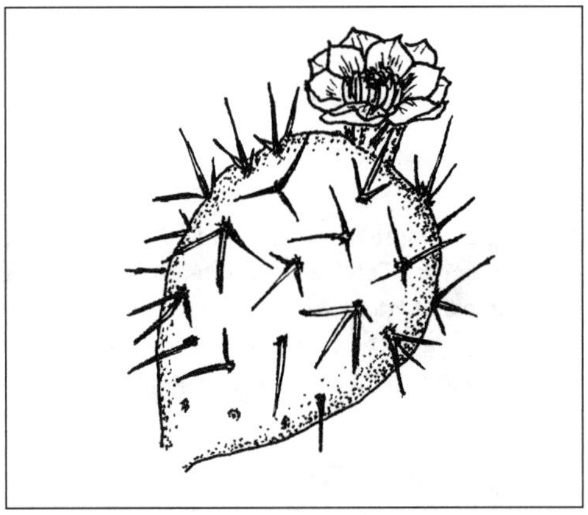

FIGURE 44. Prickly pear (*Opuntia* spp.).

Prickly Pear (genus *Opuntia*)

BRITTLE PRICKLY PEAR (*O. fragilis*)

PLAINS PRICKLY PEAR (*O. polyacantha*)

Description and Habitat. Prickly pears need very little introduction. The species in the northern Rocky Mountains are succulent herbs with fibrous roots and stems that are flat or cylindrical. The leaves when present are small, fleshy, and awl-shaped. Many *Opuntia* species have glochids: minute, nearly invisible barbed hairs that grow in clusters in areoles. They easily become embedded in skin or clothing and, because of their light tan or yellowish color and barbed surface, are almost impossible to remove. Prickly pear cacti occur in dry soils at lower elevations.

FIELD KEY TO THE PRICKLY PEARS

1. Stem segments are oval to cylindrical, about half as thick as wide, and easily detached .... *O. fragilis*
1. Stem segments are flattened, much wider than thick, and not easily detached .... *O. polyacantha*

Natural History Notes and Uses. These cacti have provided Native Americans with food, medicine, dyes, and a variety of other uses for thousands of years. They have also been credited with saving many lives by supplying both food and water to people stranded in the desert. Since eating too many cactus fruits can cause constipation, they should be eaten in moderation. The cactus also contains oxalic acid, which interferes with calcium absorption.

The pads, especially younger pads, make an excellent cooked vegetable. Harvest the young pads by grasping them with tongs and slicing them at the stem joints. Hold over a flame to singe off the spines and glochids and scrape off the remaining ones with a knife. Rinse well. Slice into thin strips and boil for at least 10 minutes. Drain off water, rinse to remove the slippery gum, and they're ready to eat. To use the older pads, slice away the more fibrous sections and cook accordingly.

After the spines are removed, the cactus fruits (also known as prickly pears) can be peeled and the pulp eaten raw, or it can be boiled and then fried or stewed. One solution for removing the spines is to burn them off; another is to split the fruit into two halves and eat the insides. The pulp can also be sun- or fire-dried for future use. High in protein and oil, the nutritious seeds may be eaten, added to soups, or dried and ground into flour.

Prickly pear cacti pads have been used as a soothing poultice for wounds and bruises. A tea made from the flowers was said to increase urine flow, and a tea from stems was used as a wash to ease headaches, eye troubles, and insomnia. The fruits are high in calcium, potassium, and vitamin C. Some Native Americans used large baskets baited with crushed prickly pear pads to catch sardines. Another tribe roasted *Opuntia* stems, then soaked them. The resulting extract was used to improve the plasticity and cohesion of clay for making pottery.

Archaeologists working in Texas discovered purses made from prickly pear pads. According to Tull (1987), the dried pads were hollowed out to form a small container. A dye can be obtained from the juice of the uncooked fruit. Bryan and Young (1940) suggest letting wool soak for about a week in the fermenting juice. The color ranges from pink to magenta and appears to fade when exposed to sunlight.

Within the genus Opuntia are various species with common names like "prickly pear," "beaver tail," and "cholla." Here is one way of distinguishing them:

- prickly pear: flat pads with spines
- beaver tail: flat pads with no spines
- cholla: cylindrical stems

SIMPSON HEDGEHOG CACTUS, MOUNTAIN CACTUS (*Pediocactus simpsonii*)

Description and Habitat. Hedgehog cactus is a common and beautiful cactus, a globular type that reaches up to 6 inches in diameter and is strongly tubercled. It flowers from early May to June; the flowers are borne in the center of the cactus and are usually a brilliant pink but can sometimes be whitish in color. Interestingly, the flowers are generally pink on plants on eastern slopes and yellowish on those on western slopes. The flowers usually remain closed on cloudy days.

Natural History Notes and Uses. This genus contains eight species, seven of which are rare plants native to the Colorado Plateau region of Utah, Colorado, New Mexico, and Arizona. *Pediocactus* is derived from the Greek words *pedinos*, meaning "plain" or "level," and *kaktos*, meaning "thistle." The species was first named by George Engelmann for army engineer James H. Simpson under the name *Echinocactus simpsonii*. Simpson led an expedition in Colorado, and Engelmann named the species "in honor of the gallant commander" of the expedition. The name of *Pediocactus simpsonii* was set forth by Britton and Rose, who described the species in 1913, the genus name meaning "from the plains"—even though this cactus grows in the mountains.

WATER STARWORT FAMILY (CALLITRICHACEAE)

Members of this family are small annual or perennial herbs with slender, usually lax stems. The leaves are simple, entire, and opposite or whorled. The minute unisexual flowers are borne in the axils of the leaves. There are no sepals or petals. The small, four-lobed fruit splits into four sections on maturity. The plants are inconspicuous in standing water or drying mud. The family has only one genus (*Callitriche*) and approximately 40 species.

Water Starwort (genus *Callitriche*)

NORTHERN WATER STARWORT (*C. hermaphroditica*)

LARGER WATER STARWORT (*C. heterophylla*)

VERNAL WATER STARWORT (*C. palustris*)

Description and Habitat. Water starworts are so called because of the shape of their floating apical rosettes. They are most common in slow-flowing waters where muddy and silty sediments predominate, although they can grow in still and fast-flowing conditions up to 3 feet deep. However, they are intolerant of inorganic pollution. They have slender stems and long, thin, submerged leaves that expand when they reach the surface. The submerged leaves are characterized by notched ends. The genus is extremely polymorphic, taking on different leaf shapes in different environmental conditions. Classification and distinction between species is often possible only by examination of the flowers and seeds.

FIELD KEY TO THE WATER STARWORTS

1. Leaves are linear; flower bracts are absent .... *C. hermaphroditica*
1. Upper leaves usually broader; flower bracts present.................... 2

2. Fruits with slight winged margin on top .... *C. palustris*
2. Winged margins not present.......................... *C. heterophylla*

Natural History Notes and Uses. This is the only genus of flowering plants in which aerial, floating, and subsurface pollination systems have all been reported. Water starworts are unique because their flowers lack petals, but in some species they are enclosed in modified leaves. The genus name is from the Greek *kallos*, which means "beautiful," and *trichos*, which means "hair," referring to the slender stems. Strike (1994) indicates that the Native Americans in California used *Callitriche* to relieve urinary problems, but the method is not reported.

BLUEBELL FAMILY (CAMPANULACEAE)

Worldwide there are over 70 genera and 2,000 species in this family. Twelve genera are native to the United States. Bluebells are annual or perennial herbs usually with milky juice. The flowers are typically 5-parted with the calyx divided into separate sepals, and the corolla is 5-lobed and bell-shaped. The family is of little economic importance, but some species are cultivated as ornamentals.

FIELD KEY TO THE BLUEBELL FAMILY

1. Corolla regular; filaments and anthers distinct 2
1. Corolla irregular; filaments and anthers united in a tube, 2 of the anthers smaller than the others . . . . *Porterella*

2. Plants perennials; corolla campanulate . . . . *Campanula*
2. Plants annuals; corolla various 3

3. Corolla campanulate and shallowly lobed. . . . . *Heterocodon*
3. Corolla rotate, deeply lobed (to at least below the middle) . . . . . *Triodanis*

Bellflower, Harebell (genus *Campanula*)

RAMPION BELLFLOWER (*C. rapunculoides*)

BLUEBELL BELLFLOWER (*C. rotundifolia*)

ARCTIC BELLFLOWER (*C. uniflora*)

Description and Habitat. Bellflowers are perennial herbs from a rhizome. The blue (occasionally white) flowers are tubular, bell-shaped, or cup-shaped. The genus name is from the Latin for "bell," and the common name, harebell, may allude to an association with witches, who were believed to transform themselves into hares, bringers of bad luck when they crossed a person's path. The species can be found in open, dry, or rocky areas from low elevations to above timberline.

FIGURE 45. Campanula (*Campanula rotundifolia*).

FIELD KEY TO THE BELLFLOWERS

1. Flowers number more than 7, with the pedicels about half as long (or less) than the flowers . . . . *C. rapunculoides*
1. Flowers usually solitary, the pedicels as long as or longer than the flowers 2

2. Plants alpine subalpine, usually less than 6 inches tall; flowers occur singly on each stem . . . . *C. uniflora*
2. Plants not as above . . . . *C. rotundifolia*

Natural History Notes and Uses. The leaves and shoots of at least bluebell bellflower can be used in salads or cooked as a potherb. The roots can also be boiled and eaten and have a nut-like taste. Rampion bellflower is also supposedly edible (Harrington 1967), though we have not sampled it.

FLESHY PORTERELLA (*Porterella carnosula*)

Description and Habitat. Porterella is a succulent annual with alternate sessile leaves. The blue to purple flowers with a yellow throat are axillary. This locally common plant is found at the margins of drying ponds.

Natural History Notes and Uses. The genus is named for botanist Thomas C. Porter (1822–1901), who was born in Pennsylvania and made his first visit to the Rocky Mountains in 1869. For five years he explored and collected in the central Rocky Mountains. As a result of these collections and with valuable help from the likes of J. M. Coulter and others, Porter wrote the *Synopsis of the Flora of Colorado*, the first Colorado flora, in 1874. This book firmly established the range of many plant species throughout the Rocky Mountains.

Porter collected the type specimen (the specimen used to scientifically describe the plant/species) of *Melica porteri* at Glen Eyrie, Colorado (near Colorado Springs). Other species first described by Porter now bear his name: *Aster porteri* (renamed in honor of Porter by Asa Gray), *Calamagrostis porteri*, and *Muhlenbergia porteri*.

He made his last publication at the age of seventy-eight. His lifelong wish to publish a flora of Pennsylvania was made possible by a provision in his will. In 1903 J. K. Small of the New York Botanical Garden produced the flora of Porter's home state. Porter is said to have been a cautious and generous man, possessing wit, impatience, and a thorough knowledge of plant ecology. Among his favorite pastimes were German literature and poetry.

CLASPING VENUS' LOOKING-GLASS (*Triodanis perfoliata*)

Description and Habitat. This upright annual reaches a height of 2½ feet. It grows in disturbed soil or moist areas such as ditches and along roadways. The blue flowers range from ¼ to ¾ inch in diameter. An older scientific name for this plant is *Specularia perfoliata*.

Natural History Notes and Uses. When it is sunny, the flowers open up during the morning and remain open for the rest of the day. They are attractive but rather small. The common name of this plant probably refers to the shiny seeds of a related European species. The seeds of Venus' looking-glass are too tiny to appear shiny to the unaided human eye.

The species was used by the Cherokee to treat indigestion. Additionally, an infusion of the roots was taken and used as a bath for dyspepsia (Harrington 1967). Venus' looking-glass is sometimes used in dried flower arrangements.

HEMP FAMILY (CANNABACEAE)

The hemp family consists of annual herbs or climbing perennial herbs. The leaves are opposite and simple or compound. There are two genera (*Cannabis* and *Humulus*) with three to five species distributed in the north temperate zone, and they are widely cultivated. *Humulus* is native to the United States. The family is a source of hempen fiber, oils, edible seeds, hops, and tetrahydrocannabinols (THC), the psychoactive compound in *Cannabis*. Some references place the family in the Moraceae (mulberry family).

COMMON HOPS (*Humulus lupulus*)

Description and Habitat. The genus was formerly included in the mulberry family (Moraceae). Common hops is a strongly twining, herbaceous vine with stems up to 15 feet long. The stems and leaves are rough to the touch. The leaves are opposite, serrate, 3–7-lobed, with heart-shaped bases. The undersides of the leaves are glandular. The flowers are small, green, and unisexual. This is a widely cultivated plant from Europe and Asia.

Natural History Notes and Uses. Hops are primarily grown for their fruits, used in brewing to give ale and beer a distinctive bitter taste. Additionally, the young shoots can be prepared as potherbs. They can be boiled in water for about 3–5 minutes, then boiled again in fresh water until tender.

A tea from the fruits was traditionally used as a sedative, antispasmodic, and diuretic and as a remedy for insomnia, cramps, coughs, and fevers. Externally, the tea was used for bruises, boils, inflammations, and rheumatism. Recently, clinical studies have disproved the sedative qualities of hops and found that it has no physiological effect on the nervous system; nevertheless, anyone who drinks much of the tea tends to fall asleep or become groggy (Moore 1979).

Caution: The plant is known to cause dermatitis when handled.

CAPER FAMILY (CAPPARACEAE; FORMERLY KNOWN AS CAPPARIDACEAE)

Members of this family are shrubs, trees, or rarely herbs. They have simple or palmately compound leaves. The flowers have 4 sepals and 4 petals, and the fruit is a capsule or berry. Worldwide there are about 46 genera and 800 species distributed in tropical and subtropical areas. Eight or nine genera are native to the United States. The family is of economic importance as a source of ornamentals and capers, a salad seasoning.

FIELD KEY TO THE CAPER FAMILY

1. Stamens 6; fruit stipitate .... *Cleome serrulata*
1. Stamens 8–22; fruit sessile...................... *Polanisia dodecandra*

ROCKY MOUNTAIN BEEPLANT (*Cleome serrulata*)

FIGURE 46. Beeplant (*Cleome* spp.).

Description and Habitat. This erect, showy plant grows up to 40 inches tall and has alternate leaves divided into 3 lance-shaped, entire leaflets. The reddish purple to pink flowers are arranged in a dense, narrow, terminal inflorescence. The petals are separate but the sepals are united. The fruits are long-stalked, pendulous capsules, linear to lance-shaped in outline. Beeplant is found in disturbed areas such as roadsides and railroad rightsof-way at the lower elevations. A second species, *C. lutea* (yellow beeplant), also occurs in dry, sandy flats and along roadsides.

Natural History Notes and Uses. An important food for many western Native Americans, beeplant was extensively used as a potherb. The young tender shoots and leaves and the flowers are preferred. The plant has an unpleasant odor, especially when older, and a pungent taste much like that of the mustards. We found it necessary to cook the plants in at least two changes of water to remove the bitter taste. The seeds can also be collected and ground into flour.

The Blackfeet Indians used the whole plant to make a medicinal tea to alleviate fever (Hart 1996). Beeplant may have been used by Native Americans to treat stomachaches. A source of dye, the plants were collected in quantity and boiled down for several hours until a thick, fluid residue was produced. The water was then drained off and the plants allowed to dry and harden into cakes. When black dye or paint was needed, a piece of the cake was soaked in hot water.

ROUGHSEED CLAMMYWEED (*Polanisia dodecandra*)

Description and Habitat. This sticky, hairy annual has simple stems and a strong, rank odor. The leaves are about 2 inches long and bear 3 leaflets about 1 inch long. About 20 flowers are clustered at the top of the plant. The flowers are about ½ inch long and white with purple bases, and they produce numerous stamens that overtop the petals. The fruits consist of slender capsules about 1–2 inches long that are filled with many tiny, dark seeds.

Natural History Notes and Uses. The generic name was compounded from the Greek *polys*, "many," and *anisos*, "unequal," ways in which the genus differs in stamen characters from another genus in the family. There are about 30 species in this widely distributed genus. The specific epithet means "having twelve stamens" in botanical Latin.

Clammyweed was first described for science by the great Swedish naturalist Carl von Linné (Linnaeus), the father of modern plant taxonomy, in his famous *Species Plantarum* of 1753. *Clammyweed* refers to the glandular, sticky pubescence that covers the plant: touching the foliage gives the fingers a "clammy" feeling. Bees visit the flowers for nectar, and although there are flies that feed on the pollen, they do little to pollinate the flowers. The plants are not known to be toxic even though they emit a foul odor.

HONEYSUCKLE FAMILY (CAPRIFOLIACEAE)

There are 15 genera and 400 species in the honeysuckle family. Of the 15 genera, 7 are native to the United States. They are woody plants with opposite leaves. The flowers are 5-merous, with the petals fused and an inferior ovary. Many genera in this family are cultivated as ornamentals.

FIELD KEY TO THE HONEYSUCKLE FAMILY

1. Flowers densely clustered in terminal cymes or umbels; flowers more or less flat 2
1. Flowers in pairs or short racemes or in axillary clusters; flowers tubular or funnel-shaped 3

2. Leaves pinnate . . . . *Sambucus*
2. Leaves simple or lobed . . . . *Viburnum*

3. Plants with slender, creeping stems and glossy, evergreen leaves . . . . *Linnaea*
3. Plants upright shrubs or vines; leaves deciduous, not glossy 3

4. Flowers nearly regular; berries white . . . . *Symphoricarpos*
4. Flowers irregular; berries red or black . . . . *Lonicera*

TWINFLOWER (*Linnaea borealis*)

Description and Habitat. Twinflower is a slender, trailing, mat-forming evergreen with short, leafless branches that divide into two at the top. At the top of these small branches arise small, bell-shaped pink or white flowers, hence the name twinflower. The evergreen leaves are oval or round and about ½ inch long. Twinflower is found from middle to subalpine elevations and is associated with conifers and moss-covered sites that also support *Pyrola*, *Clintonia uniflora*, and *Chimaphila umbellata*.

FIGURE 47. Twinflower (*Linnaea borealis*).

Natural History Notes and Uses. The genus name honors Carolus Linnaeus of Sweden, the person largely responsible for developing the binomial (two-name) system of naming plants and animals. It is said that twinflower was Linnaeus's favorite flower. The species name means "northern."

Twinflower was used medicinally by some Native Americans. As an orthopedic aid, the plant was mashed for inflammation of the limbs, and a poultice of the whole plant was applied to the head for headaches. The Algonquin Indians of Quebec made an infusion of the entire plant for menstrual difficulties and for pregnant women to ensure "good health of the child" (Turner and Kuhnlein 1991). The Iroquois made a decoction of the plant and used it as a sedative for crying children and for children with cramps or fever (Herrick 1977).

Honeysuckle (genus *Lonicera*)

SWEETBERRY HONEYSUCKLE (*L. caerulea*)

TWINBERRY HONEYSUCKLE (*L. involucrata*)

UTAH HONEYSUCKLE (*L. utahensis*)

Description and Habitat. This genus is composed of shrubs and woody vines with entire, opposite leaves (upper leaf pair fused in vines). Inflorescences are borne on either 2-flowered axillary stalks or terminal clusters in which the uppermost flower blooms earliest. The fruits are fleshy, several-seeded berries. Honeysuckles can be found in a variety of habitats from the foothills up to the alpine zone. The genus is named for Adam Lonitzer, a German naturalist who lived from 1528 to 1586.

FIELD KEY TO THE HONEYSUCKLES

1. Bracts broad and leaf-like. . . . . *L. involucrata*
1. Bracts narrow or absent 2

2. Leaves pale; fruit blue-black . . . . *L. caerulea*
2. Leaves green; fruit red . . . . *L. utahensis*

Natural History Notes and Uses. The berries of honeysuckle are seedy but can be eaten raw or dried for future use. The bark and twigs of *L. involucra* were used for a variety of medicinal preparations, ranging from digestive tract remedies to contraceptives. Additionally, the juice from the stems was used as an antidote for bee stings (Wilford et al. 1916).

The berries provide a black pigment. The long stems of honeysuckle were used as basket foundation material by a number of Native American tribes. The hairy stems were also peeled and split as wrapping material for coiled baskets.

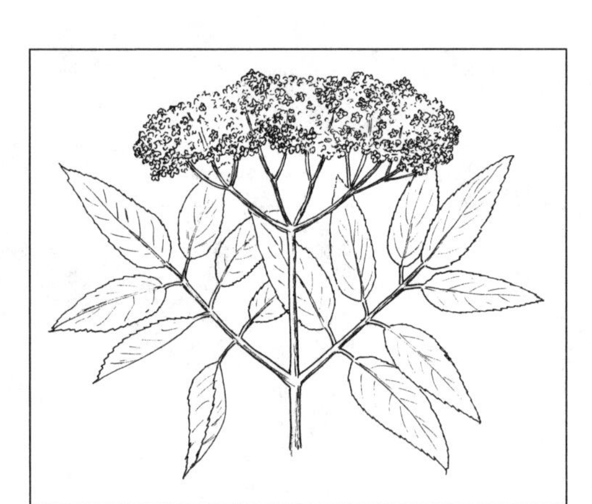

FIGURE 48. Elderberry (*Sambucus* sp.).

SCARLET ELDERBERRY (*Sambucus racemosa*)

Description and Habitat. Elderberries are shrubs with pithy stems. The species here have large, compound leaves with serrated leaflets. The white flowers are arranged in dense clusters. The fruits may be red or blue-black. Elderberries can be found in open areas, on hillsides, and in riparian habitats in the montane zone. The genus name comes from the Greek *sambuke*, a musical instrument made from the hollow stem.

Natural History Notes and Uses. The berries of blue, black, and desert elderberries are edible raw or can be made into excellent jams, jellies, and wines. They can also be dried and stored for winter use. The seeds contain hydrocyanic acid and if eaten in quantity can cause diarrhea and nausea. It is best to cook the berries or strain the seeds before use. The red-berried species contains much higher concentrations of these compounds and should be considered **poisonous**.

The blossoms can be added to pancakes to lighten the batter and add flavor. The dried flowers were also ground and added to flours and baking mixes. Flower buds can be pickled or steamed as a potherb. Both the flowers and fruits are a rich source of vitamin C.

The fresh flowers can be used externally as a decoction for an antiseptic wash. A tea made from the flowers contains a natural estrogen and is often effective for relieving menstrual cramps. The leaves were used as poultices for sprains and skin irritations. The leaves and flowers were common ingredients in skin salves for piles, burns, and boils. Recent studies of elderberry have confirmed that the berries possess antiviral properties that may be useful against influenza (Tilford 1997).

Elderberry stems can be cut and dried for use as musical instruments. After the stems are dry, holes can be bored into them to make flutes. During the drying process, the poisons are said to dissipate. The stems can also be used to make arrow shafts

Hydrocyanic acid, also known as prussic acid, is a colorless volatile liquid with a peach-blossom odor. It is an extremely deadly poison. Prussic acid does not occur freely in normal, healthy plants. Instead, certain sugar compounds called cyanogenic glycosides contain the cyanide ion and form prussic acid only when degraded by certain enzymes. Living plant tissues can contain both cyanogenic glycoside (called *dhurrin* in *Sorghum* species) and enzymes (beta-glycosidase or emulsin) in separate cells. When plant tissues are damaged, such as by freezing, chopping, or chewing, enzymes can come in contact with the cyanogenic glycoside and produce prussic acid. Bacterial action in the rumen of cattle and sheep can also release prussic acid from glycosides. There are approximately 1,000 plant species in 250 genera that are known to be cyanogenic. The most commonly known ones are the sorghums, sudangrass, and Johnsongrass as well as plants of the *Prunus* genus such as wild cherry and chokecherry.

for hunting small game. The odorous leaves can be used in water and sprayed on plants to repel aphids. The pith of the stem is used by watchmakers to absorb grease and oil. The leaves, with chromium as a mordant, yield a green hue. The berries, with alum and cream of tarter, yield a crimson dye.

Caution: The seeds, leaves, bark, and roots contain hydrocyanic acid and an alkaloid called sambucine. They are toxic and cause acute emetic and laxative effects. Berries should be consumed when ripe and used for food after cooking and removal of seeds.

Snowberry (genus *Symphoricarpos*)

COMMON SNOWBERRY (*S. albus*)

WESTERN SNOWBERRY (*S. occidentalis*)

WHORTLELEAF SNOWBERRY (*S. oreophilus*)

Description and Habitat. Snowberries are erect shrubs with elliptical to egg-shaped leaves. The flowers are white to pink and bell-shaped, accompanied by two small bracts. The fruits are berry-like and white. The three snowberry species in the area are found in dry soils at various elevations.

FIELD KEY TO THE SNOWBERRIES

1. Corolla long and narrow, bell-shaped, longer than wide .... *S. oreophilus*
1. Corolla short, broad and short-campanulate, not longer than wide 2

2. The style is exerted past the corolla lobes; pith of stem web-like, not hollow .... *S. occidentalis*
2. The style is not exerted; older stems hollow.................... *S. albus*

Natural History Notes and Uses. The white, tasteless berries are edible raw or cooked and are said to be emetic and cathartic in large amounts. Saponins are found in the leaves and can be used as a natural cleaning agent. A decoction of the pounded roots has been used for colds and stomachache. An infusion made from *S. rivularis* (= *S. albus* var. *laevigatus*) was used to cure sores and skin lesions, and a root decoction was used to alleviate colds and stomach ailments (Moerman 1986). Additionally, the bark of the current year's growth can be peeled and used for cordage.

FIGURE 49. Snowberry (*Symphoricarpos oreophilus*).

HIGHBUSH CRANBERRY (*Viburnum edule*)

Description and Habitat. Highbush cranberry is a deciduous shrub or small tree. The leaves are opposite, roundish in outline, lightly 3-lobed above the middle, irregularly serrate, glabrous or slightly pubescent beneath, and palmately veined. The flowers are white with a 5-parted calyx and corolla, all perfect and of equal size, and the 5 stamens are shorter than the corolla. The fruit is a red, 1-seeded drupe with seeds like flattened stones.

Natural History Notes and Uses. The tart, clustered berries were harvested in late summer and early fall often while still green but also after the first frost, and they were stored in boxes with water and oil. The berries will remain on the shrub well into winter. In some areas the berry patches themselves were owned by families, and only the owners were entitled to harvest the fruits; ownership passed from generation to generation. It is said that these berries mixed half and half with commercial cranberries make an excellent Thanksgiving cranberry sauce.

The bark is an effective antispasmodic for relief of menstrual and stomach cramps as well as asthma. The muscle relaxant properties are due to the presence of the bitter glucoside viburnine. Native Americans used decoctions of the bark for eye medicines, relief for lung colds, and as an astringent to wash infected cuts. The bark was often chewed and the juice swallowed for lung colds.

PINK FAMILY (CARYOPHYLLACEAE)

The pink family has approximately 80 genera and 2,000 species, found in the north temperate zone. About 20 genera are native to the United States. In general, they are annual or perennial herbs with opposite, simple leaves. The stems are often swollen at the joints. The flowers are 5-merous, and the calyx is tubular or has distinct sepals. The petals are often deeply notched, appearing like 10 petals. Some species are cultivated as ornamentals, and several genera are regarded as weedy.

FIELD KEY TO THE PINK FAMILY

1. Petals lacking . . . . *Paronychia*
1. Petals usually present 2

2. Sepals united; petals clawed 3
2. Sepals separate; petals without claws. 7

3. Ribs of calyx twice an many as the calyx teeth, ending in both apices and sinuses. . . . . *Silene*
3. Ribs of calyx as many as the calyx teeth 4

4. Calyx scarious between green midveins ending in teeth 5
4. Calyx green overall 6

5. Petals pink; calyx urn-shaped . . . . *Vaccaria*
5. Petal pink or white; calyx bell-shaped . . . . *Gypsophila*

6. Flowers immediately subtended by 1 to several pairs of tapering, often united bracts. . . . . *Dianthus*
6. Flowers not immediately subtended by bracts. . . . . *Saponaria*

7. Leaves with prominent membranous or papery stipules at the nodes where the leaves join the stem. . . . . *Spergularia*
7. Leaves without stipules at the nodes 8

8. Petals entire or slightly wavy on the margins 9
8. Petals 2-lobed at the tip 10

9. Plants prostrate or ascending; styles usually 5 . . . . *Sagina*
9. Plants usually erect; styles usually 3. . . . . *Arenaria*

10. Mature seed capsule egg-shaped . . . . *Stellaria*
10. Mature seed capsule cylindrical . . . . *Cerastium*

Sandwort (genus *Arenaria*)

Description and Habitat. The species in this genus are generally annual or perennial herbs with opposite leaves. The white flowers are borne in open to congested, flat-topped inflorescences. The species are found in various habitats from the foothills to above timberline in dry, rocky, and open areas, as well as moist open forests. The genus name is from the Latin *arena*, referring to sand, the habitat of many species. This and the related genus *Minuarta* are recognized by their linear leaves, undivided petals, and often matted growth habits.

Natural History Notes and Uses. The sandworts were used medicinally by several Native American tribes. A decoction of the root of a related species, *A. aculeata* (prickly sandwort), was used by the Shoshone as an eyewash, and a poultice of steeped leaves of *A. congesta* was applied to swellings (Strike 1994).

FIGURE 50. Sandwort (*Arenaria* spp.).

Mouse-ear Chickweed (genus *Cerastium*)

FIELD CHICKWEED (*C. arvense*)

BERING CHICKWEED (*C. beeringianum*)

COMMON CHICKWEED (*C. fontanum*)

Description and Habitat. The species in this genus are annual or perennial herbs with entire-margined leaves opposite on the stem. The herbage is usually hairy and often sticky. The few to several (rarely solitary) flowers are borne in an open inflorescence. There are 5 petals and 5 sepals, and the petals are white and deeply lobed at the tip. The genus name is from the Greek *keras*, meaning "horn," referring to the tapered capsule, which in some species is bent slightly like a cow's horn.

FIELD KEY TO THE MOUSE-EAR CHICKWEEDS

1. Size of petals about equal to size of sepals; plants of disturbed habitats .... *C. fontanum*
1. Petals about 1½ to 3 times as long as sepals; native perennials usually in undisturbed habitats 2

2. Bracts of inflorescence without papery margins .... *C. beeringianum*
2. Bracts of inflorescence with papery margins .... *C. arvense*

Natural History Notes and Uses. *Cerastium* is frequently confused with *Stellaria media* (chickweed), but to the general forager there is no danger. The tender leaves and stems of most *Cerastium* can be added to a salad, but we found they are better if boiled first and served as greens.

Pink (genus *Dianthus*)

DEPTFORD PINK (*D. armeria*)

SWEETWILLIAM (*D. barbatus*)

Description and Habitat. *Dianthus* is an introduced genus with about 300 species of annual and perennial herbs with opposite leaves and fragrant flowers. Both species in the area are found at the lower elevations.

FIELD KEY TO THE PINKS

1. Flowers many in a dense head; plant not hairy .... *D. barbatus*
1. Flowers several in a short, open cluster; plant short-hairy, at least in the inflorescence .... *D. armeria*

Natural History Notes and Uses. Most of the cultivated forms are doubles and bear many petals. Several species have given rise to some of the most important garden and cultivated flowers. In fact, there are more than 30,000 cultivars recorded in the *International Dianthus Register* and supplements. The name *carnation* is applied to cultivars whereas *pink* is loosely applied to any member of this genus.

BABY'S-BREATH (*Gypsophila paniculata*)

Description and Habitat. Baby's-breath is a non-native, perennial species with much-branched stems. The genus contains approximately 125 species of annual and perennial herbs with simple, linear, and opposite leaves. The flowers have 5 sepals, 5 petals, and 10 stamens. This species is found in open forests at about 6,500 feet.

Natural History Notes and Uses. Baby's-breath is an escaped ornamental in the United States. It was originally introduced from Eurasia in the late 1800s and is still widely sold for ornamental landscaping and floral arrangements. It is an extremely hardy perennial plant with a very deep taproot up to 12 feet long. Baby's-breath spreads in a tumbleweed fashion by abscising just above the ground and releasing seed as it rolls. Large infestations are generally found where skeletons collect along fence lines and in ditches and ravines. Several species are grown as ornamentals, and some are reported to have medicinal qualities.

YELLOW NAILWORT (*Paronychia sessiliflora*)

Description and Habitat. This small, perennial, mat-forming plant, usually no more than 3 inches tall, resembles coarse moss. The plants have a taproot topped by a much-branched stem base (caudex) at ground level. The tiny, sharp-tipped leaves with white stipules are crowded along the short stems. The small, yellowish flowers are numerous and lack petals, but their 5 yellow sepals in combination with the leaves give the plants a yellowish green color. At maturity each flower produces a tiny round fruit called a utricle. Each utricle contains a single seed.

Natural History Notes and Uses. The generic name is the Greek word for a "whitlow" or "felon," a disease of the nails, and for plants with whitish scaly parts once supposed to cure it. The specific epithet means "stalkless flowers" in botanical Latin. Yellow nailwort was first described for science in 1818 by the famous naturalist, botanist, and ornithologist Thomas Nuttall (1786–

1859). The *Paronychias* are sometimes called "whitlowworts" or "nailroots." They have been used in the Mediterranean region as well as by the North American Kiowa for diuretic and aphrodisiac teas.

TUBER STARWORT (*Pseudostellaria jamesiana*)

Description and Habitat. Tuber starwort is a weak-stemmed, glandular perennial with lanceolate leaves and many few-flowered cymes of white flowers that have slightly 2-lobed petals. This species is found about meadows and damp places and flowers from May to July.

Natural History Notes and Uses. A number of plants have *pseudo* or "false" in their name (such as *Pseudocymopterus*, *Pseudotsuga*, false Solomon's seal) to indicate that although they may resemble another plant, that resemblance is superficial. In this case, *Pseudostellaria* refers to starwort's resemblance to the genus *Stellaria*.

The species name is for Edwin P. James (1797–1861), an American naturalist and botanical explorer in the Rocky Mountains. He studied medicine, then learned botany from Professor John Torrey, and in 1820 became the naturalist-surgeon on the federal government's Yellowstone expedition, which explored the Rockies all the way south into New Mexico. He and two colleagues were the first white Americans to ascend Pike's Peak, and he was the first plant collector to explore the high alpine regions of the Rocky Mountains.

The tuber-like swellings of this starwort can be eaten raw or dried in the sun. They have a thin, light brown rind and a tender, rather mealy texture inside, similar to that of a potato.

ARCTIC PEARLWORT (*Sagina saginoides*)

Description and Habitat. This hairless, matted perennial has slender, ascending stems and a basal rosette of linear leaves. There are 5 sepals and petals and 10 stamens. The capsule is conical.

Natural History Notes and Uses. This plant is common in moist areas, especially on mud flats. The genus name is from the Greek *sagina* ("fodder"), the name for a related plant that was fed to sheep. There are about 20–25 species in this genus, usually tufted annual and perennial herbs. Several species are grown as rock-garden ornamentals.

BOUNCING-BET, SOAPWORT (*Saponaria officinalis*)

Description and Habitat. This erect perennial herb has sessile or nearly sessile leaves. The flowers are showy and usually pale pink. Soapwort can be found along roadsides, in disturbed areas, and in waste places at the lower elevations. The plant has escaped from cultivation. The genus name is Latin for "soap," since the juice of the plant lathers with water. Soapwort flowers from June to September.

SAPONINS AND SOAP

Many plants contain saponin, which, once extracted, can be used as soap. Saponins are found in soapwort (*Saponaria* spp.), clematis (*Clematis* spp.), snowberry (*Symphoricarpos* spp.), elderberry (*Sambucus* spp.), and other species. Cuts, wounds, and rashes need to be cleansed of bacteria to prevent infection; soap is also useful for eliminating human scent when hunting, to avoid alerting game animals to our presence. Food and cooking gear also need to kept clean. On the other hand, soaps can be detrimental to some organisms living at the surface of water in ponds or streams. In water, soaps break down the surface tension, making life more difficult for organisms that are dependent on this tension.

Natural History Notes and Uses. The plant contains saponins and will irritate the digestive tract if eaten. The crushed green plant and roots can be used as a soap substitute.

Campion, Catchfly (genus *Silene*)

MOSS CAMPION (*S. acaulis*)

SLEEPY SILENE (*S. antirrhina*)

MENZIES' CAMPION (*S. menziesii*)

OREGON SILENE (*S. oregana*)

PARRY'S SILENE (*S. parryi*)

PINK CAMPION (*S. repens*)

Description and Habitat. There are many species of *Silene*. They are annual, biennial, or perennial herbs with opposite leaves. The sepals are united and often inflated into a 5-lobed tube. The petals are lobed at the tip and have appendages at the point where the broader upper portion (blade) joins the narrower lower segment (claw). The various species are found in a wide range of habitats.

FIELD KEY TO THE CAMPIONS

1. Plants annual, of disturbed habitats .... *S. antirrhina*
1. Plants perennial, of undisturbed habitats 2

2. Alpine plant with cushion-like growth, less than 2 inches tall .. *S. acaulis*
2. Plant not cushion-like, generally taller than 2 inches 3

3. Petals less than ⅔ inch long; flowers in open leafy cymes .... *S. menziesii*
3. Petals longer than ½ inch 4

4. Petals with 4–6 appendages .... *S. oregana*
4. Petals with 2 appendages 5

5. Calyx purple .... *S. repens*
5. Calyx not purple .... *S. parryi*

Natural History Notes and Uses. The young shoots of *S. acaulis* can be used as potherbs. The sap of *S. antirrhina* was used by native people in California to paint designs on the faces of young girls (Barrett and Gifford 1933). The designs were cosmetic, not ritualistic.

RED SANDSPURRY (*Spergularia rubra*)

Description and Habitat. This small, prostrate plant has small flowers that are pink with petals equal in length to or shorter than the sepals. The flowers occur on long stalks arising from the leaf axils. The leaves are narrow, needle-like, and arranged in whorls about the stem. The plant grows to 2–6 inches in height.

Natural History Notes and Uses. *Spergularia* is the Latin derivative of *Spergula* (a genus in Caryophyllaceae), and *rubra* is Latin meaning "red." The genus *Spergularia* can be distinguished from other similar small caryophyllacious (of the family Caryophyllaceae) plants by the presence of stipules, the small appendages at the bases of the leaves. The tiny seeds are sometimes eaten but contain saponin. They were gathered, ground up, and mixed with flour to make bread.

Starwort, Chickweed (genus *Stellaria*)

BOREAL STARWORT (*S. borealis*)
NORTHERN STARWORT (*S. calycantha*)
CURLED STARWORT (*S. crispa*)
LONGLEAF STARWORT (*S. longifolia*)
LONGSTALK STARWORT (*S. longipes*)
GARDEN CHICKWEED (*S. media*)
ROCKY MOUNTAIN CHICKWEED (*S. obtusa*)
UMBRELLA STARWORT (*S. umbellata*)

Description and Habitat. The starworts are mostly low, annual or perennial herbs with flowers in an open inflorescence in the leaf axils or at the ends of stems. The 5 sepals are separate to the base. The petals are white and deeply lobed or lacking.

FIGURE 51. Chickweed (*Stellaria media*).

FIELD KEY TO THE STARWORTS

1. Plants annual .... *S. media*
1. Plants perennial 2

2. Leaves ovate to ovate-lanceolate; flowers solitary in leaf axils; usually there are no petals 3
2. Leaves linear to lanceolate; flowers few to many in terminal and axillary inflorescences; petals present or missing 5

3. Leaves mostly less than ½ inch long and often twice as wide or less .... *S. obtusa*
3. Leaves sometimes over ½ inch long and often over twice as long as wide 4

4. Leaf margins flat; stems terminating with a flower .... *S. calycantha*
4. Leaf margins curled and wavy; stems terminating with a pair of young leaves .... *S. crispa*

5. Petals none or vestigial and very much shorter than the sepals 6
5. Petals are longer or only slightly shorter than the sepals 7

6. Flowers in axils of green bracts that look like stem leaves .... *S. borealis*
6. Flowers in axils of scarious or scarious-margined bracts .... *S. umbellata*

7. Leaf margins smooth 8
7. Leaf margins finely bumping and rough to the touch .... *S. longifolia*

8. Petals as long as or longer than sepals .... *S. longipes*
8. Petals shorter than the sepals .... *S. borealis*

Natural History Notes and Uses. While the uses of other starworts are unknown, the young shoots of *Stellaria media* have been used as salad herbs or potherbs if cooked like spinach. Although they are edible raw, we prefer to boil them for a few minutes before eating. Since the plants are usually quite small and only the youngest parts are good, chickweed can be tedious to collect. The greens are low in calories and packed with copper, iron, phosphorus, calcium, potassium, and vitamin C, valued in the prevention and treatment of scurvy (Schofield 1989).

Medicinally, *S. media* can be used as a tonic, in large quantities a laxative, and a diuretic. For itchy skin, make a strong tea and wash the area. A poultice of the plant has been used to treat skin sores, ulcers, and infections as well as eye infections and hemorrhoids.

COW SOAPWORT (*Vaccaria hispanica*)

Description and Habitat. This native of Europe is a garden escapee and crop weed, recorded from wasteland and crops in the Southwest. An erect annual, up to 24 inches tall, it has pink petals and hairless, blue-green leaves. It flowers in spring.

Natural History Notes and Uses. This herb contains a variety of saponins, mainly vascegoside, through the hydrolysis of which vaccaroside is obtained, through the further hydrolysis of which yuccagenin is obtained. It also contains the vaccaric flavonoid glycoside, alkaloids, and such compounds as coumarin. The seed is anodyne (relieves pain), discutient (disperses morbid matter), diuretic, emmenagogue (promotes menstrual discharge), galactogogue (promotes lactation), styptic (contracts or binds), and vulnerary (heals wounds). A decoction is used to treat skin problems, breast tumors, menstrual problems, deficiency of lactation, and sluggish labor.

STAFF TREE FAMILY (CELASTRACEAE)

Members of the staff tree family are shrubs with small inconspicuous flowers borne in the axils of the leaves. The 4–5 sepals are united at the base, and the petals are separate. The fruit is a capsule. There are 60 genera and 850 species distributed in tropical and temperate regions of the world; 10 genera are native to the United States. Several species are cultivated as ornamentals.

BOXLEAF MYRTLE (*Paxistima myrsinites*)

Description and Habitat. Boxleaf myrtle is a low, dense, evergreen shrub. The thick, leathery leaves are opposite and oval to elliptic, and the toothed margins are slightly rolled under. The small flowers are maroon. The plant is found in coniferous forest, in rocky openings, and on dry mountain slopes from low to middle elevations. The genus is often spelled *Pachistima*.

Natural History Notes and Uses. The fruits were reportedly eaten by some Native Americans (Strike 1994), but other sources indicate that the plant is inedible. The Maidu Indians of California used the boiled leaves as a poultice for pain or inflammation (Strike 1994).

HORNWORT FAMILY (CERATOPHYLLACEAE)

The hornwort family has only one genus, *Ceratophyllum*, with three species distributed worldwide. These aquatic herbs occur in lakes, ponds, and slow streams and have no roots.

COON'S-TAIL (*Ceratophyllum demersum*)

Description and Habitat. This rootless aquatic forb is submersed or free-floating and has slender, lax, and much-branched stems. The sessile leaves are in whorls of 5–12, and the blades are dissected into linear, filamentous segments whose shape varies with the position on the plant. The minute flowers have no petals and are borne in the axils of the leaves. Coon's-tail is a common plant in standing or slowly flowing water of rivers, sloughs, and ponds to about 7,000 feet. It flowers from June to August.

Natural History Notes and Uses. *Ceratophyllum* comes from the Greek *keras*, "horn" or "antler," and *phyllon*, "leaf" or "foliage," describing the plant's "horn-like leaf." The specific name is from the Latin *demersus*, meaning "underwater."

This is an important habitat plant for young fish, small aquatic animals, and aquatic insects. Some waterfowl eat the seeds and foliage, although coon's-tail is not considered an important food source. It is often used in cool-water aquaria and pools.

Coon's-tail often forms monospecific populations and is found to depths of 30 feet as individual, very slow-growing plants. In the aquarium, on the other hand, it prefers relatively high light, and it does not tolerate transportation for extended periods. It excretes substances toxic to algae (allelopathic behavior), and under good growth conditions it efficiently inhibits most algae growth. Because it tolerates a certain amount of salinity, it is also an ideal plant for brackish tanks.

Native Americans in California used coon's-tail to make a soothing lotion for sore or inflamed skin (Strike 1994).

GOOSEFOOT FAMILY (CHENOPODIACEAE)

Approximately 102 genera and 1,500 species are in the goosefoot family worldwide. Fourteen genera are native to the United States, mostly in the West. This family includes several food plants (e.g., beets and spinach) and weeds (e.g., Russian thistle).

Caution: Some plants in the Chenopodiaceae can accumulate high amounts of nitrates, which, when ingested, are reduced to toxic nitrites. Nitrites cause the hemoglobin (red pigment) to transform into methemoglobin, which is dark brown in color and makes the blood unable to transport oxygen to the body tissues. Additionally, some plants in this family can become toxic through the accumulation of selenium absorbed from the soil.

FIELD KEY TO THE GOOSEFOOT FAMILY

1. Leaves scale-like; stems and branches fleshy, jointed; flowers sunk in depressions of the spike .... *Salicornia*
1. Leaves usually well developed; stems and branches not fleshy and jointed; flowers not as above 3

2. Leaves spine-tipped and dry; flowers perfect................... *Salsola*
2. Leaves not spine-tipped and dry.................................... 3

3. Flowers solitary in the axils of the leaves or bracts .... *Bassia*
3. Flowers usually not all solitary in the axils of leaves or bracts 4

4. Inflorescence of 1–3-flowered axillary clusters; leaves terete or narrowly linear . . . . *Suaeda*
4. Inflorescence not as above 5

5. Sepals 1, minute; stamen 1; plants mostly prostrate . . . . *Monolepis*
5. Sepals more than 1 or none; plants seldom prostrate 6

6. Flowers perfect 7
6. Flowers either male or female 8

7. Perianth often becoming winged or keeled in fruit . . . . *Kochia*
7. Perianth not winged in fruit . . . . *Chenopodium*

8. Pistillate calyx present . . . . *Sarcobatus*
8. Pistillate calyx lacking 9

9. Plant densely tomentose . . . . *Krascheninnikovia*
9. Plant not densely tomentose 10

10. Bracts of pistillate flowers united to form a sac with a narrow opening; spiny shrub . . . . *Grayia*
10. Bracts of pistillate flowers not forming a sac, but more or less united . . . . *Atriplex*

COPING WITH HIGH SALT CONCENTRATIONS

In arid or semi-arid environments, saline and alkaline soils (pH > 7) are common and the excess accumulations often appear as a white crust on the soil surface. High soil concentrations of soluble salts interfere with the osmotic balance needed by plants for normal uptake of water by the roots. Plants adapted to grow in salty or alkaline soils are called halophytes. Some plants, such as saltbush, solve the problem of excess salt by absorbing it, so that the concentration of salt in their cell sap exceeds that of their soil-moisture supply. Under severe conditions of soil-moisture stress, they can continually adjust the internal salt concentration to cope with the increasing salinity of the soil. Additionally, the concentration of salt in plant tissues may function as a sort of antifreeze during the winter, protecting the plant by limiting the development of ice crystals within its cells.

Saltbush (genus *Atriplex*)

FOURWING SALTBUSH (*A. canescens*)

GARDNER'S SALTBUSH (*A. gardneri*)

Description and Habitat. Saltbushes are annual or perennial herbs or shrubs with alternate leaves and glabrous or scaly herbage. The flowers are unisexual, and individual plants have one or both sexes. The various species are found at the lower elevations in valleys, disturbed areas, or dry, alkaline soils.

FIELD KEY TO THE SALTBUSHES

1. Fruiting bracts usually 4-winged lengthwise . . . . *A. canescens*
1. Fruiting bracts not winged . . . . *A. gardneri*

Natural History Notes and Uses. There are many uses for these plants, from food to medicine and dye, as well as soap and spice. The young leaves of many species can be cooked and eaten as greens and have a distinctive salty taste. We've often added them to otherwise bland foods to make our wild meals less boring. Adding the leaves to meats while they are cooking will help spice them up. The seeds were parched, ground into flour, and made into mush. They can also be soaked in water for a few minutes to make a rather pleasant-tasting drink. The Navajo used the flowers to make puddings (Bailey 1940). The ashes of *A. canescens* (fourwing saltbush) make a good substitute for baking soda.

Medicinally, the various species had many uses. The Navajo used the leaves of *A. argentea* (silverscale saltbush) as a fumigant to relieve pain (Willard 1992), and the Zuni made a poultice from the chewed root for application to sores and wounds. A warm poultice made from the pulverized root of fourwing saltbush was used to treat toothaches.

The leaves and roots of many species were used as a soap. They were rubbed in water for lather and used for washing clothing and baskets. Native Americans carved arrowheads from the wood. The seeds of some species were used to make a black dye.

SUMMER CYPRESS, FIVEHOOK BASSIA (*Bassia hyssopifolia*)

Description and Habitat. Fivehook bassia, a summer annual, is a weedy plant found most frequently on saline soils. The seeds germinate in early spring, and the plant matures in late summer. The fruit is 5-lobed with a hook on each lobe. The inconspicuous flowers occur in clusters along the ends of branches and bases of leaves.

Natural History Notes and Uses. The plant in seedling stage is difficult to differentiate from *Kochia scoparia* (Russian thistle). The seedling leaves are long and grayish with soft white hairs, and the leaves on very young plants grow in a rosette. Mature plants branch from a main stem but are not as branched as the similar-looking Russian thistle. The leaves are blue-green, flat, and narrow.

Goosefoot (genus *Chenopodium*)

LAMBSQUARTERS (*C. album*)

PINYON GOOSEFOOT (*C. atrovirens*)

PITSEED GOOSEFOOT (*C. berlandieri*)

BLITE GOOSEFOOT (*C. capitatum*)

FREMONT'S GOOSEFOOT (*C. fremontii*)

NARROWLEAF GOOSEFOOT (*C. leptophyllum*)

DESERT GOOSEFOOT (*C. pratericola*)

RED GOOSEFOOT (*C. rubrum*)

Description and Habitat. The goosefoots are annual or perennial herbs or shrubs with alternate leaves and glabrous or scaly herbage. The flowers are unisexual, and individual plants have one or both sexes. The various species are found at the lower elevations in valleys, disturbed areas, or dry, alkaline soils. The genus name comes from *cheno*, meaning "goose," and *podium*, meaning "foot," because the triangular leaves resemble the shape of a goose's foot. Oil of chenopodium, distilled from the fruits, contains a broad-spectrum vermifuge that is widely used in veterinary medicine.

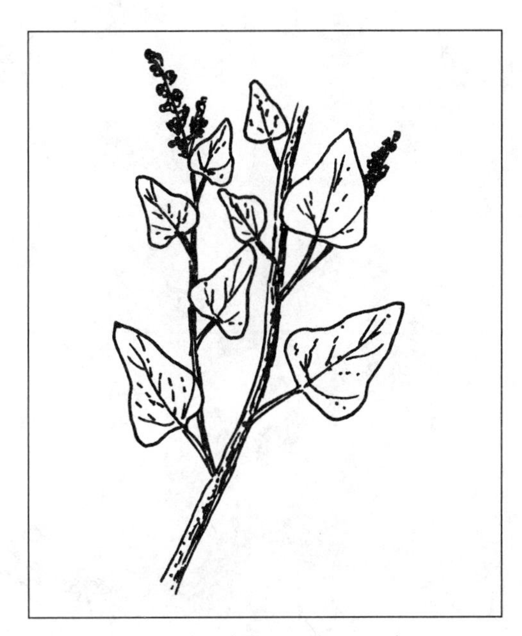

FIGURE 52. Goosefoot (*Chenopodium* spp.).

FIELD KEY TO THE GOOSEFOOTS

1. Sepals becoming bright red and fleshy in fruit. . . . . *C. capitatum*
1. Sepals not becoming fleshy 2

2. Leaves 5–10 times longer than wide, less than ¼ inch wide, smooth-edged. 3
2. Leaves at least ½ inch wide, not more than 4 times as long as wide 4

3. Plants little branched if at all . . . . *C. pratericola*
3. Plants branched to much branched . . . . *C. leptophyllum*

4. Flower 3- to 4-parted, seeds flattened on sides lengthwise .... *C. rubrum*
4. Flower 5-parted 5

5. Leaves 2–3 times long as wide, with 1–3 near parallel veins, smooth-edged............................ *C. atrovirens*
5. Leaves over ½ inch wide, mostly lobed or divided, not parallel-veined... 6

6. Leaves thin and papery when dry, about as long as wide, mostly hairless .... *C. fremontii*
6. Leaves thick, usually 1–2 times longer than wide 7

7. Seeds pitted on side.................................. *C. berlandieri*
7. Seeds smooth ....*C. album*

Natural History Notes and Uses. Leaves, tops, and seeds of all species can be used as an emergency or basic food and are quite tasty and nutritious. High in protein, the greens are a good source of vitamins A and C, iron, and potassium and are extremely rich in calcium. Since they do not become bitter with age, both young and old plants can be used. Leaves may be used raw in salads or boiled in water like spinach. The water can be saved and used as a yellow dye. The leaves were also eaten to treat stomachaches and prevent scurvy. A leaf poultice was used on burns. The flower buds and flowers can be used as potherbs. A single plant can produce up to 70,000 seeds, which can be ground as flour for use in bread or cooked as mush. The seeds can also be eaten without grinding or incorporated into pinole (flour made from a mixture of seeds of small plants). The seeds contain about 15 percent protein and 55 percent carbohydrates, more than is found in corn. The seeds can also used as a coffee substitute.

Large quantities of the plant should not be eaten as many species contain high levels of oxalic acid, which tends to bind calcium and prevent its proper absorption into the body. Cooking or freezing of *Chenopodium* apparently breaks down the oxalic acid. Additionally, *Chenopodium* has been known to accumulate toxic levels of nitrates and may cause livestock poisoning. But because large quantities of the plant must be consumed to cause problems, this type of poisoning may be unlikely.

The hard root of some *Chenopodium* species was stored until needed, then grated on a rock to make soap. The leaves were also used to make soap but are not as effective as the roots.

Warning: *Chenopodium* greens contain oxalic acid and its salts, which can reduce calcium absorption if eaten in large quantities.

SPINY HOPSAGE (*Grayia spinosa*)

Description and Habitat. This low, mealy-appearing shrub has stiff, spreading, spine-tipped branches and slightly fleshy leaves with gray tips becoming pinkish with age. The flowers occur in heads that are borne in terminal or axillary spikes or panicles. The fruits are closely subtended by a pair of attractive rose-purple, thin, flat-winged bracts that are united to the middle or higher.

Natural History Notes and Uses. The genus *Grayia* is named for American botanist Asa Gray and contains a single species. Thanks to a deep root system, this plant is extremely drought tolerant. It serves as a good food source for browsers, especially in the dry months when other plants have dropped their leaves. Some Native American peoples ground parched seeds of spiny hopsage to make pinole.

COMMON KOCHIA (*Kochia scoparia*)

Description and Habitat. Common kochia is a bushy annual with stems up to 3 feet tall. The leaves are alternate, narrowly lance-shaped, and tapered at both ends. The herbage may or may not be covered with hairs. The flowers are solitary or in clusters in spikes. The species is common in open, disturbed habitats at low elevations.

Natural History Notes and Uses. Common kochia is native to Europe and was introduced into the United States as an ornamental. It has since escaped, becoming well established. In Japan, China, and other parts of Asia common kochia was cultivated for its seeds. The tips of the young shoots can be prepared as potherbs (Clarke 1977). The seeds can be eaten raw or cooked or can be ground into meal and used in making bread. The genus is named for William Koch, a German botanist of the late eighteenth and early nineteenth centuries.

WINTERFAT (*Krascheninnikovia lanata*)

Description and Habitat. In many older references winterfat is also known as *Eurotia lanata*. Winterfat is a small shrub found at the lower plains and foothills elevations, often in saline or alkaline areas. The leaves are alternate, narrow, and entire; the flowers occur in heads or spikes in the axils of the leaves. The genus name is from the Greek *eurotios*, meaning "moldy," and refers to the dense hairiness of the plant.

Natural History Notes and Uses. Although it is unknown whether people may safely eat this species, it is an important forage plant for horses and other livestock. Medicinally, the plant has been used by many Native American tribes. For example, the Hopi Indians used the powdered root for burns, and a decoction of the leaves was used for fevers. The Navajo made a poultice of the chewed leaves and applied it to a poison ivy rash. The Navajo also used the stems and leaves of this plant in sweathouse ceremonies, placing them on hot rocks for the Mountain Chant (Wyman and Harris 1941).

NUTTALL'S POVERTYWEED (*Monolepis nuttalliana*)

Description and Habitat. Povertyweed is a low-growing winter annual with prostrate or ascending stems. The leaves are somewhat succulent and lance-shaped, broadened and lobed at the base. The flowers are borne in dense clusters at the leaf bases, and the solitary sepal is reddish in color. The seeds are dark brown

ASA GRAY

Asa Gray (1810–88) was one of America's leading botanists and taxonomists. He was born in New York and in 1842 was a professor of natural history at Harvard, where he was the teacher of many eminent botanists. Through his voluminous writings in periodicals and his well-known textbooks, he helped popularize the study of botany. Together with John Torrey he explored the western United States and helped revise the taxonomic procedure of Linnaeus on the basis of a more natural classification. Gray's *Manual of Botany* was the standard reference work for the flora of the United States east of the Rocky Mountains. He also initiated the quarterly *Gray Herbarium Card Index*, listing all the vascular plants of the Western Hemisphere described since 1873. Among his many other writings, which are still highly valued, are *Structural Botany* (6th ed., 1879) and *The Elements of Botany* (1887).

THOMAS NUTTALL

Throughout this book there are plants named after or by Thomas Nuttall (1786–1859). To many, he is known as the father of western botany. He was not only an authority in the botanical field but also a well-trained ornithologist, naturalist, and printer. He accompanied several scientific expeditions to the Mississippi and Missouri valleys and the Pacific coast and published his findings in *The Genera of North American Plants* (2 vols., 1818) and *A Manual of the Ornithology of the United States and of Canada* (1832). In his many years of western travels Nuttall visited the Sweetwater River of southern Wyoming and the Black Hills of southeastern Wyoming. He traversed the South Pass at the border of Colorado and Wyoming and followed parts of the Oregon Trail.

Nuttall's contributions were many. He wrote papers in geology, botany, and zoology, and there is still an ornithological society named in his honor. It is difficult to travel anywhere in the American West without seeing a plant that was not named or collected by him. He was the first to champion the use of a natural system of classification in the United States, he authored a textbook on botany, and he revised a magnificent silva, or guide to forest trees, with illustrations that are impossible to match even today. He was first and foremost a field botanist, and as such he changed the direction of botany.

and have a pitted covering. The plant is found in open disturbed habitats at the lower elevations.

Natural History Notes and Uses. There are only about six species of *Monolepis* worldwide. The generic name is derived from the Greek *monos*, "one," and *lepis*, "scale," alluding to the single sepal found on the flowers of members of this genus. The specific name was dedicated to the plants' original describer, Thomas Nuttall (1786–1859). The plant was finally placed in its current taxonomic position in 1891 by Edward Greene (1843–1915), who was the first professor of botany at the University of California. The aboveground parts of povertyweed may be eaten as a potherb. The seeds are also edible.

RED SWAMPFIRE (*Salicornia rubra*)

Description and Habitat. These fleshy, hairless, herbaceous annual plants have leafless, jointed stems bearing opposite branches. The flowers are borne in fleshy, cylindrical spikes with the flowers sunk in groups of 3 to 7 in cavities on opposite sides of the joints. The species is found in saline or alkaline soils and marshy ground. The genus name is Latin *sal*, meaning "salt," and *cornu*, meaning "horn" and referring to its form.

Natural History Notes and Uses. This and other *Salicornia* species are succulent and add a salty taste to salads. The young stems and branches can be pickled but must be boiled first.

PRICKLY RUSSIAN THISTLE (*Salsola tragus*)

Description and Habitat. Russian thistle is not a true thistle (*Cirsium*) but a many-branched annual with purplish striped stems up to 3 feet tall in a rounded form. The lower leaves are thread-like; the upper leaves are awl-like and spine-tipped. The plant may or may not be hairy.

Natural History Notes and Uses. When mature, the whole plant becomes rigid, breaks off at ground level, and becomes a "tumbleweed" blowing across the open plain. The flowers are solitary in the leaf axils and are subtended by spiny bracts. Russian thistle is common in open, disturbed habitats, particularly around agricultural areas at low elevations. It was introduced to the United States from Europe. Fortunately, Russian thistle is not an aggressive competitor and does not appear to replace native plant species. However, it is still considered a noxious weed because of its distributional pattern and spines.

This unsavory-looking plant is edible. The young parts of the plant may be boiled and eaten as a potherb or chopped raw into a salad. On older plants, clip the tender branch tips that are green. We find that the taste of the plant greatly improves when cooked in butter and lemon. In Europe the ashes of the plant were once used in the production of carbonate of soda, known as barilla (Clarke 1977).

Warning: The older parts of the plants contain significant quantities of nitrates and oxalates and may be toxic if eaten in quantity.

BLACK GREASEWOOD (*Sarcobatus vermiculatus*)

Description and Habitat. This perennial native is a long-lived shrub with spreading, rigid branches that often bear spines. The leaves are linear, succulent, and pale green with entire margins. Some of the leaves may be opposite and some alternate, but all are shed in winter. The plants usually have both male and female flowers on the same plant (monoecious), but they can also occur on separate plants (dioecious). The male flowers are catkin-like spikes on the ends of branches, and the female flowers usually occur singly in the axial of leaves and form the fruits, which are surrounded by a green, membranous wing.

Natural History Notes and Uses. *Sarko* is from the Greek word for "flesh," and *batos* is from the Greek word for "bramble," alluding to the succulent leaves and spiny branches. Black greasewood is used as wood for fuel, and the sharpened spines were used for painting by Native Americans. Native Americans used the seeds and leaves, which have a salty taste, for food (Elmore 1944). The Hopi and other Native Americans use greasewood for fuel and for planting sticks. In Chaco Culture National Historical Park (New Mexico) greasewood was used for construction, especially of lintels, and for fuel, being a preferred wood for Pueblo kiva fires.

The seeds, leaves, and new leaders are also consumed by a variety of small mammals. Black greasewood is an important browse plant for cattle, sheep, and big game animals in the winter and provides good cover and food for small mammals and birds. Sheep have been poisoned by rapidly consuming large amounts of new leader growth, which contains high levels of soluble oxalate (Kingsbury 1964). The numerous seeds are wind-dispersed and help to reestablish the plants after fire, although greasewood is only slightly harmed, if at all, by fire and will resprout.

Sea Blite (genus *Suaeda*)

SEEPWEED (*S. calceoliformis*)

SEA BLITE (*S. nigra = S. mcquinii*)

Description and Habitat. Sea blites are annual or perennial herbs or small shrubs. The leaves are alternate, linear, and square or flattened in cross section. The flowers are perfect and occur in axils of small bracts. *Suaeda* is an ancient Arabic name.

FIELD KEY TO THE SEA BLITES

1. Perianth lobes equal . . . . *S. nigra*
1. Perianth lobes unequal . . . . *S. calceoliformis*

Natural History Notes and Uses. Identification of *Suaeda* specimens is achieved most successfully when based on material with flowers (for ovary shape) and mature calyces (for lobe shape) containing seeds. Because of the succulent nature of most specimens, fresh material may appear quite different than dried material, especially in the accentuation of calyx features when dry.

Plants of *Suaeda* are found in saline or alkaline wetlands or occasionally in upland habitats. Some species have been cultivated and eaten as a vegetable; the seeds of some have been ground and eaten by Native Americans; and some species are used as a source for red or black dye.

MORNING GLORY FAMILY (CONVOLVULACEAE)

Members of the morning glory family are herbs, shrubs, or trees. In some species a milky latex is present. The flowers are usually 5-merous with 5 united petals. There are approximately 50 genera and 1,400 to 1,700 species, distributed in tropical and temperate regions. Nine genera are native to the United States. The family is of some economic importance because of the sweet potato (*Ipomoea batadas*), several weeds, and ornamentals.

FIELD BINDWEED (*Convolvulus arvensis*)

Description and Habitat. Bindweed is a perennial with trailing or twining stems. It spreads from a deep and brittle rhizome. The flowers are white or pinkish and funnel-shaped, arising from the axils of the arrowhead-shaped leaves. This beautiful but pernicious European weed is well established throughout North America and is frequently encountered in roadcuts and fields at low elevations. It is difficult to eradicate because of its deep rhizome and low growth. The genus name comes from the Latin *convolvere*, meaning "to entwine."

Natural History Notes and Uses. Native Americans used a cold leaf tea as a wash on spider bites. A tea from the flowers was used for fevers and wounds (Foster and Duke 1990). A tea was also made from the leaves and stems by Kashaya and Pomo women to stop excessive menstruation (Strike 1994). In European folk use the flower, leaf, and root teas were considered a laxative. The root is considered a strong purgative, cathartic, and diuretic. The powdered rootstalk was used as a laxative in ancient and modern China.

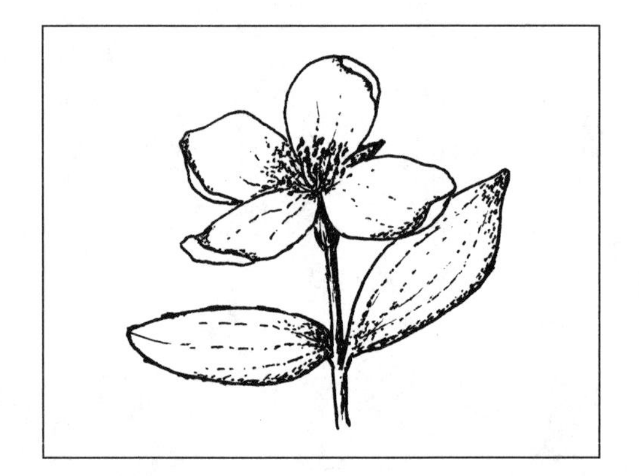

FIGURE 53. Dogwood (*Cornus* spp.).

DOGWOOD FAMILY (CORNACEAE)

There are 12 genera and 100 species in the dogwood family, including many ornamentals. A single genus, *Cornus*, occurs in the United States. Dogwoods are trees or shrubs, often with tiny flowers surrounded by petal-like bracts that resemble a single large flower. The leaves are opposite and simple.

Dogwood (genus *Cornus*)

BUNCHBERRY DOGWOOD (*C. canadensis*)

REDOSIER DOGWOOD (*C. sericea*)

Description and Habitat. Dogwoods are shrubs or semi-woody perennials with simple leaves that are opposite or whorled. The flowers mature into red or white drupes. Dogwoods prefer partial shade in moist mountain and foothill forests up to the subalpine zone.

FIELD KEY TO THE DOGWOODS

1. Plants shrubby, usually more than 8 inches tall . . . . *C. sericea*
1. Plants herbaceous or woody only at the base; plants less than 8 inches tall . . . . *C. canadensis*

Natural History Notes and Uses. The fruits of *C. sericea* were sometimes consumed by Native Americans. We have found them to be extremely bitter. In fact, in large quantities they may be toxic. The Blackfeet Indians used the bark of redosier dogwood as a laxative. Other Native Americans smoked the inner bark as part of a ceremonial herb blend. Pounded twigs can be used as a toothbrush. The wood for both species was used for bows and arrows, fishing hooks, and other implements. The bark was boiled and used to make a brown dye. The fruits of *C. canadensis* may be eaten raw or cooked. Since the berries are rather bland, we like to mix them with other, more tasty fruits. The unripe berries may cause stomachaches or act as a laxative. The chewed berries have been used as a poultice to treat local burns. A cold and fever remedy can be made by boiling dried root or bark (the root is more potent). The feathered bark can be used as a toothbrush. Fresh bark is a cathartic. Leaf tea was used for aches and pains, kidney and lung ailments, coughs, and fevers and as an eyewash. Dogwood has earned a reputation as an anti-inflammatory and general analgesic due to the presence of cornine and other flavonoid compounds. Researchers are studying these properties for use as an anti-cancer agent. The current interest by pharmaceutical companies may stem from the fact that Native Americans used dogwood as an antidote for a variety of poisons (Pojar and MacKinnon 1994).

STONECROP FAMILY (CRASSULACEAE)

Members of this family are succulent herbs or shrubs. There are 35 genera and over 1,500 species worldwide, of which 9 genera are native to the United States. They are of no real economic importance except as ornamentals.

CAM photosynthesis is one of three variations of the photosynthetic process known to occur in plants. Known as crassulacean acid metabolism, or CAM, it is named for the succulent plant family Crassulaceae. This photosynthetic process requires more energy than the other processes (called C4 and C3). The advantage of CAM photosynthesis for these plants is that it allows them to close their stomates (minute epidermal pores in a leaf or stem through which gases and water vapor can pass) during the day in order to reduce water loss. In this way the plants carry on photosynthesis using stored water and carbon dioxide. The energy demands of this process mean that CAM plants have very slow growth rates.

Stonecrop (genus *Sedum*)

ORPINE STONECROP (*S. debile*)

LEDGE STONECROP (*S. integrifolium = Rhodiola integrifolia*)

SPEARLEAF STONECROP (*S. lanceolatum*)

REDPOD STONECROP (*S. rhodanthum = Rhodiola rhodantha*)

WORMLEAF STONECROP (*S. stenopetalum*)

Description and Habitat. The Latin word *sedere* means "to sit," possibly referring to the tendency of many species to grow low to the ground. Stonecrops are well adapted to survival in shallow soil or on rocky outcroppings. The succulent leaves and stems have a waxy coating to help reduce water loss. The reddish color of the foliage in some species is enhanced by sunlight and occurs most often in plants in hot exposed sites.

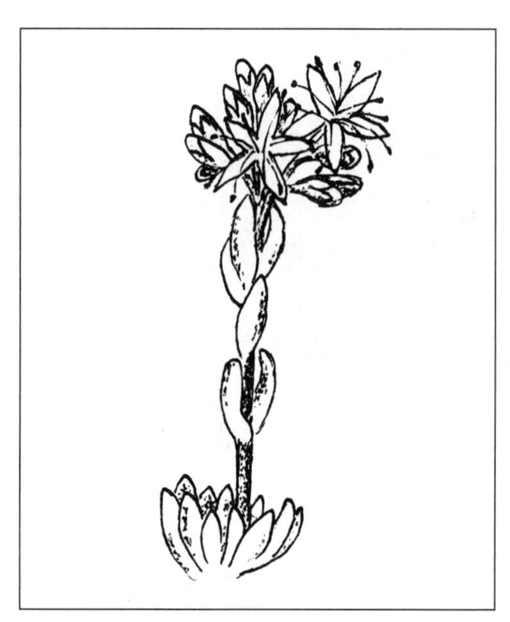

FIGURE 54. Stonecrop (*Sedum* spp.).

FIELD KEY TO THE STONECROPS

1. Flowers greenish white, pink, or purple; leaves on flowering stems 2
1. Flowers yellow, occasionally drying pink; most leaves are basal......... 3

2. Petals ¼ to ½ inch long .... *C. rhodanthum*
2. Petals less than ¼ inch long .... *C. integrifolium*

3. Leaves of flowering stems opposite or alternate and oval, egg- to spatula-shaped or broadest beyond the midpoint .... *S. debile*
3. Leaves of flowering stems alternate and linear to lance-shaped......... 4

4. Leaves strongly keeled to a point at tip .... *S. stenopetalum*
4. Leaves not keeled or pointed.......................... *S. lanceolatum*

Natural History Notes and Uses. The young leaves and stems of all species can be eaten as a salad or boiled as a potherb. We find them slightly tart and crisp, a wonderful addition to salads and trail snacks. However, some species have emetic and cathartic properties and can cause headaches. In an emergency, stonecrop can be eaten raw to allay hunger and thirst. The plants are best when collected before flowering since they tend to become bitter and fibrous in late summer. The green fleshy leaves are high in vitamins A and C. The tubers can also be boiled and eaten.

Sedum is also reported to be slightly astringent and mucilaginous (Wilford et al. 1916). It is valuable in the treatment of wounds, ulcers, lung disorders, and diarrhea. The juice can be used as a field remedy for minor burns, insect bites, and other skin irritations; just squeeze it onto the affected area. Decoctions of the plant were used for sore throats and colds and as an eyewash.

GOURD FAMILY (CUCURBITACEAE)

Members of the gourd family are annual or perennial herbs that are climbing or prostrate, with spirally coiled tendrils. The leaves are alternate, often palmately lobed. The fruit is a berry (often referred to as a *pepo*) with a leathery or hard exocarp. There are approximately 100 genera and 850 species distributed in the warmer regions of the Old and New Worlds. Fourteen genera are native to the United States. The family is economically important as a source of many food plants and ornamentals.

WILD CUCUMBER (*Echinocystis lobata*)

Description and Habitat. Wild cucumber is a high-climbing vine with angular, grooved stems. The leaves are alternate and palmately 5-lobed, with the lobes triangular-shaped and acute. The flowers are small and greenish white, and the fruits are egg-shaped and prickly. Widely distributed across North America, it is usually found at low elevations.

Natural History Notes and Uses. Native Americans used the bitter root tea as a tonic for stomach troubles, kidney ailments, rheumatism, chills, and fevers. The roots were also pulverized and then poulticed for headaches (Foster and Duke 1990).

DODDER FAMILY (CUSCUTACEAE)

Members of the dodder family are leafless, rootless, parasitic herbs that lack chlorophyll. The stems are thread-like and often yellowish. The small flowers have 4 or 5 distinct sepals and 4 or 5 united petals. The fruit is a dry or fleshy globose capsule. The family has one genus (*Cuscuta*), which is native to the United States and has approximately 170 species. The genus may cause great losses to crop plants. The family was once included in the Convolvulaceae (morning glory family).

Dodder (genus *Cuscuta*)

ALFALFA DODDER (*C. approximata*)

CLOVER DODDER (*C. epithymum*)

Description and Habitat. Dodders are leafless, twining perennials with slender stems that are colored pink, whitish, or yellowish, never green. Both the leaves and pink-to-white flowers are highly reduced. Dodder species can normally be identified only with a microscope or hand lens. The many species of *Cuscuta* parasitize different flowering plant hosts at low elevations.

Natural History Notes and Uses. Dodders have a unique life cycle. The small seeds usually germinate in the soil and produce slender stems without seed leaves (cotyledons). Unless the slowly rotating plant encounters a host plant within a short period of time, the dodder seedling will wither and die. However, if the seedling encounters the living stem of a susceptible host plant, the dodder will twine around it, developing suckers that penetrate the tissue of the host. Nutrition is received through these suckers. The dodder then loses all contact with the soil. After a period of growth, small flowers develop and large amounts of seeds are produced to start the process all over again (Frankton and Mulligan 1987).

Dodder was called "love vine" and "vegetable spaghetti" by some Native Americans, but they are generally not considered edible and may cause digestive upset. However, the seeds of *C. californica* (chaparral dodder) were parched and eaten by native people in California.

Chaparral dodder, when brewed as a tea, was considered an antidote for black widow bites. However, only the dodder from *Eriogonum fasciculatum* (California buckwheat) was used for this purpose. Other Native Americans chewed a mass of dodder and stuffed it in their noses or pulverized the plant and sniffed the powder to stop nose-bleeds.

Dodder stems were used by the Cherokees as a poultice for bruises. In China the stems of some species of dodder are used in lotions for inflamed eyes. Moore (1979) indicates that a rounded teaspoon of the chopped dodder is a good laxative-cathartic. In smaller quantities, and drunk every few hours, it is said to aid in spleen inflammations, lymph node swellings, and "liver torpor." Additionally, handfuls of dodder can be gathered and used as scouring pads for cleaning.

TEASEL FAMILY (DIPSACACEAE)

The teasel family is made up mostly of herbaceous plants with opposite leaves. There are approximately 10 genera and 270 species, found mostly in the Old World. None are native to the United States, although *Dipsacus* is widely naturalized and has become quite weedy.

FIELD SCABIOSA (*Knautia arvensis*)

Description and Habitat. The flowers of this plant are pink to lavender, and the petals have 4 lobes. The flowers grouped into a composite-like head, and the petals are interspersed with fine hairs. The stems and leaves have fine hairs, and the upper leaves are pinnately lobed; the lobes become less defined farther down the stem, and the basal leaves are unlobed. Field scabiosa grows 1–3 feet tall and is native to Europe, where it is most common in dry, grassy areas.

Natural History Notes and Uses. This plant could be regarded as a potential oilseed crop. The seeds contain about 25 percent oil, of which some 33–40 percent is in the form of saturated caprylic and capric acids (similar to what occurs in coconut oil). These acids are being increasingly used in high-performance oils for jet engines and in other lubricants of high quality, and also in the preparation of some valued dietary fats. At present, the only directly available sources of these acids are coconut and palm kernel oils, which are imported principally for their content of lauric acid, used in soaps and detergents.

Because of its high proportion of saturated fats, coconut oil is one of the most stable oils and is highly resistant to rancidity. (It is the unsaturated fats in the common seed oils that are easily oxidized and susceptible to rancidity.) Coconut oil is mild on the skin and is widely used in the tropics to protect both skin and hair from the harsh effects of the sun. Film-forming qualities allow it to act as a skin moisturizer and a means to protect from moisture loss. Its natural detergency and lathering capabilities give it a double purpose as a cleansing agent for soaps and shampoos.

FIGURE 55. Sundew (*Drosera anglica*).

SUNDEW FAMILY (DROSERACEAE)

Members of the sundew family are insectivorous herbs growing in acidic bogs. The leaves are covered with sticky glandular hairs on which insects become trapped. There are approximately four genera and 100 species, all of which grow in very nutrient-poor soil conditions. *Drosera*, a native of the United States, is cosmopolitan in its distribution, whereas the other genera are monotypic and restricted in distribution. The family is of no economic importance except that several species are cultivated as novelties because of their insectivorous habit.

ENGLISH SUNDEW (*Drosera anglica*)

Description and Habitat. These perennials have basal rosettes of leaves covered with viscid, stalked glands that trap and digest small insects. Few to several short-stalked flowers are borne on one side at the end of an erect, naked stem. Each flower has 5 petals and sepals that are separate to the base, or nearly so. There are 4–20 stamens and 3–5 deeply divided styles. They are usually found in bogs in association with *Sphagnum* moss.

Natural History Notes and Uses. The juice of some *Drosera* species has been used to curdle milk. In the Mediterranean region sundews are mixed with brandy, raisins, and sugars and allowed to ferment into a drink called *rossolis.* However, the main use of *Drosera* has been medicinal.

The leaves of a related species, *D. rotundifolia,* apparently have antispasmodic and expectorant properties. They have been used in the treatment of whooping cough, bronchitis, asthma, and other respiratory problems. To relieve a bad cough, the leaves were tinctured or made into a tea and sipped throughout the day. The plant also contains an antibiotic substance that, in pure form, is effective against streptococcus, staphylococcus, and pneumococcus (Lust 1987).

Schofield (1989) reports that sundew is helpful in a tinctured form for nausea associated with seasickness, although some drowsiness was associated with the tincture. Sundew juice has been used for removing warts and has been blended with milk to lighten freckles.

Caution: Although there are no reports of human poisoning from the ingestion of sundew, these plants contain corrosive, irritating substances and should be used only in small doses. Cats have been poisoned from daily doses of the plants.

OLEASTER FAMILY (ELAEAGNACEAE)

Plants in this family are shrubs or trees with alternate or opposite, silvery-gray leaves. The fruits are drupe- or berry-like. There are three genera and approximately 45 species distributed in North America, southern Europe, Asia, and eastern Australia.

FIELD KEY TO THE OLEASTER FAMILY

1. Leaves and branches alternate. . . . . *Eleagnus*
1. Leaves and branches opposite. . . . . *Shepherdia*

SILVERBERRY (*Eleagnus commutata*)

Description and Habitat. This shrub grows 3–9 feet tall and has alternate, silvery-scale leaves. The flowers have 4 stamens and are silvery on the outside and yellowish within. The fruits are also silvery in color.

Natural History Notes and Uses. Silverberry is found along gravel river bars and water courses. Silverberries may be eaten raw but are dry and mealy and not highly regarded. Native Americans usually fried them in moose fat or some other grease. They should be considered an emergency food. The flowers were used in making perfumes and other cosmetics.

FIGURE 56. Buffaloberry (*Shepherdia argentea*).

Buffaloberry (genus *Shepherdia*)

SILVER BUFFALOBERRY (*S. argentea*)

RUSSET BUFFALOBERRY (*S. canadensis*)

Description and Habitat. Buffaloberry is a low-spreading shrub with opposite leaves, each with a dark green upper surface and lighter underside that is covered with tiny, brown scales. The inconspicuous flowers are yellow-green; the fruits range in color from yellow to bright red. Buffaloberry grows along streams from 3,500 to 6,500 feet and is also found in recently burned areas. It blooms from April to May. The genus is named after John Shepherd (1764–1836), curator of the Liverpool Botanic Garden.

FIELD KEY TO THE BUFFALOBERRIES

1. Leaves green above, brown-scaly below; branches not spine-tipped . . . . *S. canadensis*
1. Leaves silvery on both surfaces; branches often spine-tipped . . . . *S. argentea*

Natural History Notes and Uses. The berries contain a significant amount of saponin, and another common name for russet buffaloberry is soapberry. The saponin not only gives the plant its bitter taste but also whips it up into a frothy mass called "Indian ice cream."

Native Americans used the berries of *S. canadensis* extensively, both fresh and dried. The berries are at first pleasant, but then the soap-like bitterness prevails. We enjoy cooking them with sweeter-tasting berries such as thimbleberries and serviceberries, plus a large amount of sugar. We found the berries to be somewhat unattractive for general use but a valuable consideration in emergencies. They taste better in the fall after a few good frosts. They can also be used in the making of pemmican or jelly. Dried into cakes, the berries can be stored for winter. The fruits of *S. argentea* are also edible and do not contain as much saponin as those of soapberry.

Infusions of the stems and leaves were drunk as a tonic beverage (Schofield 1989). The berries can be crushed and made into a tea for use as a liquid soap. Native Americans used the tea to relieve constipation. Hart (1967) indicates that the Flathead and Kootenai Indians made solutions from the bark of buffaloberry for eye troubles.

"INDIAN ICE CREAM"

To make Indian ice cream, Native Americans placed a small number of soapberries into a bowl with a little water, then used a special stick with some grass tied on one end to beat the fruit. The result was foamy concoction. Nowadays sugar is added to improve the taste. Care must be taken in picking and preparing the berries so that they do not come into contact with oil or grease of any kind, or they will not whip. Indian ice cream is still served in many households, especially at parties and family gatherings in the Pacific Northwest.

WATERWORT FAMILY (ELATINACEAE)

Members of the family have opposite or whorled leaves and small flowers with 2–5 overlapping petals. Waterwort (*Elatine hexandra*) and two similar species, *E. hydropiper* and *E. macropoda*, sometimes are grown in aquariums.

THREE-STAMEN WATERWORT (*Elatine triandra = E. rubella*)

Description and Habitat. This small, smooth-looking, annual plant is typically found growing along the shores of lakes, ponds, and slow-moving streams. The plants are light green and have matted, prostrate stems with delicate, upright branches. The leaves are oblong and oppositely arranged with tiny notches at the tips. Minute flowers occur at the leaf bases. Several other species of *Elatine* occur in the West; details of the tiny seeds must be examined to distinguish them.

Natural History Notes and Uses. *Elatine* is from the Greek *elatinos*, meaning "of the fir," used by Pliny to refer to a plant of the genus *Antirrhinum* (snapdragons). The specific epithet is from the Latin *tri*, referring to "three," and the Greek *andros*, for "male"; hence, "three stamens." The plant is considered beneficial for stabilizing shorelines.

HEATH FAMILY (ERICACEAE)

Members of the heath family, or as they are sometimes referred to, ericaceous plants, are mostly lime-hating or calcifuge plants that thrive in acid soils. This family includes numerous plants from mostly temperate climates: cranberry, blueberry, heath, heather, huckleberry, azalea, and rhododendron are well-known examples. Recent genetic research by the Angiosperm Phylogeny Group (an international group of botanists whose aim is to establish a consensus on the taxonomy of flowering plants in light of the rapid rise of molecular systematics) has resulted in the inclusion of the formerly recognized families Empetraceae, Epacridaceae, Monotropaceae, Prionotaceae, and Pyrolaceae in Ericaceae.

There are about 50 genera and 2,500 species in the heath family, and in the United States approximately 25 genera are indigenous. Economic products provided by this family include food plants, oil of wintergreen, and many ornamentals. Some herbaceous members are mycotrophic, that is, they depend on fungi for nutrient uptake and lack chlorophyll.

SAPROPHYTIC HEATHS

A number of species in the heath family possess little or no chlorophyll to capture sunlight from which to derive nutrition and water. To compensate, they have an intimate association with mycorrhizal fungi, depending on or associating with the fungi in various degrees.

FIELD KEY TO THE HEATH FAMILY

1. Plants usually having green leaves 2
1. Plants saprophytic, brown to red, pinkish, white, or yellowish but not green 12

2. Plants are woody, or herbaceous with leafy stems.................... 3
2. Plants are herbaceous, leafy only at or near base of plant 10

3. Petals separate 4
3. Petals united.................................. 5

4. Some or all of the leaves are whorled and toothed .... *Chimaphila*
4. The leaves are alternate and entire............................*Ledum*

5. Ovary inferior; petals united almost to tip; fruit a berry ...... *Vaccinium*
5. Ovary superior.................................. 6

6. Leaves opposite .... *Kalmia*
6. Leaves alternate 7

7. Plants are shrubs usually over 2 feet tall ....*Menziesia*
7. Plants are prostrate and creeping, mostly less than 12 inches tall 8

8. Leaves linear, looking much like the needles of a fir tree...... *Phyllodoce*
8. Leaves broader.................................. 9

9. Leaves oblanceolate to obovate....................... *Arctostaphylos*
9. Leaves ovate or orbicular .... *Gaultheria*

10. Flowers solitary and terminal .... *Moneses*
10. Flowers several in a raceme 11

11. Flowers occurring on one side of stalk .... *Orthilia*
11. Flowers on all sides of stalk .... *Pyrola*

12. Petals united almost to the tip; anthers awned on back....... *Pterospora*
12. Petals separate; anthers not awned.......................... *Hypopitys*

Bearberries, Kinnikinnick (genus *Arctostaphylos*)

KINNIKINNICK (*A. uva-ursi*)

BOG BEARBERRY (*A. rubra*)

Description and Habitat. These shrubs have reddish to brown stems that root at the nodes. The spoon- to lance-shaped leaves are leathery. The flowers are urn-shaped, and the bright red berries often persist through the winter. They are found growing in open areas with dry to well-drained soils from low to high elevations. The genus name means "bear grape" and refers to the fondness shown by bears for the fruits of these shrubs, many of which are known as bearberry.

FIGURE 57. Kinnikinnick (*Arctostaphylos uva-ursi*).

FIELD KEY TO THE BEARBERRIES

1. Leaves entire; young twigs usually hairy; fruit mealy .... *A. uva-ursi*
1. Leaves toothed; young twigs not hairy; fruit fleshy............. *A. rubra*

Natural History Notes and Uses. The berries of all *Arctostaphylos* are edible. They may be eaten raw, but large quantities may be hard to digest. Constipation or indigestion are common maladies of eating too many. The berries can also be stewed or dried and ground into meal and cooked as mush. A cider can also be made from the berries. The seeds alone can be collected and ground into meal, too.

The leaves and bark of kinnikinnick when dried can be smoked as a tobacco substitute. According to Harrington (1967), its general effect was intoxication. Owing to their tannic acid, the leaves are astringent and have been used to tan hides. The leaves can also be chewed to stimulate saliva, particularly when one is thirsty.

The leaves can also be boiled in water, allowed to cool, and the decoction applied to stop the itching and spread of poison ivy (Angier 1978). The internal consumption of the leaf tea often results in urine becoming alkaline and bright green. This effect is caused by the urinary antiseptic hydroquinolone and is relatively harmless. Hydroquinolones (particularly arbutin) are strongly antibacterial and are effective against *Klebsiella* and *E. coli*, which are often associated with urinary infections.

BEARBERRY CIDER

To make bearberry cider, simply crush the ripe or green berries in a container, then pour an equal volume of scalding water over them. After the mixture has cooled and the solids have settled, decant the liquid and drink. The cider will be a little dry to the taste.

PIPSISSEWA (*Chimaphila umbellata*)

Description and Habitat. Pipsissewa is a short evergreen semishrub (woody only at the base) that originates from a long creeping rootstalk. The leaves are whorled and leathery. It is often placed in the family Pyrolaceae.

Natural History Notes and Uses. The roots and leaves of pipsissewa may be boiled, and the liquid cooled for a refreshing drink that is high in vitamin C. The leaves may also be nibbled raw, but because of their astringency and tough texture we found them unappealing.

Pipsissewa was an important Native American remedy for rheumatism. A tea from the leaves was used to treat rheumatism and kidney problems (Foster and Duke 1990). The plant contains quinone glycosides, such as that found in *Arctostaphylos*, but is less astringent and more diuretic, making it better for long-term use. The plant was also mixed with tobacco for smoking. Pipsissewa produces a natural antibiotic (Moore 1979; Willard 1992). Hot infusions can be taken to induce perspiration in the treatment of typhus, and the berries can be eaten for stomach disorders (Moore 1979).

Pipsissewa is a "secret ingredient" in certain popular soft drinks. In the Northwest these plants, as well as certain species of *Pyrola*, are under commercial harvesting pressure and may be slowly disappearing.

ALPINE SPICY WINTERGREEN (*Gaultheria humifusa*)

Description and Habitat. Alpine spicy wintergreen is a dwarf evergreen shrub that forms small mats with leaves that are broadly egg-shaped to elliptical. The bell-shaped flowers are white to pink, and the berries are red.

Natural History Notes and Uses. The small, red fruits are edible raw or cooked and can be made into jams, wines, or pies. The young tender leaves are suitable as greens and have a wintergreen flavor. The fresh leaves of a related species, *G. hispidula* (creeping snowberry), can be used to make a tea, and the berries are also edible.

In Native American medicine the plants were used for treating aches and pains and to help with breathing while hunting or carrying heavy loads. The leaves of these species yield an oil on steam distillation. This "oil of wintergreen" (methyl salicylate) is a folk remedy for body aches and pains and is known for its astringent, diuretic, and stimulant properties.

Warning: The wintergreen flavor in the plants is due to the presence of oil of wintergreen, which if taken in excess can be toxic, especially to children. In small amounts, such as in wintergreen tea, there is little danger. Children who are allergic to aspirin (a related drug) should not eat the plant or berries or even handle the plant.

PINESAP (*Hypopitys monotropa = Monotropa hypopithys*)

Description and Habitat. Pinesap is a fleshy herb without chlorophyll and is pink- to straw-colored. The leaves are scale-like. The flowers are 4-merous, and the petals are saccate at the base, usually hairy on both surfaces. Stamens are twice the number of the petals. *Monotropa* is Greek for "one turn," referring to sharp recurving of top of the stem. *Hypopithys* means "under pines."

Natural History Notes and Uses. This fleshy tawny or reddish saprophytic herb resembles the Indian pipe (*Monotropa uniflora*) and grows in woodlands; in some classifications it is placed in a separate genus, *Hypopitys*. It is often placed in the family Monotropaceae. The plants are edible raw or cooked.

ALPINE LAUREL (*Kalmia microphylla*)

Description and Habitat. This species is a branched, evergreen shrub that spreads by short rhizomes and layering. The leaves are opposite, lance-shaped to elliptical, and have in-rolled margins. The flowers are rose-colored and bowl-shaped. The plant has 10 stamens that, when triggered by an insect landing on the flower, spring out of pockets in the petals and discharge pollen. The species occurs in the middle to upper elevations, usually in moist to wet, acidic soils, often along creeks.

Natural History Notes and Uses. The toxicity of *Kalmia* is legendary. Some Native Americans used it as a suicide plant. Game birds and livestock may be poisonous to eat if they have ingested the leaves. According to Peter Kalm (1715–79), after whom the genus is named, "sheep are especially susceptible, while deer are unharmed. Though the flesh of affected animals is apparently not contaminated, the intestines will cause poisoning if fed to dogs so that they become quite stupid and as it were intoxicated and often fall so sick that they seem to be at the point of death."

Warning: All *Kalmia* species should be considered ***poisonous.*** They contain andromedotoxin, which causes a slow pulse, low blood pressure, lack of coordination, convulsions, progressive paralysis, and death. The honey made by bees from these plants is also ***poisonous*** (Kingsbury 1965).

ANDROMEDOTOXIN

The major toxic substance in the Ericaceae appears to be andromedotoxin. This compound is known to occur in *Rhododendron, Leucothoe, Menziesia, Ledum,* and *Kalmia* and is probably more widespread than is now known. The leaves, twigs, flowers, and pollen grains of these genera all contain andromedotoxin. The course of poisoning includes watering of the mouth, eyes, and nose, followed by loss of energy, vomiting, slow pulse, low blood pressure, lack of coordination, convulsions, and slow and progressive paralysis of the arms and legs until death. Humans can be poisoned by chewing on the leaves and twigs, brewing "tea" from the leaves, or sucking nectar from the flowers of these plants. It has long been known that bees produce poisonous honey after visiting large stands of *Rhododendron*. Fortunately, the honey is so bitter that very little of it can be eaten.

WESTERN LABRADOR TEA (*Ledum glandulosum*)

Description and Habitat. Labrador tea is an evergreen shrub with short, fine hairs and glands on young branches and lower leaf surfaces. The leaves are elliptical to egg-shaped and are clustered near the stem tips, giving them a whorled effect. The white flowers are in rounded clusters at the tip of the stem. All parts of the plant smell like turpentine when crushed. Look for the plants in the middle to upper subalpine elevations, particularly in permanently wet or moist, acidic soils.

Natural History Notes and Uses. Labrador tea contains ledol, which is a narcotic toxin that causes drowsiness, delirium, cramps, paralysis, heart palpitations, and even death if taken in excess. Prolonged cooking extracts large doses of ledol. Otherwise the tea is slightly laxative and is recommended for "camper's distress" (e.g., constipation). The leaves are astringent and useful in facial creams.

Andromedotoxin is also found in the leaves of these plants (see *Kalmia* above), and therefore they should be considered ***poisonous.*** The leaves of *L. groenlandicum* (bog Labrador tea) make a mild but agreeable tea when steeped in hot water. Willard

(1992), Densmore (1974), and Foster and Duke (1990) indicate that the tea was used for colds, rheumatism, scurvy, and stomach ailments, but it is ***not recommended.*** A strong decoction of the leaves was used as a wash to get rid of lice. As an insect repellent, it was said to be quite effective against mosquitoes.

Another common name for this species is "trapper's tea." The name apparently originated not because trappers drank it but because they boiled their traps in it to de-scent them.

RUSTY MENZIESIA (*Menziesia ferruginea*)

Description and Habitat. This deciduous shrub has shredding, gray-brown bark. The leaves are crowded on the stem, giving it a whorled effect. The plant has a skunk-like odor when crushed. The cinnamon-pink flowers are small and urn-shaped. It can be found in wooded areas, usually on north- and east-facing slopes from the middle to upper elevations.

Natural History Notes and Uses. The twigs and leaves of rusty menziesia were used by some Native Americans to make a tea, but the plant contains some poisonous alkaloids. Additionally, a fungus (*Exobasidium* sp. affin. *vaccinii*) growing on the leaves of this shrub was eaten by some aboriginal peoples (Pojar and MacKinnon 1994).

SINGLE DELIGHT (*Moneses uniflora*)

Description and Habitat. The leaves of this plant are basal, thin, ovate, and sharply serrulate. The flowers are solitary, and the petals are white to pink in color. The anthers are 2-horned and the stigma 5-lobed. This is an uncommon plant.

Natural History Notes and Uses. A poultice of the leaves was applied to draw out pus from boils. An infusion of the dried plants was used for coughs and colds, and the plants were chewed for sore throats. A poultice of the chewed or pounded plant was applied to pains. The fruit was used as food (unspecified) by Montana Indians (Hart 1996).

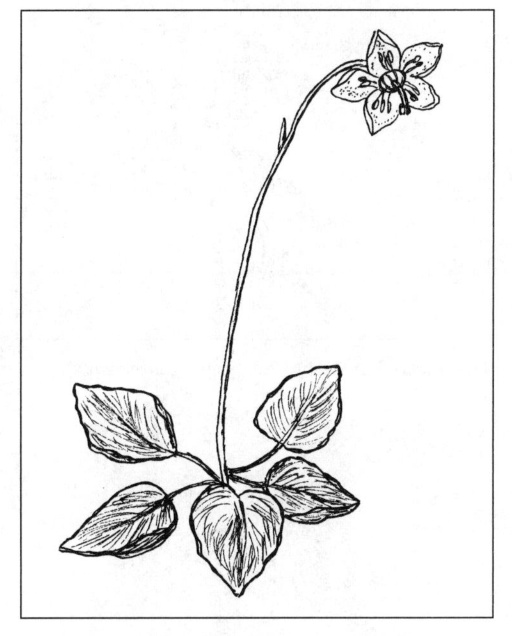

FIGURE 58. Single delight (*Moneses uniflora*).

SIDEBELLS WINTERGREEN (*Orthilia secunda*)

Description and Habitat. This species was once in the genus *Pyrola* (*P. secunda*). This low perennial has flowers dangling on one side of the stem and grows in dry shady woods from 3,000 to 10,500 feet. It flowers from July to September.

Natural History Notes and Uses. A strong decoction of the root was used as an eyewash.

Mountain Heath (genus *Phyllodoce*)

PINK MOUNTAIN HEATH (*P. empetriformis*)

YELLOW MOUNTAIN HEATH (*P. glanduliflora*)

Description and Habitat. These plants are the common heathers of subalpine meadows, although not the same as the classic heather of Europe. These low shrubs grow as spreading mats, with many short, upright stems. The evergreen leaves have glandular margins recurved and the underside appearing grooved. The flowers occur at the tops of the stems and usually nod on glandular stalks. The fruit is a roundish, dry capsule.

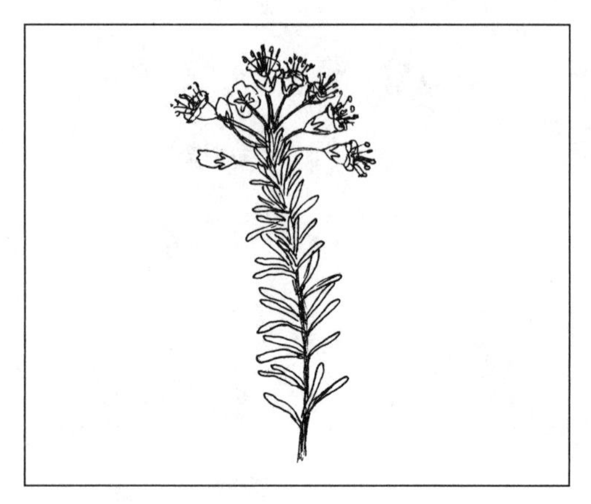

FIGURE 59. Mountain heath (*Phyllodoce* sp.).

FIELD KEY TO THE MOUNTAIN HEATHS

1. Corolla about 2 times as long as the calyx, pink to rose-colored and hairless outside . . . . *P. empetriformis*
1. Corolla not 2 times as long as calyx, yellowish to greenish white, and outer surfaces are glandular and hairy. . . . . *P. glanduliflora*

Natural History Notes and Uses. Pink mountain heath was used by the Thompson Indians of southwestern British Columbia as a tuberculosis remedy. Apparently, a decoction of the plant was taken over a period of time for tuberculosis and spitting up blood (Perry 1952).

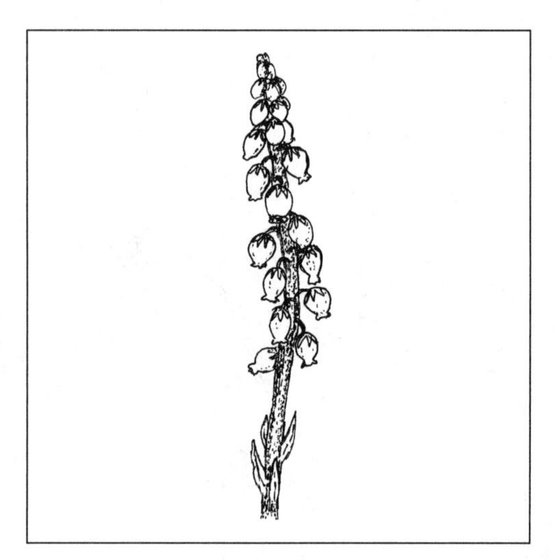

FIGURE 60. Pinedrops (*Pterospora andromedea*).

PINEDROPS (*Pterospora andromedea*)

Description and Habitat. Pinedrops is a brownish red plant with sticky stems up to 3 feet tall with pale yellow flowers. It is found in the deep humus of coniferous forests between 2,500 and 8,500 feet, usually associated with ponderosa pine (*Pinus ponderosa*). It flowers from June to August.

Natural History Notes and Uses. Foster and Duke (1990) indicate that Native Americans used a cold tea made from the pounded stems and fruits to treat bleeding from the lungs. As a dry powder, the plant was used as a snuff for nose-bleeds.

Wintergreen (genus *Pyrola*)

LIVERLEAF WINTERGREEN (*P. asarifolia*)

GREEN-FLOWERED WINTERGREEN (*P. chlorantha*)

SNOWLINE WINTERGREEN (*P. minor*)

WHITE-VEINED WINTERGREEN (*P. picta*)

Description and Habitat. In general, wintergreens are low, smooth perennial herbs with shiny, leathery leaves that are clustered at the base. The flowers are waxy and nodding. *Pyrola* stems from *pyrus*, for "pear," probably since the leaves of many species resemble pear leaves. Wintergreen is the common name for all members of this genus and refers to the persistent nature of the leaves into winter.

FIELD KEY TO THE WINTERGREENS

1. The style is nearly straight, less than ⅛ inch long . . . . *P. minor*
1. The style is curved, usually longer than ⅛ inch 2

2. Leaves prominently white-mottled along the main veins of the upper surface . . . . *P. picta*
2. Leaves not as above. 3

3. Corolla pink to purplish . . . . *P. asarifolia*
3. Corolla white or greenish white to yellowish . . . . *P. chlorantha*

Natural History Notes and Uses. A tea made from the whole plant was used to treat epileptic seizures in babies. A leaf tea was gargled for sore throats and canker sores, and a tea from the root was a tonic. A poultice from the mashed leaves was used for tumors, sores, and cuts and to relieve the itch of insect bites. The plant is also an excellent astringent and disinfectant for urinary tract infections. The plants contain ursolic acid and the glycosides arbutin and ericolin, which were used in the treatment of kidney problems and skin eruptions.

Pyrola is also used as an ingredient in popular soft drinks. It is said to be an excellent substitute for *Chimaphila umbellata* (pipsissewa). In some areas *Pyrola* may be overharvested for commercial purposes.

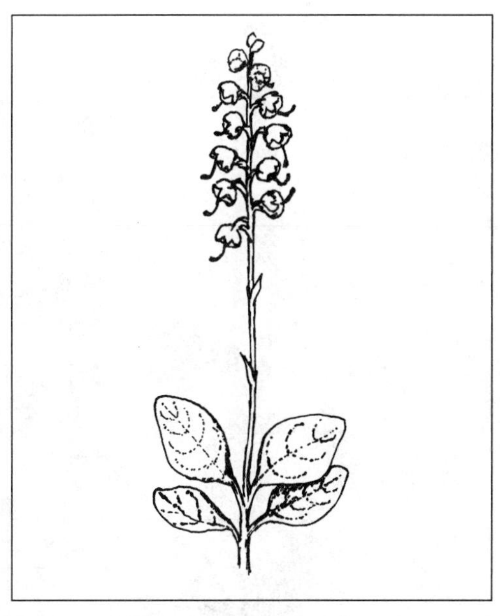

FIGURE 61. Wintergreen (*Pyrola* spp.).

Blueberry, Huckleberry (genus *Vaccinium*)

DWARF BLUEBERRY (*V. cespitosum*)

HUCKLEBERRY (*V. membrabaceum*)

WHORTLEBERRY (*V. myrtillus*)

BLUEBERRY (*V. occidentale = V. uliginosum*)

GROUSE WHORTLEBERRY (*V. scoparium*)

Description and Habitat. The species of *Vaccinium* in the area are small to midsized shrubs with deciduous leaves. The twigs are often angled. The small flowers are urn-shaped, and fruits are many-seeded berries. They can be found on well-drained sites from wet meadows and around lakes up to timberline.

FIELD KEY TO THE BLUEBERRIES

1. Leaves entire; flowers 1–4 per axil .... *V. occidentale*
1. Leaves mostly toothed; flowers usually 1 per leaf axil.................. 2

2. Leaves obovate or oblanceolate; twigs inconspicuously angled .... *V. cespitosum*
2. Leaves elliptic, oval, ovate, or lanceolate; twigs conspicuously angled....................................... 3

3. Plants usually 12 inches or taller.................... *V. membrabaceum*
3. Plants usually under 12 inches tall 4

4. Leaves ½ to 1½ inches long, about ½ inch wide; young twigs finely hairy; mature fruits bluish .... *V. myrtillus*
4. Leaves ¼ to ¾ inch long, less than ½ wide; young twigs not hairy; mature fruit reddish............... *V. scoparium*

Natural History Notes and Uses. *Vaccinium* berries can be eaten raw or be dried in the form of cakes for future use. The various species we have sampled range in taste from sweet to tart. Hybridization between the species is known to occur, but the fruits are still edible. The berries have also been used as fish bait because they look very similar to salmon eggs. The leaves can be dried to make a tea. The leaves and berries are high in vitamin C.

SPURGE FAMILY (EUPHORBIACEAE)

The spurge family has about 290 genera and 7,500 species distributed worldwide. Among the valuable products of the family are rubber, castor and tung oils, and tapioca. Most members are ***poisonous*** and have milky sap that will irritate eyes and mouth.

Spurges (genus *Euphorbia*)

HORNED SPURGE (*E. brachycera*)

WOLF'S MILK (*E. esula*)

SPOTTED SANDMAT (*E. maculata* = *Chamaesyce m.*)

SNOW-ON-THE-MOUNTAIN (*E. marginata*)

THYMELEAF SANDMAT (*E. serpyllifolia* = *Chamaesyce s.* subsp. *serpyllifolia*)

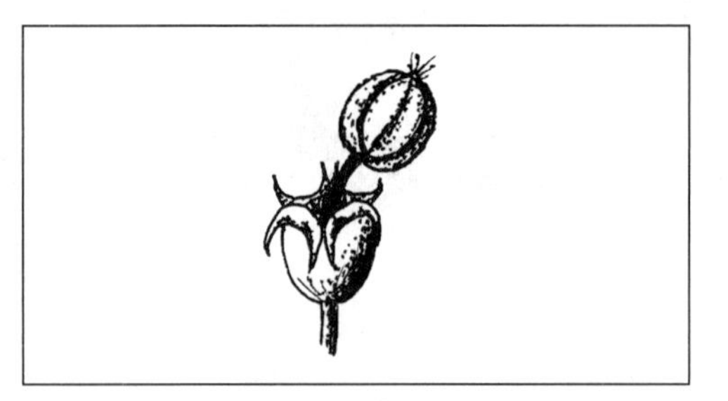

FIGURE 62. Spurge (*Euphorbia* spp.) flower: cyathium.

Description and Habitat. Spurges are annual or perennial herbs with milky juice. The flowers are borne in a complex, cup-like structure called a *cyathium*, which contains several male flowers and a single female flower. They are found in disturbed habitats in low elevations. The genus name is from the Greek *euphorbion*, a plant named after Euphorbos, a celebrated Greek physician of the first century B.C.

FIELD KEY TO THE SPURGES

1. Leaves all opposite	2
1. Leaves, at least lower, alternate	3
2. Plants hairy	*E. maculata*
2. Plants glabrous	*E. serpyllifolia*
3. Plants taprooted annuals	*E. marginata*
3. Plants perennial with woody base	4
4. Leaves mostly over 5 times as long as wide, linear or oblong to narrowly elliptic	*E. esula*
4. Leaves mostly less than 5 times as long as wide, not linear or oblong	*E. brachycera*

Natural History Notes and Uses. *Euphorbia* contains toxic constituents that will cause severe ***poisoning*** if ingested in quantity. Most species contain carcinogenic, highly irritant diterpene esters and are strong purgatives. The white sap can cause skin irritations and blisters (Willard 1992). Bean and Saubel (1972) report that the Cahuilla Indians in California used both the native and introduced species as a medicine for reducing fever and as a cure for chicken pox and smallpox. The plant was boiled and the afflicted person bathed in the decoction.

PEA FAMILY (FABACEAE = LEGUMINOSAE)

The pea family is one of the largest plant families in the world, with approximately 600 genera and 13,000 species worldwide. In economic importance, this family is second only to the grass family (Poaceae), which produces all our grains and cereals. The beans and peas that we eat for dinner, as well as the traditional peanuts at baseball games, are found in this family. But before taking a bite of the next legume you see, be aware that the family also contains a number of highly toxic members. The various species of locoweeds and milkvetches (*Oxytropis* and *Astragalus*) have caused much loss of livestock.

The pea flower is referred to as *papilionaceous,* meaning "butterfly-like." The flowers are bilaterally symmetrical, consisting of 5 petals. The largest upper petal is called the *banner*, the two lateral ones are the *wings*, and the two lowest ones are fused at the lower margins to form a boat-like structure called the *keel.*

The ripe seeds of various members of the pea family are good protein sources, especially when mixed with cereals (Poaceae). The amino acids of their respective proteins combine, which markedly increases their nutritional efficiency. Legumes are also rich in carbohydrates.

FIELD KEY TO THE PEA FAMILY

1. Leaves even-pinnate 2
1. Leaves odd-pinnate, simple, or palmate 3

2. Flowers pale yellow; style long-hairy only at the tip; leaflets more than ¾ inch broad . . . . *Lathyrus*
2. Flowers purple-blue; style long-hairy along the whole upper side; leaflets mostly less than ½ inch broad. . . . . *Vicia*

3. Leaflets 3 only. 4
3. Leaflets 5 or more, or the leaves simple. 9

4. Leave palmate, the terminal leaflets not stalked or jointed 5
4. Leaves pinnate, the terminal leaflets stalked or jointed 7

5. Flowers golden yellow, the banner circular in outline . . . . *Thermopsis*
5. Flowers white to pink or purple, banner not circular in outline. 6

6. Leaflets usually toothed; flowers in heads, pink or white. . . . . *Trifolium*
6. Leaflets entire; flowers not in heads. . . . . *Astragalus*

7. Flowers in heads; corolla persistent; fruit straight. . . . . *Trifolium*
7. Flowers in racemes; corolla not persistent; fruit curved, straight, or coiled 8

8. Pods curved or coiled; inflorescence not over 2 inches long; petals yellow or blue-purple. . . . . *Medicago*
8. Pods straight; inflorescence is a loose raceme more than 2 inches long; petals white or yellow . . . . *Melilotus*

9. Leaves palmately compound, with 5–11 leaflets. . . . . *Lupinus*
9. Leaves pinnately compound 10

10. Herbage gland-dotted. . . . . *Glycyrrhiza*
10. Herbage not gland-dotted 11

11. Margin of leaflets toothed; corolla persistent. . . . . *Trifolium*
11. Margin of leaflet usually entire; corolla deciduous 12

12. Flowers in umbels . . . . *Coronilla*
12. Flowers in racemes or cymes. 13

13. Keel petals much longer than the wings; fruit a flat loment . . *Hedysarum*
13. Keel petals about equal to or shorter than the wings; fruit a legume . . . 14

14. Keel petal narrowed to a slender beak. . . . . *Oxytropis*
14. Keel without a beak. 15

15. Flowers red-orange when fresh. . . . . *Sphaerophysa*
15. Flowers pink, pink purple, lavender, or white . . . . *Astragalus*

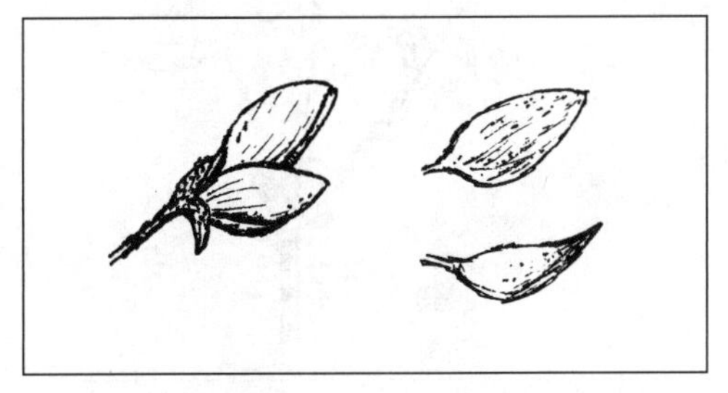

FIGURE 63. Pea family flower expanded.

Milkvetch, Locoweed (genus *Astragalus*)

Description and Habitat. Milkvetches are perennial herbs with odd-pinnate leaves that have leafy stipules. This is a difficult genus of perhaps 1,600 species, making it the largest genus in the pea family. The name comes from the ancient Greek name for a plant in the pea family.

Natural History Notes and Uses. Although the roots, pods, and peas of some species were reported to be eaten by American Indians, this genus is ***not recommended*** for consumption. All milkvetches either produce a toxic alkaloid substance or accumulate selenium from the soil or both. Selenium poisoning of livestock has the following characteristics: lethargy, diarrhea, loss of hair, breakage at the base of hoof, excessive urination, difficulty breathing, rapid and weak pulse, and coma. Death usually results from failure of the lungs and heart (Turner and Szczawinski 1991).

An interesting side benefit has developed from discovery of plants that grow only in selenium soils: scientists can use the plants to map areas high in selenium for the purpose of mining the valuable element. Some *Astragalus* species are also good indicators of uranium ore and copper-molybdenum deposits.

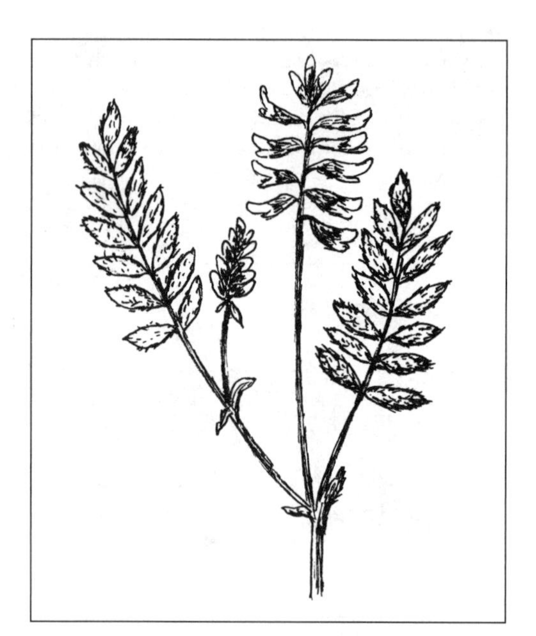

FIGURE 64. Astragalus (*Astragalus drummondii*).

PURPLE CROWN-VETCH (*Coronilla varia*)

Description and Habitat. Purple crown-vetch is a perennial herb with trailing to ascendant glabrous stems 11–40 inches long. The leaves are alternate, odd-pinnate, and 2–4 inches long. Each leaf bears 9–25 obovate to oblong leaflets. The pea-like, pink-to-white flowers occur in umbels at the end of extended peduncles that arise in the axils of leaves. The fruit, a loment, is linear, 4-angled with 3–7 segments, ¾ inch to 2 inches long.

Natural History Notes and Uses. Crown-vetch is native to Europe, southwest Asia, and northern Africa. It was introduced into the United States for use in erosion control along highway embankments. It has escaped and is now widely distributed throughout North America. Crown-vetch grows rapidly and reproduces prolifically, out-competing and displacing native species. The seeds are reported to be poisonous.

AMERICAN LICORICE (*Glycyrrhiza lepidota*)

Description and Habitat. Licorice is a perennial herb growing up to 3 feet tall. The flowers are greenish white in dense racemes. The mature fruits are a conspicuous pod up to ½ inch long and densely covered with hooked spines. This species is usually found in moist, sandy soils and on riverbanks at the lower elevations.

Natural History Notes and Uses. The plant contains glycyrrhizin, sugar, and other chemicals used in medicine as a mild laxative, a demulcent, and a flavoring to mask the taste of other drugs. It is also used in confections, root beer, and chewing tobacco. Licorice root has been used in the treatment of asthma, stomach ulcers, bronchitis, and urinary tract disorders. The plants were chewed by Native Americans and used as a flavoring.

Licorice root contains glycyrrhizin, the source of most of the pharmacological effects of licorice and rhizome. Glycyrrhizin is about 50 times sweeter than sugar, and it has a cortisone-like effect that may result in minor "poisoning" if consumed in very large amounts. Glycyrrhizin increases extracellular fluid and plasma volume and induces sodium retention and loss of potassium, which often leads to edema or water retention. Too much licorice can cause cardiac depression and edema.

Researchers using highly refined licorice extract suggest that chemicals in glycyrrhizin called triterpenoids may be effective against cancer. They block the production of prostaglandin (a hormone-like fatty acid that may be responsible for stimulating the growth of cancer cells) and help get rid of cancer-causing invaders. Triterpenoids have been shown in test tubes to stunt the growth of rapidly multiplying cells such as cancer cells, and they may even help precancerous cells return to normal (Miller 1973).

Warning: Continual use of this plant in large doses may cause water retention and elevated blood pressure.

Sweetvetch (genus *Hedysarum*)

ALPINE SWEETVETCH (*H. alpinum*)

NORTHERN SWEETVETCH (*H. boreale*)

WESTERN SWEETVETCH (*H. occidentale*)

WHITE SWEETVETCH (*H. sulphurescens*)

Description and Habitat. Sweetvetches are perennial herbs that are sparsely branched and slightly hairy. The stipules are usually brown, large, and sheath-like on the lower stem and narrow and pointed on the upper stem. The typical pea-like flower has a much longer keel than wings and banner. The species occur in open forests, meadows, and rocky ledges from the middle to higher elevations. The botanical name is from the Greek *hedys*, meaning "sweet," and *aroma*, for "smell." The flowers are quite fragrant. Harvest sweetvetches only if you can positively identify the species you are gathering.

FIGURE 65. Hedysarum (*Hedysarum sulphurescens*).

FIELD KEY TO THE SWEETVETCHES

1. Flowers yellowish white .... *H. sulphurescens*
1. Flowers pink to purple 2

2. Sepals about ¼ inch long; leaflet veins hidden .... *H. boreale*
2. Sepals less than ¼ inch long; leaflets with definite lateral veins 3

3. Flowers less than ½ inch long .... *H. alpinum*
3. Flowers more than ½ inch long .... *H. occidentale*

Natural History Notes and Uses. Unlike other genera, in which some species are better-tasting than others but all are still harmless, *Hedysarum* does not follow that pattern in terms of safety for eating. A person wishing to harvest these plants must make a special effort to notice the subtle variations in leaf and flower structure between species. A misidentification can easily lead to some badly upset stomachs. Early settlers and Native Americans used the roots of *H. boreale* as a licorice substitute.

Caution: If you intend to consume any *Hedysarum*, it is important to correctly identify it to species first. *Hedysarum* is also sometimes confused with *Astragalus* but can be easily distinguished when the pods are present: *Hedysarum* is distinguished by the presence of a loment, and the keel is longer than or equal in length to the wings and banner.

Sweet Pea (genus *Lathyrus*)

THICK-LEAF PEAVINE (*L. lanszwertii*)

FLAT PEAVINE (*L. sylvestris*)

Description and Habitat. Sweet peas are vines, climbing or supporting themselves on other vegetation. The plants have tendrils at the ends of their leaves. Sweet peas are found in a variety of habitats from the foothills up to the subalpine.

FIELD KEY TO THE SWEET PEAS

1. Leaflets 2; stems winged . . . . *L. sylvestris*
1. Leaflets 4 or more; stems angled but not winged . . . . *L. lanszwertii*

Natural History Notes and Uses. Some species of *Lathyrus* have a history of poisoning humans. Kirk (1975) indicates that an exclusive diet of some species from 10 to 30 days can bring on partial or total paralysis, and Willard (1992) suggests avoiding these plants entirely. Strike (1994) says that the greens and raw seeds of *Lathyrus* were eaten by Native Americans. Some of the seeds were parched and made into pinole, which could be stored for winter use. Weedon (1996) says that the fruits of many species are edible in small amounts but may cause paralysis and several secondary disorders if eaten in large quantities over time.

Caution: It is best to assume that these plants are poisonous if ingested.

FIGURE 66. Lupine (*Lupinus* spp.) leaf.

Lupine (genus *Lupinus*)

SILVERY LUPINE (*L. argenteus*)

PACIFIC LUPINE (*L. lepidus*)

VELVET LUPINE (*L. leucophyllus*)

BIGLEAF LUPINE (*L. polyphyllus = L. burkei*)

RUSTY LUPINE (*L. pusillus*)

SILKY LUPINE (*L. sericeus*)

Description and Habitat. The many species of lupine are showy perennial or annual herbs with palmately compound leaves. The flowers are blue, violet, rose, or rarely white in an elongated narrow inflorescence. The pods are flattened and usually hairy. They are found on open slopes and in meadows up into the alpine zone. The Latin name comes from *lupus*, meaning "wolf," alluding to the belief that this plant wolfed nutrients and caused poor soil conditions. To the contrary, lupines are nitrogen fixers that greatly improve soil conditions. Lupines have challenged generations of taxonomists, and the disposition of many species has varied tremendously.

FIELD KEY TO THE LUPINES

1. Plant annual; seed pod with 2 seeds . . . . *L. pusillus*
1. Plants not annual; seed pods with more than 2 seeds 2

2. Leaves are almost all basal, their height often surpassing that of the flowers .... *L. lepidus*
2. Leaves mostly from the stem, flowers surpassing height of leaves 3

3. Most of back surface of banner lacking or with inconspicuous hairs 5
3. Back surface of banner prominently hairy 4

4. Flowers in loose racemes; flowers stalks often greater than ⅛ inch long .... *L. sericeus*
4. Flowers densely clustered in spike-like racemes; flowerstalks less than ⅛ inch long .... *L. leucophyllus*

5. Banner only slightly reflexed from wings to form a narrow V-shaped opening of about 45 degrees .... *L. argenteus*
5. Banner greatly reflexed from wings and forms a wide V-shaped opening of 60 degrees or more .... *L. polyphyllus*

FIGURE 67. Lupine (*Lupinus* spp.).

Natural History Notes and Uses. Pliny listed 35 uses for species in this genus, and Thoreau once wrote of the lupine, "The earth is blued with it." At night or during dark stormy days the leaflets of lupines "sleep" by folding down along the petiole or by standing upright and close together.

The pea-like seeds have been wrongly recommended by some authors of edible plant books as a substitute for peas. Lupines possess many complex alkaloids and should be considered ***poisonous***. Nevertheless, some species have been safely consumed. For example, Weedon (1996) and Scully (1970) indicate that the young leaves and unopened flowers were steamed and eaten with soup by some Native American tribes. But because of hybridization, the edible species can concentrate toxic alkaloids that could result in an unhealthy game of "lupine roulette."

It appears, however, that some of the alkaloids found in lupines are removed by cooking and that toxins intensify with age. The toxic principle of lupines is excreted by the kidneys, and the poisoning is not cumulative. That is, a lethal dose must be eaten at one time to cause death. The poisonous effects produced by lupines are referred to as lupinosis, with nervousness, labored breathing, convulsions, and frothing at the mouth the obvious signs (Muenscher 1962). *But until documentation on Yellowstone and Grand Teton species is established, using lupines as a food source is **not recommended**.* Many people use the larger, hairy-leaved species as an excellent toilet paper substitute.

Warning: Lupine seeds contain alkaloids and are toxic.

ROOT NODULES AND SYMBIOTIC BACTERIA

Many soils are deficient in nitrogen and other nutrients. Some plants have evolved special structures on their roots that contain nitrogen-fixing bacteria. These single-celled organisms convert atmospheric nitrogen into water-soluble nitrate compounds that can be used by the plants. In return, the plant furnishes the bacteria with nutrients and water. The nodules on the roots of lupine contain a bacterium known as *Rhizobium*. During the early stages of this symbiotic relationship the bacteria gain entrance to the root hairs and are entirely parasitic until nodule formation has been completed. Another example of this type of relationship is seen with mountain-mahogany (*Cercocarpus*) and the filamentous bacterium *Actinomycetes*.

Medic (genus *Medicago*)

BLACK MEDIC (*M. lupulina*)

ALFALFA (*M. sativa*)

Description and Habitat. Medics are hairless, branching perennials or annual herbs with leaves divided into 3 leaflets. The terminal leaflet is evidently longer than the other two. The pods are twisted. These plants are usually found in disturbed areas at the lower elevations.

FIELD KEY TO THE MEDICS

1. Plants annual; flowers less than ¼ inch long. . . . . *M. lupulina*
1. Plants perennial; flowers greater than ¼ inch long . . . . *M. sativa*

Natural History Notes and Uses. Alfalfa can cause bloat in livestock when it constitutes a high percentage of their diet, especially when young before flowering. Saponins found in the leaves may contribute to the problem. Humans should, therefore, use this plant in moderation. The dried and powdered young leaves and flower heads of alfalfa are nutritious and can be steeped in hot water to make a bland tea. The tender leaves can also be added to salads and are rich in vitamins A, D, and K. Alfalfa also supplies calcium, magnesium, and phosphorus (Smith 1973). Alfalfa sprouts are a popular salad addition, and the seeds are available from health food stores. Nectar from the flowers produces a good honey. In addition to uses as food and medicine, alfalfa seeds contain an oil used in paints and varnishes. Paper makers have used the stem fibers in their craft, and wool dyers extract a yellow dye from the seeds. The seeds of black medic, *M. lupulina*, can be parched and eaten or ground into flour.

SWEETCLOVER (*Melilotus officinalis* = *M. albus*)

Description and Habitat. Sweetclovers are strongly taprooted perennial or annual herbs. The leaves are divided into 3 fine-toothed, wedge-shaped leaflets. The white or yellow flowers are loosely arranged in an inflorescence, and the pods are thickly spindle-shaped. They are usually found in disturbed habitats at the lower elevations. The genus name is from the Greek *mel*, meaning "honey," and lotus flower.

Natural History Notes and Uses. The young leaves (before the flowers appear) may be eaten raw or boiled. The fruit may be used as seasoning for soups. The older leaves are toxic and should be avoided. The dried flowering plant of *M. officinalis* was used in teas for neuralgic headaches, nervous stomach, diarrhea, and aching muscles. Elias and Dykeman (1982) indicate that improperly dried yellow sweetclover will easily mold and in the process produce dicoumarol, a potent anticoagulant that can cause severe bleeding and death. Molding yellow sweetclover mixed in hay has killed many cattle. Dicoumarol, which is extremely poisonous in excess, is used in rat poisons.

The plants are sweet-scented due to coumarin and become more pleasant when dried. They have been used to scent clothes and protect them from moths as an alternative to moth balls. Sweetclover has also been a traditional flavoring additive in smoking tobacco and snuff.

Stemless Locoweed (genus *Oxytropis*)

COLD MOUNTAIN CRAZYWEED (*O. campestris*)

HANGPOD CRAZYWEED (*O. deflexa*)

HARESFOOT POINTLOCO (*O. lagopus*)

PARRY'S CRAZYWEED (*O. parryi*)

STALKEDPOD LOCOWEED (*O. podocarpa*)

SILVERY OXYTROPE (*O. sericea*)

Description and Habitat. Stemless locoweeds are generally perennial herbs that are stemless or with a short leafy stem. The flowers are white to reddish or purple. The keel is prolonged into a point or tooth or a straight-to-curved beak. The name is from the Greek *oxys*, meaning "sharp," and *tropis*, meaning "keel," in reference to the beaked keel. They may be found in a variety of habitats from the lower elevations to subalpine.

FIELD KEY TO THE STEMLESS LOCOWEEDS

1. Pods pendulous; stipules slightly adnate (attached) to the petioles	*O. deflexa*
1. Pods erect or spreading; stipules adnate to the petioles	2
2. Corolla white	3
2. Corolla purplish, pinkish, or bluish	4
3. Mature pods leathery	*O. sericea*
3. Pods papery	*O campestris*
4. Racemes 6- to many-flowered	*O. lagopus*
4. Racemes 1–5-flowered	5
5. Pods papery and inflated, ellipsoidal	*O. podocarpa*
5. Pods leathery, not inflated, cylindrical or oblong in shape	*O. parryi*

Natural History Notes and Uses. None of the species is known to be poisonous (loco producing), but it's probably best if they're avoided and treated as potentially toxic. However, locoweeds are poisonous to livestock if eaten in large amounts over a long period of time. Extensive grazing of *O. sericea* induces a chronic poisoning called locoism. It is strongly recommended that these plants be avoided for the purposes of human consumption.

LOCOWEEDS (*ASTRAGALUS* AND *OXYTROPIS*)

There are as many as 300 species of "locoweeds" in North America, although not all are toxic. The two genera are *Astragalus* and *Oxytropis*, and identification requires a trained specialist. Locoweeds contain one of three toxic fractions: miserotoxin, swainsonine, and selenium. The fresh plants are toxic, and the toxicity is gradually lost as the plant dries, except in the case of selenium accumulators, which are not affected by drying. Swainsonine poisoning occurs after about two weeks or more of ingestion. Cattle and horses may become habituated to locoweed and seek it out even when good forages are available. Swainsonine may be passed in milk.

ALKALI SWAINSONPEA (*Sphaerophysa salsula*)

Description and Habitat. Swainsonpea is a perennial plant that is much branched from the base and has a woody taproot and rhizome. The leaves are pinnately compound and alternate on the stem. Each leaf is composed of 9–25 leaflets, which have silvery hairs. The flowers are brick red, drying to purple, and occur on axillary stems, each with 4–8 flowers. The fruits of this legume are inflated pods with seeds the size of alfalfa seeds.

Natural History Notes and Uses. Swainsonpea is a long-lived perennial legume found throughout the western United States. It may have been introduced from Asia for forage or soil stabilization. Like many other legumes, the seeds are extremely hard and may be viable in the soil for many years.

One species that superficially resembles swainsonpea is wild licorice (*Glycyrrhiza lepidota*). However, unlike swainsonpea, wild licorice has yellowish white to greenish white flowers, gland-dotted leaves, and bur-like fruits covered with hooked prickles.

In addition, many native locoweeds (*Astragalus* spp.) have leaves and inflated pods that closely resemble those of swainsonpea. However, locoweeds have glabrous stigmas and styles, and their pods lack the stalk-like base above the calyx; also, most do not have red flowers.

MOUNTAIN YELLOW PEA (*Thermopsis montana*)

Description and Habitat. Mountain yellow pea is a perennial up to 3 feet tall. Leaves are compound pinnate, with the leaflets linear-elliptic to broadly ovate-elliptic in shape and as long as 4 inches. The flowers are yellow, borne in small to long racemes as much as 12 inches long. Five to fifty flowers may make up the inflorescence. The pods are erect to somewhat spreading, straight, somewhat hairy, and 2–5-seeded. The plant varies in hairiness overall, from nearly glabrous to densely pubescent.

Natural History Notes and Uses. Mountain yellow pea is also known as "false lupine" because the pea-shaped flowers of this species and lupine are similar. The genus name even recognizes this similarity: *thermos* is Greek for "lupine," and *opsis* is Greek for "similar." The specific name means "pertaining to the mountains." A cold decoction of the leaves of a related species (*T. macrophylla*) was used by the Pomo Indians of California as a wash for sore eyes (Barrett 1917). A tea made from the leaves, roots, or bark was used by Kashaya (California) women to slow their menstrual flow (Strike 1994).

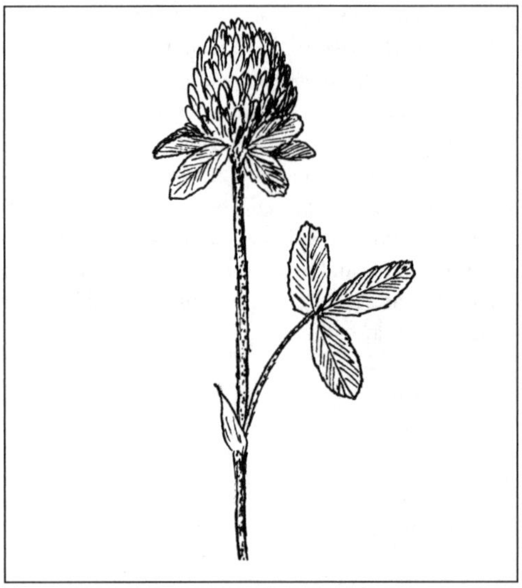

FIGURE 68. Clover (*Trifolium pratense*).

Clover (genus *Trifolium*)

ALPINE CLOVER (*T. dasyphyllum*)

HAYDEN'S CLOVER (*T. haydenii*)

ALSIKE CLOVER (*T. hybridum*)

DWARF CLOVER (*T. nanum*)

PARRY'S CLOVER (*T. parryi*)

RED CLOVER (*T. pratense*)

WHITE CLOVER (*T. repens*)

COWS CLOVER (*T. wormskioldii*)

Description and Habitat. Clovers are annual or perennial plants from rhizomes with leaves that are divided into 3 or more leaflets. The flower colors range from white to pink, yellow, red, or purple, and the seed pods are round to elongated. They are found in various habitats at all elevations. The genus name refers to the three leaflets.

FIELD KEY TO THE CLOVERS

1. Plants having a conspicuous upright, leaf-bearing stem or the stem is creeping and usually rooting at the nodes; internodes conspicuous 2
1. Plants mostly growing in dense tufts with all leaves basal or almost; internodes obscure 5

2. Flowers subtended by a true involucre or a false involucre of stipules . . . 3
2. Flowers not subtended by an involucre 4

3. Flowers subtended by a false involucre of stipules from a leaf or leaves; of disturbed areas .... *T. pratense*
3. Flowers subtended by a true involucre .... *T. wormskioldii*

4. Calyx glabrous; flowers mainly white .... *T. repens*
4. Calyx with hair between lobes; flowers usually pink......... *T. hybridum*

5. Calyx hairy .... *T. dasyphyllum*
5. Calyx glabrous 6

6. Heads mostly 1–3-flowered .... *T. nanum*
6. Heads mostly 5- or more flowered 7

7. Heads subtended by an involucre........................... *T. parryi*
7. Heads not subtended by an involucre .... *T. haydenii*

Natural History Notes and Uses. All species are nutritious and high in protein, but the flower heads and tender young leaves are hard to digest raw and may cause bloating. To improve digestibility of the plants, soak them in salt water for several hours or overnight. Leaves prepared this way may be dried and stored for future use. The dried flower heads and seeds can be ground into a flour substitute or extender.

Trifolium was an important food source for many Native Americans. In the spring, explorers and settlers saw them in the meadows picking and eating large quantities of clover. This was an annual event for the natives, who relished the greens of the spring season. Unfortunately, the non-natives, in their ignorance, compared the natives to grazing animals, just one of many disparaging comments that arose from misunderstood Native American behavior (Erichsen-Brown 1979).

A tonic tea can be made from the dried flowers. Made strong, the tea can be used as a gargle for a sore mouth or throat, and a mild sedative. The tea can also be used as a wash for skin ailments. The dried leaves can be smoked.

AMERICAN VETCH (*Vicia americana*)

Description and Habitat. Vetches are annual or perennial herbs with trailing to climbing stems. The leaves are pinnately divided with tendrils in place of terminal leaflets. American vetch is a trailing, climbing perennial that has sparsely pubescent stems and grows in open or waste places at lower elevations. It flowers from April to June. *Vicia* closely resembles *Lathyrus* (Sweet pea), and identification requires careful examination of the stipules. The stipules of *Vicia* are usually cut into narrow lobes, whereas the stipules of *Lathyrus* are entire to dentate.

Natural History Notes and Uses. Many species contain toxic compounds and therefore should be considered ***poisonous***; however, Kirk (1975) and Craighead et al. (1963) state that the young stems and seeds can be boiled or baked. The seeds of some species contain compounds that produce toxic levels of cyanide when digested.

Caution: Because of the poisonous compounds found in vetch, it is not recommended for eating.

FUMITORY FAMILY (FUMARIACEAE)

There are 16 genera and 450 species in the fumatory family. The family name is derived from *fumus*, the Latin word for "smoke," and refers to the climbing, purple-flowered European species *Fumaria officinalis*, whose finely divided leaves resemble a cloud of smoke over the ground. The members of this family are of little economic importance, except that a few are cultivated as ornamentals. Members of this family are sometimes included in the Papaveraceae (poppy family).

FIELD KEY TO THE FUMITORY FAMILY

1. Flowers solitary; leaves basal or may be lacking .... *Dicentra*
1. Flowers in racemes; leaves on stems .... *Corydalis*

SCRAMBLED EGGS (*Corydalis aurea*)

Description and Habitat. Plants in this genus are annual or perennial herbs with dissected leaves. Their flowers are generally yellow or white to pinkish. This species is found on rocky or sandy soils along lake or ponds or in open woods.

There are about 100 species of *Corydalis* worldwide. The name is derived from the Greek *korydalis*, "crested lark." The species name means "golden" in botanical Latin. Scrambled eggs was first named for science by Carl Ludwig Willdenow (1765–1812), director of the Berlin Botanical Garden and producer of the fourth edition of Linnaeus's *Species Plantarum*, the first accepted book of botanical nomenclature.

Natural History Notes and Uses. The plants are considered ***poisonous*** and contain several different alkaloids. Native Americans apparently used a tea made from *C. aurea* for painful backaches, diarrhea, menstruation, bronchitis, sore throats, and stomachaches and inhaled the fumes of burning roots for headaches (Foster and Duke 1990). The plant is reputed to be used in Mexico as a tea for women recovering from childbirth. The roots of some closely related Asian and European species are used as food or tonic medicines.

LONGHORN STEERS HEAD (*Dicentra uniflora*)

Description and Habitat. *Dicentra* is a genus of about 20 species of herbaceous flowering plants native to Asia and North America. The common name *bleeding heart* is used for many of the species. This name comes from the appearance of the pink flower, which resembles a heart with a drop of blood descending. In general, they are perennial herbs from tubers with dissected leaves, and the flowers are white to purplish or rose-colored. They are usually found growing in well-drained soils from the foothills to the subalpine.

Natural History Notes and Uses. The plants are considered ***poisonous*** and contain several different alkaloids (Kingsbury 1964). These alkaloids are found throughout the plant and can cause trembling, staggering, convulsions, and labored breathing. Large quantities can be fatal. A poultice from *D. cucullaria*

(Dutchman's breeches) was apparently made to treat skin diseases (Foster and Duke 1990).

GENTIAN FAMILY (GENTIANACEAE)

There are 70 genera and approximately 1,100 species of this family worldwide. Thirteen genera are native to the United States. They are mostly annual or perennial herbs with bitter juice. Several species are cultivated as ornamentals.

FIELD KEY TO THE GENTIAN FAMILY

1. Corolla rotate, the lobes equal to or longer than the tube .... *Swertia*
1. Corolla tubular, the lobes shorter than the tube 2

2. Corolla with conspicuous folds (plaits) between lobes......... *Gentiana*
2. Corolla lacking plaits .... *Gentianella*

Gentian (genus *Gentiana*)

PLEATED GENTIAN (*G. affinis*)

WHITISH GENTIAN (*G. algida*)

WATER GENTIAN (*G. aquatica* = *G. fremontii*)

PYGMY GENTIAN (*G. prostrata*)

Description and Habitat. *Gentiana* is a large genus of annual, biennial, or perennial herbs from fleshy roots or rhizomes; most species grow in moist or wet soil. The flowers are 4- or 5-lobed, tubular or funnel-shaped. The four species in the area can be found from the foothills to alpine meadows. The genus honors King Gentius of Illyria, an ancient country on the east side of the Adriatic Sea, who is reputed to have discovered medicinal virtues in gentians.

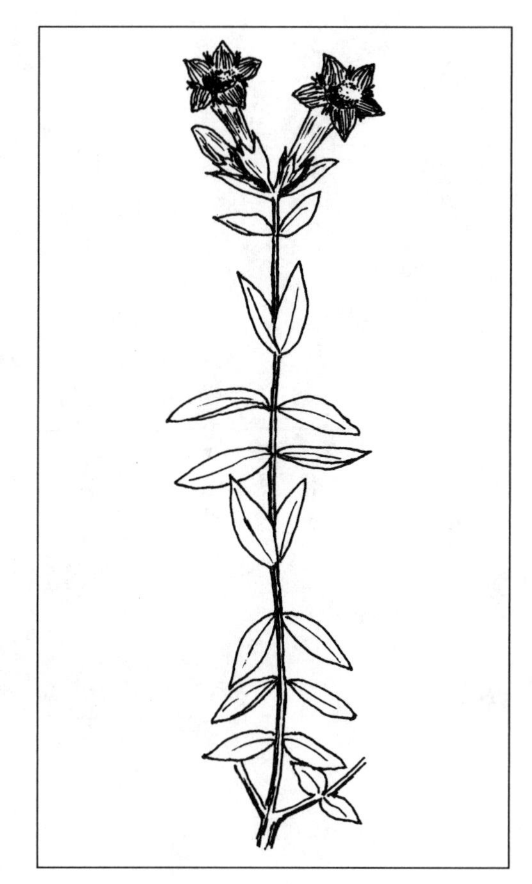

FIGURE 69. Gentian (*Gentiana* spp.).

FIELD KEY TO THE GENTIANS

1. Flowers solitary and terminal 2
1. Flowers usually more than 1 3

2. Flowers greenish outside, white inside .... *G. aquatica*
2. Flowers blue .... *G. prostrata*

3. Flowers yellowish white, spotted or streaked with purple....... *G. algida*
3. Flowers blue to purple .... *G. affinis*

Natural History Notes and Uses. Moore (1979) suggests that gentians are perhaps the best stomach tonics. As a bitter, gentians excite the flow of gastric juices, thereby promoting an appetite and aiding in digestion. The root or chopped herb is steeped and drunk before a meal. The herbage and roots of most species are bitter. Craighead et al. (1963), in discussing *G. calycosa*, mention that European and Asian gentians were used medicinally and that early American settlers used them in much the same way as a tonic. Gentians contain some of the most bitter compounds known, against which the bitterness of other substances is scientifically measured.

Fringed Gentian (genus *Gentianella*)

AUTUMN DWARF GENTIAN (*G. amarella*)

PERENNIAL FRINGED GENTIAN (*G. barbellata = Gentianopsis b.*)

WINDMILL FRINGED GENTIAN (*G. detonsa = Gentianopsis thermalis*)

FOUR-PART DWARF GENTIAN (*G. propinqua*)

DANE'S DWARF GENTIAN (*G. tenella*)

Description and Habitat. Fringed gentians are glabrous annuals with basal and cauline (stem) leaves. The calyx tube is shorter than the lobes, and the corolla has spreading lobes that are shorter than the tube.

FIELD KEY TO THE FRINGED GENTIANS

1. Corolla lobes usually 4 and fringed on margins; corolla usually over ¾ inch long, not fringed in throat from base of lobes 2
1. Corolla lobes 4 or 5, margins entire or almost; corolla less than ¾ inch, fringed in throat from base of lobes 3

2. Plants perennial .... *G. barbellata*
2. Plants annual .... *G. detonsa*

3. Base of corolla lobes not fringed on inner surface .... *G. propinqua*
3. Base of corolla lobes fringed on inner surface 4

4. Flowers usually crowded; calyx not lobed to base........... *G. amarella*
4. Flowers on long, naked pedicels; calyx lobed to base.......... *G. tenella*

Natural History Notes and Uses. These gentians have a long history of use as an herbal bitter in the treatment of digestive disorders. They are especially useful in states of exhaustion from chronic disease and in all cases of debility, weakness of the digestive system, and lack of appetite. They are one of the best strengtheners of the human system and an excellent tonic to combine with a purgative in order to prevent its debilitating effects.

G. propinqua was used as a cold remedy. A decoction of the leaves, stems, and flowers was also taken for colds and cough. Another species (*G. quinquefolia*) provided an infusion for diarrhea, and the liquid from the root was used for hemorrhages (Gottesfeld 1992).

G. amarella has also been used as a source of the medicinal gentian root. The root is an anthelmintic, anti-inflammatory, antiseptic, bitter tonic, cholagogue, emmenagogue, febrifuge, refrigerant, and stomachic. It is harvested in the autumn and dried for later use. The roots of plants that have not flowered are likely the richest in medicinal properties.

Swertia (genus *Swertia*)

STAR GENTIAN (*S. perennis*)

SHOWY OR GREEN FRASERA (*S. radiata = Frasera speciosa*)

Description and Habitat. Swertias are perennial herbs with opposite or whorled leaves. The flowers are bell-shaped and are densely aggregated into pyramid-shaped panicles. The two species in the area can be found in dry, open areas or meadows up to the subalpine zone. The genus was named for Emanuel Sweerts (1552–1612), a Dutch botanist, artist, and author of a florilegium, or book about flowering plants.

FIELD KEY TO THE SWERTIAS

1. Corolla lobes 4, green to white or yellow .... *S. radiata*
1. Corolla lobes 5, blue or purplish........................... *S. perennis*

Natural History Notes and Uses. The fleshy root of showy frasera can be eaten raw, roasted, or boiled. But because the root is very bitter, we suggest mixing it with salad greens. Medicinally, a poultice of the powdered root is applied externally to reduce fever. In small doses the root is a good laxative, though large doses may be fatal. Lard and ground roots were applied to the head for killing lice. An infusion of a related species, *Frasera albicaulis,* was used to treat infected sores.

GERANIUM FAMILY (GERANIACEAE)

The geranium family has 11 genera and 800 species distributed worldwide. *Geranium* and *Erodium* are native to the United States. Members of this family are 5-merous plants (5 petals, 5 or 10 stamens, pistil of 5 parts). The seed pod resembles the head and beak of a stork or crane, hence the common name "storksbill": the seeds are in the short, thickened "head," and the style is elongated into the pointed "beak." The family is of no real economic importance, except as a source of ornamentals, primarily from the cultivated geranium (*Pelargonium*), a tropical genus well developed in South Africa.

FIELD KEY TO THE GERANIUM FAMILY

1. Leaves pinnately compound; stamens 5....................... *Erodium*
1. Leaves palmately divided or lobed; stamens 5 or 10 .... *Geranium*

RED-STEMMED STORKSBILL (*Erodium cicutarium*)

Description and Habitat. Red-stemmed storksbill is a low-growing annual with mostly basal, finely dissected, fern-like, pinnately divided leaves. The flowers are small and pink and mature into the distinctive "stork's bill" fruit. This is an introduced plant that is widespread on disturbed sites at low to middle elevations.

Natural History Notes and Uses. In fruit the styles form tails that are 1–2 inches long and become spirally twisted. By twisting and untwisting with varying amounts of moisture, these tails serve to drive the fruits into the soil.

The leaves can be eaten raw in salads or cooked as a potherb. They are particularly palatable when picked young and have a parsley-like taste. We find that it nicely complements an otherwise bland wild salad and provides a good source of vitamin K.

It is uncertain whether other species of *Erodium* are edible, and sampling them is not recommended.

The species has a reputation for being a diuretic, astringent, and anti-inflammatory. The entire plant was used in a warm-water bath to relieve the pains of rheumatism. Leaves were also used in a hot tea to increase urine flow and to increase perspiration.

Wild Geranium (genus *Geranium*)

BICKNELL'S CRANESBILL (*G. bicknellii*)

RICHARDSON'S GERANIUM (*G. richardsonii*)

STICKY GERANIUM (*G. viscosissimum*)

FIGURE 70. Wild geranium (*Geranium* spp.).

Description and Habitat. Geraniums are annual or perennial herbs that are hairy. The leaves are mostly basal, and the flowers are showy, with 5 petals and sepals and 10 stamens. The mature fruits are spirally coiled. They can be found in dry, open forests or wet meadows. The name *geranium* is derived from the Greek word *gernion,* meaning "crane."

FIELD KEY TO THE WILD GERANIUMS

1. Plants annual or biennial; petals less than ½ inch long	*G. bicknellii*
1. Plants perennial from an enlarged woody rootstalk; petals more than ½ inch long	2
2. Petals white with dark veins; stems and leaves are not sticky	*G. richardsonii*
2. Petals normally rose-purple to pink in color; stems and leaves are sticky at least above and sometimes throughout	*G. viscosissimum*

Natural History Notes and Uses. The anthers mature earlier than the stigmas—the outer 5 first and then the inner 5. After the pollen has been shed, the anthers drop off. Only then does the pistil elongate and spread apart the 5 sticky stigmas, to be pollinated by insects with pollen from other flowers, thereby preventing self-pollination. At maturity the 5 parts of the fruits separate from the central axis at the base and curl elastically upward so suddenly as to discharge the seeds.

The leaves and flowers of most species can be eaten, but because of their astringent properties and texture, they are not a choice edible. We find that they are best when tossed in with other greens in salads or steamed as potherbs. In any case, the leaves are better treated as a filler to stretch supplies of other, more tasty and less abundant greens. The leaves can also be chopped and added to soups, blending flavors and making the leaves more acceptable. The leaves toughen with age but are still palatable in stews. Geranium leaves look similar to those of monkshood (*Aconitum* spp.), all parts of which are poisonous, so positive identification of the flowerless plants is important. Harvest leaves and roots from plants identified with flowers.

A leaf or root tea of *G. richardsonii,* which is one of the most widespread western species and frequently hybridizes with other species, can be used as a gargle for a sore throat. The root sliced fresh can be used as first aid for gum or tooth infections when applied directly on the area of pain.

The herbaceous part of *G. viscosissimum* was used as an astringent and styptic and internally for diarrhea and hemorrhages. The plant is high in tannins, providing astringent remedies important in traditional medicine for the emergency treatment of injuries and diarrhea. A hot poultice of boiled leaves was used for bruises and skin problems. The green crushed leaves can be applied to relieve pain and inflammation.

CURRANT AND GOOSEBERRY FAMILY (GROSSULARIACEAE)

This family consists of a single genus (*Ribes*) with approximately 150 species. All are shrubs with palmately lobed leaves. Some species are armed with spines. The family is a source of ornamentals and edible fruits. In many old field guides *Ribes* is included as a member of the Saxifragaceae (saxifrage family).

Currant, Gooseberry (genus *Ribes*)

AMERICAN BLACK CURRANT (*R. americanum*)

GOLDEN CURRANT (*R. aureum*)

WAX CURRANT (*R. cereum*)

NORTHERN BLACK CURRANT (*R. hudsonianum*)

WHITESTEM GOOSEBERRY (*R. inerme*)

PRICKLY CURRANT (*R. lacustre*)

GOOSEBERRY CURRANT (*R. montigenum*)

CANADIAN GOOSEBERRY (*R. oxyacanthoides*)

STICKY CURRANT (*R. viscosissimum*)

Description and Habitat. The many species of *Ribes* are shrubs. The species that have prickles on the stems and bristles on the fruit are commonly called gooseberries. Those without prickles on the stem or bristles on the fruit are currants. The leaves are palmately veined and shallowly or deeply lobed. The 5 petals are smaller than the sepals and usually narrowed to a claw-like base. The fruit is a berry.

FIGURE 71. Gooseberry (*Ribes* spp.).

FIELD KEY TO THE CURRANTS AND GOOSEBERRIES

1. Stems with spines or prickles (gooseberries) 2
1. Stems without spines or prickles (currants) 5

2. Racemes 1–4-flowered; calyx tube cylindrical or bell-shaped. 3
2. Racemes several-flowered; calyx tube salverform 4

3. Leaves glabrate or downy; fruit smooth . . . . *R. inerme*
3. Leaves closely pubescent; fruit with stiff hairs. . . . . *R. oxyacanthoides*

4. Leaves glabrate; fruit glandular-bristly, deep purple or black . . . . *R. lacustre*
4. Leaves pubescent, maybe glandular; fruit glandular-bristly, reddish . . . . *R. montigenum*

5. Flowers yellow . . . . *R. aureum*
5. Flowers not yellow 6

6. Berry glandular-bristly . . . . *R. viscosissimum*
6. Berry smooth 7

7. Calyx tube cylindrical, with dilated base . . . . *R. cereum*
7. Calyx tube bell-shaped 8

8. Racemes drooping . . . . *R. americanum*
8. Racemes erect or ascending . . . . *R. hudsonianum*

Gamma-linolenic acid is an essential fatty acid (EFA) in the omega-6 family that is found primarily in plant-based oils. EFAs are essential to human health but cannot be made in the body. For this reason, they must be obtained from food. EFAs are needed for normal brain function, growth and development, bone health, stimulation of skin and hair growth, regulation of metabolism, and maintenance of reproductive processes.

Natural History Notes and Uses. The berries of almost all species of *Ribes* are edible raw, and none are known to be poisonous. However, we have come across some unpalatable species, berries with an unpleasant odor and a taste to match. The berries are high in vitamin C and one of the richest plant sources for copper. One method of collecting them in bulk is to shake the bushes over sheets of plastic or blankets. Those that are too sour or spiny become more palatable if they are cooked or dried. For the fruits with bristles, one can also roll the berries on hot coals in a basket until the bristles have been singed off. When dried, the berries are a great trail snack. The dried berries can also be mixed with meat to make pemmican. The berries contain enough natural pectin to make jelly. The seeds also contain large quantities of gamma-linolenic acid, and many herbalists use this oil to treat skin conditions, asthma, arthritis, and premenstrual syndrome. The nectar-filled flowers are considered good trail snacks. The wood is ideal for making arrow shafts. The leaves of currants and gooseberries may be added to herbal tea blends. The leaves should be fresh or thoroughly dried, as wilted leaves may be toxic.

WATER-MILFOIL FAMILY (HALORAGACEAE)

This family occurs throughout the world but mostly in the Southern Hemisphere. Its members are generally aquatic herbs with simple or pinnatifid leaves that are opposite, alternate, or whorled.

Water-milfoil (genus *Myriophyllum*)

ANDEAN WATER-MILFOIL (*M. quitense*)

WHORLLEAF WATER-MILFOIL (*M. verticillatum*)

Description and Habitat. The genus name is from the Greek *myrios* ("many") and *phyllon* ("leaf"), referring to the finely divided leaves. There are about 40 species of these submerged aquatic and terrestrial herbs with pinnate leaves and spikes of small, wind-pollinated flowers.

Natural History Notes and Uses. The rhizomes were frozen for future use, eaten raw, fried in grease, or roasted. The rhizomes are sweet and crunchy and a much-relished food. They were also important during famine periods.

MARE'S-TAIL FAMILY (HIPPURIDACEAE)

The family consists of a single genus, *Hippuris*. The genus was at one time assigned to the Haloragaceae (water-milfoil family), but it is not closely related.

MARE'S-TAIL (*Hippuris vulgaris*)

Description and Habitat. Mare's-tail is a common plant in the main mountain chains of the western United States. The plant at first glance resembles an immature horsetail (*Equisetum* spp.), but the two are unrelated. Horsetails reproduce by spores and have stems that can be quickly pulled apart. The flowers of mare's-tail are small and inconspicuous. The plant is found at the margins of shallow waters in ponds, lakes, and streams. It can also be found in marshy and swampy areas and along roadsides and irrigation ditches.

Natural History Notes and Uses. The whole plant is edible when prepared as a potherb. The plant parts are tender and can be gathered in any stage, even in winter. Ancient herbalists are said to have employed mare's-tail for internal and external bleeding (Harrington 1967).

WATERLEAF FAMILY (HYDROPHYLLACEAE)

The 20 genera and 270 species in the waterleaf family are distributed worldwide, except for Australia. The western United States appears to be the main center of diversity. Only a few members in the family are cultivated.

FIELD KEY TO THE WATERLEAF FAMILY

1. Flowers arranged singly or in a few-flowered, terminal inflorescence .... *Nemophila*
1. Flowers not solitary... 2

2. Flowers densely arranged in a head-like cluster below the leaf blades; plants with fibrous roots.......... *Hydrophyllum*
2. Flowers rarely ball-like when expanded and always above the leaves; plants taprooted..................... *Phacelia*

FIGURE 72. Ballhead waterleaf (*Hydrophyllum capitatum*).

BALLHEAD WATERLEAF (*Hydrophyllum capitatum*)

Description and Habitat. These somewhat fleshy perennial herbs have leaves that are pinnately divided. The flowers are white to bluish. The plants are found in moist soils from the foothills to the alpine environment.

Natural History Notes and Uses. The young shoots, leaves, and flowers of *Hydrophyllum* can be eaten raw or, along with the roots, may be cooked and eaten. We find them exceptionally good in salads or when eaten as a trail nibble, although their texture takes some getting used to.

The leaves can be used as a protective dressing for minor wounds and are slightly astringent. A poultice can be used for insect bites and other minor skin irritations.

BASIN NEMOPHILA (*Nemophila breviflora*)

Description and Habitat. Nemophilas are mostly annuals with stems that are diffuse, weak, and sometimes prostrate. The leaves are mainly opposite, and the flowers occur in the upper axils of the leaves. The style is deeply bifid. The genus name is from the Greek for "grove loving," referring to the woodland habitat of many species in this genus.

Natural History Notes and Uses. Some native tribes used the roots to prepare a decoction to cure asthma.

Phacelia (genus *Phacelia*)

FRANKLIN'S PHACELIA (*P. franklinii*)

SILVERLEAF PHACELIA (*P. hastata*)

VARILEAF PHACELIA (*P. heterophylla*)

SILKY PHACELIA (*P. sericea*)

Description and Habitat. Phacelias include herbaceous annuals, biennials, and perennials with various degrees of hairiness. The flowers are 5-parted and spirally coiled, with stamens extending beyond the corolla. The flowers bloom progressively from base to tip.

FIELD KEY TO THE PHACELIAS

1. Leaves almost all entire or some with 1 or 2 pairs of entire lobes or leaflets below middle 2
1. Leaves coarsely toothed or pinnately lobed or divided. 3

2. Plants biennials or short-lived perennials from a taproot, usually a single erect stem, usually over 18 inches tall . . . . . *P. heterophylla*
2. Plants perennials from a taproot bearing a branched caudex with several stems, usually less than 18 inches . . . . *P. hastata*

3. Plants annual or biennial; usually with single stem or the central stem surrounded by smaller stems; corolla glabrous within; filament scarcely exserted . . . . *P. franklinii*
3. Plants perennial, usually several-stemmed; corolla hairy within; filaments long exserted . . . . *P. sericea*

Natural History Notes and Uses. At least one species, *P. ramosissima* (branching phacelia), can be cooked and used as greens. Strike (1994) suggests that the stems and leaves of *Phacelia* may have been eaten raw but were most likely cooked. The boiled roots of *P. ramosissima* were used to cure coughs and colds and to alleviate lethargy. A decoction was also used as an emetic and to relieve stomachaches. Medicinally, the whole plant of *P. heterophylla* (varileaf phacelia) was dried and pulverized, then used to poultice wounds (Strike 1994).

ST. JOHN'S-WORT FAMILY (HYPERICACEAE = CLUSIACEAE)

There are 40 genera and 1,000 species worldwide. The family is of little economic importance in North America. A few species are used as ornamentals.

St. John's-wort (genus *Hypericum*)

ST. JOHN'S-WORT (*H. formosum* = *H. scouleri*)

COMMON ST. JOHN'S-WORT (*H. perforatum*)

FIGURE 73. St. John's-wort (*Hypericum* spp.).

Description and Habitat. The two perennial species in the area have yellow flowers and small, translucent glands on the leaves and petals. They can be found in moist areas at various elevations.

FIELD KEY TO THE ST. JOHN'S-WORTS

1. Flowers in a flat-topped cluster; leaves narrowly oblong . . . *H. perforatum*
1. Flowers in axillary cymes; leaves broadly elliptic. . . . . *H. formosum*

Natural History Notes and Uses. The largest and most widespread of the species is *Hypericum perforatum*. Weedon (1996) indicates that the leaves may be eaten fresh or may be dried and ground to a flour that can be used like acorn meal. However, only a small amount of the herbage should be consumed.

Despite its reputation as a weed, St. John's-wort may have much to offer as a medicinal plant. A number of clinical studies strongly suggest the plant may be effective in treating depression. Laboratory studies reveal that St. John's-wort has at least two compounds, hypericin and pseudohypericin, that are active against retroviruses. Consequently, the plant is being closely investigated in acquired immune deficiency syndrome (AIDS) research (Tilford 1997). The fresh flowers of St. John's-wort, in tea, tincture, or olive oil, were once a popular domestic medicine for the treatment of external ulcers, wounds, sores, cuts, and bruises. The ancient alleged magical properties of St. John's-wort were partly due to the fluorescent red pigment, hypericin, which oozes like blood from the crushed flowers. The red dye and extracts are used in cosmetics.

Caution: Craighead et al. (1963) indicate that white-skinned animals feeding on these plants develop scabby sores and a skin itch. Apparently, the plants contain photosensitive toxins and alkaloids. Therefore, ingestion is not recommended in large quantities.

MINT FAMILY (LAMIACEAE = LABIATAE)

The mint family has approximately 180 genera and 3,500 species worldwide, with the Mediterranean region being the chief area of diversity. All have epidermal glands that exude an odor when rubbed, not necessarily pleasant and mint-like. Labiatae refers to the corolla shape, which resembles 2 lips (*labia*). This family is of considerable economic importance as a source of numerous ornamentals, aromatic oils, and medicines; it also has a few weedy genera.

FIELD KEY TO THE MINT FAMILY

1. Anther-bearing stamens 2 2
1. Anther-bearing stamens 4 4

2. Connective of anther laterally elongated........................ *Salvia*
2. Connective short.. 3

3. Flowers large, in one or a few dense capitate clusters.......... *Monarda*
3. Flowers small, in many axillary clusters .... *Hedeoma*

4. Stems white-woolly; calyx teeth 10 .... *Marrubium*
4. Stems not white-woolly; calyx teeth 5 or fewer 5

5. Calyx 2-lipped, lips being entire, and calyx with a crest-like swelling on the upper side .... *Scutellaria*
5. Calyx not 2-lipped, or if so, lips are lobed or toothed; there is no crest present.. 6

6. Upper pair of stamens as long as lower pair................... *Mentha*
6. Upper pair of stamens not the same length as lower pair 7

7. Upper pair of stamens longer than lower pair........................ 8
7. Upper pair of stamens shorter than lower pair 10

8. Anther sacs parallel or almost............................. *Agastache*
8. Anther sacs divergent 9

9. Plant perennial; corolla well exserted from calyx .... *Nepeta*
9. Plant annual or biennial; corolla barely exserted........ *Dracocephalum*

10. Calyx 2-lipped, closed in fruit.............................. *Prunella*
10. Calyx almost equally 5-toothed 11

11. Leaves small, about ½ inch long .... *Monardella*
11. Leaves large, 1½ to 2 inches long............................. *Stachys*

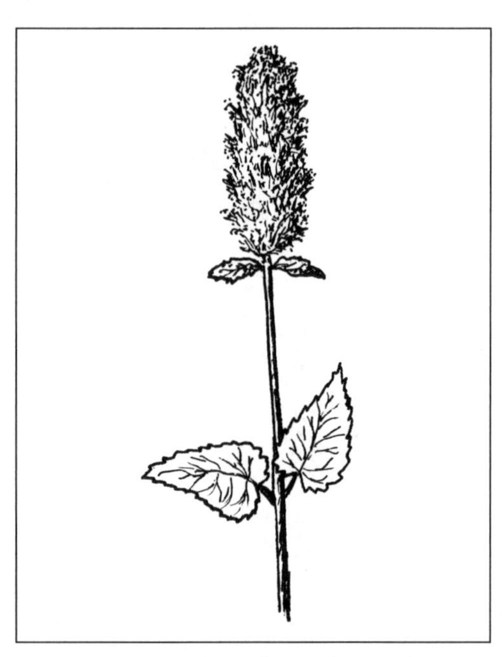

FIGURE 74. Nettleleaf giant hyssop (*Agastache urticifolia*).

NETTLELEAF GIANT HYSSOP (*Agastache urticifolia*)

Description and Habitat. This perennial grows 3–6 feet tall. The leaves are opposite, ovate in shape, 1–3 inches long and 1½ inches wide, and coarsely toothed on the margins. The flowers occur in dense whorls and form a terminal spike up to 6 inches long. The calyx is green or rose, and the corolla is 2-lipped, rose or violet. Giant hyssop grows in moist places and flowers from June to August. *Agastache* is from the Greek *agan*, meaning "much," and *stachys*, meaning "ear of grain," referring to the flower cluster.

Natural History Notes and Uses. The seeds of giant hyssop may be eaten raw or cooked, and the leaves can be used as a tea or for flavoring stews. The Miwoks of California drank an infusion of the leaves to relieve rheumatic pain, indigestion, and stomach pains. The plant is said to have mild sedative qualities. Mashed leaves were made into a poultice for swellings.

AMERICAN DRAGONHEAD (*Dracocephalum parviflorum*)

Description and Habitat. This biennial or short-lived perennial grows from a taproot. The lower stem leaves are small and wither early, and the stalked upper leaves are lance-shaped with spine-tipped teeth. The purple flowers are small in a dense, spike-like inflorescence. Dragonhead occurs in open, low-elevation, moist sites. The plant requires disturbance, such as fire or scarification, to germinate.

Natural History Notes and Uses. Kirk (1975) indicates that the Havasupai Indians of northern Arizona made a flour from the seeds.

False Pennyroyal (genus *Hedeoma*)

DRUMMOND'S FALSE PENNYROYAL (*H. drummondii*)

ROUGH FALSE PENNYROYAL (*H. hispida*)

Description and Habitat. False pennyroyal has axillary clusters of small, tubular, lavender or purplish flowers; the calyx is 2-lipped, with 3 short and 2 longer teeth; the flowers appear from June to October. The whole plant is pleasantly aromatic. The generic name *Hedeoma* is derived from the Greek *hedys*, "sweet," and *osme*, "scent." There are about 25 species of this genus in the Western Hemisphere.

FIELD KEY TO THE FALSE PENNYROYALS

1. Plant a hispid-pubescent annual . . . . *H. hispida*
1. Plant an ash-grayish perennial . . . . *H. drummondii*

Natural History Notes and Uses. False pennyroyal was used as a carminative, diaphoretic, emmenagogue (to promote menstruation), antispasmodic, mild sedative, sudorific, stimulant, and aromatic by Native Americans. An infusion of *H. hispida* was used for colds and as a tonic appetizer for the sick (Gilmore 1919). *H. drummondii* was used in soup and as a remedy for flu (Rogers 1980; Vestal 1952).

HOREHOUND (*Marrubium vulgare*)

Description and Habitat. This woolly perennial herb has bitter sap. The leaves are wrinkled and toothed. The flowers are small and white and occur in dense whorls. Horehound is a weed of waste places and fields at the lower elevations.

Natural History Notes and Uses. Horehound is listed as a stimulant, tonic, expectorant, and diuretic. The plant was highly valued by ancient Egyptian priests and Romans, the former calling it the "Seed of Horus" and "eye of the star."

The most famous use of this plant is horehound candy, which is used to soothe sore throats and coughs. A tea from the dried leaves and flowers is also used, but because of the extreme bitterness of the herb, it is obvious why it tastes better in the form of a candy. Other medicinal uses of the plant include making a warm infusion that promotes perspiration and the flow of urine. When taken cold, this infusion will expel worms.

Mint (genus *Mentha*)

WILD MINT (*M. arvensis*)

SPEARMINT (*M. spicata*)

Description and Habitat. All species are distinctly aromatic perennial herbs with rhizomes. The flowers are arranged in whorls. *Mentha* comes from Mintho, the mistress of Pluto, ruler of Hades. Pluto's jealous queen, Proserpine, on learning of Mintho, trampled her, transforming her into a lowly plant forever to be walked upon. Pluto made this horrible fate more tolerable by willing that the more the plant was trampled, the sweeter it would smell.

FIELD KEY TO THE MINTS

1. Flowers in dense axillary whorls.......................... *M. arvensis*
1. Flowers in terminal spikes............................... *M. spicata*

HERB TEAS

Plants are the source of practically all the beverages consumed by humankind (water and milk are the exceptions). Plants provide flavor, color, and aroma for endless variety and pleasure. They also provide nutrients for health. Most herbal "teas" are infusions made by pouring boiling water over herb leaves or flowers and allowing them to steep 5–10 minutes to release the herb's aromatic oils. A general rule is one teaspoon of dried herb, or 3 teaspoons of fresh crushed herb, per cup of water. To make a stronger tea, add more of the herb rather than steeping the tea longer (long steeping makes the tea bitter). Experiment by combining various herb teas for interesting flavor results.

Natural History Notes and Uses. The fresh or dried leaves of both *M. arvensis* and *M. spicata* can be steeped in hot water for a tea. They have also been used as flavoring agents for soups, meat, and pemmican. The young leaves can also be added to salads and soups. The plants are high in vitamins A, C, and K and the minerals iron, calcium, and manganese. It is an appetite stimulant and digestive aid. The leaf tea is considered medicinal and was used for colds, stomachaches, fevers, headaches, insomnia, and nervous tension. Crushed leaves were used by some Native Americans to poultice swellings and bruises.

Pure mint oil is a multimillion dollar industry. It is added to shampoos, massage oil, salves, and soaps, as well as medicines, foods, and liqueurs. It takes approximately 300 pounds of mint to yield 1 pound of oil.

WILD BERGAMOT BEEBALM (*Monarda fistulosa*)

Description and Habitat. Beebalm grows from a spreading rhizome, and the leaves are sharp-tipped and glandular. The flowers are lavender or rose-purple and aggregated in dense terminal heads. The sepals are fused into a ribbed tube, the top of which has 5 short, spiny lobes covered with dense, white hairs. It is found on open, dry to moist slopes and rockslides and in forest openings at the lower elevations.

Natural History Notes and Uses. *Monarda* is a North American genus of about 15–20 species. It was dedicated by the famous Swedish botanist Carl von Linné (Linnaeus) to Nicholás Monardes (1493–1588), a Spanish physician-botanist and the author of many tracts about useful New World plants. The specific name, *fistulosa*, means "tubular" in botanical Latin, in reference to the flowers.

The young leaves and leaf buds of beebalm can be used as a seasoning in salads, much as you would use fresh oregano. Moderation, however, is advised. Although the older plants can also be used, they are very bitter and it would be best to dry them first.

A tea can be made and inhaled to ease bronchial complaints and colds. Beebalm contains thymol, which has been used as a stimulant and to relieve digestive flatulence and nausea. Thymol is a white crystalline aromatic compound derived from thyme oil and other oils or made synthetically and used as an antiseptic, a fungicide, and a preservative.

PACIFIC MONARDELLA (*Monardella odoratissima* = *M. glauca*)

Description and Habitat. Monardellas are perennial herbs with leafy stems. The flowers are commonly rose-purple in dense terminal, globose clusters. The bracts are leaf-like but usually colored. The genus name honors Spanish botanist and physician Nicholás Monardes (1493–1588). This species is also known as mountain pennyroyal.

Natural History Notes and Uses. The leaves and stalks of *M. odoratissima* were eaten, and a thirst-quenching tea was made from the leaves and flower heads. Medicinally, a tea made from the inflorescence was used for colds and fevers, digestive upset, and blood purification. A tea made from the whole plant, used in the first stages of a cold when fever is present, is said to help relieve elevated temperatures and toxins through sweat. Leaves and flowers rubbed on exposed skin repel mosquitoes and other biting insects (Camazine and Bye 1980).

FIGURE 75. Pacific monardella (*Monardella* spp.).

PENNYROYAL INSECT REPELLENT

You will need the following ingredients:

- 2 cups mountain pennyroyal leaves and flowers
- ½ cup sagebrush (*Artemisia*) leaves
- ½ cup fresh yarrow (*Achillea*) leaves and flowers
- 2 cups rubbing alcohol
- ½ cup water (preferably distilled)
- 1 tsp. jojoba oil

Place all the ingredients except for the water and oil in a quart-sized jar and allow it to stand for 2–3 weeks. Strain the liquid concoction through a coffee filter and discard the solid material. Then add the water and jojoba oil into the strained liquid. Pour the liquid into a spray bottle. To use, shake well and apply as needed.

CATNIP (*Nepeta cataria*)

Description and Habitat. Catnip is a taprooted perennial that feels like felt. The leaves are triangular-shaped and coarsely toothed. The flowers are blue or yellowish white in terminal, spike-like inflorescences. This introduced species from Europe is now widespread across North America and occurs in waste and disturbed places and along irrigation canals at the lower elevations.

Natural History Notes and Uses. Catnip is a harmless "high" for felines. Although many cats will eat it, researchers suggest that they may be reacting to the smell rather than the taste. Cats will bite, chew, rub against, and roll in catnip to release the volatile oil trapped in the leaves. About 80 percent of adult cats, including lions, pumas, and leopards(!), react to this irresistible, intoxicating, analgesic soporific. However, the tendency to like or ignore catnip is an inherited trait, and some cats appear to be immune to its influence. Catnip is harvested when this essential oil production reaches its peak, and the leaves and fragrant flowers are carefully air-dried to preserve essential oils at their best.

The nutritious young leaves and buds can be added to salads. The dried leaves and flowers make an excellent tea and are high in trace minerals and vitamins. Taken as a hot infusion, catnip promotes sweating and is beneficial for colds, flu, fevers, and infectious childhood diseases. It is soothing to the nervous system and calming to the stomach. It aids with flatulence, diarrhea, and colic. It is sometimes used as an enema to cleanse and heal the lower bowel (use in diluted form). Catnip helps prevent miscarriage and premature birth and relieves morning sickness.

COMMON SELFHEAL (*Prunella vulgaris*)

Description and Habitat. This perennial grows 4–20 inches tall and has opposite, lanceolate-ovate leaves that are 1–2 inches long. The herbage is glabrous to short-pubescent. The flowers are in a dense, terminal spike in the axils of round, membranous, purple-tinged bracts. The calyx is purplish, and the corolla is 2-lipped, violet, and ⅜ to ¾ inch long. Self-heal is common in moist woods from May to September at various elevations. It is also called Hercules' all-heal because it is supposed that Hercules learned the herb and its virtues from Chiron.

Natural History Notes and Uses. The entire plant is edible, raw or cooked. However, we found that the young and tender plants collected in the early spring are best. The crushed leaves can be used fresh or dried to make a tea.

Historically, as the common name implies, the plant has been used as a medicine for almost everything. It is effective as an astringent, antispasmodic, tonic, and styptic. The tea of the dried plant was also used as a gargle for sore throat. Fresh plants can be made into an antiseptic poultice for bruises and scrapes, effective because of the high tannin content (Scully 1970).

Sage (genus *Salvia*)

WOODLAND SAGE (*S. nemorosa*)

LANCELEAF SAGE (*S. reflexa*)

Description and Habitat. *Salvia* has about 900 members, which can be annual or perennial herbs or shrubs. They have opposite, simple to pinnately divided leaves and whorls of tubular, 2-lipped flowers. The stamens work on a rocker mechanism: pollen is pressed onto the head of a visiting bee as it pushes against a sterile projection from the anther.

FIELD KEY TO THE SAGES

1. Plant perennial, more than 12 inches tall .... *S. nemorosa*
1. Plant annual, less than 12 inches tall .... *S. reflexa*

Natural History Notes and Uses. The genus name is from the Latin *salvare* ("to save" or "to heal"), alluding to the medicinal properties of many of the species. The seeds of perhaps all species of *Salvia* may be eaten raw or can be parched and ground into meal. The seeds can also be soaked in water for a flavorful drink. Leaves of any fragrant sage can be used as a tea or spice for soups and meats. Sage contains moderate amounts of vitamins A and C and can be added fresh to salads and sandwiches.

Tilford (1993) indicates that the above-ground parts are antiseptic, astringent, hemostatic, alterative, and tonic; they make a strong, effective topical disinfectant and cleansing wash for abrasions, contusions, and chafed skin (Moore 1979). They are also an effective gargle for sore throat and congested sinuses (Moore 1979).

Lanceleaf sage is considered a toxic plant. Reported cases of poisonings in the United States are limited to cattle and horses that have consumed hay containing the plant. Experimental feeding trials have shown that sheep are also susceptible. It is not known how much of the plant material must be eaten to cause toxicity; in one confirmed case, alfalfa hay contained about 10 percent lanceleaf salvia.

Note: *Salvia* and some species of *Artemisia* are often mistakenly considered related, both being called "sage." Both are aromatic, but they belong to different families. Sage (*Salvia*) is a member of the mint family, whereas wormwood (*Artemisia vulgaris*) and sagebrush (*Artemisia tridentata*) are in the sunflower family. Mints have opposite leaves, whereas these *Artemisia* species have alternate leaves.

MARSH SKULLCAP (*Scutellaria galericulata*)

Description and Habitat. The genus name comes from the Latin *scutella* and means "small dish," referring to the appearance of the sepals in the fruit. There are about 300 species of rhizomatous herbs and subshrubs. The flowers are curved, tubular, and 2-lipped.

Natural History Notes and Uses. *Scutellaria galericulata* has nervine-related therapeutic properties and has been used for general restlessness and in acute or chronic cases of nervous tension or anxiety. The calming effects are said to be mild but reliable. A strong tea was also made of *S. lateriflora* (blue skullcap) and was used as a sedative, nerve tonic, and antispasmodic for all types of nervous conditions. All the species contain scutellarin, the primary active compound, whose sedative and antispasmodic properties have been confirmed. Other species may have similar qualities.

MARSH HEDGE-NETTLE (*Stachys palustris* = *S. pilosa*)

Description and Habitat. Hedge-nettle has no stinging hairs, as do the true nettles (*Urtica* spp.), but resembles them before flowering. The genus name is from the Greek *stachys* ("spike") and refers to the inflorescence. There are about 300 species of perennial herbs and shrubs with wrinkled leaves.

Natural History Notes and Uses. The leaves and flowers are edible, but because of their fuzzy texture and bitter taste, we find them unpleasant. The tubers can be eaten raw, cooked, or pickled and are best if collected in the autumn. Other species may be edible, but this has not been confirmed. Strike (1994) states generally that *Stachys* tubers were eaten.

The leaves may be soaked in water for a few minutes and used as a poultice (Vizgirdas 2003a). Additionally, an infusion of fresh leaves can be used as a wash for sores and wounds.

BLADDERWORT FAMILY (LENTIBULARIACEAE)

There are five genera and 300 species in this family worldwide. Two genera, *Utricularia* and *Pinguicula*, are native to the United States. They are described as annual or perennial herbs of moist and aquatic habitats. The carnivorous species in this family, which trap their prey by means of sticky leaves and bladders, are sometimes cultivated as oddities.

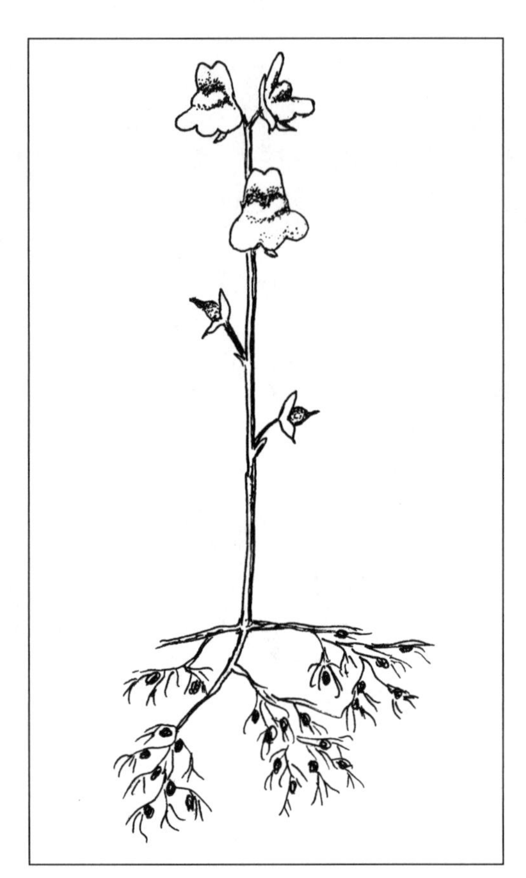

FIGURE 76. Bladderwort (*Utricularia minor*).

INSECTIVOROUS PLANTS

Insectivorous plants entrap and digest insects to supplement their nutrition. They usually grow in such habitats as acid swamps, calcareous springs, and bogs, which are often deficient in essential nutrients such as nitrogen. Nitrogen is obtained from the protein in the bodies of the prey; the passage of this element into the plant tissues has been confirmed by experiments with radioisotope tracers. Insectivorous plants in the area include *Drosera* and *Utricularia*.

Bladderwort (genus *Utricularia*)

COMMON BLADDERWORT (*U. macrorhiza*)

LESSER BLADDERWORT (*U. minor*)

Description and Habitat. About 250 species of bladderwort occur in the United States. They are aquatic or bog plants with submersed stems. The leaves are finely dissected, and the yellow flowers are strongly 2-lipped with a spur at the base. They are found growing in ponds, lakes, and sluggish streams at the low to middle elevations.

Natural History Notes and Uses. Bladderworts are carnivorous plants that entrap small insects in their bladders. The bladders are closed at the narrow end by valve-like doors that have stiff trigger hairs on the outer surface. When set, the bladders have a partial vacuum, and when a passing animal touches the bristles, the doors open, the walls of the bladder immediately expand, and the sudden inrush of water captures the prey. The process has been timed at 1⁄460 of a second. Enzymes digest the trapped victim.

The edibility and uses of bladderworts is unknown, but several species in this genus are reputed to have diuretic values and have been used to treat dysentery (Coon 1974).

LIMNANTHUS FAMILY (LIMNANTHACEAE)

The two genera and 12 species in this family are restricted to North America. They are annual herbs found in wet places.

FALSE MERMAID-WEED (*Floerkea proserpinacoides*)

Description and Habitat. False mermaid-weed is a slender annual with succulent stems. The leaves are pinnately divided into 3–5 oblong leaflets. The white flowers are stalked and borne singly in the leaf axils. The plant is found in moist, shaded habitats, especially under shrubs. The genus is named after a German botanist, H. G. Floerke. This inconspicuous little plant is easy to overlook except where it occurs in sizable colonies along the woodland floor. The delicate foliage is attractive and resembles moss or a bedstraw (*Galium*).

Natural History Notes and Uses. Because the flowers are inconspicuous, it isn't immediately obvious when false mermaid-weed is blooming; careful inspection at the right time during the spring will reveal the green sepals, the tiny white petals, and the stamens. It is fairly easy to identify this plant because each flower has only 3 sepals and 3 petals, and the alternate leaves are pinnately compound.

The root system consists of a slender branching taproot. This plant spreads by reseeding itself, and it often form colonies at favorable sites. The small inconspicuous flowers can attract flower flies and small bees. The foliage is not known to be toxic and is edible to mammalian herbivores, although little is known about floral-faunal relationships for this species. We have sampled the stems and leaves of this plant and found them to be rather spicy and an acceptable addition to wild salad. In Connecticut false mermaid-weed is considered an endangered species.

FLAX FAMILY (LINACEAE)

There are about 12 genera and 300 species in this family worldwide. The family is of some economic importance for its flax fibers, linseed oil, and ornamentals.

Flax (genus *Linum*)

KING'S FLAX (*L. kingii*)

PRAIRIE FLAX (*L. lewisii*)

Description and Habitat. Flaxes are much-branched annuals that have blue or rarely white flowers and alternate, sessile, and linear leaves. Flax has had value through the ages for its many uses, such as for thread, fabric, oil, paper money, and cigarette paper.

FIELD KEY TO THE FLAXES

1. Flowers blue; sepals without marginal glands . . . . *L. lewisii*
1. Flowers yellow; sepals with marginal glands . . . .*L. kingii*

Natural History Notes and Uses. Flax seeds contain a cyanide compound but are edible after roasting. The seeds also have high levels of a beneficial oil that contains essential fatty acids, plus they add an agreeable flavor to cooked foods. The crushed seeds have been used as a poultice for irritation, boils, and pain. An infusion of the stems is said to relieve stomachaches or intestinal disorders. The roots were also steeped to make an eye medicine. The stems are the source of linen fabric.

When the fibers of *Linum* are twisted together (spun), it is called yarn. It is strong and durable, and resists rotting in damp climates. It is one of the few textiles that has a greater breaking strength wet than dry. It has a long "staple" (individual strand length) relative to cotton and other natural fibers. The fiber in its unspun state is called flax. After it is spun into yarn, it becomes linen.

The standard measure of bulk linen yarn is the *lea*. A yarn having a size of 1 lea will give 300 yards per pound. The fine yarns used in handkerchiefs, for example, might be 40 lea, giving 12,000 yards (40 × 300) per pound. A characteristic often associated with linen yarn is the presence of *slubs*, or small knots that occur randomly along its length. However, these are actually defects associated with low quality. The finest linen has a very consistent diameter with no slubs. When being washed for the first time, linen shrinks significantly.

LOASA FAMILY (LOASACEAE)

There are about 15 genera and 250 species in this family, occurring chiefly in South America and the warmer parts of the North America. Various species of *Mentzelia* are endemic in the western United States.

Blazing Star (genus *Mentzelia*)

WHITE-STEM BLAZING STAR (*M. albicaulis*)

TEN-PETAL BLAZING STAR (*M. decapetala*)

BUSHY BLAZING STAR (*M. dispersa*)

SMOOTH-STEM BLAZING STAR (*M. laevicaulis*)

Description and Habitat. Members of this family have alternate, entire, or pinnately lobed leaves. The fruit is a capsule that opens at the top. There are many species of *Mentzelia* in the western United States. It is also called "stick-leaf" because of the leaves' barbed hairs, which readily cling to fabric.

FIELD KEY TO THE BLAZING STARS

1. Annuals with inconspicuous flowers	2
1. Biennials or perennials with showy flowers	3
2. Flower bracts mostly egg-shaped and inflorescence congested; leaves mostly entire to merely toothed	*M. dispersa*
2. Flower bracts mostly linear to lance-shaped and inflorescence not congested	*M. albicaulis*
3. Petals apparently 10; the inner 5, actually modified stamens, are slightly narrower than the outer 5	*M. decapetala*
3. Petals 5; 5 of outer stamens often flattened, much narrower than true petals	*M. laevicaulis*

Natural History Notes and Uses. *Mentzelia* was considered an important food source in many parts of the West. The seeds are edible after being parched and ground into flour and were often stored for future use. Murphey (1990:27) describes a type of "gravy" made from the seeds of *M. laevicaulis*: "the red seed is put into a hot frying pan and when the seeds turn a darker red, warm water is added and it is stirred till it thickens."

The Hopi Indians in the Southwest parched and ground the small, oily seeds of *M. albicaulis* into a fine, sweet meal and ate it in pinches (Hough 1897).

LOOSESTRIFE FAMILY (LYTHRACEAE)

Members of the loosestrife family are herbs, shrubs, or trees. The leaves are opposite or whorled. There are approximately 25 genera and 550 species widely distributed around the world. Seven genera are native to the United States. The family is a source of dyes and ornamentals.

PURPLE LOOSESTRIFE (*Lythrum salicaria*)

Description and Habitat. This erect perennial grows 20–72 inches tall and has pale green, glabrous stems. The leaves are alternate, linear to linear-oblong in shape, entire, and ⅜ to 1½ inches long. The flowers are purple, with cylindrical bases, and solitary in the leaf axils. There are 6 petals, each ¼ inch long, and 4–12 stamens. Loosestrife grows in moist places below 6,000 feet. It flowers from April to October.

Natural History Notes and Uses. Purple loosestrife is a Eurasian species that has become an aggressive wetland weed. It is also cultivated in gardens by some who are unaware of its potential. It is often called the "beautiful killer" because it can take over wetlands and displace native species. It appears to have some efficacy against gnats and flies and was reported to calm quarrelsome beasts of burden at the plow if placed on the yoke (Pojar and MacKinnon 1994).

A tea made from whole flowering plant of purple loosestrife, fresh or dried, is a European folk remedy for diarrhea and dysentery and a gargle for sore throats. It was also used as a cleansing wash for wounds. Experiments have shown that extracts of the plant stop bleeding and kill some bacteria. Other species appear to have been used by Native Americans. For example, *L. californicum* was used by the Kawaiisu Indians of California as a medicine and a dermatologic aid. The method, however, is not reported (Zigmond 1981). A related species, *L. hyssopifolia*, was used by Native Americans in California to expedite healing and to reduce inflammation of mucous membranes. It was also used as a shampoo, but the method is not reported (Zigmond 1981).

MALLOW FAMILY (MALVACEAE)

There are some 85 genera and 1,500 species in the mallow family, most of which occur in the tropics. Twenty-seven genera are native to the United States. The distinctive feature of this family is the uniting of the numerous stamen stalks to form a tube around the pistil that resembles a tree trunk, with the anthers and nonfused filaments corresponding to the branches and leaves. This "stamen tree" in the center of the flowers is almost a guaranteed characteristic of this family. The family is of moderate economic importance because of cotton fibers derived from the seeds of *Gossypium*, several ornamentals, and a few food plants.

Mallow is from the Greek word meaning "soft" and may refer to the soft, fuzzy leaves characteristic of so many plants in this family or to the sticky, soothing juice obtained from the roots of some species.

FIELD KEY TO THE MALLOW FAMILY

1. Flowers solitary or clustered .... *Malva*
1. Flowers in racemes or panicles 2

2. Flowers rusty red; stems mostly less than 12 inches high..... *Sphaeralcea*
2. Flowers lavender or whitish; stems mostly over 12 inches high 3

3. Upper stem leaves parted to the base or almost;
   stamens in 2 distinct rings around carpels .... *Sidalcea*
3. Upper stem leaves lobed less than halfway to base;
   stamens all united to form 1 ring around carpels .... *Iliamna*

STREAMBANK GLOBE MALLOW (*Iliamna rivularis*)

Description and Habitat. *Iliamna rivularis* is a perennial herb up to 4 feet tall. All green parts of the plant have star-shaped hairs. The flowers are lavender to pink and occur in dense clusters. The plant is usually found along streams, on moist sites, and on disturbed sites such as clearcuts.

Natural History Notes and Uses. One often sees this gorgeous plant in colonies, especially on burned or clearcut sites, but it spreads only by seed, not by rhizomes. The seeds have a very hard coat and can remain viable in the soil for centuries, until germination is triggered by wildfire.

The eight species of *Iliamna* in North America have a taxonomically complex history. After its conception in 1906, the genus was not recognized for some time, several species were placed into other genera, and the status of a few species was questioned. Six species are located in western North America, and two are found isolated in the East. Species in *Iliamna* are very similar morphologically with only a few characters distinguishing several as separate entities. Of the western species, four overlap in distribution (*I. crandalii*, *I. grandiflora*, *I. longisepala*, and *I. rivularis*), and their recognition as independent species has been questioned. The stems of some species were chewed as gum (Craighead et al. 1963).

Mallow (genus *Malva*)

COMMON MALLOW (*M. neglecta*)

CLUSTER MALLOW (*M. verticillata* = *M. crispa*)

Description and Habitat. The plants are distinguished by their fruit and seeds rather than their leaves and flowers. They are introduced annual or biennial herbs that are usually found in waste places at the lower elevations.

FIELD KEY TO THE MALLOWS

1. Stems prostrate spreading from the caudex; leaves obscurely lobed .... *M. neglecta*
1. Stems erect; leaves definitely lobed .... *M. verticillata*

Natural History Notes and Uses. The entire plant of *M. neglecta* is edible. The young leaves are particularly good in salads or cooked up as a potherb. The plant is often used to thicken soup, but because it is very mucilaginous, it may take a little getting used to. Eaten in large amounts, it may cause digestive disorder. The immature fruits (which look like cheese) can also be eaten raw or added to soups.

Medicinally, the bruised leaves of a related species, *M. parviflora*, can be rubbed on the skin to treat skin irritations. As a headache remedy, leaves or the whole plant can be mashed and placed on the forehead (Willard 1992). Leaf or root tea can be used for angina, coughs, bronchitis, and stomachaches (Foster and Duke 1990). The fresh or dried leaves were used as a soothing poultice (Moore 1979).

OREGON CHECKER MALLOW (*Sidalcea oregana*)

Description and Habitat. Checker mallows are annual herbs with lobed or divided leaves. The flowers are white to deep pinkish lavender, in terminal clusters. The genus name is a combination of two Latin words for mallow: *sida* and *alcea*.

Natural History Notes and Uses. Early settlers applied the whole plant of *S. glaucescens* as a poultice to ease the pain of insect stings and draw out thorns and splinters. *S. malvaeflora* was used as greens. The dried, mashed leaves were used to flavor manzanita berries. Kirk (1975) says *S. neomexicana* (New Mexico checker mallow) is edible as greens after cooking. Additionally, the thick sap of a species in the eastern United States was mixed with sugar and used to make marshmallows (Peterson 1978).

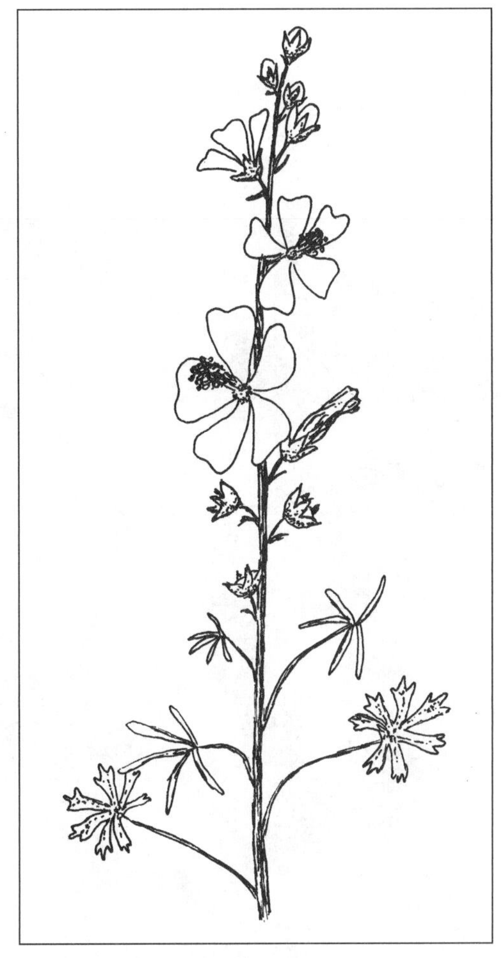

FIGURE 77. Checker mallow (*Sidalcea* spp.).

Desert Mallow (genus *Sphaeralcea*)

SCARLET GLOBE MALLOW (*S. coccinea*)

MUNRO'S GLOBE MALLOW (*S. munroana*)

Description and Habitat. Desert mallows are perennial herbs that have star-shaped hairs on the leaves and stems. Flower colors range from red to pink. They can be found in open areas at the lower elevations. The genus name comes from the Greek *sphaira* and *alkea*, meaning "spherical mallow."

FIELD KEY TO THE DESERT MALLOWS

1. Inflorescence racemose, rarely more than 1 flower per node . . *S. coccinea*
1. Inflorescence thyrsoid, usually with more than 1 flower per node . . . . *S. munroana*

Natural History Notes and Uses. *S. coccinea* was chewed and applied to inflamed sores and wounds as a cooling, healing salve. It was also used as a pharmaceutical aid: the entire plant was ground and steeped in water for a sweet-tasting tea that was mixed with bad-tasting medicines to make them more palatable. The Navajo Indians used the plant as a lotion to treat skin diseases, as a tonic to improve appetite, and as a medicine for rabies.

BUCK BEAN FAMILY (MENYANTHACEAE)

Members of the buck bean family are perennials with thick rhizomes and are usually found in aquatic habitats. The leaves are simple or divided into 3 sessile leaflets. There are five genera and 30–40 species distributed worldwide.

COMMON BUCK BEAN (*Menyanthes trifoliata*)

Description and Habitat. Buck bean is a perennial marsh herb with creeping rootstalks. The leaves have long petioles and are all basal and divided into 3 leaflets. The whitish flowers are small, star-shaped, and crowded into a short inflorescence. The species can be found in bogs and lakes. It is sometimes placed in the Gentianaceae (gentian gamily).

Natural History Notes and Uses. The herbage and rhizome of the plant are bitter, but we found that the rhizome can be made palatable when collected early in the season and boiled in several changes of water. A nutritious flour can also be made from the rhizome by drying, crushing, and leaching it thoroughly. The fresh plant eaten raw may cause vomiting.

The dried leaves are tonic and diuretic and are esteemed for their high content of vitamin C, iron, and iodine. Buck bean tea was used to relieve fever and migraine headaches, to ease indigestion, to promote a healthy appetite, and to eliminate intestinal worms. A poultice of the leaves can be applied to skin sores, herpes, glandular swelling, and sore muscles. Fresh leaves are an emetic: therefore, dry them well before use unless you intend to induce vomiting. The plant contains a bitter glycoside, menyanthine, which stimulates gastric juices. The leaves have been used in facial steams for those troubled with acne. Add the tea to a bath or use as a rinse for oily hair. The genus name is from the Greek for "month flower." The fruit has no known use. The leaves are a common ingredient in herbal smoking blends.

FOUR-O'CLOCK FAMILY (NYCTAGINACEAE)

There are 30 genera and about 300 species in this family, found in the tropics and the New World. Fifteen genera are native to the United States and chiefly centered in Arizona and Texas. In general, the plants are trees, shrubs, and herbs with opposite leaves; the flowers have bracts that mimic the sepals, and the sepals look like petals. Except for being a source of ornamentals, the family is of little economic importance.

FIELD KEY TO THE FOUR-O'CLOCK FAMILY

1. Involucral bracts separate to base or almost .... *Abronia*
1. Involucral bracts united to about one-third or more their length .... *Mirabilis*

Sand Verbena (genus *Abronia*)

YELLOWSTONE SAND VERBENA (*A. ammophila*)

SNOWBALL SAND VERBENA (*A. fragrans = A. elliptica*)

WHITE SAND VERBENA (*A. mellifera*)

SAND VERBENA (*A. micrantha = Tripterocalyx micranthus*)

Description and Habitat. Sand verbena is a genus of some 35 species of sprawling herbaceous annuals or perennials from coastal and desert areas of western North America. *Abronia*s have branching, usually sticky-hairy stems and thick, toothless leaves occurring in pairs. The flowers are more or less salverform, with 5 lobes, a thread-like style, and 3–5 unequal stamens on the tube of the perianth and not protruding from it. The fruit is winged. *Abro* is Greek for "delicate" or "pretty," referring to the flowers.

Natural History Notes and Uses. Although there are no recorded uses of these species, the Diegueño Indians in California used an *Abronia* species as a diuretic (unspecified). Other species have also been used as food and medicine.

Yellowstone sand verbena is restricted to stabilized sand sites that lie just above the maximum splash zone along the shoreline of Yellowstone Lake. A 1998 survey of the entire population found little more than 8,000 plants, most of which were seedlings. Historical collections suggest that this species was more widely distributed around the lake in the early years after the park's establishment. The high level of human activity on the beaches, especially along the northern shoreline of the lake, may have resulted in the extirpation of sand verbena from significant portions of its original range. The long-term survival of Yellowstone sand verbena is in doubt if the remaining sites are adversely affected (Whipple 1999).

NARROW-LEAF FOUR-O'CLOCK (*Mirabilis linearis*)

Description and Habitat. Narrow-leaf four-o'clock is perennial from a deep woody taproot crowned with a branching structure called a *caudex*. Plants are up to 3 feet tall and usually branched above. The opposite leaves are about 3–4 inches long but only about ¼ inch wide. The flowers occur in groups of 2–4 on stalks. These flowers have no corolla, the group of petals that normally color flowers. Instead, the ½-inch-long supporting structure for the corolla (calyx) is colored purplish red to pink and bears long hairs. At maturity the calyx swells and ripens into a 5-angled fruit containing yellowish brown achenes (seeds).

Natural History Notes and Uses. The genus *Mirabilis* is from the Latin for "wonderful" and contains about 40 species. The specific epithet *linearis* pertains to the long narrow leaves. Narrowleaf four-o'clock was first described for science under the genus *Allionia* in the 1800s by Frederick Pursh. Work published in 1901 by the Austrian botanist Anton Heimerl (1857–1942), a specialist in the Nyctaginaceae, convinced the International Botanical Congress that the plant belongs in its current taxonomic position.

Other members of the genus *Mirabilis,* also sometimes called the "umbrella-worts," are used in the tropics and Far East to make medicines, cosmetics, and jelly dyes, but economic uses for narrow-leaf four-o'-clock remain to be discovered.

WATER LILY FAMILY (NYMPHAEACEAE)

There are six genera and 68 species found throughout the world in aquatic habitats. Four of the genera are native to the United States and are a source of food for birds and aquatic animals. The family name (Nymphaeaceae) means "water nymph" or "water virgin." It was once believed that the plants in this family had anti-aphrodisiacal properties, and thus they were used in art to represent virginity. A few species are used in cultivation.

YELLOW POND-LILY (*Nuphar lutea* = *N. polysepalum*)

Description and Habitat. This large, aquatic lily is native to North America, Japan, and Europe. It has large, green, heart-shaped pads that either float or rise above the water. These plants usually obtain a spread of up to 8 feet; the rhizomes are large, and the roots can grow many feet long. The flowers, rather small when compared with the pads, are yellow and cup-shaped. They rise above the water and look like small yellow balls, appearing only half-opened when in full bloom. Several yellow pond-lilies that were once considered separate species (including *N. variegatum*, *N. rubrodisca*, *N. microphyllum*, and *N. advena*) have recently been lumped into *Nuphar lutea*.

FIGURE 78. Yellow pond-lily (*Nuphar lutea*).

Natural History Notes and Uses. Yellow pond-lily is easier to identify than to harvest. The rhizomes are prime during the early spring and fall. These starchy rootstalks can be boiled and then peeled and eaten or placed in soup or stew, or dried, ground into meal, and used as flour. The plant reproduces by seeds and rhizomes and is very easy to culture.

The seeds can be collected and, when dry, will keep indefinitely. They can also be treated like popcorn. Simply pop them and eat, or grind them into meal. The seeds can also be steamed as a dinner vegetable or cooked like oatmeal (1 part seeds to 2 parts water). In Turkey the flowers of another species are distilled into a beverage called *pufer cicegi* (Saunders 1976).

The leaves and stalks have been used as poultices for boils, ulcerous skin conditions, and swelling. An infusion of the root is useful as a gargle for mouth and throat sores. Some aboriginal people in western North America still use a root medicine for numerous illnesses, including colds, internal pains, rheumatism, chest pains, and heart conditions. This aquatic plant (like some others, including rice) gives off alcohol instead of carbon dioxide as it takes in oxygen.

EVENING-PRIMROSE FAMILY (ONAGRACEAE)

There are about 20 genera and 650 species worldwide, with a dozen genera native to the United States. A few are considered ornamentals, but otherwise the family is of little economic importance. Oil of evening-primrose, said to be the world's richest source of natural unsaturated fatty acids, is obtained from this family. The oil is helpful in cases of obesity, mental illness, heart disease, and arthritis and is advertised widely in natural food publications.

FIELD KEY TO THE EVENING-PRIMROSE FAMILY

1. Sepals, petals, and stamens 2; leaves opposite . . . . *Circaea*
1. Sepals and petals 4, stamens usually 8 2

2. Seeds with tuft of hairs at one end 3
2. Seeds naked or pubescent overall 4

3. Flowers pink, lavender, white, or yellowish . . . . *Epilobium*
3. Flowers scarlet . . . . *Zauschneria*

4. Flowers pink to red-orange or white, often irregular . . . . *Gaura*
4. Flowers usually white or yellow, regular or nearly so 5

5. Flowers minute. . . . . *Gayophytum*
5. Flowers larger. 6

6. Anthers attached near one end; petals pink to lavender . . . . *Clarkia*
6. Anthers attached near the middle, or if attached at the end, the petals are not white; otherwise, petals white, yellow, or lavender 7

7. Stigmas 4-cleft . . . . *Oenothera*
7. Stigmas hemispheric or globose, entire or merely 4-lobed. . . . *Camissonia*

Camissonia (genus *Camissonia*)

SUNDROPS EVENING-PRIMROSE (*C. andina*)

FEW-FLOWER EVENING-PRIMROSE (*C. breviflora*)

SMALL EVENING-PRIMROSE (*C. minor*)

LEWIS RIVER SUNCUP (*C. parvula*)

BARESTEM EVENING-PRIMROSE (*C. scapoidea*)

DIFFUSE-FLOWER EVENING-PRIMROSE (*C. subacaulis*)

Description and Habitat. The cammisonias are annual plants with basal or alternate leaves. The inflorescences are bracted and nodding, and the 4 sepals are reflexed. The white or yellow flowers usually fade to red.

FIELD KEY TO THE CAMISSONIAS

1. Plants perennial, leaves all basal; flowers yellow 2
1. Plants annual, stem leaves present; flowers yellow, white, or lavender . . . 3

2. Leaves entire or lobed only near the base; herbage glabrous . . . . *C. subacaulis*
2. Leaves irregularly pinnatifid; herbage short-hairy to almost glabrous. . . . . *C. breviflora*

3. Capsules with short pedicels, straight to curved; seeds in 2 rows per locule . . . . *C. scapoidea*
3. Capsules sessile or nearly so, straight, curved, or contorted; seeds in 1 row per locule. 4

4. Flowers white, opening in the evening . . . . *C. minor*
4. Flowers yellow, opening in the morning 5

5. Capsules up to ½ inch long, thickened near the base; leaves are congested at the ends of the stems. . . . . *C. andina*
5. Capsules usually longer than ½ inch, not thickened at the base; leaves are not clustered at the ends of the stems. . . . . *C. parvula*

Natural History Notes and Uses. Although there are no recorded uses for these species, other species have been used as food. The leaves were used as greens, eaten raw, boiled, or steamed.

SMALL ENCHANTER'S NIGHTSHADE (*Circaea alpina*)

Description and Habitat. This low, slender perennial has opposite leaves. The flowers occur in a bractless raceme. There are 2 sepals that are turned back, and the petals are notched and white to pink. This species grows in deep woods below 8,000 feet. It flowers from June to August.

Natural History Notes and Uses. A related species, *C. lutetiana* (broadleaf enchanter's nightshade), was used by the Iroquois as a dermatological aid on wounds. They also made an infusion as a wash for injuries (Chamberlain 1901).

PINK FAIRIES (*Clarkia pulchella*)

Description and Habitat. These annuals have brittle stems and purple or red showy flowers. They are usually found on dry slopes at the lower to middle elevations. The genus honors Captain William Clark of the Lewis and Clark expedition to the Northwest in 1806.

Natural History Notes and Uses. The seeds of *Clarkia* were among the most highly prized foods of some Native Americans in the West (Chatfield 1997). When ripe, the tops of the plants were tied in bundles and dried on rocks. After they were dry, the plants were unbundled and the seeds were dislodged by beating with a stick. The seeds were then parched and ground into meal, which was eaten dry or mixed with acorn meal. The roots of many *Clarkia* species were also eaten.

Willow-herb, Fireweed (genus *Epilobium*)

ALPINE WILLOW-HERB (*E. anagallidifolium*)

FIREWEED (*E. angustifolium = Chamerion a.*)

AUTUMN WILLOW-WEED (*E. brachycarpum*)

HAIRY WILLOW-HERB (*E. ciliatum*)

CLUB-FRUIT WILLOW-HERB (*E. clavatum*)

SMOOTH WILLOW-WEED (*E. glaberrimum*)

GLANDULAR WILLOW-HERB (*E. halleanum*)

HORNEMANN'S WILLOW-HERB (*E. hornemannii*)

MILKFLOWER WILLOW-WEED (*E. lactiflorum*)

DWARF FIREWEED (*E. latifolium = Chamerion l.*)

SLENDERFRUIT WILLOW-HERB (*E. leptocarpum*)

MARSH WILLOW-HERB (*E. palustre*)

ROCKY MOUNTAIN WILLOW-HERB (*E. saximontanum*)

SHRUB WILLOW-HERB (*E. suffruticosum*)

During the blitz of London and other English cities in World War II, fireweed was one of two plants to appear en masse among the devastation and ruin, comforting and cheering the confused and weary residents.

Description and Habitat. Willow-herbs are annual and perennial plants that have willow-like leaves. The flowers are white or lavender with petals that are often notched. The fruits are long, narrow pods that open by 4 slits to release the numerous small, densely hairy seeds. The roots and pods are often needed to make positive identification of the many species. The genus name is from the Greek, meaning "on a pod," describing the elongated ovary bearing the other flower parts on its top. The common name refers to the tufts of hairs at the end of the seed, which are similar to those on willow seeds.

FIELD KEY TO THE WILLOW-HERBS

1. Petals more than ¼ inch long, completely separate, not lobed 2
1. Petals less than ¼ inch long, united at the base, lobed at the tip 3

2. Inflorescence leafy with mostly 1–7 flowers .... *E. latifolium*
2. Inflorescence not leafy, flowers numerous .... *E. angustifolium*

3. Petals yellow; stigma of 4 long lobes .... *E. suffruticosum*
3. Petals white to purple; stigma round or only slightly lobed............. 4

4. Plants annual; leaves alternate....................... *E. brachycarpum*
4. Plants perennial; leaves opposite 5

5. Inflorescence, especially fruit, white-hairy.................. *E. palustre*
5. Inflorescence glandular-hairy or glabrous, not white-hairy 6

6. Stems glabrous..................................... *E. glaberrimum*
6. Stems hairy, often in lines from leaf bases 7

7. Turions (small, scaly, bulb-like offshoots) generally present........... 8
7. Turions lacking.. 11

8. Inflorescence without glands......................... *E. leptocarpum*
8. Inflorescence, especially fruits, at least partially glandular 9

9. Leaves clasping.................................. *E. saximontanum*
9. Leaves not clasping, with petioles 10

10. Basal rosette or large turions present....................... *E. ciliatum*
10. Compact turions present, no basal rosette................ *E. halleanum*

11. Petals white to pink, less than ¼ inch long; seeds smooth .... *E. lactiflorum*
11. Petals pink to purple, more than ¼ inch long; seeds bumpy 12

12. Plants more than 6 inches tall; leaves broadly lance-shaped; petals ¼ to ½ inch long .... *E. hornemannii*
12. Plants less than 6 inches tall; leaves narrowly lance-shaped; petals less than ¼ inch long .... *E. clavatum*

FIGURE 79. Fireweed (*Epilobium angustifolium*).

Natural History Notes and Uses. The dozens of species of *Epilobium* are all reported to be edible in emergency situations, but *E. angustifolium* and *E. latifolium* are the best-known and most commonly consumed species. Food, drink, tinder, twine, and medicine are all provided by these abundant herbs. There are many small and "weedy" species. In general, they are survivors in landscapes that have been ravaged by manmade and natural forces (e.g., fires, clearcuts). Soil conditions appear to affect their flavor. Many Native Americans "owned" good patches of fireweed that were passed on to subsequent generations. The most distinctive identifying feature of fireweed is the unique leaf venation: the veins do not terminate at the edges of the leaves, as in other plants, but rather join together in loops inside the outer margins (Vizgirdas 1999b).

The young shoots and leaves of fireweed may be boiled like asparagus but are better when mixed with other raw greens for a salad. The leaves, green or dry, make a good tea and are useful in settling an upset stomach. Be careful, though, as the leaves are slightly laxative. The unopened flower buds can be used in the same manner as the leaves and stems. The young fruits can be boiled like green beans and are tasty before the seed fibers form. Mature plants tend to become tough and bitter (Vizgirdas 1999b).

The pith of the stems can be scraped out and eaten as a snack or used as a thickener for soups. If consumed in large amounts, fireweed is a gentle but effective laxative. The plant contains a relatively high content of vitamin C and beta-carotene. The raw roots are a popular food among Siberian Eskimos. A poultice made from the roots of fireweed can be used on skin inflammations, boils, ulcers, and rashes.

BETA-CAROTENE

In addition to providing a safe source of vitamin A, beta-carotene is an antioxidant that works with other natural protectors to defend the body's cells from harmful free radical damage caused by highly reactive substances that either form in the body or are acquired from the environment (air pollution, cigarette smoke, smoke carcinogens, etc.). Dark green leafy vegetables, yellow and orange vegetables, and fruits are excellent sources of beta-carotene.

The fibrous inner bark can be used as cordage and tinder material, but I found the fibers brittle for cordage. The seeds have cotton-like hairs and are ideal for fire starting and insulation. Many western Indians used the fluffy seed cotton as a wool substitute, mixing it with mountain goat wool or duck feathers. Willow-herb fluff lacks the qualities of a really fine fiber, however. The flowers can also be rubbed into rawhide to repel water.

The seeds of *E. densiflorum* can be gathered, parched, pulverized, and then eaten. The Native Americans in California also rubbed the plant on their heads to relieve headaches.

Gaura (genus *Gaura*)

SCARLET GAURA (*G. coccinea*)

VELVETWEED (*G. parviflora* = *G. mollis*)

Description and Habitat. Gauras are annual, biennial, or perennial herbaceous plants with large, pinnately lobed basal leaves and spikes of irregular or regular, tubular flowers that bear their parts in 4's. The generic name *Gaura* stems from the Greek *gauros*, meaning "superb." *Coccinea* means "scarlet" in botanical Latin.

FIELD KEY TO THE GAURAS

1. Plants mainly 6–15 inches tall, perennial, and clump-forming; flowers are modestly showy . . . . *G. coccinea*
1. Plants 19–60 inches tall or more, annual or biennial, not clump-forming; flowers are tiny and inconspicuous . . . . *G. parviflora*

Natural History Notes and Uses. There is little information available regarding the biology and ecology of these species. They are typically native grassland perennial herbs that occupy dry sites and do not appear to be negatively affected by grazing. Scarlet gaura was first collected by the eminent botanist Thomas Nuttall and officially described for science by Frederick Pursh in his monumental *Flora Americae Septentrionale*, published in 1814.

The common name for *G. coccinea* (scarlet gaura) refers to the dye produced from galls on the plant. The Lakota are recorded to have chewed this plant and rubbed it on their hands to aid in catching horses. The Navajo used a cold infusion of this plant to settle children's upset stomachs. Some Native American tribes used the roots of *G. mollis* to treat snakebite.

Ground Smoke (genus *Gayophytum*)

SPREADING GROUND SMOKE (*G. diffusum*)

DWARF GROUND SMOKE (*G. humile*)

BACKFOOT GROUND SMOKE (*G. racemosum*)

PINYON GROUND SMOKE (*G. ramosissimum*)

Description and Habitat. The ground smokes are slender-stemmed annuals with alternate leaves, the lower ones often being opposite. The small flowers have distinct, reflexed sepals, white to pink petals, and 8 stamens. The fruits are linear to club-shaped capsules. The various species are found on dry slopes and on the edges of meadows.

FIELD KEY TO THE GROUND SMOKES

1. Capsules with small pedicels, usually constricted between the seeds; plants branched above	2
1. Capsules sessile, only slightly constricted between the seeds; plants branched at the base	3
2. Pedicels sharply reflexed	*G. ramosissimum*
2. Pedicels spreading or erect	*G. diffusum*
3. Seeds fewer than 10	*G. racemosum*
3. Seeds 15–20	*G. humile*

Natural History Notes and Uses. An infusion of *G. ramosissimum* was used to soothe irritated skin (Strike 1994).

Evening-primrose (genus *Oenothera*)

TUFTED EVENING-PRIMROSE (*O. caespitosa*)

NUTTALL'S EVENING-PRIMROSE (*O. nuttallii*)

PALE EVENING-PRIMROSE (*O. pallida*)

HAIRY EVENING-PRIMROSE (*O. villosa*)

Description and Habitat. Evening-primroses are annual, biennial, and perennial herbs. The flowers are white or yellow, often opening at night. There are 8 stamens, 4 petals, and 4 sepals, and the stigma is globe-shaped to deeply 4-lobed. The various species can be found in a range of habitats up to the subalpine zone. The many species within this genus are extremely variable, described by botanists as a hopelessly confused and freely hybridizing group (Fernald and Kinsey 1958).

FIGURE 80. Evening-primrose (*Oenothera* spp.).

FIELD KEY TO THE EVENING PRIMROSES

1. Leaves all basal, no apparent stem visible	*O. caespitosa*
1. Stems obvious and erect	2
2. Flowers yellow when fresh, rarely purple in age	*O. villosa*
2. Flowers white to pinkish, reddish, or purplish	3
3. Inflorescence glandular-pubescent	*O. nuttallii*
3. Inflorescence may be pubescent but not glandular	*O. pallida*

Natural History Notes and Uses. Most handbooks on edible plants indicate that at least *O. hookeri* (= *O. elata* subsp. *hookeri*) (Hooker's evening-primrose) and *O. biennis* (common evening-primrose) have edible roots. They can be cooked and eaten as a vegetable when young but become tough and somewhat spicy or peppery with age. The leaves of *O. biennis* are also edible as cooked greens but are not exceptional unless mixed with bland greens to make a more acceptable salad. Harrington (1967) suggests that all species would stand a trial as none are known to be poisonous. The various species are known to hybridize easily, making identification challenging at times.

We have cooked and eaten the young seed pods of several species and found them to have an acceptable taste. Olsen (1990) also suggests that many species have seeds that are edible after being parched or ground into meal. Strike (1994) states that seeds and leaves of *Oenothera* were eaten by native people.

The leaves, stems, and crushed seeds have an astringent quality and can be used as a poultice for wounds, bruises, and piles. The seeds are also high in essential oils and have been shown in clinical studies to be effective for heart disease, asthma, arthritis, alcoholism, and other fatty acid problems. The medicinal uses of the oil in these plants are a recent discovery following scientific research in the 1980s that demonstrated their effectiveness for a wide range of intractable complaints. The oil contains gamma-linoleic acid (GLA), an unsaturated fatty acid that assists in the production of hormone-like substances. Evening-primrose oil, in the form of gelcaps, is becoming popular in the natural supplements marketplace (Tilford 1993). Additionally, the stringy bark makes good cordage material.

ZAUSCHNERIA (*Zauschneria garrettii* = *Epilobium canum* subsp. *zauschneria*)

Description and Habitat. Zauschneria is a sprawly, shrubby perennial. Older plants can spread to 2 feet wide and grow to 12 inches tall. The flower is about an inch long, and its brilliant orange-red flowers are a welcome sight. The leaves are also about an inch long and about ¼ inch wide, clasping at the stem and narrowly acute at the far end. Their color varies from green to grayish green.

Natural History Notes and Uses. All the members of the genus *Zauschneria* are now considered one species in the genus *Epilobium*.

PARASITIC PLANTS

Parasitic plants obtain all their nourishment from another living organism. They have no chlorophyll and cannot make their own food. The broomrapes (*Orobanche*) are fleshy, and their leaves are reduced to scale-leaves, as they have lost their photosynthetic function. The plants attach to the roots of various host plants by means of suckers (haustoria), with which they obtain water and nutrients. Another group of parasitic plants are the dwarf mistletoes (*Arceuthobium*).

BROOMRAPE FAMILY (OROBANCHACEAE)

Thirteen genera and 180 species are found worldwide, with four of the genera being native to the United States. Members of the broomrape family are herbaceous, lack chlorophyll, and are parasitic on the roots of other flowering plants. The family is of no direct economic importance. The family name comes from the Greek *orogos*, meaning "a clinging plant," and *acho*, "to strangle."

Broomrape (genus *Orobanche*)

CLUSTERED BROOMRAPE (*O. corymbosa*)

CLUSTERED BROOMRAPE (*O. fasciculata*)

ONE-FLOWERED BROOMRAPE (*O. uniflora*)

Description and Habitat. Broomrapes parasitize the roots of other plants. These fleshy annuals are nearly white to brownish or purplish in color and lack chlorophyll. The leaves are reduced to scales. Broomrapes are usually found in dry soils, associated with such genera as *Artemisia* and *Eriogonum*. A single broomrape flower produces enormous numbers of small seeds, as many as fifty thousand in some species, which can remain dormant in the soil for 10–15 years or until suitable germination conditions are met. A fourth species, Louisiana broomrape (*O. ludoviciana*), has been recorded from the area.

FIELD KEY TO THE BROOMRAPES

1. Flowers sessile or on pedicels less than 1½ inches long and in addition to the subtending bract have a pair of bractlets just below the sepals . . . . *O. corymbosa*
1. Flowers all somewhat long-stalked and lacking small subtending bracts 2

2. Flowers no more than 2–3, usually 1, the short stems remaining underground; petals rounded; sepals narrow and slender . . . . . *O. uniflora*
2. Flowers more than 3; the top of the stems emerging from the ground; petals pointed; sepals triangular . . . . *O. fasciculata*

Natural History Notes and Uses. The entire plant of broomrape, roots and all, can be eaten raw. Being succulent plants, they answer for food and drink and are often called "sand food." We found them to be better-tasting when roasted in the hot ashes of a campfire. Strike (1994) indicates that the roots of *O. californica* (California broomrape) and the entire plant of *O. fasciculata* (clustered broomrape) were also eaten.

The decocted blanched or powdered seeds are said to ease joint and hip pain. They can also be used as a toothache remedy. Moore (1979) states that the whole plant is astringent and makes an excellent poultice. Broomrape is also mildly laxative and sedative. The stalks, with the white inner portions removed, have been used as pipes. *Orobanche uniflora* was used to treat numerous ailments, including bronchial problems, intestinal upset, toothaches, and rheumatic pain. A decoction of *O. fasciculata* was used as a skin wash to kill lice.

FIGURE 81. Broomrape (*Orobanche* spp.).

OXALIC ACID

The tart, lemony taste of wood sorrel (*Oxalis* spp.), cacti (*Opuntia* spp.), lamb's quarters (*Chenopodium* spp.), amaranth (*Amaranthus* spp.), knotweed (*Polygonum* spp.), dock (*Rumex* spp.), and other species is due to the presence of soluble oxalic acid. Without proper preparation these plants, when eaten in substantial amounts, should be considered toxic. However, when properly prepared, they are an excellent food source.

The soluble oxalic acid, also known as salt of lemon, is what makes the plants tasty as well as dangerous. Oxalic acid is dangerous because of its solubility and its affinity for calcium. The solubility allows the acid to enter the bloodstream, where it promptly combines with calcium to form nonsoluble calcium oxalate. This substance precipitates in the kidneys, where it both plugs the tubules and "burns" all cells in contact with it, potentially leading to renal failure and death.

Oxalic acid is readily dissolved in heated water and will combine with calcium as readily in that water as in the bloodstream. Adding bone fragments, eggshells, or some other source of calcium to cooking water will transform the oxalic acid to nonsoluble calcium oxalate in the pot, retaining the full flavor but rendering the acid harmless. If you have no bone fragments or eggshells, just pour out the first water after boiling for a time and replace with fresh water.

OXALIS FAMILY (OXALIDACEAE)

There are seven genera and over 1,000 species distributed worldwide. Only *Oxalis* is native in the United States. In general, they are small plants with leaf blades divided into 3 heart-shaped segments. The flowers are 5-merous and yellow or purple. The seed pods split explosively, scattering seeds some distance from the plant. The family is of little economic importance.

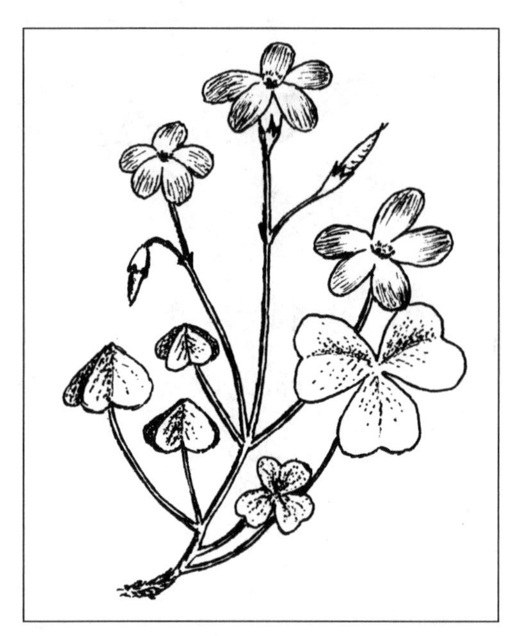

FIGURE 82. Wood sorrel (*Oxalis* spp.).

VIOLET WOOD SORREL (*Oxalis violacea*)

Description and Habitat. Wood sorrel is a herbaceous perennial with basal, alternate 3-parted leaves. The flowers are mostly bisexual, bilaterally symmetrical, and borne solitary or in a short, broad, and somewhat flat-topped inflorescence. The flowers are blue, pink, or lavender. The calyx and corolla have 5 petals and sepals and 10 stamens. The fruit is a capsule with 5 lobes and chambers.

Natural History Notes and Uses. The genus name is derived from the Greek word *oxys*, meaning "sour." The leaves and stems of *Oxalis* may be eaten raw. To make a tasty dessert, collect a mass of the plants and allow them to ferment for a while. *Oxalis* contains a high percentage of oxalic acid; therefore, it is recommended that one eat the plants sparingly until accustomed to them. One symptom of too much oxalate is painful or swollen taste buds. The plants are also high in vitamin C and were used as a remedy for scurvy. A refreshing drink can be made by steeping the leaves in hot water, then chilling and sweetening it.

PEONY FAMILY (PAEONIACEAE)

The family has only one genus with approximately 33 species.

FIGURE 83. Brown's peony (*Paeonia brownii*).

BROWN'S PEONY (*Paeonia brownii*)

Description and Habitat. This spring annual has palmately divided leaves. The flowers are large and showy, and the petals are red-brown. The flowers are often subtended by reduced leaves. This plant occurs in woodsy, shaded areas in the lower elevations. The genus name is from the Greek for Paeon, the physician of the gods, who supposedly used the plant medicinally.

Natural History Notes and Uses. The leaves are edible when cooked as greens. We have found it best to boil them in several changes of water until the bitterness is removed.

Native Americans picked the young leaves before the blossoms appeared in the spring, boiled them, and then placed them in a cloth sack and weighed the sack down in the river with a stone. Allowing the water to run through the sack overnight removed the bitterness. Medicinally, an infusion of sliced, oven-baked roots was taken for indigestion by Native Americans in California. The genus is also important in Chinese medicine.

POPPY FAMILY (PAPAVERACEAE)

Most plants in this family are annual or perennial herbs and sometimes shrubs. The sap is often milky or colored. There are 26 genera and 200 species distributed in the subtropical and temperate areas of the Northern Hemisphere, particularly in western North America. Thirteen genera are native to the United States. The family is of little economic importance except for *Papaver somniferum*, which yields opium and its many derivatives, including morphine and heroin. A few species are cultivated as ornamentals.

ALPINE POPPY (*Papaver kluanense*)

Description and Habitat. Alpine poppy is a short, tufted perennial herb with stems up to 6 inches high and covered with soft, spreading hairs. The leaves are all basal, the blades blue-green, sparsely hairy, and divided into 3–5 lobes (the lobes sometimes notched). The flowers are solitary at the ends of stems, with 2 sepals that fall off shortly after the flower opens and 4 yellow to greenish yellow petals. The capsules are covered with stiff, brownish hairs. It is found in alpine meadows, on talus slopes, and in alpine fellfields (a slope environment, usually alpine or tundra, where the dynamics of frost and of wind give rise to characteristic plant forms).

Natural History Notes and Uses. Alpine poppy ranges from southern Alaska south along the Rocky Mountains to northern New Mexico. In Wyoming it is known from alpine areas of the Absaroka, Bighorn, and Wind River ranges, between 10,800 and 12,300 feet. There are at least nine extant occurrences of alpine poppy in Wyoming and one historical record. Four populations have been observed since 1990. Additional populations have been reported for Fremont and Johnson counties.

LATEX

The plants in the poppy family are laticiferous. All parts contain a well-developed duct system (called laticifers), producing a milky or watery yellow or red juice (latex). Latex, as found in nature, is a milky sap that coagulates on exposure to air. It is a complex emulsion containing proteins, alkaloids, starches, sugars, oils, tannins, resins, and gums. In most plants latex is white, but some have yellow, orange, or scarlet latex. Latex can also be made synthetically by polymerizing a monomer that has been emulsified with surfactants.

Latex has been attributed to many plant functions. Some botanists regard it as a form of stored food, while others consider it an excretory product in which waste products of the plant are deposited. Still others believe it functions to protect the plant in case of injuries, drying to form a protective layer that prevents the entry of fungi and bacteria. Similarly, it may provide protection against browsing animals, since in some plants latex is very bitter or even poisonous. It may be that latex fulfills all these functions to varying degrees in the numerous plant species in which it occurs.

Latex has many uses, but its first and foremost is rubber. Chicle, widely used as a base for chewing gum, is another latex product. Latex paint uses synthetic latex as a binder. Finally, poppy latex is a source of opium and its many derivatives.

Some people are seriously allergic to latex, and exposure to latex or rubber products such as rubber gloves or condoms can cause anaphylactic shock. Since latex has a protein found also in bananas, care should be taken to ensure that people are not allergic to both. Guayule latex is hypoallergenic and is being studied as a substitute for the allergy-inducing *Hevea latexes*. Latex can be used for clothing as well. For some people latex is a sort of fetish. They like the feeling and smell of the material.

PLANTAIN FAMILY (PLANTAGINACEAE)

Three genera and 270 species of this family are found worldwide. *Plantago* is widespread in the United States. In general, the family is of little economic importance, but several species are weeds and one (*P. psyllium*) is the source of seeds used to make a commercial laxative.

Plantain (genus *Plantago*)

LONGLEAF PLANTAIN (*P. elongata*)

REDWOOL PLANTAIN (*P. eriopoda*)

NARROWLEAF PLANTAIN (*P. lanceolata*)

COMMON PLANTAIN (*P. major*)

WOLLY PLANTAIN (*P. patagonica*)

TWEEDY'S PLANTAIN (*P. tweedyi*)

Description and Habitat. Plantains are short-stemmed annual or perennial herbs with basal leaves. The flowers are greenish or purplish. Many of the species were introduced from Europe and can be found at the lower elevations, particularly in fields and waste places. *Plantago* means "sole of foot" and refers to the sole-shaped leaves of plantain, which lie close to the ground as though stepped on.

FIELD KEY TO THE PLANTAINS

1. Plants annual; leaves linear to filiform, rarely more than ½ inch wide . . . 2
1. Plants biennial or perennial; leaves lanceolate or broader, more than ½ inch wide . . . 3

2. Stamens 2; corolla lobes erect and closing over the capsule . . . *P. elongata*
2. Stamens 4; corolla lobes spreading or reflexed . . . *P. patagonica*

3. Leaf blades broadly ovate, abruptly contracted to the petiole.... *P. major*
3. Leaf blades lanceolate, oblanceolate, lance-oblong, or elliptic, gradually tapering into the petiole 4

4. Stamens conspicuously exserted .... *P. lanceolata*
4. Stamens not conspicuous..5

5. Crown of plant conspicuously woolly-hairy at base of petiole; leaves thick.................................... *P. eriopoda*
5. Crown of plant not woolly; leaves thin .... *P. tweedyi*

Natural History Notes and Uses. Because of their reputation as weeds, plantains are a forgotten edible. In fact, a lot of effort is spent trying to get rid of the plants from gardens. *Plantago major* and *P. lanceolata* were brought over by European settlers for use as potherbs and medicine. The Native Americans called the plants "white man's foot" because they followed the settlers west. The native species of plantain are uncommon in comparison, and it is suggested that they only used only when large populations are found.

The young leaves of common plantain and narrowleaf plantain were eaten fresh or cooked. They contain calcium and other minerals. One hundred grams of plantain is said to furnish as much vitamin A as a large carrot. The older leaves may be too fibrous and bitter, but they are usable if one is able to remove the fibers. The seeds are tedious to collect in quantity but can be ground and used as flour substitute or extender (Doebley 1984).

The leaves and seeds of many species were used medicinally. The foliage contains tannins and iridoid glycosides, notably aucubin, which stimulates uric acid secretion from the kidneys. The crushed leaves of common plantain provide an astringent juice that can be used to soothe wounds, sores, insect bites, and the rash of poison ivy. Plantain juice is a traditional treatment for earaches. The seeds contain up to 30 percent mucilage, which swells in the gut to act as a bulk laxative and soothes irritated membranes. Rubbing the leaves on one's skin works as a natural, moderately effective insect repellent.

PHLOX FAMILY (POLEMONIACEAE)

There are about 18 genera and 320 species in this family, found chiefly in North America and particularly in western United States. The family includes annual and perennial herbs and low-growing, woody-based subshrubs. The leaves are opposite or alternate and undivided with entire margins to compound or variously divided. Both sepals and petals are 5-parted and partially fused; the 5 stamens are partially fused to the corolla tube. The fruits are capsules. The family is a source of a few ornamentals.

FIELD KEY TO THE PHLOX FAMILY

1. Leaves opposite, simple, and entire; filaments attached to corolla at different levels 2
1. Leaves alternate or basal, or if opposite, then lobed or compound 4

2. Calyx tube of nearly uniform texture; calyx lobes somewhat triangular in shape. . . . . *Collomia*
2. Calyx tube usually with green costae alternating with hyaline intervals; calyx lobes needle-shaped. 3

3. Plants annual; upper leaves usually alternate . . . . . *Microsteris*
3. Plants perennial; leaves often all opposite or densely crowded. . . . . *Phlox*

4. Plants annual with flowers in leafy bracted heads. . . . . *Navarretia*
4. Plants not as above 5

5. Leaves sessile, palmately divided to near base 6
5. Leaves not sessile and not palmately divided to near base. 8

6. Plants annual . . . . . *Linanthus*
6. Plants perennial. 7

7. Plants woody nearly throughout or may be matted; leaves prickly . . . . . *Leptodactylon*
7. Plants woody only at base, not matted; leaves soft, not prickly. . . . . *Linanthus*

8. Calyx tube of nearly uniform texture; calyx lobes without needle-like tips 9
8. Calyx tube usually with green costae alternating with hyaline intervals. 10

9. Leaves pinnately compound . . . . . *Polemonium*
9. Leaves simple and mostly entire. . . . . *Collomia*

10. Corolla over ½ inch long . . . . . *Ipomopsis*
10. Corolla ¼ to ½ inch long. . . . . *Gilia*

Mountain Trumpet (genus *Collomia*)

ALPINE MOUNTAIN TRUMPET (*C. debilis*)

NARROWLEAF MOUNTAIN TRUMPET (*C. linearis*)

DIFFUSE MOUNTAIN TRUMPET (*C. tenella*)

Description and Habitat. Mountain trumpets are annual or perennial herbs with simple or branched stems. There are approximately 15 species in western North America and South America. The flowers are funnel-shaped or tubular with throats that abruptly flare into an expanded limb. The genus name is from the Greek *kolla,* meaning "glue," because of the mucilaginous layer on the seeds of most species.

FIELD KEY TO THE MOUNTAIN TRUMPETS

1. Plants perennial; alpine . . . . . *C. debilis*
1. Plants annual from taproot; lower elevation to alpine 2

2. Flowers solitary in leaf axils; leaves linear. . . . . *C. tenella*
2. Flowers in capitate, terminal clusters; leaves elliptic to lanceolate. . . . . *C. linearis*

Natural History Notes and Uses. From the roots of large-flowered collomia an infusion was made for high fevers. Additionally, an infusion of the leaves and stalks was taken for constipation and to "clean out the system."

Narrowleaf collomia was used as a dermatological aid by the Gosiute. They made a poultice of the mashed plant and applied it to wounds and bruises (Chamberlain 1911).

The seed coat becomes mucilaginous when wet. This "glue" is a mechanism to help keep the germinated seeds from drying out, storing water between the first autumn rains and those that may not come for several weeks.

Gilia (genus *Gilia*)

DELICATE GILIA (*G. tenerrima*)

TWEEDY'S GILIA (*G. tweedyi*)

Description and Habitat. There are many species of *Gilia*, and they are characterized as annual, biennial, or perennial plants. The leaves are mostly alternate and are lobed or dissected with the tips acute. The seeds are sticky when wet. They are found in a range of habitats at various elevations. The genus was named for Spanish botanist Philipp Salvador Gil. The gilias are related to the phloxes and, like them, have funnel- or salverform flowers.

FIELD KEY TO THE GILIAS

1. Leaves entire	*G. tenerrima*
1. Leaves strongly toothed to pinnatifid	*G. tweedyi*

Natural History Notes and Uses. Strike (1994) indicates that *Gilia* seeds were eaten by many Native American tribes.

Ipomopsis (genus *Ipomopsis*)

SKYROCKET GILIA (*I. aggregata*)

DWARF GILIA (*I. pumila*)

SPIKED GILIA (*I. spicata*)

Description and Habitat. These annual and perennial herbs have basal rosettes of pinnately cut leaves and tubular flowers. The genus name is from the Greek *ipo* ("to impress") and *opsis* ("appearance"), referring to the showy flowers.

FIELD KEY TO THE IPOMOPSIS

1. Corolla tubes ¾ to 2 inches long	*I. aggregata*
1. Corolla tubes less than ½ inch long	2
2. Plants annual	*I. pumila*
2. Plants perennial	*I. spicata*

FIGURE 84. Skyrocket gilia (*Ipomopsis aggregata*).

Natural History Notes and Uses. *Ipomopsis aggregata* was used by Native Americans to make glue, to treat blood troubles, and, as a tea, to relieve colds. In Nevada the principal use of this plant was for the treatment of venereal diseases. The whole plant was boiled for the purpose, and a solution was taken as a tea or used as a wash. The whole plant was also boiled by the Ute Indians in Utah to make a glue. A blue dye can be extracted from the roots (Gifford 1967).

Leptodactylon (genus *Leptodactylon*)

MAT PRICKLY GILIA (*L. caespitosum*)

GRANITE PRICKLY GILIA (*L. pungens*)

WATSON'S PRICKLY GILIA (*L. watsonii*)

Description and Habitat. The leptodactylons are low shrubs with prickly leaves. The flowers are funnel-form to salverform and often very showy. The species are found in dry, rocky places in the lower to high elevations.

FIELD KEY TO THE LEPTODACTYLONS

1. Leaves alternate .... *L. pungens*
1. Leaves opposite 2

2. Flowers 6-merous; plants forming clumps .... *L. watsonii*
2. Flowers 4-merous; plant phlox-like .... *L. caespitosum*

Natural History Notes and Uses. *Leptodactylon pungens* was used as a decoction to bathe swellings, sore eyes, and scorpion stings (Strike 1994).

Linanthus (genus *Linanthus*)

NUTTALL'S DESERT TRUMPETS (*L. nuttallii*)

NORTHERN LINANTHUS (*L. septentrionalis*)

Description and Habitat. These low annuals have opposite leaves that are palmately parted into slender segments or reduced to linear blades. They are found on dry, open slopes.

FIELD KEY TO THE LINANTHUS

1. Plants annual .... *L. septentrionalis*
1. Plants perennial .... *L. nuttallii*

Natural History Notes and Uses. An infusion was made from *L. ciliatus* by the West Coast Indians to treat children's coughs and colds. An unheated decoction was drunk instead of water to purify the blood.

SLENDER PHLOX (*Microsteris gracilis = Phlox m.*)

Description and Habitat. Slender phlox is an annual with linear to elliptical-linear, sessile leaves. The lower leaves are opposite, and the upper leaves are alternate. The flowers are white, pink, or lavender. This is a very common and variable species. It is found in a variety of mostly open areas at the low to middle elevations.

Natural History Notes and Uses. Slender phlox was eaten by some native people as greens. The plant was also used as a poultice on bruises and wounds (Strike 1994).

NAVARRETIA (*Navarretia intertexta*)

Description and Habitat. Navarretia is a low, rather stout, usually branched herb with pinnately lobed leaves, with the

POLLINATION ECOLOGY 101

Successful reproduction is defined by an individual's passing on its genes to the next generation. The showy wildflowers of grasslands and meadows are insect- or hummingbird-pollinated. The array of shapes, colors, and fragrances of these flowers is for the sole purpose of attracting pollinators and maintaining species' isolation. For its services, the pollinator receives nectar, pollen, or oil.

This relationship between plant and pollinator is the result of a long and intimate coevolutionary process. The animal pollinators have distinct color, shape, and food-type preferences in the flowers they visit, and have specific anatomical features that maximize their abilities to collect pollen or nectar. For example, beetles have brush-like mouth parts for collecting pollen; bees have hollows, or pollen baskets, on their hind legs for storing pollen; hummingbirds have long, narrow bills and tongues for gathering nectar; and butterflies have tubular, bristled proboscises for nectar drinking.

The patches, streaks, and spots on flowers are called nectar guides and often direct the pollinator to the nectar glands. These guides are situated so that the pollinator must brush against the sex organs of the flower on its way in. Many of these guides are conspicuous, such as the lines on the interior of some penstemon flowers; others are produced by specialized plant tissues that strongly reflect ultraviolet light and are seen only by the pollinators.

lobes spiny-tipped. The white flowers occur in dense, head-like clusters. It is found in meadows, on slopes, and along pond margins. The seeds are usually mucilaginous when wetted. The genus name honors Spanish physician Fr. Ferd. Navarrete.

Natural History Notes and Uses. The seeds of a related species, skunkweed (*N. squarrosa*), were gathered and dried in the sun and then stored. When needed, the seeds were parched and then eaten dry. Medicinally, skunkweed was used as a tonic, fever reducer, laxative, and dye. The seeds of *N. leucocephala* were also eaten, and a decoction of the plant was used to reduce swellings.

Phlox (genus *Phlox*)

SPINY PHLOX (*P. hoodii*)

LONG-LEAVED PHLOX (*P. longifolia*)

FLOWERY PHLOX (*P. multiflora*)

PHLOX (*P. muscoides* = *P. hoodii*)

CUSHION PHLOX (*P. pulvinata*)

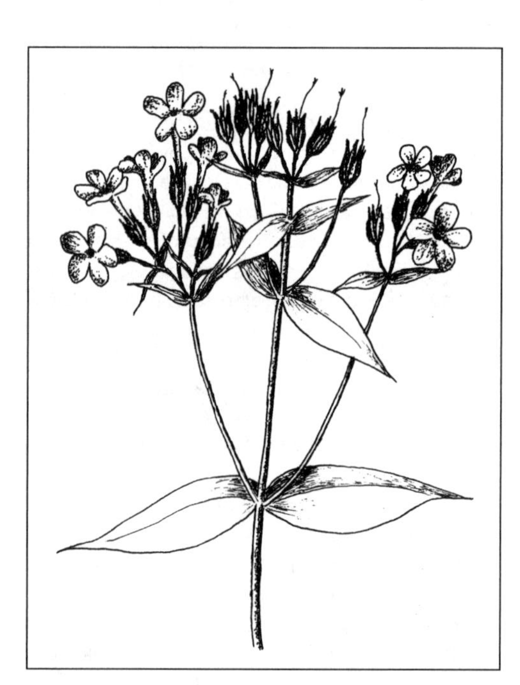

FIGURE 85. Phlox (*Phlox* spp.).

Description and Habitat. The plants in this genus are low shrubs, perennials, or annuals with opposite leaves. The flowers are salverform in shape. The many species can be found in various habitats at all elevations.

FIELD KEY TO THE PHLOXES

1. Plants growing more or less erect; leaves more than 1 inch long	*P. longifolia*
1. Plants growing cushion-like; leaves short and crowded, usually less than 1 inch long	2
2. Leaves narrow-linear, up to ¼ inch long, sometimes with soft woolly hairs	*P. hoodii*
2. Leaves wider and usually with no soft hairs	3
3. Calyx glandular-hairy	*P. pulvinata*
3. Calyx not glandular-hairy	*P. multiflora*

Natural History Notes and Uses. A decoction from the roots of *P. longifolia* and other species were used by some Native Americans as an eyewash. The scraped roots were soaked in water or steeped or boiled to make the wash (Train et al. 1957).

Jacob's Ladder (genus *Polemonium*)

ANNUAL POLEMONIUM (*P. micranthum*)

WESTERN POLEMONIUM (*P. occidentale*)

SHOWY POLEMONIUM (*P. pulcherrimum*)

STICKY POLEMONIUM (*P. viscosum*)

Description and Habitat. These perennials grow from woody rootstalks. The flowers occur in terminal or axillary cymes. One species has small, white flowers, and the other three are blue-flowered. All are malodorous to various degrees.

FIELD KEY TO THE JACOB'S LADDERS

1. Plants annual; petals shorter than or equal to the sepals . . . *P. micranthum*
1. Plants perennial; petals longer than sepals 2

2. Leaflets are crowded on the stem and deeply 2–5-cleft as to appear whorled; petals longer than wide, and lobes definitely shorter than the tube. . . . . *P. viscosum*
2. Leaflets not crowded and usually not lobed; petals about as wide as long or wider, and the lobes about as long as the tube or longer 3

3. Stems erect, plants more than 12 inches tall. . . . . *P. occidentale*
3. Stems lax, plants less than 12 inches tall . . . . *P. pulcherrimum*

Natural History Notes and Uses. The genus name is thought to honor Polemon, an early Greek philosopher. The common name of Jacob's ladder is based on the arrangement of the leaflets and comes from the story in Genesis 28:12 of Jacob's dream of a ladder connecting heaven to earth. Medicinally, a decoction of showy polemonium (*P. pulcherrimum*) was used as a wash for the head and hair.

FIGURE 86. Showy polemonium (*Polemonium pulcherrimum*).

BUCKWHEAT FAMILY (POLYGONACEAE)

The buckwheat family has 40 genera and 800 species, of which 15 genera are native to the United States. The family is best represented in the western states. The economic products include food plants and a few ornamentals. In general, this family is safe to eat.

FIELD KEY TO THE BUCKWHEAT FAMILY

1. Plants are annuals found in the alpine zone and are less than 1½ inches tall . . . . *Koenigia*
1. Plants not as above 2

2. Leaves without sheathing stipules . . . . *Eriogonum*
2. Leaves with sheathing stipules (ochrea) 3

3. Sepals 5, all similar and erect in fruit . . . . *Polygonum*
3. Sepals 4 or 6, the outer 3 turned back, the inner 3 erect in fruit 4

4. Sepals 6; styles 3 . . . . *Rumex*
4. Sepals 4; styles 2 . . . . *Oxyria*

Wild Buckwheat (genus *Eriogonum*)

MATTED BUCKWHEAT (*E. caespitosum*)

NODDING BUCKWHEAT (*E. cernuum*)

YELLOW ERIOGONUM (*E. flavum*)

CUSHION BUCKWHEAT (*E. ovalifolium*)

SULPHUR WILD BUCKWHEAT (*E. umbellatum*)

Description and Habitat. Wild buckwheats are annual or perennial herbs; some species are woody at the base. The flowers are small and usually bright-colored. The many species of *Eriogonum* can be found in various habitats at all elevations. The genus name is from the Greek *erion*, meaning "wool," and *gony*, meaning "knee" or "joint," referring to the hairy stems of many species.

FIELD KEY TO THE WILD BUCKWHEATS

1. Plants annual or biennial with a slender taproot	*E. cernuum*
1. Plants perennial with a thick taproot, a caudex, or mat-forming	2
2. Bracts usually scale-like and 3-parted; perianth not appearing to form a stipe	*E. ovalifolium*
2. Bracts leafy in texture, 2 to several; perianth appearing to form a stipe	3
3. Involucral bracts erect, lobes usually less than half as long as the tube	*E. flavum*
3. Involucral bracts generally reflexed or spreading, lobes at least as long as the tube	4
4. External surface of perianth with hairs	*E. caespitosum*
4. External surface of perianth glabrous	*E. umbellatum*

Natural History Notes and Uses. None of the species is known to be poisonous. The flowering stems can be eaten raw or cooked before they have flowered. The seeds can be collected (though the process is tedious) and ground into flour. A tea from the root of *Eriogonum* was used to treat headaches and stomach problems. The plants are mildly astringent and were used as a gargle for sore throats.

ISLAND PURSLANE (*Koenigia islandica*)

Description and Habitat. *Koenigia* is a glabrous, reddish-stemmed annual herb less than 2 inches tall. The leaves are oblanceolate to obovate and blunt, sessile, and essentially opposite to whorled. The flowers are 1 to few and borne in axillary or terminal clusters. The perianth consists of 3 (rarely 4) greenish white calyx lobes. Flowers have 3 stamens and 2 styles and produce a triangular, dark brown achene.

Natural History Notes and Uses. This plant has a circumpolar distribution, extending south in North America to scattered alpine summits in the Rocky Mountains as far south as Colorado. In Wyoming it is known from the Beartooth and Wind River ranges in Fremont and Park counties. It grows in wet nival (snow) basins, along streambanks and lake shores, and in areas of patterned ground at 10,000–12,500 feet.

ALPINE MOUNTAIN SORREL (*Oxyria digyna*)

Description and Habitat. Alpine mountain sorrel is a low perennial with simple, roundish leaves clustered at the base of the stem. The flowers are small and red or greenish. The plant is found in cold, wet places among rock crevices between 9,400 and 10,700 feet. It flowers from July to September. The plant resembles a miniature rhubarb, with small, rounded leaves. It has always been highly esteemed in arctic regions as a "scurvy grass" with an agreeable sour taste.

Natural History Notes and Uses. Perhaps one of the most refreshing plants one encounters in the high country is the alpine mountain sorrel. The new growth up to flowering time can be eaten raw, when it tastes like a mild rhubarb. The stems and leaves can be used in salads or prepared as a potherb. Some aboriginal peoples have been known to ferment mountain sorrel as a kind of sauerkraut. This is accomplished by simply letting the plants sit in water for a while. The sauerkraut can then be stored for winter use (Schofield 1989). The plants were also dried in the sun for traveling. The plants are high in vitamin C and can be used to prevent and cure scurvy. Large amounts could, however, cause oxalate poisoning.

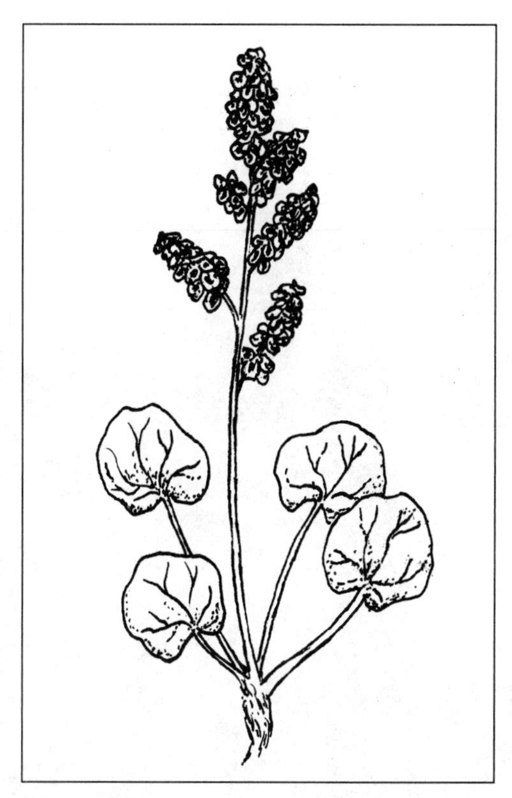

FIGURE 87. Alpine mountain sorrel (*Oxyria digyna*).

Knotweed, Smartweed (genus *Polygonum*)

WATER KNOTWEED (*P. amphibium*)

PROSTRATE KNOTWEED (*P. aviculare*)

AMERICAN BISTORT (*P. bistortoides*)

BLACK BINDWEED (*P. convolvulus*)

DOUGLAS' KNOTWEED (*P. douglasii*)

BROADLEAF KNOTWEED (*P. minimum*)

MILKWORT KNOTWEED (*P. polygaloides*)

BUSHY KNOTWEED (*P. ramosissimum*)

ALPINE BISTORT (*P. viviparum*)

Description and Habitat. Knotweed is a highly variable genus that includes annual or perennial forbs and shrubs. Some species are viny. The leaves are usually simple and alternate. The pink, green, or white flowers have jointed stalks, and the stems have swollen nodes. The flowers can be either perfect or imperfect. The fruit is a 3- or 4-angled achene. Some species have rhizomes or taproots. About 150 species of knotweed or smartweed are known, and most are weeds of disturbed, moist areas. Mature achenes are often needed for positive identification. Annuals are predominantly self-pollinating, sometimes creating an array of locally distinctive forms.

FIELD KEY TO THE KNOTWEEDS

1. Flowers and fruits numerous in terminal, elongate clusters 2
1. Flowers and fruits 1–4 in axils of leaves or leaf-like bracts 4

2. Plants aquatic or stems prostrate and rooting at nodes .... *P. amphibium*
2. Plants not aquatic and rooting and nodes; stems erect 3

3. Lower flowers converted to bulbs........................ *P. viviparum*
3. Bulbs lacking in inflorescence.......................... *P. bistortoides*

4. Stems twining or creeping; leaf blade arrow-shaped ...... *P. convolvulus*
4. Stems erect to prostrate but stiff; leaf blades linear to elliptic 5

5. Stems with 8–16 longitudinal ribs 6
5. Stems 3- to 5-angled but not ribbed................................ 7

6. Stems prostrate .... *P. aviculare*
6. Stems erect or nearly so .... *P. ramosissimum*

7. Leaves linear 8
7. Leaves elliptic 9

8. Mature achene black, shiny .... *P. douglasii*
8. Mature achene light brown to dull black .... *P. polygaloides*

9. Flowers erect or spreading; plants of subalpine and alpine areas .... *P. minimum*
9. Older flowers nodding; plants subalpine and below .... *P. douglasii*

Natural History Notes and Uses. Experimentation may be the rule for *Polygonum* as none of the species is known to be poisonous. They vary in degree of palatability, however. Tannins are found in the plants, and large amounts might cause digestive upset and possible kidney damage, but in moderate quantities the genus is generally regarded as safe. In our experiments with various species, we found that some have peppery-tasting leaves that can be used in flavoring foods. Others have starchy roots that may be eaten raw, boiled, or roasted. Still others have young foliage that makes good salads or potherbs. In our opinion, of all the species, *Polygonum bistortoides* (bistort) tastes the best. This species is very common in mountain meadows.

The seeds have been used whole or ground into flour. They are described as a prehistoric food source and are frequently found in archaeological remains.

A decoction of the roots can be made for a sore mouth or gums. The root can also be used as an astringent, diuretic, antiseptic, and alterative. The roots were eaten by maritime explorers to prevent scurvy. A traditional European "Easter pudding" is made of bistort, nettle, and dock, all of which are high in vitamin C (Schofield 1989).

Dock, Sorrel (genus *Rumex*)

COMMON SHEEP SORREL (*R. acetosella*)

WESTERN DOCK (*R. aquaticus*)

CURLY DOCK (*R. crispus*)

GOLDEN DOCK (*R. maritimus*)

FEW-LEAVED DOCK (*R. paucifolius*)

WILLOW DOCK (*R. salicifolius*)

VEINY DOCK (*R. venosus*)

Description and Habitat. Docks are annual or perennial herbs. They have small flowers that are greenish and aggregated in a large terminal inflorescence. They can be found in many habitats in the mountains.

FIELD KEY TO THE DOCKS

1. Flowers imperfect, with staminate and pistillate flowers on separate plants 2
1. Flowers perfect, or if imperfect, the male and female flowers are on the same plant 3

2. Inner sepals becoming winged;
leaves with a large lobe at the base on either side . . . . *R. acetosella*
2. Inner sepals not winged; leaves without lobes at base . . . . . . *R. paucifolius*

3. Inner sepals very large, ½ to 1 inch broad;
rose-colored and showy . . . . *R. venosus*
3. Inner sepals less than 1.2 inch broad 4

4. Inner sepals not bearing tubercles;
leaves wavy-margined, the lower one very large . . . . *R. aquaticus*
4. Inner sepals with a swelling or tubercle on the back. 5

5. Inner sepals toothed or fringed. . . . . *R. maritimus*
5. Inner sepals entire, not toothed or fringed. 6

6. Leaves flat . . . . *R. salicifolius*
6. Leaves strongly wavy-curled . . . . *R. crispus*

Natural History Notes and Uses. The young leaves of dock can be used as greens, and we have found that the flavor varies from species to species. The young leaves are best when collected before the flowerstalk emerges. Also, because the leaves become watery when cooked, use very little water and don't overcook them. In most cases, the older leaves are too bitter for use. Euell Gibbons (1966) found that the leaves of dock are high in vitamin C and contain more vitamin A than carrots. Native Americans ground dock seeds and used the meal to make breads. However, removing the papery seed cover involves a lot of work and, depending on the species, is probably more work than it is worth. The distinctive sour taste of these plants is due to oxalic acid. As with other species that contain oxalic acid, docks should be used in small portions as they can cause calcium deficiency.

Poisoning from *Rumex* has been recorded in livestock, but only after large quantities were eaten. Medicinally, the crushed leaves can be applied to boils and the juice of the leaves used to treat ringworms and other skin parasites. The juice of the plant and a poultice of the leaves have also been applied to the rash and pain caused by stinging nettles. A poultice of leaves was used for nervous or allergic hives. The fresh roots were boiled in water to provide a decoction for use internally as a laxative. The powdered yellow roots have been used as a tooth cleanser, laxative, astringent, and antiseptic (Lewis and Elvin-Lewis 1977). Some *Rumex* roots contain as much as 35 percent tannin and were used for tanning animal hides.

PURSLANE FAMILY (PORTULACACEAE)

Nineteen genera and 600 species occur worldwide, of which nine genera are native to the United States. The family is particularly well represented along the Pacific Coast. It is of little economic importance but includes several ornamentals.

FIELD KEY TO THE PURSLANE FAMILY

1. Stamens 3; petals 4; stigmas mostly 2; sepals thin, dry-membranous, not green .... *Spraguea*
1. Stamens 3–50; petals 5–18; stigmas 3–8; sepals green, herbaceous 2

2. Plants with slender taproots, rhizomes, or stolons 3
2. Plants with thick, fleshy taproots or globe- or egg-shaped corms 5

3. Petals yellow; fruits (capsules) opening in a transverse line bisecting the long axis................................. *Portulaca*
3. Petals white, pink, rose, or lavender; capsules opening by valves running longitudinally.................... 4

4. Stem leaves mostly 2 and opposite or even, joined at their bases to form a disc surrounding the stem at the base of the inflorescence...................................... *Claytonia*
4. Stem leaves more than 2, opposite or alternate .... *Montia*

5. Petals and stamens usually 5; capsule dehiscent from the summit by 3 valves .... *Claytonia*
5. Petals and stamens often more than 5; capsule circumscissile near the base .... *Lewisia*

Springbeauty, Miner's Lettuce (genus *Claytonia*)

LANCE-LEAF SPRINGBEAUTY (*C. lanceolata*)

ALPINE SPRINGBEAUTY (*C. megarhiza*)

MINER'S LETTUCE (*C. perfoliata*)

Description and Habitat. The 15–25 species in this genus are succulent herbs with simple leaves and racemes of white flowers. The genus is named for Dr. John Clayton, an American botanist and notable plant collector of colonial days.

FIGURE 88. Lance-leaf springbeauty (*Claytonia lanceolata*).

FIELD KEY TO THE SPRINGBEAUTIES

1. Basal leaves usually 1–2; stems from a globose rhizome .... *C. lanceolata*
1. Basal leaves more than 2; plants taprooted or fibrous-rooted.......... 2

2. Stem leaves linear; alpine perennial...................... *C. megarhiza*
2. Stem leaves wider; plants annual .... *C. perfoliata*

Natural History Notes and Uses. Often called Indian potato, wild potato, or mountain potato, the small corms can be eaten raw, boiled, or roasted. For many Native Americans, springbeauty was an important root vegetable. At first, many find the corms distasteful, as they take a little getting used to. They are high in starch and, when cooked, taste like potatoes. Boil or bake the corm for 30 minutes. Most species are not plentiful, so be conservative, keeping only the largest corms and replanting the others. They can also be dried on strings for long-term storage.

The rosettes can also be eaten raw or cooked and are high in vitamins A and C. They are better when mixed with other salad plants. The leaves of *C. sibirica* (Siberian springbeauty) were soaked and applied to the head to relieve a headache (Schofield 1989).

In some places in the West, Native Americans picked miner's lettuce and placed it near the nests of red ants. The ants were allowed to crawl over the leaves and were then shaken off. The residue they left on the leaves had an acerbic flavor. A tea from the leaves was used as a laxative. A poultice made from the plant was used for rheumatic pains and to stimulate a poor appetite. Miner's lettuce is one of the few native plants of the United States that has been cultivated elsewhere. Introduced into Europe, it is used for salads and as a potherb.

Lewisia (genus *Lewisia*)

PIGMY BITTERROOT (*L. pygmaea*)

OREGON BITTERROOT (*L. rediviva*)

THREE-LEAF LEWISIA (*L. triphylla*)

Description and Habitat. Lewisias are indigenous to the western United States. They can be found clinging precariously to rocky ledges among boulders, on rock-strewn slopes, in damp gravelly places, in alpine meadows, and in near desert conditions where rainfall is seasonal and unpredictable. There are about 18 species, many evergreen, but other are bulb-like in that they are below ground for part of the year. Several species have large, showy flowers.

FIGURE 89. Bitterroot (*Lewisia rediviva*).

FIELD KEY TO THE LEWISIAS

1. Stem leaves 2–3 in a whorl; plant with rounded corms	*L. triphylla*
1. Stem leaves reduced and bract-like; root carrot-shaped	2
2. Sepals 5–9; one flower per naked flowering stalk	*L. rediviva*
2. Sepals 2; 2 to many flowers per stem	*L. pygmaea*

Natural History Notes and Uses. Although all species may be edible, *L. rediviva* is the species that has been used extensively. These plants were an important food item for many Native Americans. The root is remarkably large and thick for a small plant, and it contains highly prized, nutritious starch. The roots are dug up in spring before flowering. Once dug, the root is peeled promptly and the small red "heart" (the embryo of next year's growth) is removed to reduce the root's bitter flavor. It is then steamed, boiled, or pit-cooked. The root can also be dried and will keep for a long time. The bitterness of the root varies, and cooking is said to improve the flavor. The root turns pink when boiled to a jelly-like consistency. The pounded root was chewed for a sore throat (Hart 1996).

Though some still collect it today, bitterroot is considered a rare plant in many areas. There is little evidence, however, that harvesting by Native Americans has contributed to its decline. Overgrazing and trampling by range livestock and habitat destruction from agricultural encroachment seem to have had the most impact on *Lewisia* populations. Remember, digging the roots destroys the plant. Programs to maintain and enhance habitat for the plant are recommended.

Miner's Lettuce, Montiastrum (genus *Montia*)

WATER MINER'S LETTUCE (*M. chamissoi*)

MONTIASTRUM (*M. lineare = Montiastrum l.*)

Description and Habitat. The genus is composed of slightly succulent annual and perennial herbs. The flowers have 2 persistent sepals and 5 white or pinkish petals. Most *Montia* species grow in moist or seasonally wet areas that are partially to fully shaded.

FIELD KEY TO THE MINER'S LETTUCE

1. Plants perennial; stem leaves alternate and linear . . . . *M. lineare*
1. Plants annual or perennial; stem leaves opposite and not linear . . . . *M. chamissoi*

Natural History Notes and Uses. All species of *Montia* have stems and leaves that can be eaten raw or boiled like spinach. The roots are also edible raw or boiled. A tea from the leaves was used as a laxative. A poultice made from the plant was used for rheumatic pains and to stimulate a poor appetite (Strike 1994).

PURSLANE, LITTLE HOGWEED (*Portulaca oleracea*)

Description and Habitat. Purslane is a small, succulent annual herb found at the lower elevations. It has been used as a food for more than 2,000 years in India and Persia. In Europe it is grown as a garden vegetable. The genus name may be derived from the Latin *portula,* meaning "little gate," referring to the lid on the capsule.

Natural History Notes and Uses. The stems and leaves of purslane have a tart taste. The entire above-ground part of the plant can be boiled, steamed, fried, or pickled. The mucilaginous juice of the stems makes a good thickener for soups. Since the plant tends to hold a lot of dirt and grit, you may want to wash it thoroughly. Besides the good flavor, purslane provides vitamins A and C, iron, and calcium. The tiny black seeds are also nutritious. They can be ground and mixed with other flours.

PUSSY PAWS (*Spraguea umbellata = Cistanthe u.*)

Description and Habitat. This is a low, spreading alpine plant with dense pink and white flowers at the apex of a 5-inch stem. The blossoms are short-lived but are followed immediately by cream-colored fruits that closely resemble a blossom. The rosettes of fleshy basal leaves are green touched with rose. They grow in sand and fine gravel at high elevations in the spruce-fir and alpine tundra plant communities. *S. umbellata,* a singular and pretty plant allied to *Claytonia,* grows 6–9 inches high and has fleshy foliage and spikes of showy pinkish blossoms.

Natural History Notes and Uses. The genus name honors Isaac Sprague (1811–95), an American botanical and zoological artist. The genus contains about nine species of annual and perennial herbs with rosettes of simple basal leaves.

PRIMROSE FAMILY (PRIMULACEAE)

Worldwide there are approximately 28 genera and 800 species in this family. Eleven of the genera are native to the United States, mostly to the eastern part of the country. The family as a whole has little or no economic value. Many of the species are very interesting to the scientific observer, for the structure of their flowers is such that they are peculiarly adapted for cross-fertilization (see *Primula*).

FIELD KEY TO THE PRIMROSE FAMILY

1. At least some leaves on flowering stems . . . . *Glaux*
1. Leaves all basal or almost 2

2. Corolla lobes several times longer than tube and sharply reflexed; stamens protruding full length . . . . *Dodecatheon*
2. Corolla lobes less than 2 times the length of tube, not sharply reflexed; stamens not protruding 3

3. Plants perennial with densely tufted, small, narrow persistent leaves; flowers showy pink to violet; calyx somewhat keeled on and below the lobes . . . . *Douglasia*
3. Plants annual or perennial; if perennial and with tufted leaves, then flowers white and calyx not keeled along or below lobes 4

4. Flowers white; plant sometimes with long, grayish hairs . . . . . . *Androsace*
4. Flowers pink to lavender; plant not with gray hairs . . . . *Primula*

Rockjasmine (genus *Androsace*)

SWEETFLOWER ROCKJASMINE (*A. chamaejasme*)

FILIFORM ROCKJASMINE (*A. filiformis*)

PYGMYFLOWER ROCKJASMINE (*A. septentrionalis*)

Description and Habitat. The genus name is from the Greek *andros* ("male") and *sakus* ("buckle"), alluding to the shape of the anther. The genus contains about 100 species of annual and perennial alpine herbs with simple, often tuft-forming leaves and clusters of showy flowers. The petals are fused at the base. The genus differs from *Primula* in that the petal tube is shorter than the sepals and somewhat constricted at the mouth.

FIELD KEY TO THE ROCKJASMINES

1. Plants are mat-forming perennials; herbage has long hairs . . . . *A. chamaejasme*
1. Plants annual or biennial; herbage glabrous minutely hairy 2

2. Leaf blades abruptly narrowed to a petiole; corolla lobes reflexed . . . . *A. filiformis*
2. Leaf blades tapering to the base; corolla lobes spreading to erect . . . . *A. septentrionalis*

Natural History Notes and Uses. A compound decoction was made from the whole plant of western androsace (*A. occidentalis*) and used for postpartum bleeding. A cold infusion of northern androsace (*A. septentrionalis*) was taken for internal pain. The plant was also used as a lotion to give protection from witches.

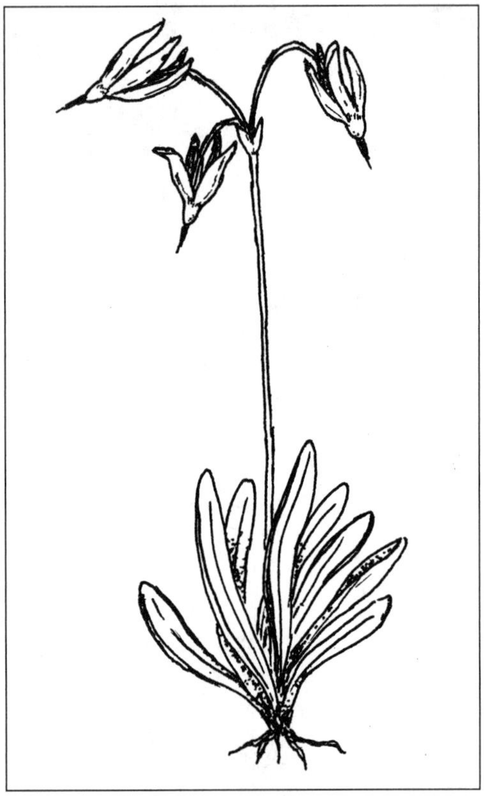

Shooting Star (genus *Dodecatheon*)

BONNEVILLE SHOOTING STAR (*D. conjugens*)

TALL MOUNTAIN SHOOTING STAR (*D. jeffreyi*)

DARKTHROAT SHOOTING STAR (*D. pulchellum*)

FIGURE 90. Shooting star (*Dodecatheon* spp.).

Description and Habitat. All leaves are basal and form a loose rosette. The flowers are located at the end of a stalk with narrow, reflexed, rose-colored petals. The distinctive shape of the shooting star comes from the swept-back petals and the forward-pointed, pollen-bearing stamens. When blooming, the flowers nod toward the ground; then, as they dry and the petals drop, the dried seed capsules turn to point straight up. Other descriptive common names are Mosquito Bills, Prairie Pointers, Roosters' Heads, and Sailor Caps. The species' habitats range from grassland to shrubland, meadows, and riparian habitats up to the alpine zone.

FIELD KEY TO THE SHOOTING STARS

1. Stigma in a dense head-like cluster, about twice as wide as the style	*D. jeffreyi*
1. Stigma not as above, a little if at all wider than the style	2
2. Leaves glandular-hairy	*D. conjugens*
2. Leaves not hairy	*D. pulchellum*

Natural History Notes and Uses. The genus name *Dodecatheon* comes from two Greek words: *dodeka*, meaning "twelve," and *theoi*, meaning "gods," referring to plants protected by the gods. The name was first used by Pliny for the primrose, which was thought to be under the care of the twelve principal Greek gods. Linnaeus used it for the shooting star, a mostly North American flower in the primrose family. There are 30–50 species of shooting stars in the world, depending on who is doing the counting. At least one is from the arctic of northeastern Asia. Most are from western North America, often of arctic or alpine tundra. *Dodecatheon meadia* is the largest, growing up to 2 feet or more in height and having up to 30 flowers in its inflorescence. Shooting stars are pollinated by bumblebees and bloom from mid-April through June, when the leaves disappear. The seed capsules turn upright, and their 5 sepals form a star opening from which a multitude of tiny seeds shake out whenever nudged by the wind or passing animals.

Since none of the species are listed anywhere as poisonous, it is likely that all the species may be edible. It is usually the texture that discourages people from using the plants. We have found that the leaves of many species have a good flavor when eaten raw. Weedon (1996) and Strike (1994) also indicate that the roots and leaves of *D. hendersonii* (mosquito bills) are edible after roasting or boiling. Scully (1970) believes that at least five species of shooting star in the Rocky Mountains were used by American Indians and that they ate the green leaves and roasted the roots. Thompson and Thompson (1972:120) provide some additional insight into their preparation of *D. jeffreyi*: "We tried eating the leaves of Shootingstar raw, but decided that their texture made them unappealing to chew on. At least they do not seem to become bitter, even after the flowers are blooming. When boiled for about 15 minutes and seasoned with butter and salt, they make a satisfactory but bland green vegetable."

Very little information regarding the medicinal uses of shooting stars could be found except for the fact that the Native Americans of the Northwest used a leaf tea as a treatment for cold sores (Pojar and MacKinnon 1994).

ROCKY MOUNTAIN DWARF PRIMROSE (*Douglasia montana*)

Description and Habitat. Dwarf primrose is a low plant with small, awl-shaped leaves that are crowded at the ends of short branches. The slender flowering stalk usually bears a solitary purple or lilac-colored flower.

Natural History Notes and Uses. This species was once placed within the genus *Androsace* (rockjasmine), a genus with about 100 species of annual and perennial alpine herbs with simple, often tuft-forming leaves and clusters of showy flowers that bear 5 basally fused petals. Many species are grown as ornamentals.

SALTWORT, SEA MILKWORT (*Glaux maritima*)

Description and Habitat. This species has small, white flowers with 5 petals that are mostly fused except for terminal petal lobes. The flowers are sessile and arise from the leaf axils. The stem is green and smooth, and the leaves are fleshy, small, ovate, sessile to the stem, and in opposite pairs. The plant grows 2–12 inches tall and in salty soils.

Natural History Notes and Uses. Sea milkwort could be superficially confused with some species in the family Euphorbiaceae. However, *Euphorbia* flowers have a complicated structure whereas sea milkwort flowers are typical white flowers with 5 petals. The young shoots are said to be edible raw. The fleshy leaves and stems can be pickled.

Primrose (genus *Primula*)

GREENLAND PRIMROSE (*P. egaliksensis*)

SILVERY PRIMROSE (*P. incana*)

PARRY'S PRIMROSE (*P. parryi*)

Description and Habitat. This large group of plants is commonly known as primroses. They are mostly hardy perennial herbs and are natives of Europe and temperate Asia, Java, and North America. The generic name is from the diminutive of the Latin word *primus*, meaning "first."

FIELD KEY TO THE PRIMROSES

1. Plant with mealy, usually whitish covering .... *P. incana*
1. Plant not mealy 2

2. Flowers reddish-purple or pinkish; limb 1.5 mm or more wide . . *P. parryi*
2. Flowers white, pink, or lilac; limb less than 1.5 mm wide. . . *P. egaliksensis*

Natural History Notes and Uses. Next time you come across a primrose, take a close look at the flowers and you may notice something very interesting. In fact, you will see something that Charles Darwin once studied. Primroses have two types of flowers: *long-pistil,* whose pistil is approximately twice as long as the stamens, and *short-pistil,* whose stamens are approximately twice as long as the pistil. This feature ensures that the plants are cross-pollinated. In nature, both flower types can be found with equal frequency. Although cross-pollination of two flowers of the same type (and even self-pollination) is possible, the cross-pollination of flowers of different types produces more viable seeds.

A number of different insects visit the flowers in search of nectar, which is located at the bottom of the flower tube. This positioning means that only long-tongued insects can actually reach the nectar at the base.

The young leaves of some primroses were used in salads. The leaves of these related species contain many vitamins. Additionally, some species are said to possess curative properties that the ancient Greeks used in the treatment of coughs, tuberculosis, rheumatism, and insomnia.

BUTTERCUP FAMILY (RANUNCULACEAE)

There are 35–70 genera and 2,000 species in the buttercup family worldwide, all in the cooler regions of the Northern Hemisphere. Twenty-one genera are native to the United States. Many plants in this family are poisonous, some are grown as ornamentals, and others provide drugs.

Caution: Only a few plants in this family were eaten. Most contain an irritating compound, protoanemonin, in the fresh leaves, stems, roots, flowers, and seeds. All must be cooked before eating.

FIELD KEY TO THE BUTTERCUP FAMILY

1. Flowers irregular, mostly dark blue or purple 2
1. Flowers regular, seldom dark blue or purple 3

2. Upper sepal spurred at the base; petals usually 4 . . . . *Delphinium*
2. Upper sepal not spurred, hooded at apex, petals usually 2 . . . . . *Aconitum*

3. Flowers with sepals or petals spurred 4
3. Flowers not spurred 5

4. Leaves simple, linear to narrowly spatulate;
plants small annuals . . . . *Myosurus*
4. Leaves 1–3 times compound; plants perennial. . . . . *Aquilegia*

5. Perianth consisting of a single whorl (called sepals
but look like petals). 6
5. Perianth of sepals and petals, but the sepals
sometimes deciduous and leaving scars below the petals 10

6. Leaves simple; plants of wet meadows. . . . .*Caltha*
6. Leaves deeply lobed or compound. 7

7. Leaves alternate; flowers greenish 8
7. Leaves opposite 9

8. Leaves simple and palmately lobed or parted............. *Trautvetteria*
8. Leaves compound .... *Thalictrum*

9. Leaves whorled and sessile or short-petioled................ *Anemone*
9. Leaves opposite, with petioles.............................. *Clematis*

10. Flowers inconspicuous, numerous, in dense
terminal racemes; fruit red or white fleshy berries .... *Actaea*
10. Flowers solitary or few; fruit an achene or follicle................... 11

11. Petals linear, narrower than the white to cream petaloid sepals... *Trollius*
11. Petals as broad or broader than the usually green sepals,
or lacking... *Ranunculus*

COLUMBIAN MONKSHOOD (*Aconitum columbianum*)

Description and Habitat. Monkshood is a perennial herb with palmately divided or lobed leaves. The flowers are usually deep blue or purple but may also be pale to white. The plant is usually found in moist, densely shaded places, often with streamside vegetation, up to timberline.

Natural History Notes and Uses. Some species of monkshood have been a source of painkillers or sedatives for nervous disorders. However, all parts of the plant are ***poisonous*** and should be considered dangerous if ingested. The drug aconitine from *A. columbianum* was used to treat pain from neuralgia, toothache, and sciatica. Aconitine is one of the most toxic plant compounds known. Nevertheless, a number of different species are used medicinally in various parts of the world, with apparently beneficial therapeutic effects. In China, processing was the key. For example, used raw, monkshood was an arrow poison, whereas steamed, it was an internal medicine to help improve the digestive system.

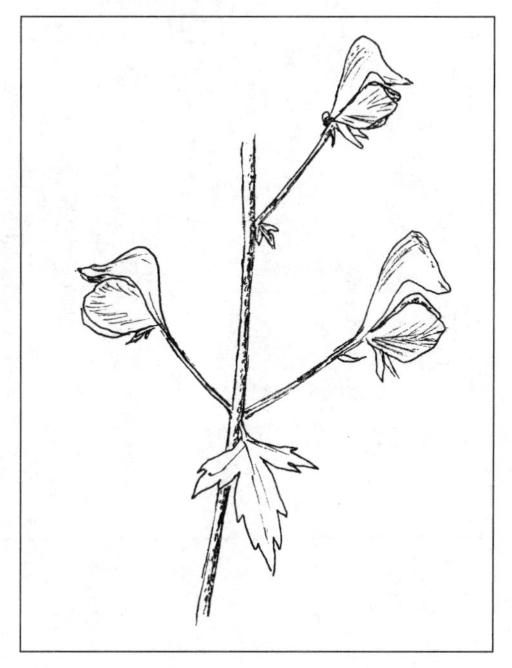

FIGURE 91. Monkshood (*Aconitum* spp.).

RED BANEBERRY (*Actaea rubra*)

Description and Habitat. Red baneberry is a perennial herb with fibrous roots. The leaves have long petioles and are 2–3 times divided into sharply toothed, lance-shaped segments. The small flowers are white and borne in a branched, congested, hemispheric inflorescence. The fruits are shiny red or white. Red baneberry is common in moist, montane forests and riparian areas, usually with some partial shade.

Natural History Notes and Uses. The entire plant, especially the berries, is ***poisonous***. The plant is sometimes confused with *Osmorhiza chilensis* (western sweetroot), which often shares the same habitat. However, unlike red baneberry, sweetroot has a strong licorice-like odor.

The roots are considered a laxative and can cause vomiting. The roots were also ground, mixed with grease or tobacco, and rubbed on the body to treat rheumatism (Bacon 1903). Ground seeds mixed with pine pitch were applied as a poultice for neuralgia. *Actaea arguta* is described by Moore (1979) as moderately

poisonous when taken internally, with cardiac arrest possible from large doses. The powdered root was mixed with hot water and applied as a counter-irritant.

Warning: If large quantities of this plant are consumed, it may cause cardiac arrest.

Wind Flower (genus *Anemone*)

LITTLE BELT MOUNTAIN THIMBLEWEED (*A. lithophila*)

PACIFIC ANEMONE (*A. multifida*)

SMALL-FLOWERED ANEMONE (*A. parviflora*)

AMERICAN PASQUEFLOWER (*A. patens = Pulsatilla p.*)

TETON ANEMONE (*A. tetonensis*)

Description and Habitat. Wind flowers are perennial herbs from a rootstalk or rhizome. The basal leaves are palmately lobed or divided, whereas the stem leaves are in whorls, each with 2–4 compound or simple leaves. There are no petals in the flowers, but the 5 to many sepals resemble petals. The plants can be found in dry to moist meadow areas from the foothills to the alpine zone.

FIELD KEY TO THE WIND FLOWERS

1. Sepals generally ¾ to 1½ inches long	*A. patens*
1. Sepals less than ¾ inch	2
2. Achenes pubescent or glabrate	*A. tetonensis*
2. Achenes densely woolly	3
3. Plant with 2 or more flowers	*A. multifida*
3. Plant with usually 1 flower	4
4. Plants arising from slender rootstalks	*A. parviflora*
4. Plants arising from a short erect caudex	*A. lithophila*

Natural History Notes and Uses. The plants are sometimes called "towhead babies" because of the fuzzy appearance of the fruiting heads. The autumn winds carry the seeds, whose feathery tails act as parachutes in the dispersal to new habitats.

Native Americans made a poultice from the leaves of related species, *A. cylindrica* (candle wind flower), to treat rheumatism and burns (Coffey 1993). The roots of *A. globosa* (= *A. multifida*) were used for treating wounds. *A. patens* contains a volatile oil used in medicine as an irritant (Craighead et al. 1963).

Columbine (genus *Aquilegia*)

COLORADO BLUE COLUMBINE (*A. caerulea*)

YELLOW COLUMBINE (*A. flavescens*)

WESTERN COLUMBINE (*A. formosa*)

JONES' COLUMBINE (*A. jonesii*)

Description and Habitat. Columbines are perennials with ternately compound leaves. There are 5 petaloid sepals and 5 petals, each ending in a spur. The genus name stems from the Latin word for "eagle." It may also come from the Latin *aqua*, "water," and *legere*, "to collect," perhaps referring to the nectar that collects at the tips of the spurs.

FIELD KEY TO THE COLUMBINES

1. Sepals are blue, purple, or white 2
1. Sepals are red or yellow 3

2. Spurs are ¾ to 2 inches long .... *A. caerulea*
2. Spurs are less than ½ inch long............................ *A. jonesii*

3. Sepals and spurs are red................................ *A. formosa*
3. Sepals and spurs are yellow .... *A. flavescens*

Natural History Notes and Uses. The flowers of western columbine are edible and have a sweet taste. They can be added to salads in small amounts. Weedon (1996) indicates that the leaves of this species are also edible but grow bitter with age.

A tea made from the roots of western columbine is said to stop diarrhea, and the fresh roots can be mashed and rubbed on aching joints. Aboriginal peoples used various parts of the plants in medicinal preparations for diarrhea, dizziness, aching joints, and possibly venereal disease. The root, boiled with *Ipomopsis aggregata* (scarlet gilia), resulted in a brew that induced vomiting (Strike 1994). Ripe seeds can be mashed and rubbed into the hair to discourage lice.

Warning: The seeds can be fatal if eaten, and most parts of columbine contain cyanogenic glycosides. Any therapeutic use of columbine is strongly discouraged.

FIGURE 92. Columbine (*Aquilegia formosa*).

WHITE MARSH-MARIGOLD (*Caltha leptosepala*)

Description and Habitat. The leafless flowering stem of this plant has 1 or 2 flowers with 5–15 white or blue-tinged, petal-like sepals. There are no petals. The leaves are dark green and basal. Marsh-marigold can be found in marshes and wet meadows to above timberline, where it blooms close to receding snowbanks. It is not related to the cultivated marigold, a member of the sunflower family. Its nickname dates back to the Middle Ages, when a plant (*Caltha*) dedicated to the Virgin Mary was widely used in church celebrations. The flowers were fermented for making wine. The genus name is from the Greek *calathos*, meaning "cup."

Natural History Notes and Uses. The young leaves can be used as a potherb, and the spaghetti-like roots can be dug up during the winter and boiled as a pasta substitute. Though the plant is poisonous when raw (it contains the poisonous glucoside protoanemonin), cooking appears to destroy the poison. Marsh-marigold also contains the deadly glucoside hellebrin, which breaks down with boiling (Mitchell and Dean 1982).

The roots have diaphoretic, emetic, and expectorant properties. The leaves are diuretic and laxative, and a tea from the leaves mixed with maple sugar was used as a cough syrup by the Ojibwa

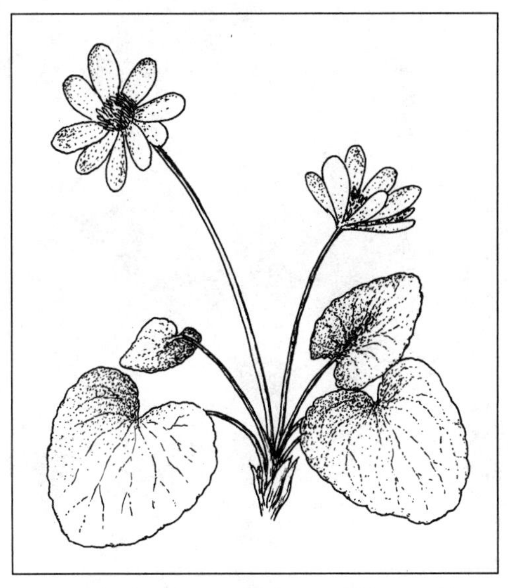

FIGURE 93. Marsh-marigold (*Caltha leptosepala*).

(Wilford et al. 1916). The tea was also used as an antispasmodic and expectorant for treating cramps and convulsions. In *Stalking the Healthful Herbs* Euell Gibbons (1966) reports that a drop of juice squeezed from the fresh leaves is caustic and will remove warts.

Clematis, Virgin's Bower (genus *Clematis*)

ROCK CLEMATIS (*C. columbiana*)

HAIRY CLEMATIS (*C. hirsutissima*)

WESTERN WHITE CLEMATIS (*C. ligusticifolia*)

WESTERN BLUE VIRGIN'S BOWER (*C. occidentalis*)

Description and Habitat. Clematis are herbaceous perennials with erect stems or woody vines. The leaves are opposite or whorled and simple to pinnately compound. The flowers, lacking petals, are solitary or borne in an open, pyramid-shaped inflorescence. The sepals are petal-like. The various species can be found from brushy slopes above creek bottoms to open areas from the low to high elevations.

FIELD KEY TO THE CLEMATIS

1. Flowers few to many in clusters; sepals white . . . . *C. ligusticifolia*
1. Flowers solitary; sepals yellow to blue or purple 2

2. Plants vine-like; leaves ternate to biternate; sepals glabrous or only very slightly hairy on dorsal side. 3
2. Plants not vine-like; leaves pinnately compound; sepals densely hairy on dorsal side . . . . *C. hirsutissima*

3. Leaves bi- or triternately compound with the ultimate segments usually sharply toothed. . . . . *C. columbiana*
3. Leaves only ternate, the leaflets bluntly toothed . . . . *C. occidentalis*

Natural History Notes and Uses. The genus is essentially composed of ***poisonous*** species. Many references list western virgin's bower as poisonous even though the stems and leaves have been chewed by Native Americans as a remedy for colds and sore throats. The plants have a peppery taste and may cause lightheadedness. Tilford (1993) also indicates that western virgin's bower is diaphoretic and diuretic and offers a unique vasoconstrictory/dilating action that makes it useful in the treatment of migraine headaches. The Thompson Indians used the plant to make a head wash for scabs and eczema, and a mild decoction was drunk as a tonic (Teit 1930). Sweet (1976) states that the white portion of the bark was used for fever, the leaves and bark for shampoo, and a decoction of the leaves for sores and cuts on horses. The fiber in the bark was used for snares and carrying nets. The dried stalks were used in fire-by-friction sets, and the feathery seed tails for tinder.

Caution: The consumption of *Clematis* may cause internal bleeding. The entire genus contains strong chemical constituents that can irritate the skin and mucous membranes.

Larkspur (genus *Delphinium*)

LITTLE LARKSPUR (*D. bicolor*)

LARKSPUR (*D. burkei*)

GEYER'S LARKSPUR (*D. geyeri*)

SIERRA LARKSPUR (*D. glaucum*)

NUTTALL'S LARKSPUR (*D. nuttallianum*)

LARKSPUR (*D. occidentale*)

Description and Habitat. Larkspurs are perennial herbs with tuberous or fibrous roots and erect stems. The leaves are roundish in outline and deeply lobed or divided. The flowers are showy, blue to partly white, containing 5 petal-like sepals with the uppermost prolonged into a spur. There are also 4 petals, 2 partly enclosed by the upper sepals, the lower 2 often hairy and lobed at the tip. They can be found in various habitats, including meadows, thickets, and open woods from low to high elevations.

FIELD KEY TO THE LARKSPURS

1. Root system shallow, fibrous to fleshy 2
1. Root system appearing deep and woody 4

2. Sepals pointed forward, not flared, less than ½ inch long . . . . . . *D. burkei*
2. Sepals flared, usually more than ½ inch long 3

3. Lower petals shallowly notched; sepals unequal, lower pair longer. . . . . *D. bicolor*
3. Lower petals deeply notched; sepals nearly equal. . . . . . . *D. nuttallianum*

4. Sepals mostly pointed forward, not flared; leaves rarely dissected into linear segments. 5
4. Sepals mostly flared; leaves often dissected into linear segments 6

5. Ovary and fruit usually glabrous . . . . *D. glaucum*
5. Ovary and fruit hairy . . . . *D. occidentale*

6. Upper petals whitish with prominent purple or blue lines; leaves glabrous . . . . *D. bicolor*
6. Upper petals whitish with blue tips, lacking prominent purple or blue lines; leaves hairy. . . . . *D. geyeri*

Natural History Notes and Uses. The generic name *Delphinium* comes from the Latin for "dolphin," which the shape of the flowers somewhat resembles. The juices or seeds of many of the delphiniums were used by various native peoples worldwide as insecticides, internal parasiticides, and to control lice and ticks; the flowers were made into green or orange dyes. Seeds of prairie larkspur (*D. virescens*) were used by the Kiowa tribe in their ceremonial rattles. Strike (1994) indicates that some *Delphinium* roots were dried, pulverized, mixed with water, and used by Native Americans as a salve for swollen limbs. Cattle and horses can contract the usually fatal disease delphinosis from eating delphiniums (Muenscher 1962). The plants should therefore be regarded as ***poisonous.***

Mousetail (genus *Myosurus*)

BRISTLE MOUSETAIL (*M. aristatus = M. apetalus*)

TINY MOUSETAIL (*M. minumus*)

Description and Habitat. About 15 species of small, annual, herbaceous (nonwoody) plants constitute the genus *Myosurus*. They generally occur in the temperate zones of both the Northern and Southern Hemispheres. Mousetails are so named for a long, slender column covered with pistils (female seed-bearing organs) that arises from the center of the flower.

FIELD KEY TO THE MOUSETAILS

1. Sepal blade definitely longer than spur; beak of achene less than ¼ inch long....................... *M. minimus*
1. Sepal blade about as long as spur; beak of achene greater than ¼ inch long................................ *M. aristatus*

Natural History Notes and Uses. Although tiny flies have been observed visiting mousetails, insects apparently are not necessary to transfer pollen. Stone (1959) noted that tiny mousetails are predominantly self-pollinating. Pollen is shed before the flower opens, when the pistils and stamens are covered by the sepals. Fertilization does not take place until 3–10 days later, which ensures that pollen will reach all the pistils that have developed. After the pollen is shed, the flower opens.

Buttercup (genus *Ranunculus*)

LITTLELEAF BUTTERCUP (*R. abortivus*)

SHARPLEAF BUTTERCUP (*R. acriformis*)

ALPINE BUTTERCUP (*R. adoneus*)

PLANTAINLEAF BUTTERCUP (*R. alismifolius*)

WHITEWATER CROWFOOT (*R. aquatilis*)

ALKALI BUTTERCUP (*R. cymbalaria*)

ESCHSCHOLTZ'S BUTTERCUP (*R. eschscholtzii*)

SPEARWORT BUTTERCUP (*R. flammula*)

SAGEBRUSH BUTTERCUP (*R. glaberrimus*)

GRACEFUL BUTTERCUP (*R. inamoenus*)

UTAH BUTTERCUP (*R. jovis*)

MACOUN'S BUTTERCUP (*R. macounii*)

NODDING BUTTERCUP (*R. natans = R. hypoboreus*)

SUREFOOT BUTTERCUP (*R. pedatifidus*)

PYGMY BUTTERCUP (*R. pygmaeus*)

HOOKED BUTTERCUP (*R. uncinatus*)

Description and Habitat. The many species of buttercup are either perennial or occasionally annual herbs with simple to compound leaves. The flowers are solitary or borne in a small inflorescence. The 5 petals are normally yellow or white and have a nectar gland at the base. They can be found in many different habitats from the lower elevations to the alpine zone. The genus name is from the Latin *rana*, "frog," and refers to the wet habitat of some species. Mature fruits as well as flowers are needed for positive identification.

FIELD KEY TO THE BUTTERCUPS

1. Plants growing in water 2
1. Plants terrestrial, maybe growing in wet places; flowers yellow 3

2. Flowers white; leaves thread-shaped . . . . *R. aquatilis*
2. Flowers yellow; leaves kidney-shaped, 3-lobed . . . . *R. natans*

3. Plants with stolons, or rooting at the nodes; plants growing in mud or water 4
3. Plants without stolons, not rooting at the nodes 5

4. Leaves linear to lanceolate . . . . *R. flammula*
4. Leaves heart-shaped, kidney-shaped, or ovate . . . . *R. cymbalaria*

5. Leaves compound 6
5. Leaves simple, entire to deeply cleft 9

6. Stems glabrous, at least at maturity 7
6. Stems pubescent 8

7. Plants small, 1½ to 4 inches tall; basal leaves missing; sepals glabrous; petals barely exceeding broader sepals; leaf divided into 3–5 narrowly oblanceolate and entire lobes . . . . *R. jovis*
7. Plants taller; petals exceed sepals in length . . . . *R. adoneus*

8. Stems appressed-pubescent . . . . *R. acriformis*
8. Sepals hirsute or hispid, spreading hairs . . . . *R. macounii*

9. Leaves all simple, entire to slightly serrate or toothed; petals longer than sepals . . . . *R. alismifolius*
9. Leaves deeply parted or divided 10

10. Some leaves entire and some 3–5-lobed 11
10. None of the leaves entire 12

11. Basal leaves elliptic to oblanceolate . . . . *R. glaberrimus*
11. Basal leaves kidney- to heart-shaped . . . . *R. abortivus*

12. Stems ¾ to 3 inches tall . . . . *R. pygmaeus*
12. Stems more than 3 inches tall 13

13. Leaves deeply 3-parted 14
13. Leaves often deeply and many-parted, but not deeply 3-parted 15

14. Stems stout, hispid, the hairs reddish-brown . . . . *R. uncinatus*
14. Stems glabrous; sepals hairy on outside . . . . *R. eschscholtzii*

15. Leaves obovate to ovate . . . . *R. inamoenus*
15. Leaves heart-shaped . . . . *R. pedatifidus*

Natural History Notes and Uses. All species are more or less ***poisonous*** when raw. The leaves and stems should be boiled in several changes of water to remove the poisonous compounds (Kingsbury 1964). The volatile toxin is also rendered harmless by drying. The seeds can be parched and ground into meal for bread or pinole. The roots can also be boiled and eaten and were an important part of some Native American diets. A yellow dye can be obtained by crushing and washing the flowers.

Meadow-rue (genus *Thalictrum*)

ALPINE MEADOW-RUE (*T. alpinum*)

FENDLER'S MEADOW-RUE (*T. fendleri*)

WESTERN MEADOW-RUE (*T. occidentale*)

FEW-FLOWER MEADOW-RUE (*T. sparsiflorum*)

VEINY MEADOW-RUE (*T. venulosum*)

Description and Habitat. Meadow-rues are rhizomatous, erect perennial herbs. The alternate leaves are 2–4 times branched into ultimate leaflets that are petiolate and shallowly lobed or toothed, closely resembling the leaves of columbine (*Aquilegia*). There are no petals, and the 4–5 sepals fall soon after opening. The fruit is a ridged or nerved achene with a persistent style (beak). Mature achenes are important for identification.

FIELD KEY TO THE MEADOW-RUES

1. Flowers usually unisexual 2
1. Flowers with both stamens and pistils 4

2. Achenes spreading to reflexed, elliptical to spindle-shaped; stigma often purple . . . . *T. occidentale*
2. Achenes usually ascending or erect, obliquely elliptical in outline; stigma not purple 3

3. Achenes flat . . . . *T. fendleri*
3. Achenes round in cross section . . . . *T. venulosum*

4. Leaves occurring on stems . . . . *T. sparsiflorum*
4. Leaves mostly basal . . . . *T. alpinum*

Natural History Notes and Uses. The dried plant of Fendler's meadow-rue was sprinkled on a fire or rolled into a cigarette and smoked to treat headaches (Pfeiffer 1922). The young leaves of *T. occidentale* (western meadow-rue) are said to be edible. A tea was made from the roots as a cure for colds and venereal disease. The roots dried and powdered can be used as a shampoo. Additionally, thalicarpine, a substance used in cancer treatment, has been isolated from *T. pubescens*, *T. revolutum*, and *T. dasycarpum*, species found in eastern North America (Mitchell and Dean 1982).

CAROLINA BUGBANE (*Trautvetteria caroliniensis*)

Description and Habitat. This erect, nearly glabrous plant is much branched above and has lower palmately lobed leaves that are long-petioled. The flowers have no petals. It is found in swamps and along streams.

Natural History Notes and Uses. This species was used by the Bella Coola as a dermatological aid, and a poultice of the roots was applied to boils.

AMERICAN GLOBEFLOWER (*Trollius laxus*)

Description and Habitat. Globeflower has showy white flowers with golden centers. The leaves are deeply 5- to 7-parted and occur at the base of the stem. Globeflower is found in wet alpine meadows and along marshy borders of higher-elevation streams.

Natural History Notes and Uses. This attractive, large-flowered plant is often associated with *Caltha leptosepala* in cold, wet sites at higher elevations. In reality, there are no petals, but rather 5–10 petal-like sepals, bright yellow stamens, and green pistils. The undersides of the sepals have a rose-green tinge, which is most easily seen when the flowers are partly closed. At this stage they look like small globes, as suggested by the common name.

BUCKTHORN FAMILY (RHAMNACEAE)

Of the approximately 60 genera and 900 species found worldwide, 10 genera are native to the United States. Economic uses of the Rhamnaceae are chiefly as ornamental plants and as the source of many brilliant green and yellow dyes. The wood of *Rhamnus* was also the most favored species to make charcoal for use in gunpowder before the development of modern propellants.

FIELD KEY TO THE BUCKTHORN FAMILY

1. Leaves with 3 prominent veins running from common origin at base to leaf margins (palmately veined); fruit a capsule..... *Ceanothus*
1. Leaves with a prominent leaf midrib and the other prominent veins lateral and ascending from all along the midrib; fruit a berry... *Rhamnus*

SNOWBRUSH CEANOTHUS (*Ceanothus velutinus*)

Description and Habitat. These shrubs have more or less leathery leaves. Three prominent veins originate near the base of the egg-shaped leaves on some species, an important distinguishing feature. The flowers are small and blue or white. Look for these plants on open and dry montane slopes. The seeds of this species may lie dormant in the soil for hundreds of years and require heat to germinate. The presence of this shrub often indicates that fire has passed through the area.

Natural History Notes and Uses. The genus has long been recognized as a substitute for commercial black tea, and the leaves and flowers can be used to make tea. The seeds can also be used as food. An infusion of the bark may be taken as a tonic. Many species contain saponin, which gives the flowers and fruits their soap-like qualities. The flowers when crushed and rubbed in water produce a light lather. The leaves can also be used as a tobacco substitute. The long, flexible shoots were used in basketry. The red roots yield a red dye.

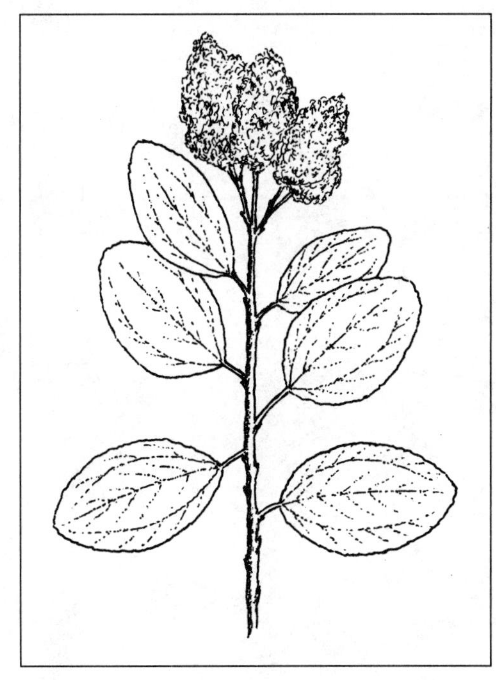

FIGURE 94. Snowbrush ceanothus (*Ceanothus velutinus*).

ALDERLEAF BUCKTHORN (*Rhamnus alnifolia*)

Description and Habitat. This shrub or small tree has flowers that are greenish yellow and 4- or 5-parted. Alderleaf buckthorn is found in wet or moist soils at low elevations. Buckthorn is also known as coffee berry.

Natural History Notes and Uses. A decoction of the leaves of this species was used by some Native Americans to soothe rashes

A number of plants were used by Native Americans to produce a laxative or purgative effect. This use may be related to the foods they ate and the resulting gastrointestinal problems they created. On the other hand, it may be related to a ceremonial or ritual requirement that required purging the body.

caused by poison ivy. The inner bark provided a purgative and a laxative. For relief from toothache, a heated piece of coffee berry root was held against the aching tooth.

The Kawaiisu used the crushed berries of coffee berry as a salve on burns, wounds, and sores to prevent infection. The berries stopped hemorrhages, counteracted poisons, and had a laxative effect.

ROSE FAMILY (ROSACEAE)

The rose family consists of approximately 100 genera and 3,000 species worldwide; the family is particularly common in Europe, Asia, and North America. About 50 genera occur in the United States. The family is of considerable economic importance because of the edible fruits (apples, pears, cherries, plums, peaches, apricots, blackberries, raspberries, and strawberries, among others) and many ornamentals.

FIELD KEY TO THE ROSE FAMILY

1. Plants herbaceous, maybe woody at the base and usually forming low mats 2
1. Plants trees or shrubs, not low and mat-forming 18

2. Leaves simple, entire or toothed, rarely with shallow rounded lobes. 3
2. Leaves compound, or simple and deeply lobed or cleft into narrow segments 6

3. Leaves white-tomentose underneath; green above; subalpine and above . . . . . *Dryas*
3. Leaves green on both sides; below alpine 4

4. Leaves usually over ½ inch long; plants not mat forming . . . . . *Spiraea*
4. Leaves about ½ inch or less long, less than ⅛ inch wide; usually forming mats on rocks or in rock crevices 5

5. Flowers solitary, not exceeding leaves or almost . . . . . *Kelseya*
5. Flowers in spike-like racemes or panicles, much exceeding leaves . . . . . *Petrophyton*

6. Leaves with 3 broad leaflets 7
6. Leaves with more than 3 leaflets or divided into narrow divisions many times longer than wide 11

7. Calyx lobes usually 5 or 6, not alternating with bracteoles; petals pink, rose-purple, or white. . . . . *Rubus*
7. Calyx lobes usually 5, alternating with bracteoles; petals yellow or white 8

8. Stamens 5 . . . . . *Sibbaldia*
8. Stamens 10 or more. 9

9. Petals white; plants with stolons. . . . . *Fragaria*
9. Petals yellow or white; plants without stolons 10

10. Ovary and achene hairy; style persistent, straight or with a hook toward the tip. . . . . *Geum*
10. Ovary and achene glabrous; style usually deciduous, straight, of 1 segment . . . . . *Potentilla*

11. Petals lacking; sepals 4 . . . . . *Sanguisorba*
11. Petals present; sepals usually 5, rarely more. 12

12. Hooked bristles present near base of the calyx lobes .... *Agrimonia*
12. Hooked bristles lacking near base of calyx lobes 13

13. Stamens 5 .... *Ivesia*
13. Stamens 10 or more 14

14. Stamens 10, the filaments broadened .... *Horkelia*
14. Stamens generally more than 10, filaments filiform 15

15. Calyx lobes not alternating with bracteoles .... *Rubus*
15. Calyx lobes alternating with bracteoles 16

16. Leaves palmately compound .... *Potentilla*
16. Leaves pinnately compound or dissected 17

17. Ovary and achene hairy; style persistent, straight or
with a hook toward the tip .... *Geum*
17. Ovary and achene glabrous; style usually deciduous,
straight, of 1 segment .... *Potentilla*

18. Leaves compound 19
18. Leaves simple 22

19. Stems spiny or prickly, the prickles or spines sometimes sparse 20
19. Stems not spiny or prickly 21

20. Stipules attached to petiole most of their length;
leaflets mostly 5 or more .... *Rosa*
20. Stipules lacking or free most of their length; leaflets mostly 3 ..... *Rubus*

21. Leaflets toothed, about 1 inch wide .... *Sorbus*
21. Leaflets entire, less than ¼ inch wide .... *Pentaphyloides*

22. Leaves palmately veined .... *Rubus*
22. Leaves pinnately veined 23

23. Branches with spines or thorns 24
23. Branches without spines or thorns 25

24. Leaves doubly toothed or lobed, at least near tip .... *Crataegus*
24. Leaves not doubly toothed or lobed .... *Prunus*

25. Leaves 3-parted or lobed at tip, otherwise entire,
tomentose beneath .... *Purshia*
25. Leaves not 3-lobed or parted, not tomentose underneath 26

26. Leaves palmately 3–5 lobed 27
26. Leaves not lobed, maybe shallowly pinnately lobed 28

27. Plants with some branched hairs .... *Physocarpus*
27. Plants with simple hairs .... *Rubus*

28. Leaf blades ovate to orbicular in shape, somewhat rounded
at the base, or if occasionally broadly elliptic, then with
acute tips and serrate margins throughout 29
28. Leaves obovate or narrowly elliptic, rarely ovate, tapering
to base, often obtuse to rounded at tip, usually coarsely toothed
or lobed or entire, the teeth not extending to base of leaf 31

29. Leaves acuminate at tip, serrate or serrulate throughout,
glabrous except along veins .... *Prunus*
29. Leaves not as above 30

30. Leaves with mostly 5-8 pair of prominent lateral veins,
glabrous or almost; pistils 3 or more .... *Spiraea*
30. Leaves with mostly 8-12 pair of prominent lateral veins,
or if fewer, then prominently hairy at least underneath;
pistils solitary .... *Amelanchier*

31. Leaves entire, the margins usually rolled .... *Cercocarpus*
31. Leaves toothed or lobed, margins not rolled 32

32. Leaves minutely glandular-toothed to near base or conspicuously toothed with the teeth not extending to base; petals present .... *Prunus*
32. Leaves conspicuously toothed, teeth usually not extending to near base; petals lacking 33

33. Leaves glabrous .... *Spiraea*
33. Leaves hairy .... *Cercocarpus*

TALL HAIRY AGRIMONY (*Agrimonia gryposepala*)

Description and Habitat. This glandular perennial has pinnately compound leaves divided into 5–9 leaflets that are evenly serrate. The flowers are yellow. The genus has about 15 species of rhizomatous herbs. The genus name is possibly from the Greek *argema*, an eye disease, because of supposed medicinal value. The species name is also Greek: *grypo*, "curved or hooked," and *sepala*, "sepal."

Natural History Notes and Uses. The plant was used to treat eye disease. Agrimony contains essential oil, bitters, and vitamins, but it is the large amounts of tannins that are responsible for most of its medicinal properties. Being astringent, it has been used to stop bleeding. It has been prescribed by herbalists in the United States and Europe for gastric problems, including gas and diarrhea, and for urinary disorders. In the 1800s doctors in the United States used it to treat incontinence (Erichsen-Brown 1979). The plant has been applied to skin irritations and cuts and used in baths. In addition to its medicinal uses, the leaves and stems make a flavorful tea, and the European species has been used to make a yellow dye.

Serviceberry (genus *Amelanchier*)

SASKATOON SERVICEBERRY (*A. alnifolia*)

UTAH SERVICEBERRY (*A. utahensis*)

Description and Habitat. Serviceberries are shrubs or small trees with simple leaves that are serrate on the terminal half. The white flowers have 5 petals, 5 reflexed sepals, and many stamens. The ovary is inferior, and the fruit is a pome. The genus name is the French Savoy word for the medlar (*Mespilus germanica*), a species that has similar fruits.

FIELD KEY TO THE SERVICEBERRIES

1. Flower petals ½ to 1 inch long; styles usually numbering 5; fruit not hairy.. *A. alnifolia*
1. Flower petals less than ½ inch long; usually 2–4 styles; fruit often hairy .... *A. utahensis*

FIGURE 95. Serviceberry (*Amelanchier alnifolia*).

Natural History Notes and Uses. All species within the genus *Amelanchier* produce edible pomes that ripen in late spring and summer. They were a major food for many native peoples. In fact, some Native Americans intentionally moved their camps to locations where the pomes could be more easily harvested. The pomes may be eaten raw, cooked, or dried. After drying, the pomes can be pounded into loaves or cakes. These in turn may be eaten after softening a piece in water or placing them in soups or stews. Prepared this way, the pomes could be kept for several years. Additionally, the dried pomes could be incorporated into pemmican.

The boiled inner bark was a Native American remedy for snowblindness. One drop of strained fluid would be placed in an afflicted eye three times daily. It was also used for eardrops and to stop vaginal bleeding. These applications probably stem from the astringency of the plant's tannic acid content.

The wood can be used for arrows, digging sticks, and other useful items. The berry juice makes a purple dye.

PEMMICAN

Pemmican is a concentrated food used by many early peoples. It is extremely nourishing and does not spoil. To make pemmican, cut meat into strips and dry it until it completely dries and crumbles. Then grind it as fine as possible, usually by pounding. Pour melted suet (fat) over the meat, and add salt for taste, along with fresh berries such as currants (*Ribes*) and serviceberries (*Amelanchier*). Knead the mixture into a paste and pack it into containers. In the "old days" the containers were intestines, but you can use plastic ware. The finished product can be eaten raw, boiled in stews and soups, or fried like sausage.

CURLLEAF MOUNTAIN-MAHOGANY (*Cercocarpus ledifolius*)

Description and Habitat. There are 13 species of *Cercocarpus* in the western United States and Mexico. They are shrubs and small trees with hard wood. The flowers are not showy but are sweet with nectar. The distinctive fruit is an achene that ends with a long terminal style covered with shiny hairs at maturity. The shrubs glisten in the sun from the mass of silvery fruits, each one a "tailed fruit," as indicated by the generic name.

Natural History Notes and Uses. The common name applied to this genus is somewhat misleading. These shrubby trees are not related to true mahogany (*Swietenia*), a valuable cabinet wood of tropical America. The dark reddish brown, mahogany-colored hardwood of *Cercocarpus* may have led to the name. Native Americans used the wood for spears, arrow shafts, and digging sticks. The inner brown bark produced a red-purple dye, as did the roots.

A tea to treat colds was prepared by peeling the bark, scraping out its inner layer, and drying and boiling it. The dried sap was pulverized and applied to the ears to treat earaches. A decoction of the bark and leaves was used for gynecological problems.

Caution: The leaves and seeds of curlleaf mountain-mahogany contain cyanogenic glycosides and should be considered toxic.

FIGURE 96. Mountain-mahogany (*Cercocarpus ledifolius*).

BLACK HAWTHORN (*Crataegus douglasii*)

Description and Habitat. Black hawthorn is a large deciduous shrub or small tree with thorns. The leaves are toothed or lobed, and the white flowers are borne in an open inflorescence. The fruits are small pomes, borne in tremendous quantity and remaining on the tree all winter. *Crataegus* is a large and varied genus containing many species that readily hybridize.

Natural History Notes and Uses. All hawthorns produce edible, albeit mealy fruits (called pomes) that may be eaten raw or cooked in small amounts, or dried and mixed into pemmican. A diet high in hawthorn pomes or drinking hawthorn tea is said to reduce weight (Willard 1992; Moore 1979). The pomes contain a nontoxic heart stimulant and should not be eaten in large amounts or without admixture. Herbal folk medicine considers *Crataegus* a heart tonic (Foster and Duke 1990). Centuries of empirical validation and many scientific studies have shown that the plant is useful in the treatment of hypertension, angina pectoris, and other heart disorders. The pomes also contain vitamin C. The thorns have many practical uses such as prongs or rakes, lances for blisters, needles for piercing ears, and fish hooks.

MOUNTAIN AVENS, WHITE DRYAD (*Dryas octopetala*)

Description and Habitat. Also known as alpine dryad, mountain avens, and white mountain avens, this dwarf matted plant, somewhat shrubby at the base, has simple toothed leaves and relatively large solitary white or yellow flowers with about 8 petals. The leaves are clothed with short white hairs on the lower surface, and the veins are prominent. This prostrate, trailing perennial shrub is usually less than 12 inches tall but can grow up to 18 inches, and it commonly forms mats, usually above timberline.

Natural History Notes and Uses. The epithets *dryad* and *dryas* derive from the mythical oak nymph Dryas because the shape of the leaves resembles that of an oak. White dryad is a pioneer species and is important in stabilizing the thin soils on mountain slopes. It spreads relatively rapidly and is popular in rock gardens. Several cultivated varieties, including hybrids, are available. Mountain avens is an important food source for ptarmigan and pikas. In Europe the leaves of this small plant have been used to make a tea. The plant contains tannins and can be used as an astringent and stomachic.

Strawberry (genus *Fragaria*)

WOODLAND STRAWBERRY (*F. vesca*)

VIRGINIA STRAWBERRY (*F. virginiana*)

Description and Habitat. These white-flowered perennial herbs are produced from rootstalks and long runners that root at the nodes. The flowers are white to pinkish and borne in cymes. The leaves are clustered at the base of the stem and are divided into 3 egg-shaped, coarsely toothed leaflets. The genus name comes from the Latin *fraga*, the classical name used for the strawberry fruit, referring to its fragrance. The common name comes from the Anglo-Saxon *streawberige* and refers to the berries "strewing" their runners over the ground.

FIELD KEY TO THE STRAWBERRIES

1. Apical tooth of the leaflets greater than those on either side; leaves yellow-green, upper surface bulged between the veins .... *F. vesca*
1. Apical tooth of leaflets smaller than those on either side; leaves blue-green .... *F. virginiana*

Natural History Notes and Uses. Strawberries do not keep well and should be dried for future use if not eaten soon after being picked. Tea made from the green or dried leaves is said to boost one's appetite. It may also be a nerve tonic and was used for bladder and kidney ailments, jaundice, scurvy, diarrhea, and stomachaches. Externally, the leaf tea can also be used as an antiseptic wash for eczema and wounds and as a gargle for sore throat and mouth ulcers. The plants contain substantial amounts of vitamins A and C and sulfur, calcium, potassium, and iron. To remove dental tartar, rub the berries on your teeth and let the juice sit for a few minutes. Afterward brush your teeth thoroughly with baking soda and water.

Avens (genus *Geum*)

YELLOW AVENS (*G. aleppicum*)

LARGELEAF AVENS (*G. macrophyllum*)

ROSS' AVENS (*G. rossii*)

PRAIRIE SMOKE (*G. triflorum*)

Description and Habitat. The avens species in the area are perennial herbs that are found in wet, open areas from low elevations to above timberline. Most of the leaves are basal and pinnately divided, but the stem leaves are small and less divided. The bell- or saucer-shaped flowers are solitary or borne in an open, few-flowered inflorescence.

FIELD KEY TO THE AVENS

1. Stem leaves, at least lower ones, not much smaller than basal leaves 2
1. Stem leaves few and greatly reduced 3

2. Terminal leaflet much larger than the two below it; lower segment of style with glandular hairs . . . . *G. macrophyllum*
2. Terminal leaflet not much larger than the two below it; lower style segment without glands . . . . *G. aleppicum*

3. Petals erect, pinkish to cream-colored; styles elongated and feather-like . . . . *G. triflorum*
3. Petals yellow and spreading; styles glabrous and short. . . . . *G. rossii*

Natural History Notes and Uses. These species were used medicinally by various Native American groups. A decoction of the root was taken for stomach pain, and a poultice of chewed or bruised leaves was applied to boils and wounds. For others, it was a panacea, and the leaves were chewed as a universal remedy.

TAWNY HORKELIA (*Horkelia fusca*)

Description and Habitat. These perennial herbs have pinnate leaves. The white to pink flowers occur in cymose panicles, and there are usually 10 stamens with dilated filaments. The genus name honors Johann Horkel (1769–1846), a German physiologist.

Natural History Notes and Uses. The Kashaya drank a tea made from the roots of a related species, *H. californica*, to purify the blood (Goodrich and Lawson 1980).

GORDON'S IVESIA (*Ivesia gordonii*)

Description and Habitat. The yellow flowers of this plant occur in a crowded, head-like cluster on each of several nearly leafless stalks growing from basal leaves; a few species of *Ivesia* have white or pink flowers. The 5 petals are narrow, on small stalks, and shorter than the narrowly triangular lobes of sepals, and there are 5 stamens. The leaves are pinnately compound, usually with more than 20 leaflets, each with 3–5 rounded segments. There are about 22 species occurring in western North America. The genus is named for Lt. Eli Ives, leader of a Pacific railway survey.

Natural History Notes and Uses. The genus is closely related to *Horkelia*, which has stamens with broad filaments. Cinquefoils (*Potentilla*) are also similar but usually do not have stalks at the base of the petals, and the 3 uppermost leaflets do not grow together as they do in many *Ivesia* species. Gordon's ivesia was used by the Arapaho. They apparently made a tonic from an infusion of the root (Arnason et al. 1981).

ONE-FLOWER KELSEYA (*Kelseya uniflora*)

Description and Habitat. This genus has only one species. One-flower kelseya is a small, cushion-forming subshrub with rosettes of simple silky leaves and small white flowers that bear 5 petals and 10 stamens.

Natural History Notes and Uses. The genus name is for Harlan P. Kelsey (1872–1958), a nurseryman of Boxford, Massachusetts. The species is often cultivated as a rock-garden plant.

SHRUBBY CINQUEFOIL (*Pentaphyloides floribunda = Potentilla fruticosa*)

Description and Habitat. This deciduous shrub has shredding, reddish brown bark. The leaves are alternate, pinnately compound, with 3–7 linear- to oblong-shaped leaflets. The flowers are yellow with 5 broad petals and appear buttercup-like when open. This plant is found in wet to dry, often rocky areas from the low elevations to the subalpine zone.

Natural History Notes and Uses. We have used the thin, dry, papery strips of bark as tinder for fire starting. A tea can be made from the leaves. This plant has been used in erosion control projects along roads, as the cuttings will root readily in the sand. Deer and other ungulates will browse on shrubby cinquefoil only when other more desirable plants missing. If you come across an area where cinquefoils have been heavily browsed by wild or domestic animals, it probably means that there are too many animals in the area and not enough food.

MAT ROCK SPIRAEA (*Petrophyton caespitosum*)

Description and Habitat. This genus has about 3 species of tuft-forming evergreen shrubs with crowded leathery leaves and dense racemes of white flowers.

Natural History Notes and Uses. This beautiful dwarf shrub flowers in late summer and fall. It grows as a crevice plant or plastered to rock surfaces. Some species are grown as ornamentals.

Ninebark (genus *Physocarpus*)

MALLOW NINEBARK (*P. malvaceus*)

MOUNTAIN NINEBARK (*P. monogynus*)

Description and Habitat. These shrubs have bark that exfoliates in stripes. The palmately veined leaves are alternate and lobed. The white flowers occur in loose or crowded terminal corymbs, and there are 20–40 stamens. The genus name is from the Greek *physa* ("bladder") and *karpon* ("fruit"), in reference to the inflated fruits.

FIELD KEY TO THE NINEBARKS

1. Leaves usually less than 1½ inches long. . . . . *P. monogynus*
1. Leaves generally greater than 1½ inches long . . . . *P. malvaceus*

Natural History Notes and Uses. Although most Native Americans considered mallow ninebark highly ***poisonous***, a tea was made from a stick with the outer bark peeled off. It was used as an emetic and purgative (Pojar and MacKinnon 1994). The Miwok in California apparently ate ninebark pods raw. The branches of these plants were straightened and used to make arrow shafts. The stems were also peeled and pounded to obtain fiber for making cordage.

Cinquefoil (genus *Potentilla*)

Description and Habitat. The many species of cinquefoil include perennial, biennial, and annual herbs and one shrub, with alternate, mostly compound leaves that have a membranous appendage (stipule) at the base of the petioles. The flowers are yellow, white, or, in one case, purple. The fruit is a cluster of achenes borne on the convex receptacle. Cinquefoils can be encountered in various habitat types at all elevations.

This is a difficult genus to work with because of its frequent asexual reproduction, and the taxonomy is far from stable. Careful examination of good material is required to separate the species. Therefore, no field key is presented here. Members of the closely related genus *Geum* have persistent styles that make the clusters of achenes appear shiny or cottony.

Natural History Notes and Uses. The large fleshy, older roots of *P. anserina* (= *Argentina anserina*) can be boiled or roasted and added to soups and stews. Prepared this way, they are quite tasty and have a nutty or a parsnip-like texture but are more woody. They were a staple among many Native Americans. Today they are seldom harvested but are greatly enjoyed by those who still use them. Silverweed is high in tannins and can be used to tan leather. Other cinquefoils are considered astringent as well. The whole plant or root of *P. arguta* (tall cinquefoil), in a tea or poultice, stops bleeding and has been used on cuts and wounds and for diarrhea and dysentery. A strong tea was used as a mouthwash and gargle for sore throats or tonsil inflammations and helped reduce gum inflammation.

Caution: In ancient times these plants were grown for food and medicine. Although there are no reports of toxic reactions from use of this genus, moderation is still advised.

Wild Plum, Cherry (genus *Prunus*)

BITTER CHERRY (*P. emarginata*)

COMMON CHOKECHERRY (*P. virginiana*)

Description and Habitat. *Prunus* has about 400 species of shrubs or trees with simple leaves. Many species have a pair of warty glands at the top of the petiole or the base of the leaf blade. The flowers are pink to white and rather showy. The fruits are drupes with 1 stone and typically embedded in the fleshy pulp. The seeds and leaves are toxic, as they contain hydrocyanic acid.

FIELD KEY TO THE WILD PLUMS

1. Flowers numerous, in unbranched, long, narrow inflorescence. . . . . *P. virginiana*
1. Flowers in hemispheric or flat-topped inflorescence . . . . . . *P. emarginata*

Natural History Notes and Uses. The fruits of these species are sour or bitter when raw, but after cooking the sourness disappears. Native Americans dried the berries whole or in cakes for use in winter. When needed, the dried fruits were soaked in water and then eaten. Members of Lewis and Clark's expedition ate western chokecherry when other foods were scarce. It seems that after drying, the fruits lose some of their bitterness, resulting in an almost sweet taste.

To make the cakes, Native Americans ground up the ripe fruits, pits and all, and dried them in the sun. When needed, the cakes, or portions thereof, could be soaked in water, mixed with flour and sugar, and made into a sauce or gravy. This sauce was eagerly traded among some Native Americans, including the Navajo, Shoshone, Arapahoe, and Ute (Bailey 1940). The only difficulty we've found in preparing cakes this way is that the pits do not grind down nicely into a fine material but leave chunks that could result in broken teeth.

In addition, the berries were incorporated into pemmican. They can also be used in making jelly, but because chokecherries are low in natural pectin, it is advisable to add pectin. The leaves of both species contain toxic amounts of cyanide, as do the seeds (pits). Cyanide is highly volatile, and the pits can be rendered safe by long-term drying, by boiling in several changes of water, or by dry roasting. Do not eat them in significant amounts even then unless you mix them with larger quantities of other foods. *Prunus* shoots, peeled and split, were used in basketry. The wood was used for implements such as digging sticks, arrows, and arrow foreshafts.

Warning: The leaves, bark, and seeds of all *Prunus* contain cyanide-producing glycosides. Therefore, eating large quantities of ripe berries with their pits could cause nausea and vomiting. In some instances it could be fatal. Cooking and drying the seeds appears to dispel most of the glycosides, and the seeds in dried, mashed chokecherries are not as significant a problem. To be safe, it is best to discard the seeds before eating the fruits.

ANTELOPE BITTERBRUSH (*Purshia tridentata*)

Description and Habitat. This fragrant shrub grows up to 8 feet tall. The leaves are deeply 3-cleft into linear lobes, glandular above and tomentose below. The leaf margins are rolled inward. The flowers are pale yellow to white, and the fruit is a pubescent, oblong achene. Bitterbrush grows on dry slopes and canyons of the desert mountains between 3,000 and 9,000 feet. It flowers from April to June. The genus was named for Frederick Pursh, author of an early flora of North America.

Natural History Notes and Uses. The leaves and inner bark of this species were used to produce a tea that was acted as an emetic or strong laxative. The tea was also used as an analgesic and to relieve menstrual cramps. The ripe seed coat yields a violet dye. Old *Purshia* stumps produce shredding bark that women would peel off, work with their hands to soften, and use as baby diapers. This bark was sometimes combined with juniper (*Juniperus*) bark.

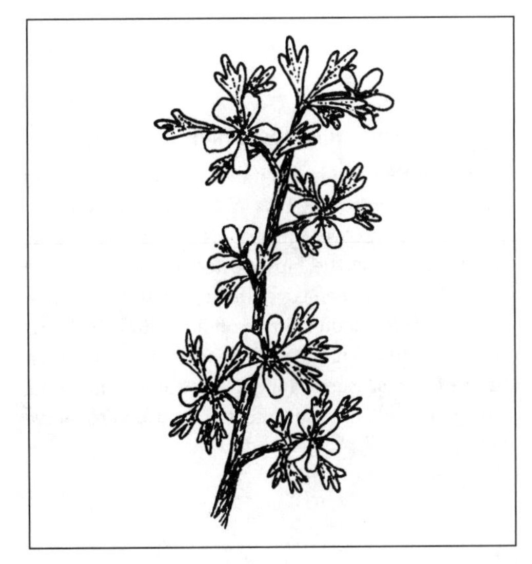

FIGURE 97. Bitterbrush (*Purshia tridentata*).

FREDERICK PURSH

The botanist Frederick Pursh was born in Tobolsk, Siberia, in 1774; he died in Montreal, Canada, in June 1820. He was educated at Dresden and in 1799 came to the United States, where he spent twelve years in botanical explorations. He visited England in 1811 and published *Flora Americaae Septentrionalis, or a Systematic Arrangement and Description of the Plants of North America* (2 vols., London, 1814). He then returned to North America and died while he was collecting materials for a flora of Canada. Until superseded by Torrey and Gray's *Flora of North America* (2 vols., 1838–43), Pursh's work was the most important on the botany of North America.

Wild Rose (genus *Rosa*)

NOOTKA ROSE (*R. nutkana*)

WOODS' ROSE (*R. woodsii*)

Description and Habitat. These shrubs have prickles and leaves that are pinnately divided into 3–11 leaflets. The large, red to pink flowers are borne singly or a few together. The fruits, called hips, are orange, red, or purplish and urn-shaped. This is a taxonomically difficult genus with many variable species that tend to hybridize.

FIELD KEY TO THE ROSES

1. Flowers often solitary; petals 1 to 1½ inches long; sepals ¾ to 1½ inch long . . . . *R. nutkana*
1. Flowers usually growing in clusters; petals ½ to 1 inch long; sepals mostly ½ to ¾ inch long . . . . *R. woodsii*

FIGURE 98. Wild rose (*Rosa* spp.).

Natural History Notes and Uses. The hips are edible raw, stewed, candied, or made into preserves. They are high in vitamin C and also contain vitamins E, B, and K, beta-carotene, calcium, iron, and phosphorus. There are many other edible parts besides the fruit. Young *Rosa* shoots in spring make an excellent potherb, and the roots and stems can be used to make a tea. The petals may be used in salads. The peeled spring shoots can also be nibbled on. Almost all parts of the plant have been made into a wash or dressing for cuts or sores to coagulate blood. One of the

WILD ROSE SYRUP

Gather rose hips in late summer or autumn. Snip off both ends, then cut the hips in half and put them into a pot and barely cover with water. Bring to a boil, cover, and simmer until the hips are tender. Strain off and retain the liquid, then cover the hips with more water and boil for 15 minutes. Strain off and retain the liquid and add it to the first batch. Measure the liquid, then add half as much sugar. Bring to a boil and simmer until the syrup thickens. Pour into sterilized bottles. Serve over pancakes.

more common methods is to sprinkle fine shavings of de-barked stems into a washed wound. The petals can be used as a dressing. A poultice of leaves can be used to relieve insect stings. In addition, the young leaves can be washed, cut into small pieces, and dried for a hot tea.

Raspberry, Blackberry (genus *Rubus*)

AMERICAN RED RASPBERRY (*R. idaeus*)

CUT-LEAF BLACKBERRY (*R. laciniatus*)

THIMBLEBERRY (*R. parviflorus*)

Description and Habitat. The *Rubus* species in the area are deciduous shrubs with arching or trailing stems covered with bristles and prickles. The flowers have white petals, and the fruit is a coherent cluster of small, 1-seeded drupes (raspberries, blackberries, dewberries, cloudberries, marionberries). The genus name is derived from the Latin *ruber*, meaning "red," in reference to the color of the fruit. This is a large and complicated group taxonomically.

FIELD KEY TO THE RASPBERRIES

1. Leaves simple . . . . *R. parviflorus*
1. Leaves compound 2

2. Leaves once compound, with the primary leaflets shallowly lobed . . . . *R. idaeus*
2. Leaves often twice compound, with the primary leaflets divided to the midrib . . . . *R. laciniatus*

Natural History Notes and Uses. All species produce edible berries. Archaeological evidence shows that *Rubus* species have formed part of the human diet from very early times (Smith 1973).

The flowers of all species can be added to salads or nibbled on when hiking. The fresh or dried leaves can be steeped for a tea, alone or in herbal blends. Do not use the wilted or molded foliage, as it may be toxic. The young shoots cut just above ground can be peeled and eaten raw or cooked. A tea from the roots was used to dry runny noses, and a tea from the bark was used to stop dysentery. The plants can also provide a uterine astringent, diuretic, laxative, and mild sedative.

SMALL BURNET (*Sanguisorba minor*)

Description and Habitat. Small burnet is an annual herb with pinnately divided leaves. The flowers are clustered in heads and have no petals. The plants are found in waste places or moist soils at the lower elevations. The generic name comes from the Latin *sanguis*, meaning "blood," and *sorbeo*, meaning "to stanch," referring to the herb's ability to stop bleeding.

Natural History Notes and Uses. The young leaves make a good salad plant, tasting somewhat like cucumber. The leaves can be chopped and blended or mixed with other herbs as a seasoning. The dried flowers and leaves can be prepared as a tea. The roots are very astringent, and a decoction was used in the treatment of internal and external bleeding and dysentery. The brew can also be used as a mouthwash for gum problems.

CREEPING SIBBALDIA (*Sibbaldia procumbens*)

Description and Habitat. Sibbaldia is a low, mat-forming perennial herb with trailing stems. The leaves have sparsely hairy petioles and are clover-like, divided into 3 oblong leaflets. The flowers are saucer-shaped with yellow petals. It can be found in sparsely vegetated soils of subalpine and alpine meadows where snow accumulates.

Natural History Notes and Uses. The genus was named for Robert Sibbald (1641–1722), first professor of medicine at the University of Edinburgh. *Procumbens* refers to its prostrate growth form. The Inuit name of this plant in northern Quebec is *arpehutik*, and the leaves were used to make a medicinal tea to relieve general stomachache (Avataq Cultural Institute 1984).

MOUNTAIN ASH (*Sorbus scopulina*)

Description and Habitat. Mountain ash is a tree or shrub with deciduous, pinnately compound leaves. The flowers are white to cream-colored and borne in densely branched, flat-topped inflorescences. The fruits are small and berry-like, ranging in color from orange to red. Mountain ash can be found in moist meadows and forest openings at the middle elevations.

Natural History Notes and Uses. The fruits may be eaten raw, cooked, or dried. They are high in vitamins A and C and carbohydrates. Unripe berries are very bitter and somewhat unpalatable. The fruits, which are pomes, are commonly processed into jams and jellies. They have a high pectin content and jell readily. To use as a coffee substitute, grind the dried, roasted seeds. The berry juice can be used as a gargle for sore throat and as an antiseptic wash for cuts. Sorbitol, the sugar in the fruit of *Sorbus*, is used commercially for sweetening candies, toothpaste, and other products.

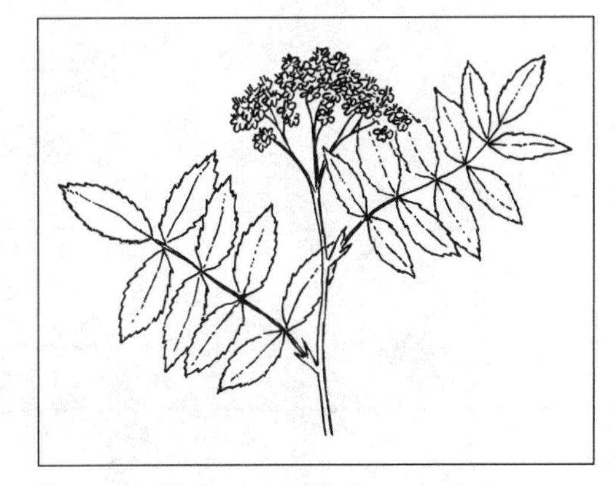

FIGURE 99. Mountain ash (*Sorbus scopulina*).

Spiraea (genus *Spiraea*)

WHITE SPIRAEA (*S. betulifolia*)

ROSE MEADOW-SWEET (*S. splendens*)

Description and Habitat. *Spiraea* is a genus of about 80–100 species of shrubs in the Rosaceae, subfamily Spiraeoideae. They are native to the temperate Northern Hemisphere, with the greatest diversity in eastern Asia. Two species of *Spiraea* can be found in the area. They are small shrubs with deciduous leaves and white to pink flowers that are densely clustered in showy, flat-topped to spike-like inflorescences. The species can be found on brushy, open slopes and in moist habitats up to timberline.

FIELD KEY TO THE SPIRAEAS

1. Flowers white	*S. betulifolia*
1. Flowers pink or rose-colored	*S. splendens*

Natural History Notes and Uses. *Spiraea* is a source of methyl salicylate, similar to the active ingredient in aspirin. Native Americans brewed a tea from the stem, leaves, and flowers of some species to use as a pain reliever (Craighead et al. 1963). The plants are astringent, and a poultice made from the leaves and bark was used to treat ulcers, burns, and tumors. The roots were also peeled and boiled until soft, mashed, and used as a poultice for burns. The wiry, branching twigs can be used to make broom-like implements for collecting tubers.

Most *Spiraea* species are indigenous to central and eastern Asia, whence come most of the popular ornamental species (e.g., bridal wreath from *S. x prunifolia*, native to Japan, and its similar hybrid, *S. vanhouttei*). In these species the fragrant, spire-like flower clusters typical of the genus are borne on long, arching branches. Spiraeas native to North America include hardhack, or steeplebush (*S. tomentosa*), a local source of astringent and tonic, and the meadowsweets (several species). The name *meadowsweet* is also applied to the related genus *Filipendula*, tall, hardy perennials (also often cultivated) formerly classified as *Spiraea* because of the similar showy blossoms. *Filipendula* includes the Eurasian dropwort (*F. hexapetala*), the queen of the meadow (*F. ulmaria*), now naturalized in the United States, and the North American queen of the prairie (*F. rubra*).

MADDER FAMILY (RUBIACEAE)

The madder family consists of approximately 500 genera and 6,000–7,000 species distributed worldwide. About 20 genera are native to the United States. The family is of economic importance because of coffee, quinine, and many ornamentals.

FIELD KEY TO THE MADDER FAMILY

1. Leaves opposite; corolla lobe elongate	*Kelloggia*
1. Leaves whorled; corolla tube short	*Galium*

Bedstraw (genus *Galium*)

STICKYWILLY (*G. aparine*)

TWIN-LEAF BEDSTRAW (*G. bifolium*)

NORTHERN BEDSTRAW (*G. boreale*)

THREE-PETAL BEDSTRAW (*G. trifidum*)

FRAGRANT BEDSTRAW (*G. triflorum*)

Description and Habitat. Despite their small flowers, the various species of *Galium* are unmistakable. They are annual or perennial herbs with 4-angled stems and whorled leaves. The small, 4-parted flowers are white or greenish, and the fruits are smooth or bristly-hairy. They can be found in various habitats from low to high elevations. *Galium* is from the Greek *gala*, meaning "milk," referring to the herb's traditional use as a milk coagulant for making cheese. The rennet (a substance that curdles milk in making cheese and junket) for this use was obtained by blending the herb with an equal amount of salt, covering it with water, and then simmering away half the fluid.

FIELD KEY TO THE BEDSTRAWS

1. Plants annual 2
1. Plants perennial 3

2. Leaves 2–4 in a whorl, lacking the abrupt pointed tips; stems smooth; flowers solitary .... *G. bifolium*
2. Leaves 5–8 in a whorl; stems rough when stroked tip to base; 3–5 flowers on axillary peduncles .... *G. aparine*

3. Fruits with hairs or bristles that are hooked at tip........... *G. triflorum*
3. Fruits without hooked hairs.. 4

4. Flowers many in a terminal, compound, much-branched inflorescence; stems erect and smooth to hairy, but not rough .... *G. boreale*
4. Flowers solitary or few in a small, rather inconspicuous inflorescence; stems are weak, and the plants tend to scramble on other vegetation; plants are rough on stem angles .... *G. trifidum*

Natural History Notes and Uses. None of the species of *Galium* is known to be poisonous. Although *G. aparine* is the most commonly used species, Tilford (1993) believes, as we do, that all other species can be used similarly. The very young leaves and stems can be used as a potherb. The small hairs on the stems make the plants difficult to swallow raw, but boiling or steaming softens them up. If the stems are too fibrous, use only the leaves. Slow roasted until dark brown and ground, the ripe fruit can be used as a coffee substitute.

Medicinally, the plants were used to increase urine flow, stimulate appetite, reduce fevers, and remedy vitamin C deficiencies. The plant has diuretic, anti-inflammatory, and astringent qualities and has been used as a lymphatic tonic. A wash made from the plant is said to remove freckles, whereas a cool tea is reported to cool sunburns. Many species of *Galium* contain asperuloside, which produces coumarin, giving it the sweet smell of new-mown hay as the foliage dries. Asperuloside can be converted to prostaglandins (hormone-like compounds that stimulate the uterus and affect the blood vessels), making *Galium* species of great interest to the pharmaceutical industry.

Dried, the foliage of bedstraw has been used as a stuffing for mattresses or as a tinder for starting fires. The roots yield a red dye, but because they are thread-like and produce little dye, collecting enough for a strong dye bath would be fairly laborious.

MILK KELLOGGIA (*Kelloggia galioides*)

Description and Habitat. Milk kelloggia is a glabrous, herbaceous perennial with clustered stems 4–24 inches tall arising from creeping rhizomes. The leaves are opposite, sessile, and narrow. The flowers have 4 fused sepals and 4 pink or white fused petals arising from the top of the ovary. The fruits are ball-like, covered with hooked bristles, and break into 2 segments at maturity.

Natural History Notes and Uses. This rare plant is known from at least three extant populations and two historical records in Wyoming. At least four populations are protected within Grand Teton and Yellowstone national parks and the North Absaroka and Washakie wilderness areas. One population near Jackson could be adversely affected by the expansion of a ski area.

WILLOW FAMILY (SALICAEAE)

This family has two to three genera and over 500 species distributed worldwide. *Salix* and *Populus* are native to the United States. The family is of little economic importance except as a source of ornamentals.

FIELD KEY TO THE WILLOW FAMILY

1. Plants mostly trees with pendulous catkins; leafbuds usually covered by several, generally resinous scales . . . . *Populus*
1. Plants trees or shrubs with mostly ascending to erect catkins; leafbuds covered by a single, nonresinous scale . . . . . . *Salix*

FIRE BY FRICTION

To early humans, the ability to start a fire to cook food and use as a tool was an important skill. In fact, a look at human history reveals starting a fire as the greatest invention of all time.

There are several methods for starting a fire by friction. The drill-and-hearth method was used by a great many Native Americans. The hearth was a flat piece of wood, usually a softer wood than the drill. It had a small hole reamed in it and a notch leading to the edge of the hearth. The drill was a piece of wood about 2 feet long and a half-inch or less in diameter. When the drill was held vertically and rapidly rotated between the palms of the hands, embers were formed in the hole. Embers spilled through the notch onto the tinder placed beside the hearth. Cradling the tinder in the hands and gently blowing on the embers produced a flame. Depending on craftsmanship, humidity, and other factors, a fire could be created in just a few seconds—or never. To help increase the friction, sand grains were sometimes placed in the hole. Also, "points" of ember-producing wood were inserted into straight cane-like sticks when appropriate wood could not be found in long straight pieces. The wood these points were made of was more efficient in starting a glowing ember. This method of making fire could be accomplished by a single person, or two people could take turns rotating the drill. Another variation used a "bow" to rotate the drill.

Some woods are better than others in starting a fire. A few choice woods in the area are sagebrush (*Artemisia tridentata*), certain species of willow (*Salix* spp.), juniper (*Juniperus* spp.), mullein (*Verbascum thapsus*), cattail (*Typha*), aspen (*Populus tremuloides*), yucca (*Yucca*), and cottonwood (*Populus* spp.).

Cottonwood, Aspen (genus *Populus*)

NARROW-LEAF COTTONWOOD (*P. angustifolia*)

BALSAM POPLAR (*P. balsamifera*)

QUAKING ASPEN (*P. tremuloides*)

Description and Habitat. These trees have sticky, resinous leaf buds and deciduous leaves. Older trees of some species have gray, rough bark; young bark is smooth and whitish. The flowers are borne in catkins that appear before the leaves. While cottonwoods are generally found along streams and in moist soils, aspens often grow in the uplands. Cottonwoods grow rapidly and are planted for quick shade or wind protection. The soft wood of some species is used for veneers, boxes, matches, excelsior, and paper.

FIELD KEY TO THE COTTONWOODS

1. Plants not usually confined to water courses; bark is white and smooth, covered with a whitish powdery bloom; bark furrowed and gray only when very old . . . . *P. tremuloides*
1. Plants usually growing along water courses or edges of lakes; bark usually gray or brown and roughly furrowed. 2

2. Leaf blades distinctly darker above than beneath . . . . *P. balsamifera*
2. Leaf blades about equally yellow-green on both sides . . . . . *P. angustifolia*

Natural History Notes and Uses. The catkins may be eaten raw or boiled in stews and are a source of vitamin C. The inner bark can also be eaten as a spring tonic or dried and ground into a flour substitute or extender. The fresh or dried plant can be used in poultices for muscle aches, sprains, or swollen joints. The primary action of *Populus* is that of an analgesic, used topically and internally. It contains varying amounts of populin and salicin, compounds related to early forms of aspirin. The leaves and bark are the most effective parts for tea and aid in diarrhea problems. The wood makes an excellent bow-and-drill fire set. Cottonwoods are considered botanical indicators of water. Trappers often used aspen as bait in beaver sets.

Willow (genus *Salix*)

Description and Habitat. Many species of willow are found in the area. They are mostly shrubs with numerous stems. The flowers are in catkins that appear before, with, or after the leaves. The various species of willows generally grow along streams or other moist habitats. Willows root easily and occasionally form dense thickets. They are often planted to reduce streambank erosion.

Mature catkins (mainly pistillate flowers) and mature leaves are useful for identification, but both are rarely available at the same time. Vegetative characters have a good deal of environmentally or developmentally induced variation. Understanding this variation is required for correct identification.

Natural History Notes and Uses. The young shoots and leaves can be eaten raw. The bitter inner bark can also be eaten raw, although it is better dried and ground into a flour substitute or extender. The plant contains salicin, which is similar to aspirin and useful as a substitute. Any part of the willow can be used to produce a tea for use as an aspirin replacement for headache and body pain. The highest concentrations of salicin, however, are found in the inner bark. Because it is not nearly as strong as aspirin, you may have to drink quite a bit of it. The leaves have astringent properties that are effective when placed on wounds and cuts. The bark was chewed as a toothache remedy. The bark, leaves, twigs, and roots produced medicinal teas, powders, washes, and poultices to relieve pain, swelling, infection, bleeding, and many other ailments. Willows, like cottonwoods, are botanical indicators of water. The branches of many willow species are very flexible, making them useful for traps, arrow shafts, and other items. The bark can also be used as crude cordage.

Willow was an important basketry plant and was often used as a foundation material and twining material for twined baskets. Other uses of the wood include frameworks for dwellings, fish dams and weirs, racks for drying and cooking food, and light hunting bows. Fiber from the bark was used for cordage, nets, and clothing. Willows root easily due to the large amounts of indole acetic acid (IAA), a plant hormone, in their stems. IAA can be extracted in cold water from 1-inch sections of stem and used to induce rooting of other species for transplanting.

SANDALWOOD FAMILY (SANTALACEAE)

Members of the sandalwood family are partially parasitic herbs. There are about 30 genera and 400 species distributed in tropical and temperate parts of the world. The most common representative in North America is the genus *Comandra*.

BASTARD TOADFLAX (*Comandra umbellata*)

Description and Habitat. Bastard toadflax is a partially parasitic perennial herb with a waxy surface and a rather woody base. The leaves are linear, and the flowers are bell-shaped. The fruit is a 1-seeded, berry-like drupe. Bastard toadflax is common and widespread in shrublands up to the subalpine zone. *Comandra* comes from the Greek *kome*, meaning "tufts of hairs," and *aner*, meaning "man," in reference to the stamens. The roots are blue when cut.

Natural History Notes and Uses. The mature, brown, urn-shaped fruits of bastard toadflax may be eaten raw and are best when slightly green. They were popular with Native Americans because of their sweet taste. The berries are rarely found in sufficient quantities for more than a pleasant tidbit, however. Consuming too many berries may cause nausea. Strike (1994) indicates that a root preparation was used to soothe sore, inflamed eyes.

SAXIFRAGE FAMILY (SAXIFRAGACEAE)

There are about 30 genera and 580 species in this family, found chiefly in cooler and temperate regions of the Northern Hemisphere. About 20 genera are native to the United States, with more species occurring in the western part of the country. Several species are cultivated as ornamentals. The family name means "breaker of rocks"—many members are found growing in rock crevices. All members of the saxifrage family are at least somewhat edible.

FIELD KEY TO THE SAXIFRAGE FAMILY

1. There are 5 stamens, and sometimes 5 sterile stamens (staminodia) 2
1. There are 10 stamens 5

2. Stamens alternate with staminodia; stigmas 4 . . . . . *Parnassia*
2. Only stamens present; stigmas usually 2 3

3. Petals divided into linear divisions . . . . . *Mitella*
3. Petals entire . . . . . *Heuchera*

4. Petals deeply 3–7-parted; styles generally 3 . . . . . *Lithophragma*
4. Petals entire or slightly lobed; styles 1 or 2 5

5. Flowers reddish or purplish; leaves kidney-shaped with rounded teeth . . . . . *Telesonix*
5. Flowers white, yellow, or greenish, or if purplish, the leaves are entire, opposite, and narrower . . . . . *Saxifraga*

Alumroot (genus *Heuchera*)

ROUNDLEAF ALUMROOT (*H. cylindrica*)

LITTLELEAF ALUMROOT (*H. parvifolia*)

Description and Habitat. In general, *Heuchera* species are perennial herbs with basal leaves. Flowers are small, saucer- to bell-shaped, and greenish, white, or pinkish in color. Their various habitats include moist soils and rocky areas up to the alpine zone. The genus is named in honor of J. H. Heucher (1677–1747), professor of botany and custodian of the botanic garden in Wittenberg, Germany. Many species of alumroot readily hybridize, making identification of some plants difficult.

FIELD KEY TO THE ALUMROOTS

1. Calyx saucer-shaped, with all lobes of equal length .... *H. parvifolia*
1. Calyx bell-shaped, with some lobes longer than others ..... *H. cylindrica*

Natural History Notes and Uses. The leaves of all species in the area are edible, although they are not choice. They have a sour taste because of the high tannin content and should therefore be boiled or steamed. Since they are rather tough, we found them to be more palatable if chopped and added to soups or salads.

Because of its high tannin content, *Heuchera* is said to be one of the strongest astringents. Tilford (1997) indicates that tannins may account for as much as 20 percent of the weight of the roots of these plants. Tannins tend to shrink swollen, moist tissues. Therefore, they are gastrointestinal irritants and have been known to cause kidney and liver failure. Ingestion of the plant should be in moderation. Otherwise, the pounded, dried roots of many species have been used as a poultice that stops bleeding and promotes healing when applied to cuts and sores. The raw root, eaten in small amounts, has been used as a cure for diarrhea. A tea from the roots can also be used as a gargle for sore throats. The powdered roots have been used as an antiseptic.

Alumroots are also commonly used as mordants, substances that make natural dyes colorfast. The alumroot of choice, however, occurs in western deserts near sulfur springs, but satisfactory substitutions probably exist in the northern Rockies.

Woodland Star (genus *Lithophragma*)

BULBOUS WOODLAND STAR (*L. glabrum*)

SMALL-FLOWER WOODLAND STAR (*L. parviflorum*)

Description and Habitat. These perennial herbs have slender, bulbet-bearing rootstalks. There are approximately 10 species in western North America. The genus name is from the Greek *lithos* ("rock") and *phragma* ("fence"). Incidentally, generic names that end in *phragma* are considered to have neuter, not feminine, gender.

NATURE'S COLORS

Plants have provided humans with color for uncounted ages. In the last 5,000 to 7,500 years, humans have learned to transfer some of nature's colors to cloth, paper, wood, leather, and wool. Natural dyes do not last as long as modern synthetics, but when you are living simply, natural dyes are readily available.

Most vegetable dyes are prone to fading and need some added treatment to become colorfast. This process is called mordanting, that is, treating the material to be dyed with other substances that serve to fix the color. Basically, mordanting involves soaking (or boiling or simmering) the material in water with the dissolved mordant, the type of which depends on the dye source and the material to be dyed. Common mordants are alum, chrome, iron, tin, copper sulfate, and ammonia. After a prescribed time the material is rinsed and allowed to dry. A dye bath is then prepared by soaking the chopped or crushed plant material in water overnight and boiling until the appropriate color is extracted. Plant material is strained out, and water is added to make 4–5 gallons of lukewarm dye bath, to which 1 pound of fabric is added. After being dyed and stirred as long as necessary for the color desired, the dyed material is passed through a series of rinses, each a little cooler than the previous one, until the rinse water remains clear. After drying, the dyed material is ready for use.

FIELD KEY TO THE WOODLAND STARS

1. Basal leaves glabrous or only very sparsely hairy............. *L. glabrum*
1. Basal leaves hairy, at least on lower surfaces .... *L. parviflorum*

Natural History Notes and Uses. The root of a related species, *L. affinis,* was chewed by the Native Americans in California to treat stomach ailments and colds.

Miterwort (genus *Mitella*)

FIVE-STAMEN MITERWORT (*M. pentandra*)

SMALL-FLOWER MITERWORT (*M. stauropetala*)

Description and Habitat. Found along mountain streams, miterwort has heart-shaped basal leaves with prominent veins and toothed, shallow lobes. The tiny, greenish yellow flowers grow along the upper half of a leafless stalk. The flowers have 5 sepals turned back at the tip, 5 petals that are divided into 5–11 thread-like segments, and 5 stamens. They grow in shady forests up to the alpine zone.

FIELD KEY TO THE MITERWORTS

1. Stamens attached directly below the base of the petals..... *M. pentandra*
1. Stamens attached below the apex of the sepals, to one side of the base of the petals .... *M. stauropetala*

Natural History Notes and Uses. There is broad ecological overlap among the various species. *Mitella* is Latin for "turban," and *stauro* is Greek for "cross-like," referring to the tiny petals. *Wort* is from the Old English *wyrt,* meaning "plant" (e.g., figwort, spiderwort, spleenwort). The name *miterwort* is a reference to the shape of that plant's seed pods, which resemble a miter, or bishop's cap.

Grass-of-Parnassus (genus *Parnassia*)

ROCKY MOUNTAIN PARNASSIA (*P. fimbriata*)

KOTZEBUE'S GRASS-OF-PARNASSUS (*P. kotzebuei*)

NORTHERN GRASS-OF-PARNASSUS (*P. palustris*)

Description and Habitat. These plants have a sessile, spade-shaped leaf near the midlength of the stem. The basal leaves have petioles up to 4 inches long and are broadly spade-shaped blades with prominent parallel nerves. Calyx lobes are elliptical with fringed tips, and the petals have 5–7 prominent, parallel veins and long, filiform fringes on both sides near the base. The genus is named after Mount Parnassus in Greece.

FIELD KEY TO THE GRASS-OF-PARNASSUS

1. Petals fringed .... *P. fimbriata*
1. Petals not fringed 2

2. Petals 1–3-nerved and equal in size to the sepals .... *P. kotzebuei*
2. Petals 5–13-nerved and longer than the sepals............... *P. palustris*

Natural History Notes and Uses. The plants can be found in wet, mossy areas along creeks, springs, and lake shores. Some Native Americans made an infusion of the powdered leaves of *P. fimbriata,* which was given to babies for dullness or stomach problems. A poultice was also made from the plant (unknown parts), and it was used as a wash for venereal disease.

Saxifrage (genus *Saxifraga*)

WEDGE-LEAF SAXIFRAGE (*S. adscendens*)

YELLOW-DOT SAXIFRAGE (*S. bronchialis*)

TUFTED ALPINE SAXIFRAGE (*S. caespitosa*)

NODDING SAXIFRAGE (*S. cernua*)

WHIPLASH SAXIFRAGE (*S. flagellaris*)

BROOK SAXIFRAGE (*S. odontoloma*)

PURPLE MOUNTAIN SAXIFRAGE (*S. oppositifolia*)

DIAMOND-LEAF SAXIFRAGE (*S. rhomboidea*)

WEAK SAXIFRAGE (*S. rivularis*)

THYME-LEAF SAXIFRAGE (*S. serpyllifolia = S. crysantha*)

YELLOWSTONE SAXIFRAGE (*S. subapetala*)

Description and Habitat. Saxifrages are glabrous or, more often, glandular hairy perennials with simple to divided leaves that are basal or alternate (opposite in 1 species) and sometimes have small bulbs in the axils. Flowers (sometimes replaced by small bulbs) are solitary or borne on stalks in a simple to branched inflorescence. The calyx is 5-lobed and saucer- to bell-shaped, and there are 5 petals. The species with all basal leaves may eventually be placed in a separate genus, *Micranthes.*

FIELD KEY TO THE SAXIFRAGES

1. Leaves all basal...2
1. Leaves occurring on stems (maybe only 1–2 leaves)..................4

2. Leaves kidney- to heart-shaped .... *S. odontoloma*
2. Leaves egg-shaped to oblong.......................................3

3. Sepals erect.. *S. rhomboidea*
3. Sepals reflexed .... *S. subapetala*

4. Stem leaves opposite................................*S. oppositifolia*
4. Leaves alternate 5

5. Flowers yellow 6
5. Flowers white...7

6. Plants with stolons.................................... *S. flagellaris*
6. Plants without stolons .... *S. serpyllifolia*

7. Leaves as broad as they are long, kidney-shaped, lobed, or entire.......8
7. Leaves longer than they are broad, entire, cleft, or parted.............9

8. Calyx bell-shaped .... *S. cernua*
8. Calyx shaped like a top.................................. *S. rivularis*

9. Leaves 3–5-cleft; plants are matted or tufted............. *S. caespitosa*
9. Leaves entire, linear to oblong in shape...........................10

10. Calyx cleft to the base . . . . *S. bronchialis*
10. Calyx lobes shorter than the tube . . . . *S. adscendens*

Natural History Notes and Uses. The generic name is from the Latin *saxum*, meaning "rock," and *frangere*, meaning "to break." It alludes to the plants' rocky habitat. Herbalists once used some species in a treatment for "stones" in the urinary tract.

The genus as a whole is regarded as a safe group of plants. The leaves can be used fresh or in stews and are high in vitamins A and C. Richard Scott (formerly of Central Wyoming College) shared with us this favorite backcountry salad of his: "When backpacking, we always carry a little bottle of vinegar, one of oil, and one of dried or powdered salad dressing. In fact, an excellent salad is based on saxifrage, particularly *S. arguta* leaves with a few alpine sorrel leaves (*Oxyria digyna*), a couple of Indian paintbrush inflorescences (*Castilleja miniata* or *C. rhexifolia*), and a couple Columbine flowers (*Aquillea caerulea*)."

In China some species were used in the treatment of nausea and ear infections. In western North America there is little documentation regarding any medicinal uses.

TELESONIX (*Telesonix heucheriformis* = *Boykinia jamesii*)

Description and Habitat. Telesonix is an uncommon plant found on limestone or dolomite cliffs. It has basal leaves that are covered with sticky, glandular hairs. The leaves are round in outline and scalloped with rounded teeth. The bell-shaped flowers are composed of 5 small, reddish purple petals that are enclosed by 5 large, brick red sepals. The flowers are also bell-shaped.

Natural History Notes and Uses. This species was once included in the genus *Boykinia*, which contains about 8 species of perennial rhizomatous herbs with alternate leaves and crowded flowers. The species was named for Edward James, a nineteenth-century American naturalist.

FIGWORT FAMILY (SCROPHULARIACEAE)

The figwort, or snapdragon, family has about 220 genera and 3,000 species distributed worldwide. About 40 of the genera are native to the United States. The family is of economic importance because of the cardiac glycosides derived from *Digitalis* (foxglove) and the many fine ornamentals.

FIELD KEY TO THE FIGWORT FAMILY

1. Flowers almost regular; anther-bearing stamens 5 . . . . *Verbascum*
1. Flowers usually irregular; anther-bearing stamens usually 2 or 4 2

2. Petals with a definite spur at the base . . . . *Linaria*
2. Petals without a spur 3

3. Anther-bearing stamens 2 4
3. Anther-bearing stamens 4 6

4. Leaves all arising from the stem and opposite . . . . *Veronica*
4. Leaves mostly basal, the stem leaves, if present, reduced in size and alternate 5

5. Leaves merely toothed .... *Besseya*
5. Leaves much dissected .... *Synthyris*

6. Flowers with a 5th sterile stamen, often rudimentary 7
6. Flowers with 4 fertile stamens and no sterile stamen 9

7. Sterile stamen conspicuous and elongate;
corolla tube- or funnel-shaped .... *Penstemon*
7. Sterile stamen appearing as a scale or gland attached to the corolla 8

8. Plants annual .... *Collinsia*
8. Plants perennial .... *Scrophularia*

9. Upper lip of corolla hooded or narrow 10
9. Upper lip of corolla not hooded or narrow 13

10. Leaves toothed to dissected, never entire; anther sacs
similar in size and position .... *Pedicularis*
10. Leaves all attached to erect stem, entire or cleft;
anther sacs of each stamen of different sizes and/or position 11

11. Calyx is 2-parted to the base; bracts not
distinctly colored .... *Cordylanthus*
11. Calyx 2- or 4-parted; bracts colored 12

12. Plants annual .... *Orthocarpus*
12. Plants perennial .... *Castilleja*

13. Leaves all basal; flowers small and inconspicuous .... *Limosella*
13. Leaves well distributed along the stem; flowers showy .... *Mimulus*

WYOMING KITTENTAIL (*Besseya wyomingensis*)

Description and Habitat. This perennial herb has somewhat hairy stems and grows to 12 inches tall. The basal leaves are toothed, and there are a few small, entire, sessile leaves on the stems. The flowers have no petals but have 2 hairy sepals and 2 long stamens with reddish purple filaments that confer the prominent color. This plant could be confused with some species of *Synthyris*, also called kittentails and members of the figwort family. Wyoming kittentail grows on dry soils from the foothills and alpine zones, blooming soon after the snow melts.

Natural History Notes and Uses. The genus name honors Charles E. Bessey (1845–1915), an American botanist, and the species name means "of Wyoming." There are about seven species of *Besseya*, mainly in the Rocky Mountains of the United States and Canada.

Paintbrush (genus *Castilleja*)

NORTHWESTERN PAINTBRUSH (*C. angustifolia*)

MOUNTAINSIDE PAINTBRUSH (*C. cristagalli*)

CUSICK'S PAINTBRUSH (*C. cusickii*)

PAINTBRUSH (*C. exilis*)

YELLOW PAINTBRUSH (*C. flava*)

WYOMING PAINTBRUSH (*C. linariifolia*)

SCARLET PAINTBRUSH (*C. miniata*)

SNOW PAINTBRUSH (*C. nivea*)

PALE PAINTBRUSH (*C. pallescens*)

PARROTHEAD PAINTBRUSH (*C. pilosa*)

BEAUTIFUL PAINTBRUSH (*C. pulchella*)

SPLITLEAF PAINTBRUSH (*C. rhexiifolia*)

SULPHUR PAINTBRUSH (*C. sulphurea*)

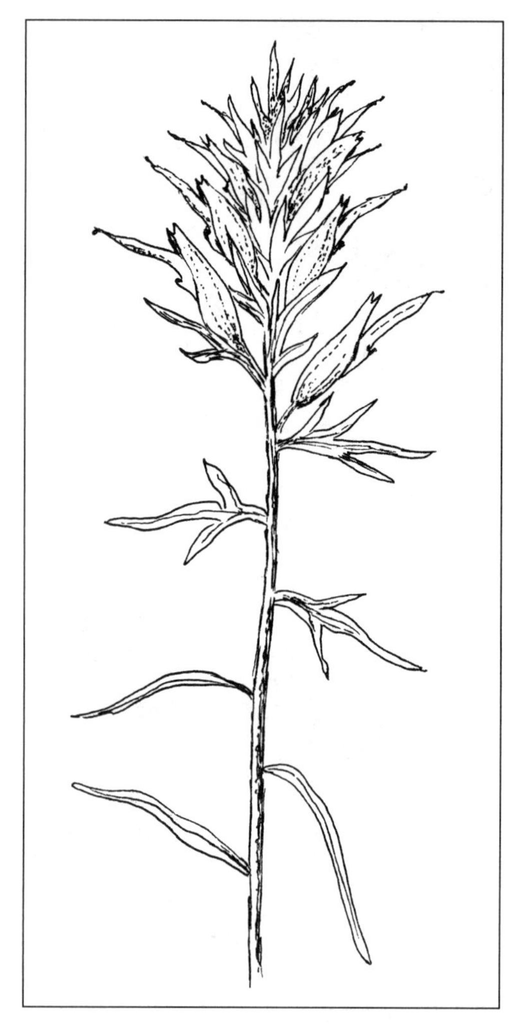

FIGURE 100. Wyoming paintbrush (*Castilleja linariifolia*).

Description and Habitat. *Castilleja* is a large genus found primarily in western North America. The many species are perennials with deeply lobed to entire leaves. The flowers are subtended by colorful, leaf-like bracts. The calyx is composed of 4 sepals that are united below. The 5 petals are united below into a tube but separate above into a small, 3-lobed lower lip and an elongated upper lip (*galea*) that encloses the 4 stamens. The genus is easily recognized, but many species are notoriously difficult to identify. In fact, most are distinguished by habitat, color, size of the petals, and other, microscopic characteristics. Paintbrushes grow in moist to dry situations in all vegetation zones. Hybridization between closely related species adds to the difficulty of identification.

FIELD KEY TO THE PAINTBRUSHES

1. Plants annual; stems solitary . . . . *C. exilis*
1. Plants perennial; stems clustered 2

2. Galea less than half the length of corolla tube 3
2. Galea more than half the length of corolla tube 10

3. Lower lip of corolla small, usually less than one-third the length of galea; bracts yellow, calyx yellow 4
3. Lower lip of corolla conspicuous, usually one-third to half the length of galea 5

4. Bracts deeply cleft . . . . *C. flava*
4. Bracts entire or with a pair of small lobes near the tip . . . . . . *C. sulphurea*

5. Calyx cleft into 4 unequal linear triangular lobes 6
5. Calyx cleft less than half as deeply laterally as in front and back 7

6. Upper stems and bracts puberulent to villous-hirsute . . . . *C. pilosa*
6. Upper stems and bracts woolly . . . . *C. nivea*

7. Calyx lobes acute; plants with fine hairs . . . . *C. pallescens*
7. Calyx lobes obtuse; plants sticky-hairy 8

8. Leaves and bracts entire or with few teeth near tip . . . . *C. sulphurea*
8. At least the upper leaves and bracts mostly deeply lobed 9

9. Bracts equaling or exceeding usually yellow flowers . . . . *C. cusickii*
9. Bracts shorter than purplish or reddish flowers . . . . *C. pulchella*

10. Calyx conspicuously more deeply cleft on side of galea, and usually more showy than bracts . . . . *C. linariifolia*
10. Calyx not equally cleft above and below or more deeply cleft on side of galea; bracts usually more showy than calyx 11

11. Leaves usually all entire 12
11. At least the upper leaves and bracts usually deeply cleft into 3–7 linear lobes 14

12. Bracts yellow . . . . *C. sulphurea*
12. Bracts red to purple 13

13. Stems less than 12 inches tall; bracts rose-purple to pink . . . *C. rhexiifolia*
13. Stems much taller than 12 inches; bracts red or orange. . . . . . . *C. miniata*

14. Plants densely hispid; most leaves lobed. . . . . *C. angustifolia*
14. Plants with or without hair, not hispid;
lower leaves not lobed. . . . . *C. cristagalli*

Natural History Notes and Uses. Some paintbrushes are partial root parasites. Their roots attach to those of neighboring plants, from which they obtain supplemental water and nutrients. For this reason, paintbrushes cannot be transplanted or easily grown from seed. The genus name honors Spanish botanist Domingo Castillejo. Wyoming paintbrush (*C. linearifolia*) is the official state flower of Wyoming. It has leafy, red bracts with protruding, yellow flowers.

Many if not all of the species have flowers and bracts that can be eaten raw. The seeds of some species were gathered, winnowed, dried, and stored for winter use. In winter they were parched, pounded, and eaten dry. The plants absorb selenium from the soil, however, and so should be taken in moderation. Symptoms in humans of selenium poisoning vary with the amount and form ingested but may include difficulty breathing, excessive urine production, loss of appetite, depression, a weak and rapid pulse, blurry vision, digestive upset, and eventually coma and death (Kirk 1975).

WYOMING STATE FLOWER

The Indian paintbrush, or painted cup (*Castilleja linariaefolia*), was adopted as the Wyoming state flower on January 31, 1917. The roots of the painted cups are partially parasitic on the roots of other green plants. Their true flowers are inconspicuous but are commonly enveloped by flower-like bracts in various shades of orange, red, and sometimes yellow. This species of *Castilleja* grows in moist areas, dry areas, and sandy prairies and attains a height of 1–2 feet. Choosing this species as the state flower was politically challenging.

The Wyoming chapter of the Daughters of the American Revolution supported the Indian paintbrush as the state flower. Dr. Grace R. Hebard, of the University of Wyoming, gets most of the credit for its adoption and ultimately drafted the state flower bill and found state legislators willing to sponsor it. However, one of the country's leading botanists, Dr. Aven Nelson (also of the University of Wyoming), strongly objected, for the following reasons: (1) Indian paintbrushes were not common throughout the state; (2) there were too many varieties and only an expert could tell them apart; (3) they were parasitic, feeding on the roots of other plants; and (4) there was no real widespread support for the plant. Dr. Nelson favored adopting a flower that was more common throughout the state and one that could be grown in gardens. He suggested the columbine (*Aquilegia*) or the schoolchildren's favorite, the fringed gentian (*Gentianella*).

SMALL-FLOWER BLUE-EYED MARY (*Collinsia parviflora*)

Description and Habitat. This small annual has slender, widely branched stems. The narrow, lanceolate, purple-tinged leaves are opposite, and the tiny, blue-and-white flowers are seldom more than ¼ inch long. The upper lip, bent upward, has 2 white lobes, while the lower lip has 3 bright blue lobes (the middle lobe is smaller than the lateral ones). It is usually found on stony slopes and rocky disturbed soils near trees and along roads in a variety of habitats up to subalpine.

Natural History Notes and Uses. The genus name honors Zaccheus Collins (1764–1831), a Philadelphia botanist. Small-flowered collinsia was used to make a horse run fast and was used externally for sore flesh. The leaves of another species, *C. heterophylla* (Chinese houses), was used by Native Americans in California as a poultice for bites from insects or snakes.

BUSHY BIRD'S-BEAK (*Cordylanthus ramosus*)

Description and Habitat. This annual herb may be a partial root parasite. The leaves are narrow or cut into narrow divisions. The flowers are tubular and 2-lipped with the upper lip incurved and enclosing the style and 2 or 4 of the stamens.

Natural History Notes and Uses. Strike (1994) indicates that *Cordylanthus* was used by California Indians as an emetic. Additionally, a powder of this species was taken with water every morning for syphilis.

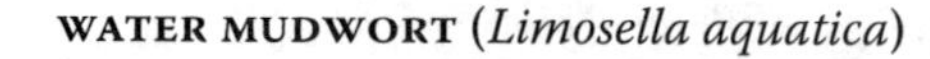

WATER MUDWORT (*Limosella aquatica*)

Description and Habitat. Water mudworts are acaulescent annual glabrous herbs with stolons and rosulate leaves. The flowers are solitary on a scape-like peduncle, and the calyx is 5-lobed and campanulate. The corolla is 5-cleft, campanulate, and nearly regular. There are 4 stamens. The genus name is Latin for "mud" and "seat." Water mudwort occurs in shallow water or on wet mud.

Natural History Notes and Uses. The Navajo used the leaves of mudwort ceremonially, rubbing them on the body for protection in hunting and from witches. A roll of washed leaves was used to plug a bullet or arrow wound (Elmore 1944).

Toadflax, Butter-and-eggs (genus *Linaria*)

DALMATIAN TOADFLAX (*L. dalmatica*)

BUTTER-AND-EGGS (*L. vulgaris*)

Description and Habitat. Both these *Linaria* species are rhizomatous perennial herbs. The flowers are funnel-shaped with 2 lips and a long, narrow, tubular projection (spur) at the base. They are introduced species found in disturbed areas at low elevations.

FIELD KEY TO THE TOADFLAXES

1. Leaves linear	*L. vulgaris*
1. Leaves ovate	*L. dalmatica*

Natural History Notes and Uses. In folk medicine a tea made from the leaves of *L. vulgaris* was said to be a laxative and strong diuretic. A "tea" made in milk was used as an insecticide (Foster and Duke 1990). The species has a long history of medicinal use and was once highly regarded as a diuretic for edema. It is seldom used now but undoubtedly merits investigation.

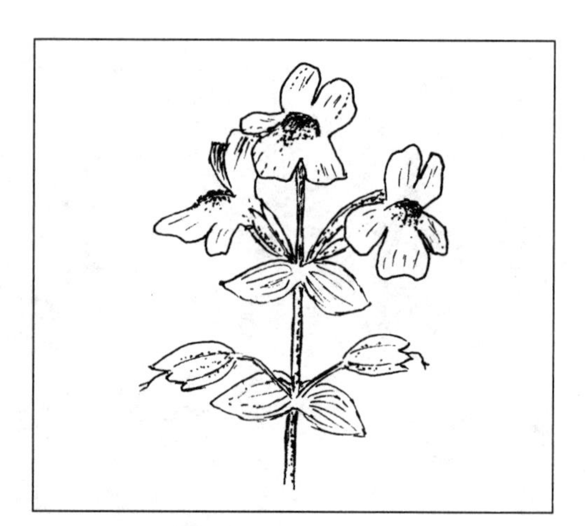

FIGURE 101. Common monkeyflower (*Mimulus guttatus*).

Monkeyflower (genus *Mimulus*)

BREWER'S MONKEYFLOWER (*M. breweri*)

MANY-FLOWERED MONKEYFLOWER (*M. floribundus*)

SEEP MONKEYFLOWER (*M. guttatus*)

LEWIS'S MONKEYFLOWER (*M. lewisii*)

MUSK MONKEYFLOWER (*M. moschatus*)

DWARF PURPLE MONKEYFLOWER (*M. nanus*)

RED MONKEYFLOWER (*M. rubellus*)

SUKSDORF'S MONKEYFLOWER (*M. suksdorfii*)

TILING'S MONKEYFLOWER (*M. tilingii*)

Description and Habitat. Many species of monkeyflower can be found in the area. They are annual or perennial herbs with opposite leaves. The flowers flare at the mouth to form 5 lobes, 2 forming the upper lip and 3 forming the lower. The perennial species occur in permanently wet soil, whereas the annuals are found in vernally moist habitats (Grant 1924).

FIELD KEY TO THE MONKEYFLOWERS

1. Plants perennial 2
1. Plants annual 5

2. Flowers reddish, rose, or pink-purple, often marked with yellow . . . . *M. lewisii*
2. Flowers usually yellow, often marked with red 3

3. Leaves pinnately veined . . . . *M. moschatus*
3. Leaves palmately veined 4

4. Stems mostly 1- to 5-flowered; petals over ¾ inch long . . . . . . *M. guttatus*
4. Stems usually more than 5-flowered; petals less than ¾ inch long . . . . *M. tilingii*

5. Flowers yellow, often marked with red 6
5. Flowers purple, pink, or red, often marked with yellow 9

6. Leaves linear to narrow lanceolate 7
6. Leaves mostly ovate or heart-shaped 8

7. Calyx teeth fringed with hairs . . . . *M. rubellus*
7. Calyx teeth not fringed with hairs . . . . *M. suksdorfii*

8. Calyx lobes distinctly unequal, the upper lobe much longer . . . . *M. guttatus*
8. Calyx lobes more or less equal . . . . *M. floribundus*

9. Flowers ¼ to 1 inch long, petals usually 2–3 times as long as sepals . . . . *M. nanus*
9. Flowers less than ½ inch long, petals less than twice as long as sepals 10

10. Margins of calyx lobes with gland-tipped hairs . . . . *M. breweri*
10. Margins of calyx lobes with nonglandular hairs . . . . *M. rubellus*

Natural History Notes and Uses. *Mimulus* is from the Latin *mimus* ("mimic" or "actor"), supposedly because of the "grinning face" of the flower. Lewis's monkeyflower (*M. lewisii*) was named in honor of Meriwether Lewis, who is credited with collecting the first described specimen. Hummingbirds seek monkeyflower nectar and are dusted with pollen from the anthers.

The young stems and leaves of *M. guttatus* have been used as salad greens. Sometimes the leaves were burned and the ash used as a salt. Weedon (1996) indicates that the young herbage of *Mimulus* species may be eaten in salads and that they grow bitter with age but remain edible. For example, the leaves of *M. primuloides* (primrose monkeyflower) were eaten by California Indians, and the young plants of *M. moschatus* (musk monkeyflower) were boiled and eaten by other Native Americans.

There are also reports of some Native Americans making a tea from monkeyflower stems, leaves, and flowers as a treatment for kidney or urinary problems and diarrhea (Moerman 1986). Monkeyflower leaves and stems were used externally to poultice wounds or internally to reduce fevers. *Mimulus* roots were used to treat fevers, dysentery, and diarrhea and to stanch hemorrhages. The raw leaves and stems can be applied as a poultice to burns and wounds, as well as to insect bites and pain from contacting stinging nettle.

Owl's Clover (genus *Orthocarpus*)

YELLOW OWL'S CLOVER (*O. luteus*)

TOLMIE'S OWL'S CLOVER (*O. tolmiei*)

Description and Habitat. Members of this genus are semi-parasitic slender plants with sessile flowers that are subtended by a large, colorful bract. The tubular flowers are 2-lipped with 2 upper lobes united to form a small, hood-shaped galea. The 3 lower lobes are partly united and nearly as long as the galea.

FIELD KEY TO THE OWL'S CLOVERS

1. Galea longer than the lower lip, with the tip ending in a short hook . . . . *O. tolmiei*
1. Galea about equal to the lower lip, tip not hooked . . . . *O. luteus*

Natural History Notes and Uses. Owl's clovers resemble paintbrushes (*Castilleja*) but can be distinguished by the annual growth form and by having a lower corolla that nearly equals the galea. The foliage of *O. luteus* was used by some Native Americans to dye small skins and feathers a reddish tan.

Lousewort (genus *Pedicularis*)

BRACTED LOUSEWORT (*P. bracteosa*)

COILED LOUSEWORT (*P. contorta*)

FERNLEAVED PEDICULARIS (*P. cystopteridifolia*)

ELEPHANTHEAD LOUSEWORT (*P. groenlandica*)

OEDER'S LOUSEWORT (*P. oederi*)

PARRY'S LOUSEWORT (*P. parryi*)

MOUNTAIN LOUSEWORT (*P. pulchella*)

SICKLETOP LOUSEWORT (*P. racemosa*)

Description and Habitat. Louseworts are perennial, partially parasitic herbs with toothed or pinnately divided leaves. The flowers are subtended by leaf-like bracts and occur in a spike-like inflorescence. The tubular flower is strongly 2-lipped at the mouth. They are found in dry or moist open habitats, including meadows up to the alpine zone. The genus name comes from the Latin *pediculus* ("little louse"), alluding to the superstition that livestock that ate the plants would suffer an infestation of lice. Although many members of this genus are attractive, they are partially parasitic on the roots of other plants, and thus none of them has been cultivated for garden use.

FIELD KEY TO THE LOUSEWORTS

1. Leaves simple and merely toothed; calyx usually 2-lobed or toothed . . . . *P. racemosa*
1. Most of the leaves deeply pinnately lobed to bipinnatifid; calyx usually 5-lobed or toothed 2

2. Galea with a beak ⅛ inch or more long, upturned or downturned, may be hidden by the lower lip 3
2. Galea without a beak or with a straight beak less than ⅛ inch long 4

3. Beak upturned, flower resembling an elephant's head .... *P. groenlandica*
3. Beak downturned, largely hidden by the lower lip .... *P. contorta*

4. Stems leafy .... *P. bracteosa*
4. Stems not leafy, or a few leaves reduced in size 5

5. Galea with tapering beak .... *P. parryi*
5. Galea without a beak or with a pair of very minute projections........ 6

6. Flowers yellow or cream-colored .... *P. oederi*
6. Flowers purple, reddish, or pinkish 7

7. Plants mostly under 8 inches tall; subalpine and alpine habitats .... *P. pulchella*
7. Plants mostly more than 8 inches tall; usually below subalpine.......................... *P. cystopteridifolia*

Natural History Notes and Uses. Tilford (1997) indicates that as long as lousewort is not attached to an unpalatable host, the fleshy roots can be prepared and eaten in moderation. He does not, however, describe how to prepare them. He also states that the leaves and stems of some species may be steamed or boiled as potherbs, but this is not recommended.

Moore (1979) describes *Pedicularis* as an effective sedative for children and tranquilizer for adults. The whole stalk, when dried and prepared as a tea, acts as a mild relaxant, quieting anxiety and tension. The fresh or dried plant is a vulnerary for minor injuries, with mild astringent and antiseptic properties.

Warning: Because of the sedative nature of this plant, its potential toxic alkaloids, and the host species it may be attached to, ingestion of lousewort is not recommended.

SEMI-PARASITIC PLANTS

The penstemons (*Penstemon*), paintbrushes (*Castilleja*), owl's clover (*Orthocarpus*), louseworts (*Pedicularis*), and bastard toadflax (*Comandra*) are semi-parasites. They look like normal plants, having green leaves, but they attach to the roots of adjacent plants by means of suckers and are probably never completely independent.

Penstemon (genus *Penstemon*)

ABSAROKA RANGE BEARDTONGUE (*P. absarokensis*)
STIFFLEAF PENSTEMON (*P. aridus*)
SULPHUR PENSTEMON (*P. attenuatus*)
CARY'S BEARDTONGUE (*P. caryi*)
WASATCH BEARDTONGUE (*P. cyananthus*)
BLUE PENSTEMON (*P. cyaneus*)
SCABLAND PENSTEMON (*P. deustus*)
BUSH PENSTEMON (*P. fruticosus*)
LARCHLEAF BEARDTONGUE (*P. laricifolius*)
MOUNTAIN PENSTEMON (*P. montanus*)
LITTLEFLOWER PENSTEMON (*P. procerus*)
MATROOT PENSTEMON (*P. radicosus*)
RYDBERG'S PENSTEMON (*P. rydbergii*)
WHIPPLE'S PENSTEMON (*P. whippleanus*)

Description and Habitat. Because there are so many species in this genus, it is sometimes customary to divide them into groups according to color, size, or some other dependable character. In general, they are perennial herbs with opposite leaves. The flower is strongly to indistinctly 2-lipped at the mouth with a 2-lobed upper lip and a 3-lobed lower lip. There are 4 anther-bearing (fertile) stamens and a single sterile stamen (staminode) that is often hairy at the tip (hence the common name "beardstongue"). The fruit is a many-seeded capsule. Penstemons occur in dry or moist meadows or forest openings up into the alpine zone.

FIELD KEY TO THE PENSTEMONS

1. Plants with anthers densely long-woolly 2
1. Plants with anthers either glabrous or pubescent, but not densely long-woolly 3

2. Plant woody (shrub); calyx glandular-pubescent .... *P. fruticosus*
2. Plants herbaceous; corolla glabrous outside .... *P. montanus*

3. Inflorescence glandular-pubescent outside 4
3. Inflorescence glabrous outside, not glandular 9

4. Corolla white 5
4. Corolla blue, violet, or dark purple 7

5. Corolla glandular-hairy inside; leaves toothed .... *P. deustus*
5. Corolla glabrous inside; leaves not toothed 6

6. Staminode glabrous or bearded only at tip .... *P. whippleanus*
6. Staminode bearded for one-third its length .... *P. attenuatus*

7. Leaves of lower stem wholly glabrous, bright or dark green 8
7. Leaves of lower stem either puberulent or pubescent throughout, or at least on veins and margins; dull or grayish green .... *P. radicosus*

8. Ovary and capsule glandular-puberulent at top .... *P. whippleanus*
8. Ovary and capsule glabrous .... *P. aridus*

9. Anthers short- or long-hairy only on the side opposite dehiscence 10
9. Anthers glabrous on the side opposite dehiscence 12

10. Anthers woolly on side opposite dehiscence .... *P. caryi*
10. Anthers short-pubescent on side opposite dehiscence 11

11. Corolla less than ¾ inch long .... *P. cyananthus*
11. Corolla more than ¾ inch long .... *P. cyaneus*

12. Corolla mostly ¼ inch long; flowers clustered .... *P. procerus*
12. Corolla more than ¼ inch long 13

13. Leaf blades linear or filiform, mostly a basal tuft, glabrous . . *P. laricifolius*
13. Leaf blades mostly broader than linear, entire .... *P. rydbergii*

Natural History Notes and Uses. *Penstemon* is the largest genus of flowering plants endemic to North America. With the exception of a few unique species, most penstemons are difficult to identify, especially those with blue to lavender or purple flowers. The genus name is from the Greek *pente*, meaning "five," and *stemon*, meaning "thread," referring to the slender 5th stamen. Several species of penstemon are cultivated, and most are easily raised from seed.

As a topical astringent, pureed or juiced, the plants can be used as a general dressing for minor irritations of the skin (e.g., insect bites). Penstemon oil is a good addition to an all-purpose salve.

LANCE-LEAF FIGWORT (*Scrophularia lanceolata*)

Description and Habitat. Lance-leaf figwort is a herbaceous perennial with a thickened root. The stem leaves are opposite and have triangular blades with toothed margins. The 2-lipped flower is yellowish green and maroon. There are 4 functional stamens, and the 5th one is reduced to a small knob on the corolla. This is an uncommon species found in moist meadows and forest openings in the low to middle elevations.

Natural History Notes and Uses. A strong tea made from the plant can be applied to fungal infections of the skin (e.g., athlete's foot) and can help eczema, rashes, burns, and hemorrhoids (Moore 1979).

FEATHER-LEAF KITTENSTAILS (*Synthyris pinnatifida*)

Description and Habitat. This early-blooming plant appears shortly after snowmelt. It has deep blue flower clusters resembling tails of upset kittens; the florets stand out in all directions on short, dense spikes over thick, powdery blue, dissected ferny leaves.

Natural History Notes and Uses. The genus name is from the Greek *syn* ("with") and *thyris* ("window"), alluding to the valves of the capsule. This genus has 14–15 species of tufted rhizomatous herbs with heart- or kidney-shaped leaves and racemes of 4-lobed, more or less bell-shaped flowers. Several species are grown as ornamentals.

COMMON MULLEIN (*Verbascum thapsus*)

Description and Habitat. Mullein is a coarse plant with stems up to 6 feet tall. The basal leaves are entire-margined or shallowly toothed, broadly lance-shaped, and up to 16 inches long. The stem leaves are smaller, and the upper ones clasp the stems, forming low wings below the point of attachment. The whole plant is densely covered with long, soft hairs. The white or yellow, saucer-shaped flowers are borne in a densely congested inflorescence. The upper 3 stamens are densely hairy. Mullein can be found in disturbed places up to the subalpine zone.

Natural History Notes and Uses. The leaves of mullein are said to be edible in small quantities when cooked. Because of their woolly texture, however, we have found the plants to be undesirable.

Native Americans smoked the dried leaves of mullein for asthma and sore throat. A tea from the leaves was used to treat colds, and the flowers contain an oil that has been used for earaches. The flowers and first-year leaves make a soothing decoction for coughs and sore throats. Boil leaves in water for 10 minutes, then strain the liquid through a cheesecloth to remove the tiny hairs. The leaves can also be used as poultices applied locally to hemorrhoids, sunburn, and inflammations. The dried stalks are ideal for use as hand drills to start fires. The flowers and leaves produce yellow dye; as a toilet paper substitute, the large fresh leaves are choice.

Caution: Mullein contains coumarin and rotenone, two substances that may be toxic in large quantities if ingested. Also, the seeds are not recommended for consumption.

Speedwell (genus *Veronica*)

AMERICAN SPEEDWELL (*V. americana*)

WATER SPEEDWELL (*V. anagallis-aquatica*)

TWO-LOBE SPEEDWELL (*V. biloba*)

SKULLCAP SPEEDWELL (*V. scutellata*)

THYMELEAF SPEEDWELL (*V. serpyllifolia*)

AMERICAN ALPINE SPEEDWELL (*V. wormskjoldii*)

Description and Habitat. Speedwells are annual or perennial herbs with only stem leaves; the leaves are opposite. The stalked flowers are borne on the end of the stem or on paired stalks that arise from the upper leaf axils. The saucer-shaped corollas are 4-lobed with the uppermost lobe the largest and the lowest one the smallest. The underside of the petals is lighter in color, often nearly white. There are 2 stamens, and the seed capsule is somewhat compressed and usually notched or 2-lobed at the top. The mature fruit is often necessary for identification.

FIELD KEY TO THE SPEEDWELLS

1. Flowers borne on branches of the inflorescence, which arises from the axils of the upper leaves 2
1. Flowers borne in a narrow inflorescence that terminates the leafy stem 4

2. Leaves all with short petioles. . . . . *V. americana*
2. At least the middle and upper leaves without petioles 3

3. Capsule evidently wider than high with a conspicuous notch on top . . . . *V. scutellata*
3. Capsule about as wide as high with an inconspicuous notch on top . . . . *V. anagallis-aquatica*

4. Plants annuals. . . . . *V. biloba*
4. Plants perennials 5

5. Lower portion of stems prostrate and rooting at the nodes. . . . . *V. serpyllifolia*
5. Stems erect or somewhat curved at base, not prostrate. . . *V. wormskjoldii*

Natural History Notes and Uses. The genus may be named after Saint Veronica, who was said to have wiped the face of Christ en route to his crucifixion. The name may also have arisen from the plant's beginning as a medicinal herb, after a shepherd observed an injured deer heal its wounds by rolling in and eating the herb. The shepherd reported this discovery to his king, who was ill. The king became well after trying the herb and showered the shepherd with riches.

The leaves and stems of all species, when collected during the spring and early summer, can be eaten like watercress, added to salads, or prepared as potherbs. The taste of the various species ranges from spicy to bitter to bland. The plants also contain moderate amounts of vitamin C and were once used to prevent scurvy. The leaves and stems can also be steeped as a tea. Care should be taken to avoid plants growing in polluted waters. If desired, the addition of halazone tablets or chlorine bleach to the wash water may kill harmful microbes. After flowering, it is better to boil the plants to eliminate the bitterness.

Medicinally, the plants were used mainly as an expectorant for respiratory problems. Infusions were also used in hair-conditioning rinses and skin-cleansing herbal steams and as an ingredient in massage oils and ointments that were added to baths for a soothing soak. Pojar and MacKinnon (1994) indicate that *V. americana* has been used for centuries to treat urinary and kidney complaints and as a blood purifier. The leaf juice from *V. serpyllifolia* has been used for earaches, its leaves were poulticed for boils, and a tea was used for chills and coughs.

NIGHTSHADE FAMILY (SOLANACEAE)

There are about 85 genera and 2,300 species in the nightshade family. About 13 genera are indigenous to the United States. The family is noted for its edible, poisonous, and medicinal plants. Some of the economically important plants are tomato, potato, chili, tobacco, and eggplant. Medicinally, two alkaloids, atropine (belladonna) and scopolamine, which can be deadly in large amounts, are obtained from this family. Eye doctors today use atropine in very exact small doses as an effective dilator when examining eyes.

FIELD KEY TO THE NIGHTSHADE FAMILY

1. Corolla rotate to broadly campanulate 2
1. Corolla salverform to funnel-form or urn-shaped 3

2. Calyx enlarging with age and more or less enclosing the fruit ... *Physalis*
2. Calyx not enlarging as fruit matures .... *Solanum*

3. Flowers and leaves sessile................................ *Hyoscyamus*
3. Flowers and leaves with pedicels............................ *Nicotiana*

BLACK HENBANE (*Hyoscyamus niger*)

Description and Habitat. Henbane grows up to 2 feet tall and has large, pale green, oval leaves with deeply toothed edges. Tiny hairs cover the stem and leaves. The flowers are bell-shaped and have mustard-yellow petals with purplish brown throats and veins. The seeds are enclosed in ½-inch-long capsules.

Natural History Notes and Uses. Native to western Asia and southern Europe, henbane is now found across much of western and central Europe and North and South America. Henbane is cultivated for therapeutic use in parts of Europe, including England, and in North America. The leaves and flowers are picked just after the plant has flowered, in the first year for the annual variety and in the second year for the biennial.

A traditional witches' brew ingredient, henbane has had a deservedly villainous reputation since ancient times. The narcotic alkaloids hyoscyamine, scopolamine, and atropine are derived from this foul-smelling, weedy plant. Therefore, all parts of the plant are considered **poisonous**, and if eaten, even small amounts cause anything from dizziness to delirium. Too much brings about a slow and painful death. In past times henbane served as a sedative to ease pain and spasms, but the determination of a safe dose has always been a tricky business, and for long periods the drug seems to have been left alone by medical practitioners.

Today henbane is used extensively in herbal medicine as a sedative and painkiller. The plant is specifically used for pain affecting the urinary tract, especially pain due to kidney stones, and is also given for abdominal cramping. Its sedative and antispasmodic effect makes henbane a valuable treatment for the symptoms of Parkinson's disease, relieving tremor and rigidity during the early stages of the illness. Henbane has also been used to treat asthma and bronchitis, usually as a "burning powder" or in the form of a cigarette. Applied externally as an oil, it can relieve painful conditions such as neuralgia, sciatica, and rheumatism. Henbane reduces mucus secretions as well as saliva and other digestive juices. Like its cousin deadly nightshade, henbane dilates the pupils. One of henbane's active components, hyoscine, is sometimes used as a substitute for opium. Hyoscine is commonly used as a preoperative anesthetic and in motion sickness formulations.

COYOTE TOBACCO (*Nicotiana attenuata*)

Description and Habitat. This annual plant grows up to 40 inches tall and has glabrous to glandular-pubescent stems. The leaves are large and ovate to ovate-lanceolate, and the upper leaves are narrowed. The flowers occur in a terminal raceme, and the calyx is 5-cleft. The corolla is white and tubular or funnel-shaped. The flowers close in the sun. Coyote tobacco grows in disturbed places and flowers from May to October.

Natural History Notes and Uses. All species of *Nicotiana* contain the highly toxic alkaloid nicotine. In addition to commercial cultivation for the leaves, this genus is grown as an ornamental plant. A relative, *N. tabaccum*, is the species used in smoking products. Although tobacco plants are distasteful, grazing animals may ingest them under some conditions. An effective insecticide against aphids can be prepared by steeping tobacco leaves in water and spraying the solution on affected parts of the plant.

Tobacco's primary medicinal use is as a topical analgesic. External preparations are useful in relieving pain and sensitivity from contusions, sprains, and other sport-accident types of injuries. Soaking in a bath made with tobacco tea relieves joint soreness, aches, and pains from a hard day's work. Topical preparations are also effective when applied to muscular spasm, whether occurring from overwork or injury. All species in the western United States have either substantial amounts of nicotine, such as in *N. attenuata*, or other alkaloids that are therapeutically active, such as nornicotine in *N. trigonophylla* and anabasine in *N. glauca*. Tobacco's alkaloid content has well-documented inhibitory activity on body-brain pain transmission. The oil or leaf bolus is soothing to hemorrhoids.

LONG-LEAF GROUNDCHERRY (*Physalis longifolia*)

Description and Habitat. Groundcherry is an herb with alternate, entire to coarsely toothed leaves. The flowers are either solitary in the axils of the leaves or in clusters of 2–5. The calyx is 5-lobed and becomes enlarged and inflated in the fruit, which is a many-seeded berry. The genus name is from the Greek, meaning "bladder" and referring to the inflated calyx. The plants are found in waste places and fields at the lower elevations.

Natural History Notes and Uses. The berries can be eaten raw or cooked and taste best when fully ripe. There is a tendency for the fruits to fall off the plants before maturation. In that case, simply gather the fruits and allow them to ripen in the husk. They will keep for a couple of weeks in this condition. The fruits have been used in preserves and pies. The unripe fruits of some species are considered ***poisonous*** when eaten in large amounts. If the fruits are bitter, sour, or otherwise strongly flavored, they are unripe. In any case, let them ripen until soft and sweet.

Nightshade (genus *Solanum*)

CLIMBING NIGHTSHADE (*S. dulcamara*)

CUT-LEAF NIGHTSHADE (*S. triflorum*)

Description and Habitat. Nightshade is a highly diverse genus containing more than a thousand species worldwide. In general, nightshades are annual or perennial herbs with flowers that resemble those of tomato or potato plants. The fruit is a many-seeded berry, surrounded in part by the persistent calyx. The various nightshades can be found from moist, open habitats to disturbed waste areas at the lower elevations. *Solanum* probably comes from the Latin *solamen*, which means "quieting," referring to the sedative properties of some species.

FIELD KEY TO THE NIGHTSHADES

1. Plants sprawling, vine-like; leaves 3-lobed; petals bright violet or blue purple .... *S. dulcamara*
1. Plants not sprawling or vine-like; leaves pinnately compound or pinnatifid; flowers white .... *S. triflorum*

Natural History Notes and Uses. The most ***poisonous*** part of the plant is the unripe fruit, but the stems, leaves, and roots are also dangerous. In fact, the alkaloid content in the plants decreases in this order: unripe fruit, leaves, stems, ripe fruit (Schofield 1989). Solanine is the predominant glyco-alkaloid, but others may be present. Although the degree of toxicity varies among and within species, solanine is highly toxic and can cause death. Solanine has been known to cause abdominal pain, headache, flushing, restlessness, irritation of the skin and mucous membranes, and tiredness. Craighead et al. (1963) report that cooking destroys the solanine, making the ripe fruit edible.

The berries of *S. nigrum* (black nightshade) were formerly used as a diuretic (Foster and Duke 1990). Native Americans of the Southwest used the crushed berries to curdle milk for making cheese, and they have also been used in various preparations for sore throat and toothaches.

S. dulcamara (climbing nightshade), when used correctly and in appropriate dosages, is said to be effective in the treatment of skin disorders, rheumatism, and bronchitis. Studies have shown that this plant possesses anti-cancer qualities (Tilford 1997).

Members of the genus *Solanum* are pollinated when an insect such as a bumblebee or hoverfly uses its wings and thorax to set up a vibration in and around the anthers. This causes the pollen to stream out onto the insect, a process called "vibrating pollen collection."

Caution: Some references describe the berries of some species of *Solanum* as edible. However, it is our recommendation that none of the plants that occur in the Greater Yellowstone Area be consumed in any manner.

NETTLE FAMILY (URTICACEAE)

There are about 45 genera and 550 species in the nettle family. Six of the genera are native to the United States and are of little economic importance.

FIELD KEY TO THE NETTLE FAMILY

1. Plant is an annual; leaves are alternate and entire . . . . *Parietaria*
1. Plant perennial; leaves are opposite and toothed. . . . . *Urtica*

PENNSYLVANIA PELLITORY (*Parietaria pennsylvanica*)

Description and Habitat. Pellitory is recognized by its long-tapering alternate leaves, the lack of stinging hairs on the stems, and the small axillary clusters of green flowers. It is an annual herb with a tuft of roots. The leaves are alternate, simple, lanceolate to oblong-lanceolate, long-tapering to a rounded tip, tapering to the base, without teeth, hairy, very thin, and up to 2½ inches long. Several flowers are crowded into small axillary clusters; some of the flowers are perfect, some male only, some female only, all often in the same cluster on the same plant. Each flower is green, surpassed by bracts.

Natural History Notes and Uses. The genus name is from the Latin *paritinae* ("old walls"), alluding to the habitat of some species. There are about 20 species of annual and perennial herbs with simple, alternate leaves and clusters of small, unisexual, green flowers.

STINGING NETTLE (*Urtica dioicia*)

Description and Habitat. This plant is an annual or perennial herb with stinging hairs. The flowers are numerous, small, and clustered on drooping branches at the base of the leaves. Stinging nettles can be found along roadsides and streams, and in moist areas and waste places in the low to middle elevations. Stinging nettle is an indicator of good soil conditions. For obvious reasons, many people consider stinging nettles obnoxious weeds. Recent taxonomic revisions have consolidated various *Urtica* species into a single species.

Natural History Notes and Uses. One of the first things a person learns about stinging nettles is their stinging effect. The intense burning and itching or stinging of the skin may persist for a long time. If you look closely at the hairs, you will see a hypodermic mechanism consisting of a very fine capillary tube with a bladder-like base that is filled with chemical irritant. When brushed against, a minute spherical tip breaks off, uncovering a sharp tip that easily penetrates the skin. The chemical is forced into the skin through the tube as the hair-like tube bends and constricts the bladder-like base. Therefore, stinging nettles should be collected with gloves.

The young stems and leaves of stinging nettle are edible after boiling or steaming and are delicious as a spinach substitute. Boiling the leaves destroys the formic acid found in the hairs. The leaves are high in vitamins A, C, and D, the last of which is rare in plants. The roots are also edible after they have been roasted. A tea made from the leaves is said to have astringent and diuretic qualities and has been used for internal bleeding and nose-bleeds as well as chest colds and internal pains. As a poultice, *Urtica* was used for headaches. Native Americans learned of stinging nettle as a food from early European travelers and settlers and possibly from Chinese immigrants.

The older stems become fibrous, which reduces their edible qualities but makes them ideal for producing strong cordage. The older leaves also contain cystoliths that can irritate the kidneys. A yellow dye may be obtained by boiling the roots.

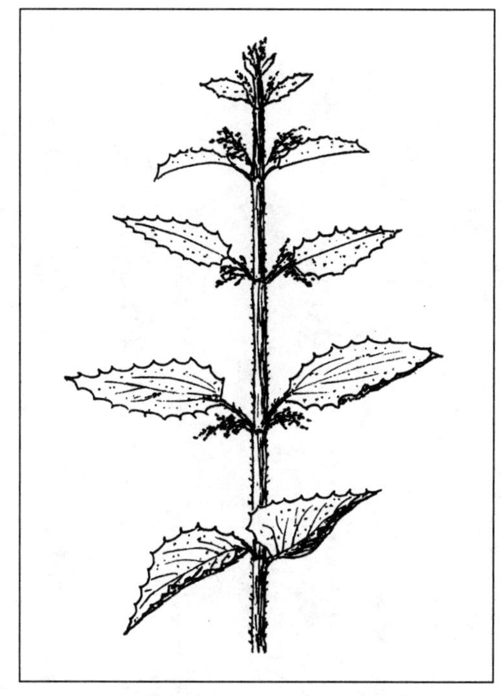

FIGURE 102. Stinging nettle (*Urtica dioicia*).

VALERIAN FAMILY (VALERIANACEAE)

The valerian family has about 13 genera and 400 species of small shrubs throughout the world, chiefly in the temperate and cold regions of the Northern Hemisphere. Three of the genera are native to the United States, but only *Valeriana* is found in the area. In general, the family is of no economic importance. However, *V. officinalis,* which is indigenous to eastern and central Europe, has become naturalized in western Europe and parts of North America after early introduction by colonists. The genus also includes species used medicinally in India (*V. wallichii*), Asia (*V. officinalis var. latifolia*, kesso root) and the Americas (*V. edulis*).

Valerian (genus *Valeriana*)

SHARP-LEAF VALERIAN (*V. acutiloba*)

MARSH VALERIAN (*V. dioica*)

EDIBLE VALERIAN (*V. edulis*)

WESTERN VALERIAN (*V. occidentalis*)

Description and Habitat. Valerians are perennial herbs with aromatic (actually ill-smelling) roots. The stem leaves are opposite and pinnately compound, and the flowers have 3 stamens. They can be found in open forests and meadows to timberline and above. The genus name comes from the Latin *valere* ("to be strong") and refers to the plant's medicinal qualities.

FIELD KEY TO THE VALERIANS

1. Plants with a stout taproot and a branched caudex; inflorescence appearing panicle-like; basal leaves tapering gradually to the petiolar base .... *V. edulis*
1. Plants with numerous, fibrous roots from a rhizome or a caudex; inflorescence like a corymb during flowering; lower leaves differentiated into blade and petiole 2

2. Corolla mostly more than ⅛ inch long, the lobes about half as long as the tube or shorter......................... *V. acutiloba*
2. Corolla mostly less than ⅛ inch long, the lobes about half as long as the tube to longer than the tube....................... 3

3. Lateral lobes of stem leaves strap-shaped..................... *V. dioica*
3. Lateral lobes of stem leaves ovate....................... *V. occidentalis*

Natural History Notes and Uses. The plants commonly contain the alkaloids chatinine and valerine. They are known to act on the central nervous system as depressants and are prescribed to calm nerves and relieve insomnia. Valerian has been used by physicians since at least the ninth century. Extracts were used as nerve tonics, and their relaxing properties may rival those of opium. Valerian was one of 72 ingredients Mithridates, king of Pontus, compounded as an antidote to poison, using poisoned slaves as test subjects.

The roots and leaves of *V. edulis* can be collected, steamed for a day or two to remove the disagreeable odor, and then used in soups as a potato substitute. However, the taste takes a little getting used to and may remain somewhat unpalatable. We found that the steamed roots are better if dried, ground into flour, and then added to other flours. The other species could probably also be used in an emergency, but they do not have the large taproot of *V. edulis*.

Valerians are best known for their calming qualities. They have been used for more than a hundred years in the area as a remedy for anxiety, muscle tension, and insomnia. The plants contain valepotriate, which is a known herbal calmative, antispasmodic, and nerve tonic and is used for hypochondria, nervous headaches, irritability, and insomnia (Tilford 1993). Research has confirmed that teas, tinctures, and extracts of this plant depress the central nervous system and act as a sedative for agitation (Foster and Duke 1990).

V. dioica was used by the Gitxsan as part of a ceremonial smudge. The whole plant was placed in bear grease, and then the grease was smeared onto the hair and face.

Valerian is a highly variable species. Polyploidy, a condition in which cells contain more than two copies of each chromosome, occurs in *V. officinalis*, and there are diploid, tetraploid, and octaploid types. English valerian is usually octaploid, and central European usually tetraploid.

VERBENA FAMILY (VERBENACEAE)

There are approximately 75 genera and 3,000 species in the verbena family worldwide, of which 14 genera are native to the United States. The family is of economic importance because of its highly prized teak wood (*Tectona grandis*) and a number of ornamentals.

Verbena (genus *Verbena*)

BIG-BRACT VERBENA (*V. hastata*)

HOARY VERBENA (*V. stricta*)

Description and Habitat. Members of this genus are perennials with opposite, toothed leaves. The tubular flowers with flaring lobes each have a subtending bract. The fruit is a cluster of 4 nutlets.

FIELD KEY TO THE VERBENAS

1. Leaves lance-shaped with a definite petiole; inflorescence of numerous spikes . . . . *V. hastata*
1. Leaves sessile or almost with elliptic blades; inflorescence of 1 to few spikes . . . . *V. stricta*

Natural History Notes and Uses. The seeds of *V. hastata* may be gathered, roasted, and ground into a bitter-tasting flour. Leaching the flour may remove the bitter taste. A tea from the boiled leaves can be used for a stomachache, and a tea from the roots was used to clear cloudy urine (Willard 1992; Foster and Duke 1990). Moore (1979) says that the plant is used as a sedative, diaphoretic, bitter tonic, and mild coagulant. It promotes sweating, relaxes and soothes, settles the stomach, and gives an overall feeling of relaxed well-being.

VIOLET FAMILY (VIOLACEAE)

The violet family has approximately 16 genera and 850 species distributed worldwide. Two genera are native to the United States. The family is of little economic importance other than as a source of ornamentals, as many species of *Viola* are cultivated. *Viola* is the classical name for violets.

Violets (genus *Viola*)

HOOKED-SPUR VIOLET (*V. adunca*)

CANADIAN WHITE VIOLET (*V. canadensis*)

SMALL WHITE VIOLET (*V. macloskeyi*)

BOG VIOLET (*V. nephrophylla*)

NUTTALL'S VIOLET (*V. nuttallii*)

DARK-WOODS VIOLET (*V. orbiculata*)

MARSH VIOLET (*V. palustris*)

CANARY VIOLET (*V. praemorsa*)

GOOSEFOOT OR PINE VIOLET (*V. purpurea*)

WHITE VIOLET (*V. renifolia*)

VALLEY VIOLET (*V. vallicola*)

FIGURE 103. Violet (*Viola* spp.).

Description and Habitat. There are many species of violets. In general, they are low-growing herbs that are perennial or annual. The leaves are spade-shaped and basal. The flowers occur singly on the ends of stems and have 5 petals. There are 2 upper and 2 lateral petals and 1 lower petal that is prolonged into a nectar-holding pouch at the base of the flower. Most species also have small, self-fertilizing flowers that do not open. Violets can be found in meadows and open forests from the foothills to above timberline.

FIELD KEY TO THE VIOLETS

1. Flowers usually yellow 2
1. Flowers generally blue, purple, or white 6

2. Leaves generally toothed or lobed; upper petals purple on the back .... *V. purpurea*
2. Leaves entire to slightly finely toothed; upper petals not purple on back 3

3. Leaves cordate, orbicular, or kidney-shaped, about as long as wide .... *V. orbiculata*
3. Leaves ovate or lanceolate to elliptic or deltoid, usually longer than wide 4

4. Leaves narrowly lanceolate, usually about 3 or more times as long as wide .... *V. nuttallii*
4. Leaves more ovate or elliptic, generally less than 3 times as long as wide 5

5. Leaves less than 2 inches long .... *V. vallicola*
5. Leaves usually over 2 inches long .... *V. praemorsa*

6. Flowers white on inner surface, may be tinged with blue 7
6. Flowers deep blue or purple 10

7. Plants with leafy stems .... *V. canadensis*
7. Leafless flowering stems arising from roots 8

8. Leaf blades hairy underneath, especially near the petiole .... *V. renifolia*
8. Leaf blades glabrous 9

9. Upper petals usually tinged with violet on the back; leaves more than 1 inch wide .... *V. palustris*
9. Upper petals white on the outside; leaves usually less than 1 inch wide .... *V. macloskeyi*

10. Leafy stems present, sometimes short; spur half as long as lower petal .... *V. adunca*
10. Leafless flowering stems arising from the roots; spur one-third as long as lower petal or shorter .... *V. nephrophylla*

Natural History Notes and Uses. The leaves, buds, and flowers of possibly all species are edible raw or cooked, with some being more palatable than others; the leaves make a good tea. Adding the leaves to soups makes them thicker. Violets are high in vitamin C and beta-carotene. Collect the plants by leaving the roots intact; since many species reproduce vegetatively, you will probably not inhibit next year's growth significantly. Many naturalists indicate that all violets are safe for consumption, but some experts insist that some yellow species may be somewhat purgative. All species have a tendency to be slightly laxative, so proceed slowly (Kirk 1975). The flowers have also been candied or made into jellies and jams.

Violet salve can be made by simmering the entire herb in lard. It was a famous remedy for skin inflammations and abrasions. Violets are also emollients (which soften the skin) and are an excellent ingredient in lotions such as night creams. The flowers and leaves of some species have been used in various herbal remedies as poultices and laxatives and to relieve cough and lung congestion.

MISTLETOE FAMILY (VISCACEAE)

Mistletoes are parasitic shrubs found on the branches of a variety of trees such as oaks, alders, conifers, and cottonwoods. The leaves are opposite and somewhat leathery. The flowers are small, composed of 2–4 tepals (petals and sepals not differentiated) inserted on a cup-like receptacle. The fruit is a drupe or berry. There are approximately 11 genera and 450 species in this family distributed worldwide. In the United States *Arceuthobium* and *Phoradendron* are native.

Dwarf Mistletoe (genus *Arceuthobium*)

AMERICAN DWARF MISTLETOE (*A. americanum*)

LIMBER PINE DWARF MISTLETOE (*A. cyanocarpum*)

DOUGLAS-FIR DWARF MISTLETOE (*A. douglasii*)

Description and Habitat. Dwarf mistletoes are parasitic flowering plants that infect conifers, producing characteristic yellow-to-orange or green-to-brown leafless aerial shoots on the host plant. They have no developed leaves, and the shoots remain short, varying in length from a few inches to several inches.

FIELD KEY TO THE DWARF MISTLETOES

1. Flowers sessile or with female ones with slender pedicels, usually 2 per node.................... *A. cyanocarpum*
1. Flowers terminal on short lateral branches, those bearing male flowers with more than 2 per node................ 2

2. Plants not parasitic on *Pinus* spp.......................... *A. douglasii*
2. Plants not parasitic on *Pseudotsuga*, *Abies*, or *Picea*...... *A. americanum*

Natural History Notes and Uses. Most dwarf mistletoes are host-specialized, parasitizing only one or a few principal host conifers. Dwarf mistletoes grow on trees and extract water, minerals, and nutrients. Major effects include witches'-brooms, loss of vigor, disfigurement due to distorted trunks, swellings and knots in branches, dieback, and death. Witches'-brooms are proliferations of shoots that result from the death of growing points or mutations of plant cells. Weakened trees are susceptible to insects and fungal diseases, which are often the direct causes of death.

Dwarf mistletoes differ from true mistletoes in that they are much smaller and are more damaging to the host. Seeds of dwarf mistletoes are forcibly discharged and infect the same tree or trees nearby, whereas seeds of true mistletoes are spread by birds and infect trees and shrubs in a large area.

Dwarf mistletoe is especially interesting because sap within the white berries develops considerable hydrostatic pressure, causing the berries to literally explode when they are ripe. Placing your warm hands near the ripe berries can hasten the forceful ejection of seeds. Seeds only 3 millimeters long may shoot up to 49 feet laterally, with an initial velocity of about 62 miles per hour. The small, sticky seeds can be felt if they strike a sensitive area of your body.

CALTROP FAMILY (ZYGOPHYLLACEAE)

The caltrop family is composed of herbs and shrubs, and rarely trees, that are xerophytic (prefer limited water) or halophytic (prefer salty soil). The leaves are usually opposite and pinnately compound. There are approximately 30 genera and 250 species, distributed primarily in tropical and subtropical areas. Six genera are native to the United States.

Guaiacum officinale, Lignum vitae, is an important economic member of this family. This small ornamental tree or shrub from Central America has a hard wood, denser than water, and is used for making mallets and bowling balls. It also produces a resin that is used in medicines and stains and as a chemical indicator. Populations of *G. officinale* are now so severely reduced in the Lesser Antilles, Puerto Rico, Barbados, Virgin Islands, and Colombia, and extinct or almost extinct in Antigua, Anguilla, and Barbuda, that this taxon is considered endangered by the World Conservation Monitoring Center. In 1998 the population was estimated at less than 2,500 mature individuals.

PUNCTURE VINE (*Tribulus terrestris*)

Description and Habitat. Puncture vine is an introduced annual plant with trailing, hairy stems and an extensive root system. The leaves are opposite and pinnate with 4–8 pairs of leaflets. The flowers are yellow and are borne singly in the leaf axils. The fruit is hard, consisting of 5 spiny nutlets or burrs that break apart into 5 tack-like sections on maturity. These burrs may injure livestock and are the bane of bicyclists. The plant is found in disturbed and waste places at the lower elevations.

Natural History Notes and Uses. Puncture vine has a 5,000-year history of medicinal uses, particularly in China and India. It was used for boosting hormone production in men and women and for urinary tract problems, itchy skin, and blood purification. The stems of the plant are considered astringent and act on the mucous membrane of the urinary tract. A tea from the aboveground part of the plant is said to be good for arthritis.

Moore (1979) indicates that studies on the seeds show they are useful in the early treatment of elevated blood fats and cholesterol. They apparently help prevent or lessen the severity of arteriosclerosis and atherosclerosis. Moreover, he says that the plant is useful in treating mild hypertension, contributing to a slower, stronger, more well-defined heart function, with greater relaxation between contractions and a lowering of the diastolic pressure.

5

Flowering Plants: Monocots

Monocots have a number of distinguishing features. The embryos of the seeds have only one cotyledon (seed leaf). The plant leaves usually have parallel veins. The vascular bundles of the stems are irregularly arranged, and the cambium is lacking. The flower parts are arranged in 3's or 6's, never in 5's (or 4's). Monocots are usually herbs, rarely shrubby.

FIELD KEY TO THE MONOCOT FAMILIES

1. Plants are small, floating on water or stranded on mud, not obviously differentiated into stem and leaves . . . . Lemnaceae
1. Plants with stems and leaves 2

2. Perianth lacking or reduced; its parts are often scales or bristles, not petal-like in color and texture 3
2. Perianth is well developed, at least the inner segments, petal-like in color and texture 14

3. Flowers in axils of chaffy or husk-like scales and more or less concealed by them, at times stamens and styles protruding; flowers in spikes, spikelets, or heads 4
3. Flowers not in axils of chaffy bracts, or if subtended by bracts, exceeding or equaling them and not concealed. 5

4. Stems round (circular in cross section) . . . . Poaceae
4. Stems 3-angled (triangular in cross section). . . . . Cyperaceae

5. Plants floating or submerged; flowers floating or submerged, or barely raised above the surface of the water 6
5. Plants of land or shallow water; usually leaves and flowers well above the surface of the water. 10

6. Flowers perfect. 7
6. Flowers unisexual 8

7. Tepals 5; stamens 4; plants of fresh water . . . . Potamogetonaceae
7. Tepals lacking; stamens 2; plants chiefly of brackish or alkaline water. . . . . Ruppiaceae

8. Flowers in conspicuous globose heads, the lower flowers female, the upper male . . . . Sparganiaceae
8. Flowers not in globose heads. 9

9. Perianth absent . . . . Najadaceae
9. Sepals 3; petals 3 . . . . Zannichelliaceae

10. Inflorescence a dense elongate spike 11
10. Inflorescence of subglobose heads, racemes, or open clusters 12

11. Plants 4–16 inches tall; male and female flowers intermingled, or flowers perfect . . . . Juncaginaceae
11. Plants 3–7 feet tall; stamens usually 3, the male flowers in a separate upper portion of the spike, the female below . . . . Typhaceae

12. Flowers unisexual, the lower heads female . . . . Sparganiaceae
12. Flowers bisexual; perianth segments distinguishable 13

13. Flowers in erect, dense, terminal spikes . . . . Juncaginaceae
13. Flowers in clusters, panicles, or heads . . . . Juncaceae

14. Carpels more or less free, 1-loculed, mostly 1-ovuled, maturing into a bunch or whorl of achenes . . . . Alismataceae
14. Carpels united into mostly 3–12-loculed ovary, maturing into a capsule or berry 15

15. Plants quite woody, with long, stiff, more or less sword-like leaves . . . . Agavaceae
15. Plants herbaceous, leaves various 16

16. Ovary superior . . . . Liliaceae
16. Ovary inferior 17

17. Aquatic plants with mostly submerged leaves . . . . Hydrocharitaceae
17. Land plants, the leaves not submerged 18

18. Flowers irregular . . . . .Orchidaceae
18. Flowers regular or almost 19

19. Stamens 3 . . . . .Iridaceae
19. Stamens 6 . . . . Amaryllidaceae

AGAVE FAMILY (AGAVACEAE)

Members of the agave family are adapted to warm, dry areas. The plants often have a rosette of leaves and produce rhizomes. Some, such as the century plant, grow vegetatively for many years, then flower and die. Suckers develop from the dying plant and start the growth cycle over again. Sissal and hemp are grown for fiber, and some plants produce a sap from which mescal, tequila, and pulque are made. Some botanists feel that the Agavaceae is not a true family but belongs in the Liliaceae.

The complex chemicals in this family have many uses. Compresses for wounds have been made from macerated agave pulp, and juices from leaves and roots were used in tonics. However, the sap from many agaves can cause severe dermatitis. The juice of the more virulent agaves has been used as fish poison and arrow poison. Agaves and yuccas are used in Mexico to make soap. Yuccas were once used to provide the foam of root beer and are still used in livestock deodorant. More recently, steroid drugs have been synthesized from extracts of several species in the family.

SPANISH BAYONET (*Yucca glauca*)

Description and Habitat. This plant, also sometimes called soapweed and beargrass, is a perennial that arises from a fibrous horizontal or upright stem bearing one or more erect crowns. Sharp-tipped leaves up to 2 feet long are rounded on the back and have in-rolled margins bearing white filaments. The leaves contain tough fibers. The flowers are greenish white and up to 2½ inches long; 10–15 of them form along a spike about 3 feet long. At maturity the large capsules enclose long black seeds. *Yucca* is a native Haitian name, and *glauca* means "blue-green" in Latin. This species was first described for science in 1813 by the famous English botanist-naturalist Thomas Nuttall. Its needle-sharp leaves have given it the common name Spanish bayonet. This species grows in sandy soils in the plains and steppe zones on the eastern side of the Continental Divide.

Natural History Notes and Uses. All species of *Yucca* provided a plentiful and dependable food source for Native Americans. The large pulpy fruits were eaten raw or roasted, or were cooked and formed into cakes or dried for future use. The flowers buds were also roasted and eaten, and their high sugar content made them a sweet treat for children. A tea can be made from the seeds, which were also pounded and cooked as mush.

There is a unique relationship between yuccas and the yucca moth (*Tegeticula maculata*). This insect is the only pollinator for the plant. The mouth parts of the female moth are such that it gathers the pollen and forms it into a ball, which is then scraped across the next stigma. In addition, the moth lays its eggs only in the seed pod of the yucca. Without each other, neither of these organisms could exist.

The leaves can be pounded to release the fibers for use in basketry and cordage making. One method used by the Diegueño Indians was to bury the leaves until the fleshy part rotted away, yielding a fine white fiber. The tough fibers were also used for making nets, hats, sandals, and mattresses (Hinton 1975).

The chopped or pounded roots yield a soap-like substance. It can be used for general washing. Paintbrushes can be made by fringing the smaller leaves.

WATER PLANTAIN FAMILY (ALISMATACEAE)

Thirteen genera and 90 species in this family are found worldwide, of which 5 genera are native to the United States. Members of this family are aquatic and marsh perennial herbs. Their long-stalked leaves and even longer flowering stems are well adapted to wetland habitats. The flowers are composed of 3 green sepals, 3 white petals, numerous stamens, and many pistils. This family is of little economic importance.

Caution: Aquatic plants such as arrowhead and water plantain are sometimes found growing in polluted or contaminated water.

FIELD KEY TO THE WATER PLANTAIN FAMILY

1. Emergent or floating leaves with arrowhead-shaped blades. . . . . *Sagittaria cuneata*
1. Emergent or floating leaves with elliptical or lance-shaped blades . . . . *Alisma*

Water Plantain (genus *Alisma*)

NARROW-LEAF WATER PLANTAIN (*A. gramineum*)

NORTHERN WATER PLANTAIN (*A. triviale*)

Description and Habitat. These perennial plants arise from fleshy, bulb-like stems. The basal leaves are long-stalked and egg-shaped, and the flowers are white. Water plantain is usually found in marshes and ponds at lower elevations.

FIELD KEY TO THE WATER PLANTAINS

1. Leaves surpassing the inflorescence .... *A. gramineum*
1. Inflorescence surpassing the leaves .... *A. triviale*

FIGURE 104. Water plantain (*Alisma triviale*).

Natural History Notes and Uses. The starchy, bulbous bases of water plantain are edible as a starchy vegetable after drying. Drying is said to remove the strong flavor. *Alisma* has a long history of use in Chinese medicine and is mentioned in texts dating back to about A.D. 200. It was also used by early herbalists as a diuretic and by the Cherokee Indians for application to sores, wounds, and bruises (Hamel and Chiltoskey 1975). It is described as a sweet, cooling herb that lowers blood pressure, cholesterol, and blood sugar levels. The root was also used as a diuretic in the treatment of dysuria, edema, distention, diarrhea, and other ailments (Foster and Duke 1990). Water plantain furnishes food for water birds and muskrats. The flowers may be dried upside-down in a cool, dark room and used in arrangements.

ARUM-LEAF ARROWHEAD (*Sagittaria cuneata*)

Description and Habitat. These aquatic perennials have glabrous stems that grow 8–48 inches high and contain a milky juice. The leaves are large and arrow-shaped, and the basal lobes spread outward. The stem is mostly leafless and is often branched. The flowers occur in whorls of 3 and are located near the ends of the stem. The sepals are ovate; the petals are white. Arum-leaf arrowhead grows in marshy places along the edges of ponds or streams or in wet meadows. It flowers in July and August. *Sagitta* is Latin for "arrow," referring to the shape of the leaves.

Natural History Notes and Uses. All species produce starchy, white tubers that can be roasted or boiled and then eaten. An important source of carbohydrates, the tubers contain a milky juice with a bitter flavor that is destroyed by heat. The Lewis and Clark expedition is said to have used these tubers extensively while exploring the Columbia River region (Hart 1996). To many Native American tribes, arrowhead was a primary vegetable. The small tubers are located at the ends of the long underwater rhizomes, perhaps a meter or more from the plant. They can be carefully removed without pulling up the whole plant. To collect the tubers, you can use your hands, a forked stick, or, if the water is deep, your feet. Wade into the pond where the plants are growing and feel around for the tubers in the mud with your toes. They should feel like round lumps varying in size from peanut to potato. After they are dislodged, the tubers usually float to the surface for easy harvesting. They are best-developed in late summer or autumn. Boil or bake them like a potato to remove the poisonous properties, then peel and eat them. They can also be dried for future use. The tubers were used in a tea for indigestion and in a poultice for wounds and sores. The leaves and stem contain alkaloids and may be poisonous.

ARUM FAMILY (ARACEAE)

These fleshy perennials have basal leaves. The flowers are borne in dense fleshy spikes (spadix), subtended or enclosed by a foliaceous or colored bract (spathe). There are more than 115 genera and 2,000 species, which are found in shady, damp, or wet places. The family is well represented in the tropics. Members include calla lily, philodendron, and dieffenbachia, which are grown as ornamentals.

AMERICAN SKUNK CABBAGE (*Lysichiton americanus*)

Description and Habitat. Skunk cabbage is a leafy perennial herb. The leaves are opposite but appear whorled. The flowers are solitary or in terminal racemes. This species is found in marshes and wet woods at lower elevations in the northern part of the state. The genus name is from the Greek *lysis*, meaning "loosening," and *chiton*, meaning "tunic," pertaining to the large spathe.

Natural History Notes and Uses. The plant contains calcium oxalate in all parts. Cooking the plant renders it edible. Native Americans roasted or dried the root and made flour from the starch. The young green leaves can be eaten after boiling in several changes of water. This plant is related to taro, a staple food of the Polynesians (Craighead et al. 1963).

SEDGE FAMILY (CYPERACEAE)

There are 90 genera and 4,000 species worldwide, particularly in the cool temperate and subarctic regions. Twenty-four genera are native to the United States. In this area sedges are grass-like plants often found growing in wet places. Sedges make up a large proportion of the plants found at higher elevations. Most of the species have triangular stalks with the flowers arranged in spikelets. Species of this family are often confused with grasses (Poaceae) and rushes (Juncaceae). Grasses generally have round, hollow stems (except at the nodes) and a characteristic flower consisting of 2 glumes subtending one to many flowers, each with a lemma or palea. Members of the rush family have round stems and small, lily-like flowers with 3 petals, 3 sepals, and 6 stamens.

The family is of little economic importance, although several members are edible and medicinal. Sedge family members, however, should not be used therapeutically without medical supervision since their properties and effectiveness are variable.

FIELD KEY TO THE SEDGE FAMILY

1. Flowers resembling a soft, feathery ball, the perianth composed of long, white, silky hairs . . . . *Eriophorum*
1. Flower clusters variable (not as above) 2

2. Achenes surrounded or enveloped by a small bract (perigynium) and subtended by a scale 3
2. Achenes not surrounded by a perigynium 4

3. Perigynium closed except at the apex, where the style protrudes . . . *Carex*
3. Perigynium split down the middle, with unsealed margins. . . . . . *Kobresia*

4. Spikelets solitary and terminal, without subtending bracts .... *Eleocharis*
4. Spikelets one to many, subtended by one to several leafy to small scale-like bracts................................ *Scirpus*

Sedge (genus *Carex*)

Description and Habitat. Numerous species within this genus occur in the area. They are normally difficult to identify in the field: a hand lens and mature fruit are often necessary. In general, they are perennial grass-like herbs with creeping rhizomes, short rootstalks, or fibrous roots; they have 3-sided stems, and the leaves are 3-ranked. Sedges occur in a great variety of habitats from the foothills to the alpine zone.

Natural History Notes and Uses. The young shoots and tender leaf bases of almost all species are sweet and furnish a tasty nibble. The fruits are also edible. *Carex* is a widespread genus with more than 1,000 species, making it available as an important emergency food in many parts of North America.

Sedge roots were used extensively as basketry material, particularly in coiled baskets. Some Native Americans would spend considerable time untangling the roots in order to remove them in one piece. In fact, some areas were cleared of other plants to allow sedge to grow long and untangled.

Spike Rush (genus *Eleocharis*)

Description and Habitat. The spike rushes are aquatic, rush-like perennials. Some species grow along the edges of ponds, and others grow completely submerged. One species, *E. acicularis* (hair grass), is a hardy variety native to North America, Asia, and Europe. It has pale green, needle-like leaves up to 8 inches in length. The genus offers many fascinating taxonomic problems. Because of the simplified body plan, there are few characters available for recognition and identification of species. It is very helpful to know whether the plant has thickened rhizomes. Also, one needs to have mature spikelets so that the color and shape of the scales, color and shape of the achenes, shape of the tubercles, and number of stigmas can be determined.

Natural History Notes and Uses. These sedges can be distinguished from other sedges by the presence of a tubercle (a small bump, like a nipple) on the achene (or fruit) and the fact that the stems grow in groups and are usually tufted. *Eleocharis* derives from the Greek words *heleios* ("marsh") and *charis* ("joy," "beauty," or "grace"). Hence the name means "marsh beauty." Common spike rush is an invaluable food source for waterfowl. Small mammals seek shelter in dense stands of this plant but do not use it for food.

Native groups did not use spike rush extensively, although they prepared a decoction of the plants for emetic purposes and occasionally used the stems for weaving. The bulbs and sap were eaten infrequently, but it is not known how they were prepared.

Cotton Grass (genus *Eriophorum*)

Description and Habitat. These perennials have solid stems, triangular or round, from rhizomes. The lower leaves are grass-like. The flowers are spirally arranged in spikes that resemble tufts of cotton at maturity. *Eriophorum* translates as "wool" (*erion*) and "bearing" (*phoros*). The species can be found in cold swamps and bogs from middle to high elevations.

Natural History Notes and Uses. The pinkish bases of these plants can be collected in the early summer and eaten raw or added to soups. The corms were dug in the early spring or fall, and boiling water was poured over them to remove the thick black outer covering. They were then eaten raw or boiled. A decoction from the rootstalk was used in treating colds and coughs, but its use was not widespread (Schofield 1989). The woolly down from the flowers was spun as fiber, although it was inferior to cotton. However, the fiber can be used as candle wicks.

Kobresia (genus *Kobresia*)

Description and Habitat. Members of this genus are grass-like perennial herbs very similar in appearance to *Carex* except that the open perigynium is diagnostic.

Natural History Notes and Uses. Kobresia grows in the Arctic and on high ridges and peaks of the Rocky Mountains in snow-free areas. In fact, it is a dominant plant on mature alpine tundra. Its uniform stands of short, close, grass-like growth can be recognized in the late summer and fall by an orange or gold color.

Bulrush (genus *Scirpus*)

Description and Habitat. The many species of bulrush are perennials with grass-like or scale-like leaves. The various species can be found in marshy areas around lakes and ponds and in other moist or wet areas from low to high elevations.

Natural History Notes and Uses. The edible rhizomes of all species are quite starchy. They may be eaten raw, baked, dried, or ground into a flour. As Olsen (1990:269) describes: "The young shoots just protruding from the mud are a delicacy raw or cooked. Furthermore, one can harvest them by wading into the water and feeling down along the plant until you come across the last shoot in a string of shoots that protrudes above the water. You then push your hands into the mud until the lateral rootstalk is encountered. By feeling along the rootstalk in a direction leading away from the last shoot, one can find a protruding bulb from which the new shoot is starting. This is easily snapped off and is edible on the spot."

The young roots when crushed and boiled yield a sweet syrup. The pollen may be gathered and pressed into cakes and baked. The seeds may be used whole, parched, ground, or in mush. The stem bases may be eaten raw and are good for quenching thirst.

Scirpus stems, roots, and leaves can be used as foundation and twining material in twined baskets. They were used extensively by Native Americans for making cordage, sandals, baskets, and mats. Waterproof "water bottles" can be made from the baskets by coating the inside with asphaltum. The stems were used for sleeping mats, padding, skirts, sandals, thatching for dwellings, and duck-shaped decoys for hunting.

The leaves of *S. acutus* (also known as *Schoenoplectus acutus*) were used to poultice wounds and burns.

FROG'S BIT FAMILY (HYDROCHARITACEAE)

This family contains about 20 genera and about 150 species of marine and freshwater aquatic herbs with more or less regular flowers that bear 3 sepals, 3 petals, and 1 to many stamens. A few species are grown in aquaria.

WATERWEED (*Elodea longivaginata = E. bifoliata*)

Description and Habitat. Plants of this genus have long, branching stems. The leaves are less than 1 inch long, have very fine teeth on the edge of the leaf, and lack a ridge on the underside. The leaves are usually in whorls of 3. Male and female flowers are separate and colorless and pollinate underwater. This plant is soft to the touch.

Natural History Notes and Uses. These plants can clog ponds and waterways if not properly controlled. The aquarium trade is one of the main reasons these plants are becoming such a nuisance in non-native waterways, because aquarium owners simply throw clippings or dump tanks into a lake or stream.

FIGURE 105. Wild iris (*Iris missouriensis*).

IRIS FAMILY (IRIDACEAE)

There are about 70 genera and 1,500 species in this family worldwide, with the chief center of distribution in South America and tropical America. Five genera are native to the United States. Members of this family are generally perennials with narrow, grass-like leaves and thick rhizomes or fibrous roots. The flowers have 3 petals and 3 petal-like sepals joined on top of the ovary. The family is of economic importance as a source of ornamentals and saffron dye.

FIELD KEY TO THE IRIS FAMILY

1. Petal-like sepals larger than the petals and reflexed; flowers iris-like . . *Iris*
1. Sepals and petals similar in size and shape; flower star-shaped . . . . *Sisyrinchium*

ROCKY MOUNTAIN IRIS (*Iris missouriensis*)

Description and Habitat. This perennial arises from a creeping rhizome. The stems are erect, and the leaves are sword-shaped or linear. The flowers are large, occurring singly or in panicles. The perianth is composed of 6 clawed segments that are united below into a tube. The outer 3 segments are spreading or reflexed whereas the inner ones are smaller and erect. Iris is the Greek word for "rainbow."

Natural History Notes and Uses. Members of the genus *Iris* contain irisin, an acrid resin concentrated mainly in the rhizomes and present in the foliage and flowers. People who raise irises sometimes develop a skin rash from handling the rhizomes. Cattle have died as a result of eating relatively large quantities of the plants. The rootstalk produces a burning sensation when chewed. If eaten in quantity, irises will cause diarrhea and vomiting. The poisonous rootstalks were used by Native Americans in a mixture of bile to poison arrow points.

Iris was an important plant for Native Americans. The leaves were woven into mats and lined with cattail "down" for use as baby diapers.

IDAHO BLUE-EYED GRASS (*Sisyrinchium idahoense*)

Description and Habitat. The pinkish purple to blue flowers have 3 petals and 3 sepals that are alike. They are found in meadows and are often inconspicuous because of their tufted, grass-like leaves. The flowers open only in bright sunshine.

Natural History Notes and Uses. *S. bellum* was known among the Native Americans in the South as "azulea" and "villela." It was made into a tea that was considered a valuable remedy for treating fevers. It was thought that a patient could subsist for many days on it alone. The uses of most *Sisyrinchium* species are unknown.

RUSH FAMILY (JUNCACEAE)

Nine genera and 400 species of rushes are found in damp and wet sites of the cool temperate and subarctic regions. Two genera are native to the United States. They are grass-like annual and perennial plants with solid, rounded, or flattened stems. The leaves are basal or alternate, and may be flat, folded, or round, and taper to a point. The flowers are small and have 6 undifferentiated sepals and petals (often termed *tepals*), 3–6 stamens, and a 3-parted ovary with many seeds. The family is of no direct economic importance, although a few members are ornamentals.

Grasses, sedges, and rushes look very similar at first glance and can be difficult to identify correctly. The best way to tell them apart is to examine the flowers. However, these parts are often very small and can be difficult to distinguish in the field. There are some vegetative features that can help distinguish one from another, such as the structure of the stems and leaves. The important parts of the stem for identification are the joints, internodes, and cross section. The important characteristics of the leaves are the ligules and the sheath orientation.

FIELD KEY TO THE RUSH FAMILY

1. Seeds many per capsule; leaves usually with open sheaths and hairless blades . . . . *Juncus*
1. Seeds 3 per capsule; leaves with closed sheaths and often with marginal hairs . . . . *Luzula*

Rush, Wire-grass (genus *Juncus*)

Description, Natural History Notes, and Uses. Many rush species occur in the area. They are annual or perennial herbs often found in water or wet places. The flowers are in heads or panicles, and the tough, fibrous stems are inedible. However, they are useful in weaving baskets and mats.

Wood Rush (genus *Luzula*)

Description and Habitat. Wood rushes are tufted perennials with slender, unbranched stems. There are about 80 species of annual or perennial herbs with grass-like leaves throughout the world. The brown-green or white flowers occur in panicles. The genus name *Luzula* comes from the Latin *lux*, or "light," and refers to the way the plant shines when morning dew covers its hairy leaves.

Natural History Notes and Uses. Some species have been used medicinally as emetics and decoctions. Additionally, *L. luciola* was used as a wick in candles. The seeds of *L. campestris* (field wood rush) are dispersed by ants because of their juicy outgrowths.

ARROW-GRASS FAMILY (JUNCAGINACEAE)

There are four genera and 26 species in the arrow-grass family. Two genera occur in the United States. The family is of no economic importance.

Arrow-grass (*Triglochin maritimum*)

Description and Habitat. This slender, grass-like plant arises from a rhizome and has fleshy basal leaves. The flowering stems are long and smooth, and the flowers occur in terminal bractless racemes. Arrow-grass grows in mountain swamps and around lakes. It flowers from April to August. Another species, *T. palustris* (marsh arrow-grass), also occurs in this area and is found in mud flats and springy habitat.

Natural History Notes and Uses. The seeds of arrow-grass can be parched and ground into flour. Roasted, they can be used as a coffee substitute. Seeds need to be parched because they contain cyanogenic toxins that have caused death in livestock (Muenscher 1962). Parching or roasting the seeds renders them safe since the poison is volatile (Harrington 1967).

The young white leaf bases are collected around April or May from the inner leaves of the basal cluster. When eaten raw at the right stage, these leaf bases have a mild, sweet, cucumber-like taste. They are generally better if cooked. In springtime the leaf bases contain few toxic compounds, but the mature leaves and flowerstalks should never be eaten. The leaves contain hydrocyanic acid, a toxin that interferes with the uptake of oxygen. Symptoms include headache, heart palpitations, dizziness, and convulsions.

Caution: These plants are toxic when fresh. Several references list them as livestock poisoners until they dry, when the cyanogenic properties evaporate, break down, or dissipate.

DUCKWEED FAMILY (LEMNACEAE)

Plants in this family are small and float on slow or stagnant waters. They have small, thread-like root hairs that obtain nutrients from the water. All duckweeds are used as food by wildlife and have been recorded from the stomachs of ducks. Two genera can be encountered in the area: *Lemna* and *Spirodela.*

FIELD KEY TO THE DUCKWEED FAMILY

1. Plants with 1 root per thallus (plant body) . . . . *Lemna*
1. Plants with more than 1 root per thallus . . . . *Spirodela*

Duckweed (genus *Lemna*)

COMMON DUCKWEED (*L. minor*)

STAR DUCKWEED (*L. trisulca*)

VALDIVIA DUCKWEED (*L. valdiviana*)

Description and Habitat. Duckweeds are small plants, often not much larger than a pinhead. The thallus is flattened with a single root. They float on the surface or are submerged in the water. Natural populations of duckweeds are usually mixtures of several species.

FIELD KEY TO THE DUCKWEEDS

1. Thalli (disks) oblong to lanceolate in shape, often with slender stalks. . . . . *L. trisulca*
1. Thalli oval and without stalks 2

2. Thalli about 2–3 times as long as broad, nerveless . . . . *L. valdiviana*
2. Thalli generally less than twice as broad, usually 3-nerved . . . . . . *L. minor*

Natural History Notes and Uses. Unlike the ordinary leaves of most plants, each duckweed frond contains buds from which more fronds may grow. These buds are hidden in pouches along the center axis of older fronds. As they grow, new fronds emerge through slits in the side of their parent fronds. Until they mature, daughter fronds may remain attached to the parent frond. Rapidly growing plants often have 3 or 4 attached fronds.

Duckweeds are an essential link in the food chain. Useful as a water crop, they can acclimatize themselves to almost all growing conditions, with some thriving in manure-rich or eutrophic waters. They reproduce quickly, extending over large surface areas, and are easily harvested. Their high fat and protein content makes them an ideal source of food for animals and poultry. Duckweeds have potential in wastewater treatment, absorbing excess nutrients such as phosphorus and ammonia from surface waters, reducing suspended solids, and reducing biochemical oxygen demand.

More cold-tolerant than other aquatic vascular plants, duckweeds can sustain temperatures as low as 7° Celsius for normal growth. Under freezing conditions, they will lie dormant on the pond bottom until warmer conditions return.

As survival food, duckweeds can provide copious and palatable material for salads. Additionally, some Native Americans used duckweed as a diuretic and a general tonic.

AQUATIC PLANTS AND POLLUTED WATER

The days of safely drinking water straight out of a mountain stream are long gone. That also goes for eating aquatic plants such as cattail (*Typha*), duckweed (*Lemna* and *Spirodela*), watercress (*Rorippa*), pond lily (*Nuphar*), and bulrush (*Scirpus*). Even in the most remote areas of the country, there are a number of disease-causing bacteria, protozoans, and viruses. A fairly common intestinal disorder among backcountry hikers is caused by the protozoan *Giardia lamblia*, which is carried in the intestines and feces of muskrats, beavers, moose, voles, and other water-loving mammals. Humans, too, are responsible for spreading *Giardia* to other areas by careless or improper disposal of human waste.

Other intestinal diseases also can be found in what appear to be clean water sources. Cases of *Campylobactor*, *E. coli*, and type A hepatitis have been traced back to drinking untreated water in the backcountry. Cities, farms, and suburbs can also experience clean water problems, resulting from oil spills, pesticides, and toxic pollutants from mining operations.

In areas where edible plants are found, it is best to treat the water and the plants that grow in it. One method is to soak the fresh greens in a disinfectant if pollution is suspected. You can use any of the commercially available water purification tablets or a teaspoon to a tablespoon of chlorine bleach in a quart of water. Then rinse the plants well and prepare. Obviously, if the waters are polluted with oils, pesticides, or other toxins, this method will not work. In any case, extreme care should be exercised.

GIANT DUCKWEED (*Spirodela polyrhiza*)

Description and Habitat. This species is frequently associated with *Lemna*. Giant duckweed is a coarse species with a purplish-tinged lower side. The thalli are egg-shaped or round.

Natural History Notes and Uses. Giant duckweed is largely made up of metabolically active cells with very little structural fiber, and thus the tissue contains twice the protein, fat, nitrogen, and phosphorus of other vascular plants. Each frond absorbs nutrients through the whole plant and not through a central root system, directly assimilating simple carbohydrates and various amino acids. Photosynthesis in this plant is mostly devoted to the production of protein and nucleic acids, giving it a very high nutritional value. As with *Lemna*, giant duckweed can provide copious and palatable material for salads. The Chinese use this species to treat hypothermia, flatulence, and acute kidney infections (Culley and Epps 1973).

LILY FAMILY (LILIACEAE)

This large and varied family contains many beautiful wildflowers. There are approximately 250 genera and 4,000–6,000 species worldwide. About 75 genera are native to the United States. The family is characterized by a perianth of 6 parts, with a superior ovary and a 3-lobed stigma. The fruit is a capsule that splits open when ripe. The family is a source of many ornamentals, several important fibers, fermented and distilled beverages, and steroidal compounds. Although some members of this family are edible, there are many poisonous species. For example, the bulbs of the true lilies (*Lilium*) can be boiled and eaten, but some genera contain highly toxic alkaloids. Nearly 200 alkaloids and numerous glycosides occur in the family.

The lily family formerly was a catch-all group that included a great number of genera that are now members of other families, and in some cases even of other orders: Agavaceae, Alliaceae, Smilacaceae, Tofieldiaceae, Trilliaceae, and Uvulariaceae.

FIELD KEY TO THE LILY FAMILY

1. Perianth segments and stamens 3 . . . . *Maianthemum*
1. Perianth segments and stamens usually 6 2

2. Perianth of well-differentiated green sepals and colored or white petals . . . . *Calochortus*
2. Perianth parts similar in appearance, at least in color 3

3. Styles 3; fruit a capsule 4
3. Styles solitary 8

4. Leaves ovate or elliptic to lanceolate . . . . *Veratrum*
4. Leaves mostly linear 5

5. Leaves stiff . . . . *Xerophyllum*
5. Leaves usually not stiff 6

6. Plants with short rhizomes and usually glandular at least above . . . . *Tofieldia*
6. Plants with bulbs or corms, usually not glandular 7

7. Perianth segments with prominent gland on inner surface toward base, flowers white to cream-colored......... *Zigadenus*
7. Perianth segments without glands, flowers brown to purplish .... *Fritillaria*

8. Leaves mostly linear 9
8. Leaves not linear 13

9. Perianth segments united half their length or more........ *Leucocrinum*
9. Perianth segments free or almost.................................. 10

10. Flowers in terminal umbel subtended by one or more bracts..... *Allium*
10. Flowers solitary or in a raceme, usually not subtended by bracts....... 11

11. Perianth yellow or orange, flowers nodding.................. *Fritillaria*
11. Perianth white or cream-colored to blue or violet, flowers not nodding 12

12. Flowers blue or violet .... *Camassia*
12. Flowers white or creamy .... *Lloydia*

13. Plants with lobed bulbs or elongated corms; fruit a capsule .... *Erythronium*
13. Plants with rhizomes; fruit a berry................................. 14

14. Stems simple; flowers in terminal racemes or panicles ... *Maianthemum*
14. Stems usually branched; flowers axillary or terminal on stem branches 15

15. Flowers in axils of leaves .... *Streptopus*
15. Flowers terminal on stem branches......................... *Disporum*

Onion (genus *Allium*)

TAPERTIP ONION (*A. acuminatum*)

BRANDEGEE'S ONION (*A. brandegei*)

SHORTSTYLE ONION (*A. brevistylum*)

NODDING ONION (*A. cernuum*)

GEYER'S ONION (*A. geyeri*)

WILD CHIVES (*A. schoenoprasum*)

TEXTILE ONION (*A. textile*)

Description and Habitat. The many species of *Allium* all arise from bulbs and have the characteristically distinct onion odor. The odor is caused by the presence of volatile sulfur compounds in all parts of the plant (causing their strong flavor and irritation to eyes). There are about 300 species of onions in the world. Some other common names of *Allium* are leeks, garlic, and chives. The small flowers are clustered together in umbels. Onions are found in a variety of habitats from low elevations to the alpine zone. *Allium* is the ancient name for garlic. Its derivation may be from the Celtic *all*, which means "pungent."

FIGURE 106. Wild onion (*Allium* spp.).

FIELD KEY TO THE ONIONS

1. Leaves rounded and hollow........................ *A. schoenoprasum*
1. Leaves flat, maybe rounded, but not hollow.......................... 2

2. Umbel nodding; stamens exserted from perianth .... *A. cernuum*
2. Umbel erect; stamens included within perianth 3

3. Bulb coats persisting as a network of fibers 4
3. Bulb coats never with a network of fibers 5

4. Flowers usually white; tips of inner perianth parts spreading... *A. textile*
4. Flowers mostly pink, tips of inner perianth parts erect .... *A. geyeri*

5. Bulbs from a rhizome, roots appearing to come from the side of the bulb near base .... *A. brevistylum*
5. Bulbs not from a rhizome, roots basal 6

6. Leaves usually more than 2 per stem, shorter than the scape and withering early .... *A. acuminatum*
6. Leaves usually 2 per stem, longer than the scape. .... *A. brandegei*

Natural History Notes and Uses. All *Allium* species are known to be edible. The bulbs may be eaten raw, boiled, steamed, creamed, and in soup and are especially good when used as a seasoning. Ingestion of large amounts of onions, including the cultivated ones, can cause poisoning or goiter but are otherwise not known to be harmful. Regardless, eating them in moderation is the key. The plants are valuable in all seasons and can be used as greens and as flavoring. The seeds and leaves can also be eaten. Onions will keep a long time because the skin dries and preserves the flesh inside. Wild onions contain large amounts of some important micronutrients, more vitamin C than an equal weight of oranges, and more than twice as much vitamin A as an equal weight of spinach (Kindscher 1987). Additionally, onions contain a significant amount of a starch called inulin, which is not easily digested by humans.

Medicinally, onions have a number of uses. Soldiers during World War I took advantage of their natural antiseptic properties by applying *Allium* juice to wounds to prevent infection. The juice of wild onions can be boiled down until it is thick and used as a treatment for colds and throat irritations. The juice was also rubbed over the body as an insect repellent. The onion smell apparently has some beneficial effects on the circulatory, digestive, and respiratory systems (Turner and Kuhnlein 1991).

Warning: Wild onions should not be confused with the so-called poison onions, or death camas (*Zigadenus*). *Zigadenus venenosus* is most likely to be confused with *Allium*. *Zigadenus* species are bulb-bearing plants with grass-like leaves, also in the lily family. They have upright, more elongated (not umbrella-like) clusters of white or cream-colored flowers, and they lack the characteristic strong odor of onions. They contain highly toxic alkaloids, and all plant parts, including the bulbs, can be fatal if ingested in any quantity.

Mariposa Lily, Sego Lily (genus *Calochortus*)

WHITE MARIPOSA LILY (*C. eurycarpus*)

GUNNISON'S MARIPOSA LILY (*C. gunnisonii*)

SEGO LILY (*C. nuttallii*)

Description and Habitat. Mariposa lilies are perennials from bulbs, with tulip-like flowers that are few and showy. The generic name *Calochortus* was compounded from the Greek *calos*, "beautiful," and *chortos*, "grass," in reference to the flowers and leaves. *Mariposa* is Spanish for "butterfly," and *sego* is a Shoshonean word for "edible bulb." These species can be found in dry open places from low to middle elevations.

FIELD KEY TO THE MARIPOSA LILIES

1. Ovaries and fruits circular to oblong in shape; basal leaf is flat and broad . . . . *C. eurycarpus*
1. Ovaries and fruits linear in shape; basal leaf linear and flat. 2

2. Hairs on face of petals branched and gland-tipped. . . . . *C. gunnisonii*
2. Hairs on face of petals not branched or gland-tipped . . . . *C. nuttallii*

Natural History Notes and Uses. Although the entire *Calochortus* plant is edible and can be used as a potherb, the highly nutritious bulb is the part usually sought. The bulbs are smaller than walnuts and may be eaten raw, boiled, roasted in hot ashes in pits, or steamed. They can also be threaded on a string and dried before or after cooking. The dried bulbs can be ground into flour or cooked and mashed into cakes for preservation. Dug in the spring, usually before flowering, *Calochortus* bulbs were an important food source for many Native Americans (Olsen 1990). When numerous bulbs were collected, they were usually pit-cooked. The flower buds can be eaten and have a sweet taste. The seeds are also edible. Sego lily is the state flower of Utah.

Note: These plants are becoming increasingly rare in some areas, primarily due to habitat destruction and overgrazing. Harvesting the corms destroys the plants. Because of the plants' rarity and beauty, and the risk of overharvesting, their use today is not recommended.

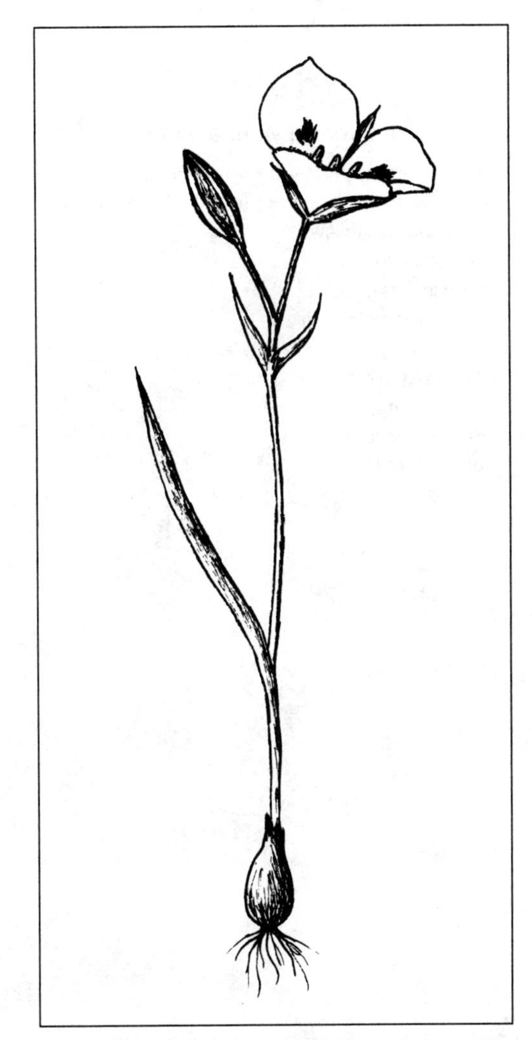

FIGURE 107. Mariposa lily (*Calochortus* spp.).

CAMAS (*Camassia quamash*)

Description and Habitat. Arising from a bulb, camas has bright blue to violet, 6-parted flowers in a showy, spike-like raceme. The leaves are basal and grass-like. The plant can be found in meadows, marshes, and fields and on grassy slopes from low to middle elevations. The plant's common name is derived from the Nootka Indian word *chamas*, which means "sweet."

Natural History Notes and Uses. Camas was perhaps the most important food of Native Americans in western North America. As Gunther (1973:24) writes, "except for choice varieties of dried salmon there was no article of food that was more widely traded in western Washington than camas."

Camas was an important staple food for many aboriginal peoples, and the bulbs were eaten wherever available. In fact, some tribes managed or "owned" areas where the plant grew. They would intentionally set fires to ensure a bountiful harvest. As testimony to the species' former abundance, Meriwether

DIGGING STICKS

The roots and bulbs of many plants are best collected with the aid of a digging stick. A digging stick is about 2–3 feet long, 1–2 inches thick, beveled at one end, and fire-hardened. To use, thrust the stick into the ground beside the plant and pry upward while pulling on the plant from above. To fire-harden your digging stick, simply hold the point a few inches above a bed of hot coals and slowly turn it as you would a skewer. Take care not to char the wood but let it turn a light brown color.

Lewis, in his journal entry of June 12, 1806, stated: "The quawmash is now in blume and from the colour of its bloom at a short distance it resembles lakes of fine clear water, so complete in this deseption that on first sight I could have sworn it was water."

Camas bulbs were dug out of the ground from late July through September with digging sticks. The black outer covering of the bulb was removed, and the white bulbs were then steamed or cooked in pits for 24 hours or more. When cooked this way, the bulbs turn dark brown, become quite moist and soft, and turn sweet. Cooking is required because the plant contains a carbohydrate called inulin, which is not very digestible or very palatable in its "raw" form. Cooking chemically breaks down the inulin into its component fructose sugar. Common in fruits and honey, fructose is both easily digested and sweet-tasting. After cooking, the camas bulbs can be mashed and dried into cakes for storage. The bulbs can also be boiled down into a syrup. Too much camas, however, is both an emetic and a purgative.

Caution: If collecting bulbs, be aware that the poisonous death camas (*Zigadenas*) may be in the area, too.

NUTRITIONAL VALUE OF CAMAS

Common camas is poor in energy and protein value. The nutrient composition of fresh bulbs (per gram of dry weight) is as follows:

- calories, 3.90
- protein, 0.13 g
- carbohydrates, 0.80 g
- lipids, 0.03 g
- calcium, 1.76 mg
- iron, 0.23 mg
- magnesium, 0.40 mg
- zinc, 0.03 mg

ROUGH-FRUIT FAIRYBELLS (*Disporum trachycarpum*)

Description and Habitat. This tall plant has lily-style leaves that clasp tightly to the stem. The leaves are wide and almost heart-shaped. The stem often forms very angular joints with leaves growing at the joints. The dangling, bell-like flowers, from which the plant takes its name, begin as closed bells but soon open and rapidly fade. Generally, the 1–4 flowers grow together in a hanging cluster. The leaves are later replaced by red, velvety berries. The berries are 6–15-seeded.

Natural History Notes and Uses. The fruits, which are strikingly reddish orange in color, have been observed remaining on the plants into late summer. True to its common name, under magnification the surface of the fruit is distinctly rough and wart-like. The velvet-skinned yellow or orange berries can be eaten raw. The Blackfoot would place the berries in the eyes overnight with an infusion of bark as a treatment for snowblindness. The dampened, bruised leaves were also used as a bandage and treatment for wounds. The fruit of a related species, *D. hookeri,* was used to relieve kidney ailments.

GLACIER LILY, DOGTOOTH LILY (*Erythronium grandiflorum*)

Description and Habitat. This perennial plant grows to 13 inches in height and arises from an underground corm. The leaves are basal, bright yellow-green, and up to 8 inches long and clasp the flowering stem base. The flowers are golden-yellow, and the 6 tepals are strongly recurved with 6 stamens and typically occur singly on a leafless, unbranched stem. The fruits are capsules an inch long.

Natural History Notes and Uses. Flowers of glacier lily are mainly pollinated by bumblebees (*Bombus*), although other bees are also important. They are occasionally pollinated by hummingbirds. The corms are a favorite food of grizzly bears, which will dig them up in the spring and fall when other foods are scarce and when the corms are most nutritious. Ground squirrels also feed on the corms, and the foliage is grazed on by large ungulates. In an Idaho study, glacier lily made up the bulk of mule deer diets during May.

Additionally, the corms were eaten by many Native Americans, even though they smelled rather bad. However, they were not eaten raw because the corms contain the complex carbohydrate inulin, which is modified by cooking into the more edible and sweet-tasting fructose. Some corms are as big as one's fist, resembling white clear tubes with the root coming out of the side. They were boiled and then either eaten fresh or dried for later use. Drying took up to two weeks. Before they were eaten, they were soaked in water to soften and the outer covering was removed. The dried bulbs were cooked in soups and stews with fish or meat, or in special "puddings" that included dried black tree lichen, Saskatoon berries, deer fat, salmon eggs, and tiger lily bulbs. As well as being a good food, the corms were said to be a good medicine for a bad cold.

Note: These beautiful wildflowers are seldom abundant. Harvesting the bulbs destroys the entire plant. They should not be used today except in an emergency.

Fritillary (genus *Fritillaria*)

SPOTTED MISSIONBELLS (*F. atropurpurea*)

YELLOW MISSIONBELLS (*F. pudica*)

Description and Habitat. Members of this genus have bell-shaped, nodding flowers that usually are solitary. A nectar gland is present at the base of each of the 6 parts of the flower. In many species the flower has a checkered appearance. The leaves alternate along the stem or are in whorls. The fruit is a 3-valved capsule with many seeds. The genus name comes from the Latin *fritillus*, meaning "dice box," in reference to the short, broad capsule characteristic of the genus. Fritillary can be found in open areas, forests, meadows, and grasslands at low and middle elevations. The flowers are hermaphrodite (have both male and female organs) and are pollinated by insects.

FIELD KEY TO THE FRITILLARIES

1. Flowers yellow, fading to red . . . . *F. pudica*
1. Flowers brown or purple, mottled with yellow . . . . *F. atropurpurea*

Natural History Notes and Uses. The bulbs of this genus have been a staple for native peoples since prehistoric times. The bulbs of all species are edible raw or cooked but are relatively rare and should be collected only in an emergency. The fruit pods of yellow missionbells can be eaten as a potherb.

These plants also contain a variety of alkaloids, including

fritimine, imperialine, chinpeimine, fritiminine, and sonpeimine. Both the alkaloid and non-alkaloid parts of the fritillary bulb have antitussive (cough suppressant) effects. Fritillary bulb liquid extract and fritimine have expectorant effects (promoting the ejection, by spitting, of mucus or other fluids) and can lower blood pressure, diminish excitability of respiratory centers, paralyze voluntary movement, and counter the effects of opium. Therefore, caution is advised.

COMMON STARLILY (*Leucocrinum montanum*)

Description and Habitat. Common starlily is a stemless perennial from short, fleshy roots. Each plant has about a dozen grass-like leaves about 6 inches long that overtop the flowers and are surrounded at the base by papery white sheaths. The fragrant white flowers are up to 1½ inches wide and include a slender tube nearly 4 inches long. At the bottom of the tube are the ovaries and seed capsules, which mature below ground. The seeds are black.

Natural History Notes and Uses. The generic name was compounded from the Greek *leuco*, "white," and *krinon*, "lily." The specific epithet means "of mountains" in Latin. Common starlily was described for science in the early 1800s by the famous English botanist-naturalist Thomas Nuttall (1786–1859). Medicinally, a poultice of pulverized roots was applied to sores or swellings by the Paiute. The plant was also used as food by the Crow.

COMMON ALPLILY (*Lloydia serotina*)

Description and Habitat. The alplily is a diminutive plant that is often overlooked. It has one to several stems arising 4–8 inches tall from bulbs on short, thick rootstalks. The needle-like basal leaves are 1½ to 4 inches long. The old, withered leaf bases remain at ground level, simulating a bulb. The 1 (usual) to 2 flowers are found atop the scape and are usually held erect. The flowers are whitish with green or purplish veins. The 6 tepals are oblong or oblanceolate in shape and are usually about ½ inch long. The 6 stamens are about half to two-thirds the length of the tepals, and the fruit (a capsule) is ovoid.

Natural History Notes and Uses. This dainty lily grows high in alpine meadows, often abundantly, and might at first be taken for springbeauty (*Claytonia*). Additionally, its long thin leaves might be taken for wild onion (*Allium*) leaves. *Lloydia* was named for Edward Lloyd (1660–1709), curator of the Museum of Oxford University and discoverer of this plant. The species name is Latin for "late."

Interestingly, populations of this plant tend to be androdioecious, that is, they have a large percentage of male flowers as well as hermaphrodites, raising some interesting questions for researchers. Initial investigations into pollen production and viability suggest that the theoretical requirements for double fertility in male flowers in androdioecious populations is not met with in *Lloydia*, but it may be possible that males are maintained in a population if they have an increased vegetative reproductive capability.

False Lily-of-the-valley (*Maianthemum = Smilacina*)

FEATHERY FALSE SOLOMON'S-SEAL (*M. racemosum*)

STARRY FALSE SOLOMON'S-SEAL (*M. stellatum*)

Description and Habitat. These annual herbs have extensive, horizontal rootstalks. The leaves are alternate and sessile or on short petioles. The genus *Maianthemum* (including *Smilacina*) consists of about 35 species and is widely distributed in the northern temperate zone as well as subtropical montane Asia and Central America.

Two genera, *Smilacina* and *Maianthemum*, have been recognized traditionally and historically. The genus *Smilacina* has recently been folded into *Maianthemum* on the basis of molecular data. *Maianthemum*, from the Latin, means "May blossom."

FIELD KEY TO THE FALSE LILIES-OF-THE-VALLEY

1. Inflorescence branched; petals shorter than the stamens	*M. racemosum*
1. Inflorescence unbranched; petals longer than the stamens	*M. stellatum*

Natural History Notes and Uses. False Solomon's-seal has its flowers at the end of the stem, which distinguishes it from the true Solomon's-seal. The plant is generally an indicator of moist environments; it also occurs on rocky, well-drained slopes.

Both species have edible berries that are not especially palatable. If eaten in quantity, they can act as a laxative. Cooking the berries removes much of the purgative elements, making them a bit more palatable. They are also high in vitamin C. The young shoots and leaves can be used like asparagus or eaten as a potherb.

False Solomon's-seals have starchy rootstalks that may be eaten, but they must be soaked overnight in lye. The roots are then boiled and rinsed several times to remove the lye (Turner and Szczawinski 1991). The Ojibwa Indians of Ontario used the white ashes from their fire pits instead of lye to remove the bitterness. A tea made from the roots was used for headaches.

A tea from starry false solomon's-seal was used by some Native Americans as a contraceptive. The powdered roots were used on wounds to stop bleeding, and a root decoction was used internally as a tonic or externally as an antiseptic wash for infected sores or wounds. The mashed root of starry false solomon's-seal was thrown into a stream to stupefy fish, making them easier to catch. A decoction of feathery false solomon's-seal was used as a contraceptive, to regulate menstrual disorders, to relieve kidney problems, to heal wounds, and as a heart tonic.

BREEDING SYSTEMS: EVOLUTION OF ANDRODIOECY

A primary goal of evolutionary biology is to understand the genetic and ecological forces controlling adaptive evolution. Because an individual's reproductive strategy has a direct effect on fitness, the forces controlling breeding-system evolution are easily shaped both theoretically and empirically, making it an important model for the study of evolution. When systems arise that appear to contradict the general assumptions of breeding-system theory, they are worth examining in detail.

Androdioecy is an uncommon form of reproduction in which males coexist with hermaphrodites, and it is thought to be difficult to evolve in species that regularly inbreed. The evolution of androdioecy is still poorly understood; however, there is evidence from several androdioecious species that it may have evolved from dioecy (males and females).

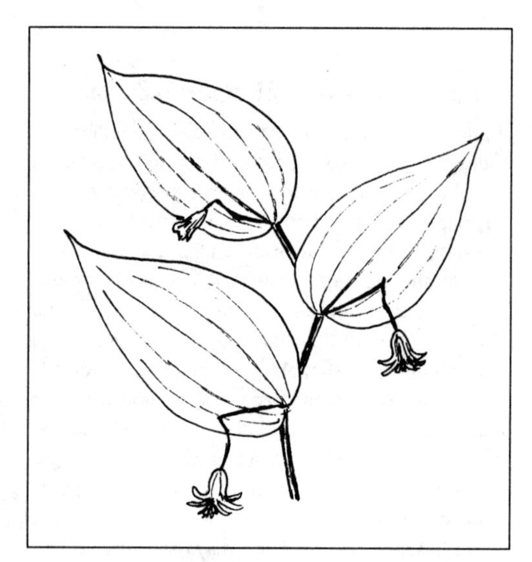

FIGURE 108. Twisted-stalk (*Streptopus amplexifolius*).

CLASPLEAF TWISTED-STALK (*Streptopus amplexifolius*)

Description and Habitat. Twisted-stalk is a perennial herb with creeping rootstalks. The sessile or clasping leaves are alternate and elliptical to ovate in shape, and the flowers are yellowish green. The 1–2 pendant flowers hang from the axils of the upper leaves on stalks that are bent in the middle. Common in moist soil and along streams and thickets in the montane and lower subalpine zone, these plants are often associated with genera such as *Smilacina* (false solomon's-seal) and *Actaea* (baneberry).

Natural History Notes and Uses. In terms of edibility, these plants have escaped mention in many guides but are indeed safe. The new spring shoots and clasping young leaves can be eaten raw or added to salads and taste somewhat like cucumbers. The berries, often referred to as watermelon berries, are somewhat laxative if eaten in excess, but may be eaten raw or cooked in soups and stews. They are sometimes called "scooter berries" because if you eat too many you can find yourself "scooting" to the bathroom. The species are easy to grow in wild gardens. The stems were used in poultices for cuts (Schofield 1989).

Warning: Anyone wishing to use the young shoots of twisted-stalk should be very careful to identify it correctly. At the shoot stage, these plants resemble the highly toxic *Veratrum* spp. (California false hellebore or corn lily).

GLUTIN TOFIELDIA (*Tofieldia glutinosa*)

Description and Habitat. This unique plant can be found near bogs and marshes. It has numerous, narrow, lily-like leaves growing in a basal clump or slightly up the stem. The erect, green stem is sticky. The dense terminal clusters of white flowers have dark anthers that extend beyond the petals.

Natural History Notes and Uses. The genus name honors Thomas Tofield (1730–79), a British botanist. The genus contains about 18 species of rhizomatous herbs that have narrow, tufted, basal leaves and racemes of small, star-shaped flowers. Some species are cultivated ornamentally.

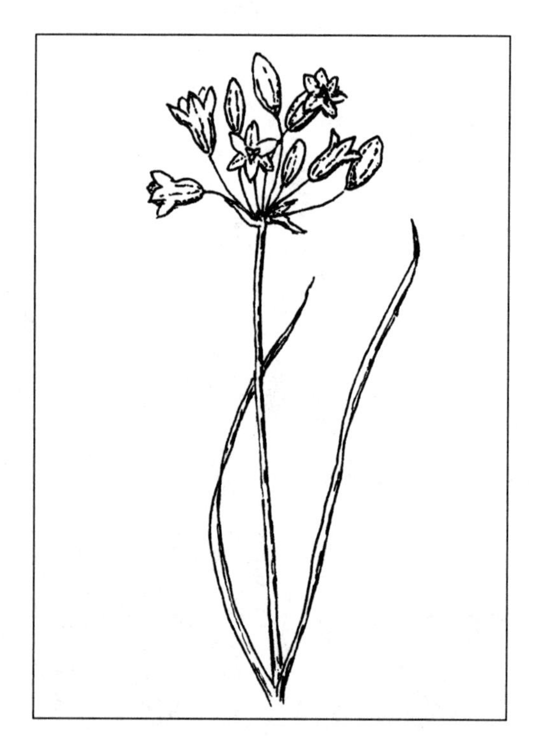

FIGURE 109. Brodiaea (*Triteleia* = *Brodiaea* spp.).

WILD HYACINTH (*Triteleia grandiflora*)

Description and Habitat. This perennial arises from a corm. The leaves are grass-like, few in number, and withering by the time the plants flower. The flowers occur in umbels. The genus name comes from the Greek *tri* ("three") and *teleios* ("perfect"), referring to the floral parts, which occur in 3's.

Natural History Notes and Uses. The corms are edible after cooking. The family Themidaceae was set up recently to hold *Triteleia* and other genera (*Brodiaea, Bessera, Dichelostemma, Milla, Muilla*) which were formerly in the Alliaceae but which were not compatible with that classification after DNA analysis. They are now considered more closely related to the Hyacinthaceae than to the alliums.

CALIFORNIA FALSE HELLEBORE, CORN LILY
(*Veratrum californicum*)

Description and Habitat. This tall, stout perennial has leafy stems and grows 3–6 feet tall. The leaves are broad and clasping, strongly veined, and ovate to elliptic. The cream to white flowers occur in a dense, terminal, branched cluster. The 6 perianth segments are all alike and have a green gland near the base. Corn lily is common in wet meadows and along streambanks, particularly at higher elevations. It is usually found below 11,000 feet, and flowers in July and August. These plants are often shredded by hail during storms.

Natural History Notes and Uses. **These plants are very poisonous if ingested and have an inconsistent mixture of several powerful alkaloids.** Some of the symptoms include depressed heart action, salivation, headache, burning sensation in the mouth, slowing of respiration, and death from asphyxia. These violent symptoms of poisoning may occur within 10 minutes. Avoid any use of the plant that involves ingestion. In some cases, just handling *Veratrum* can cause severe itchiness and irritation. The water in which the roots of *Veratrum* were boiled was considered effective in killing head lice. The powdered roots have also been used as an insecticide (Sweet 1976). Even nectar in the flowers is poisonous to insects and can cause serious losses among honeybees.

Caution: This plant should be considered **poisonous.**

COMMON BEARGRASS (*Xerophyllum tenax*)

Description and Habitat. This stout perennial grows up to 5 feet tall. The stems arise from large clumps of bluish green, wiry, saw-edged, grass-like leaves that are up to 2 feet long. The cream-colored flowers are borne in a hemispheric terminal inflorescence. Common beargrass is found in middle to lower alpine open slopes and forests. The thick, shallow rhizome can remain vegetative for many years, then flowers and dies.

Natural History Notes and Uses. The fibrous roots of beargrass can be eaten after roasting or boiling. Although the sharp leaves are not very pleasant to handle, when dried and bleached they can be used for weaving baskets and clothing. The baskets are particularly pliable and durable. Lather from the roots was used to bathe sores.

Death Camas (genus *Zigadenus*)

MOUNTAIN DEATH CAMAS (*Z. elegans*)

FOOTHILL DEATH CAMAS (*Z. paniculatus*)

MEADOW DEATH CAMAS (*Z. venenosus*)

Description and Habitat. The three species of *Zigadenus* are glabrous perennials with bulbs and grass-like leaves. The cream-colored to greenish white flowers are stalked and subtended by narrow bracts in an elongated inflorescence. Death camas

FIGURE 110. Death camas (*Zigadenus* spp.).

occurs in grasslands, meadows, and forest openings into the alpine zone.

FIELD KEY TO THE DEATH CAMAS

1. Perianth segments more than ¼ inch long, not clawed .... *Z. elegans*
1. Perianth segments less than ¼ inch long; inner segment clawed 2

2. Inflorescence usually in a panicle .... *Z. paniculatus*
2. Inflorescence usually a raceme .... *Z. venenosus*

Natural History Notes and Uses. Death camas is very **poisonous** if ingested. The alkaloids, primarily concentrated in the bulbs, can cause muscular weakness, slow heartbeat, subnormal temperature, excessive watering of the mouth, and stomach upset with pain, vomiting, and diarrhea. Death camas should not be confused with the edible camas (*Camassia*), which was a staple food for aboriginal peoples in the Northwest. It is also difficult to distinguish death camas from other edible plants, including wild onions (*Allium*), sego lilies (*Calochortus*), fritillaries (*Fritillaria*), and brodiaeas (*Brodiaea*) prior to flowering.

Crushed death camas bulbs were used by some Native Americans as poultices for boils, bruises, strains, rheumatism, and in some cases rattlesnake bites.

Caution: This plant should be considered **poisonous**.

WATER-NYMPH FAMILY (NAJADACEAE)

Only one genus is in this family of submerged aquatic herbs. The leaves are opposite and minutely serrulate. The flowers are imperfect and contain 1 stamen, 1 pistil, but 2–4 stigmas.

SOUTHERN WATER-NYMPH (*Najas guadalupensis*)

Description and Habitat. Water-nymph is a completely submerged annual plant, although it is often found as floating fragments. It has opposite leaves that are often clustered near the tips of the stems. The leaf base is much wider than the rest of the leaf blade, which helps to distinguish the water-nymphs from other underwater plants. These plants have inconspicuous flowers and fruits that are almost completely hidden by the leaf bases. Water-nymph pollination takes place underwater.

Natural History Notes and Uses. Water-nymph can form dense colonies in shallow water and hinder swimming, fishing, boating, and other forms of water recreation. It is a major problem in some areas and is reported to impede water flow in drainage and irrigation canals. However, water-nymphs are one of the most important food sources for waterfowl, which eat the entire plant. They also provide shelter for small fish and insects.

Water-nymph is a free-floating plant that may be easily grown when planted in water tanks. It will quickly fill a tank, making it ideal for breeding and fry setups. It needs good light, however, and seems to do better in harder water with a higher pH. Some aquarists have found that *Najas* is a heavy feeder that requires the presence of fish (and their wastes) to thrive. In fishless tanks it often turns pale yellow, apparently needing the additional ammonia and micronutrients provided by the fish to aid its growth.

ORCHID FAMILY (ORCHIDACEAE)

Orchid flowers are irregular in shape with 3 sepals and 3 petals. One of the petals forms a lip, sac, or pouch on the lower side of the flower. The flower structure is highly specialized for insect pollination. The family is an outstanding source of ornamentals, and although a few orchid species have utilitarian uses, the majority are rare and should be collected for use only in emergency situations.

This family is believed to contain more species than any other plant family in the world, with the possible exception of the sunflower family (Asteraceae). Orchid taxonomists can only estimate the species numbers due to the huge magnitude of the family as well as the relative inaccessibility of many species that are hidden high in the canopies of tropical forests.

AN ENDANGERED FAMILY

The entire orchid family is listed on Appendix II to the Convention on Trade in International Species of Flora and Fauna (CITES), meaning that trade in these species is restricted. Nine species of orchids are listed on Appendix I to CITES, meaning that trade is severely restricted because they are in danger of extinction. The World Conservation Union (IUCN) lists 325 species of orchids as endangered in the 1997 Red List of Threatened Plants. Orchids have been affected mainly by habitat destruction but also by collection.

FIELD KEY TO THE ORCHID FAMILY

1. Stems leafless; leaves all basal or lacking entirely 2
1. Leaves occurring on the lower third of the stem or higher 8

2. Flowers solitary . . . . *Calypso*
2. Flowers more than 1 3

3. Green leaves completely absent; stems yellow to purple . . . . . *Corallorhiza*
3. Green basal leaves present 4

4. Basal leaf solitary (rarely 2) 5
4. Basal leaves more than 1 6

5. Lip lobed and spotted . . . . *Orchis*
5. Lip with entire margins, not spotted . . . . *Platanthera*

6. Upper stem glandular-hairy; flowers without a down-turned, nectar-holding spur . . . . *Goodyera*
6. Stems glabrous; flowers with a spur 7

7. Leaves opposite, nearly orbicular, appressed to the ground . . . *Platanthera*
7. Leaves not opposite or strictly appressed to the ground, narrowly elliptic . . . . *Piperia*

8. Leaves 2, opposite each other at mid-stem . . . . *Listera*
8. Leaves usually more than 2, alternate 9

9. Flowers 1–3; lip petal more than ½ inch long, pouch-shaped . . . . *Cypripedium*
9. Flowers mostly more than 3; lip petal usually less than ½ inch long 10

10. Flowers purplish . . . . *Epipactis*
10. Flowers white or greenish white 11

11. Upper stem glandular-hairy . . . . *Goodyera*
11. Stems glabrous 12

12. Flowers without a down-turned, nectar-holding spur, spirally arranged in a spike . . . . *Spiranthes*
12. Flowers with a spur, not spirally arranged 13

13. Leaves on lower third of stem only. . . . . *Piperia*
13. Leaves present above midstem . . . . *Platanthera*

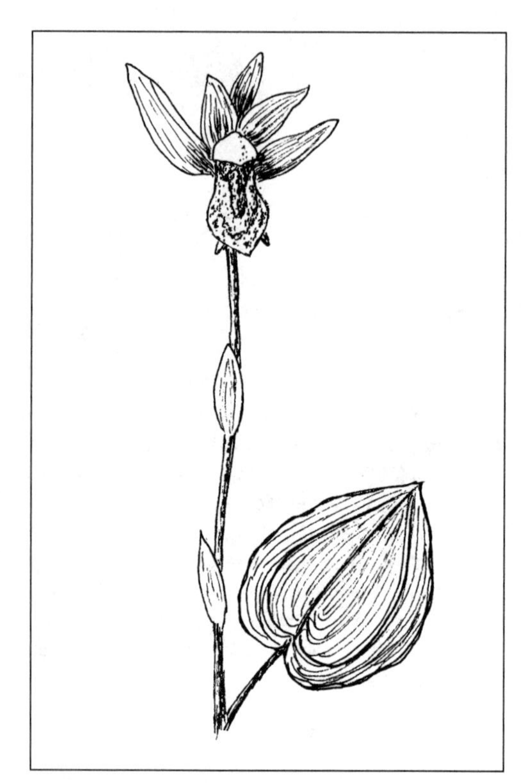

FIGURE 111. Fairy slipper orchid (*Calypso bulbosa*).

FAIRY SLIPPER ORCHID (*Calypso bulbosa*)

Description and Habitat. This orchid has a spectacular slipper-shaped flower. The pinkish brown, sheathed stem is 2–6 inches tall, with a single small, oval basal leaf that emerges after the flower has appeared. Each flower has 3 sharp-pointed sepals and 2 similar petals with purple stripes. The third petal or cup-like lip forms an apron with dark purple stripes inside and below and 3 short rows of white or yellow hairs on the upper surface. It blooms after the snow has melted in moist cool forests with decaying duff on the floor.

Natural History Notes and Uses. The genus *Calypso* is named for the beautiful nymph in Homer's *Odyssey* who waylaid Odysseus on his return to Ithaca. The specific epithet is Latin, meaning "bulbed," in reference to the small pseudo-bulb of this species.

This widespread species nearly circles the globe in the Northern Hemisphere, ranging throughout North America, Europe, and Asia. In North America calypso is found from Labrador to Alaska, south to New England, Minnesota, the Great Plains, Arizona, and along the west coast to California.

Calypso is pollinated by a number of species of bumblebees (Mosquin 1970; Boyden 1982). Like several other North American orchids, it has a deceptive pollination system. The bumblebees are attracted by the yellow bristles but on visiting the flower, find no nectar or pollen reward.

This species has nutritional as well as aesthetic value, as the mucilaginous corms were eaten by Native Americans (Correll 1950). In the *Odyssey* Calypso kept Odysseus concealed for seven years on her island. Both the beauty and rarity of calypso, as well as the seclusion of its habitats, make this a fitting name (Correll 1950).

Coralroot (genus *Corallorhiza*)

SUMMER CORALROOT (*C. maculata*)

PACIFIC CORALROOT (*C. mertensiana*)

HOODED CORALROOT (*C. striata*)

YELLOW CORALROOT (*C. trifida*)

SPRING CORALROOT (*C. wisteriana*)

Description and Habitat. Coralroots are typically yellowish to brownish red perennials with coral-like rhizomes. *Corallorrhiza* means "coral root." The rhizomes are associated with fungi that aid in the uptake of nutrients.

FIELD KEY TO THE CORALROOTS

1. Sepals and petals pink with 3–5 distinct reddish brown or purple stripes; lip margin without lobes. . . . . *C. striata*
1. Sepals and petals yellow to pink to wine red and not striped; lip often lobed. 2

2. Sepals yellow or greenish yellow, under ¼ inch long, and 1-nerved . . . . *C. trifida*
2. Sepals reddish, more than ¼ inch long, and usually 3-nerved 3

3. Lip not lobed, white, often with a few purple spots and upturned . . . . *C. wisteriana*
3. Lip usually lobed or toothed on the side, margin generally not upturned 4

4. Spur lacking or maybe a slight bulge; lip white to cream-colored, usually with reddish spots . . . . *C. maculata*
4. Spur small and visible; lip pink to reddish purple, without spots . . . . *C. mertensiana*

Natural History Notes and Uses. Consumption of the toxic rhizome can cause hyperthermia and profuse perspiration (Weedon 1996). The rhizome of *C. maculata* has been used as a diaphoretic, febrifuge, and sedative, and the dried stems were used by the Paiute and Shoshone Indians of Nevada to make a tea to build up blood in pneumonia patients (Coffey 1993; Foster and Duke 1990). Strike (1994) indicates that the plant was used to reduce fevers or as a sedative.

LADY SLIPPER (*Cypripedium calceolus* = *C. pubescens*)

Description and Habitat. This plant grows up to 14 inches tall and has 3–4 leaves. The 1–2 flowers have greenish yellow to purplish brown sepals and upper petals. The yellow slipper petal is about an inch long and has purple spots around the opening. The column is also yellow with purple spots. *Calceolus* is Latin meaning "little shoe," in reference to the slipper-like shape of the labellum. The plants of the genus *Cypripedium* are often called lady slippers but might better be called Venus' slippers, for the name comes from the Latin *Cypris*, "Venus," and *pedilon*, "shoe."

FIGURE 112. Lady slipper (*Cypripedium* spp.).

Natural History Notes and Uses. The plants are pollinated by a number of different species of small bees, primarily adrenid and halictid bees. The plants are also visited and sometimes pollinated by a variety of Diptera (flies). The flowers attract bees for pollination by giving off a nectar-like scent but do not actually produce any nectar. Once the disappointed bee is inside the "slipper" part of the flower and finds no nectar, it departs along one of two channels near the rear of the flower. While the bee is squeezing through either of these two openings, its back becomes covered with the flower's pollen, which it takes to the next flower in its search for the nonexistent nectar.

GIANT HELLEBORINE (*Epipactis gigantea*)

Description and Habitat. Giant helleborine is a native perennial plant that grows 1–3 feet tall. It has short rhizomes and leaves 2–8 inches long. It is sparsely pubescent. Only 2–3 flowers appear on the stalk at a time. The petals are reddish brown; the lower one is sac-like. The sepals are green with brownish veins. The nodding, elliptic capsule contains thousands of tiny seeds. The plants commonly form dense stands but also grow in small or large groups.

About 25 species are found worldwide, mostly in Europe (15 species) and Asia. Two species are found in the United States: *E. gigantea* and *E. helleborine*. The latter is thought to have arrived in North America from Europe in the nineteenth century and grows in the northeastern United States.

Natural History Notes and Uses. Giant helleborine appears to attract pollinators by mimicking their food supply without giving any real reward. This orchid is pollinated by flies of the family Syrphidae. They are attracted to the flowers by the aroma, which mimics the honeydew smell given off by aphids. The flies normally lay their eggs in masses of aphids, which become the food supply for their larvae. The aroma fools the fly into laying its eggs among the flowers, pollinating them in the process but not receiving any reward.

Vegetative plants may be confused with some members of the orchid genus *Platanthera*, or more likely with *Maianthemum stellatum*, in the lily family, species that can co-occur with giant helleborine. The prominently clasping leaf bases and taller habit of giant helleborine distinguish it from *Maianthemum*, and its generally more numerous and larger leaves and taller habit from *Platanthera*.

The name *Epipactis* derives from a classical name used by Theophrastus (ca. 350 B.C.) for a plant used to curdle milk, or it may refer to the Greek *epi* for "on" and *pactos* for "solid" or "firm." The specific epithet is Latin for "gigantic," referring to the large size of the plant. The Indians of northern California made a decoction of the fleshy roots for internal use when they felt "sick all over."

WESTERN RATTLESNAKE PLANTAIN (*Goodyera oblongifolia*)

Description and Habitat. Rattlesnake plantain is a perennial evergreen herb with a basal rosette of dark green leaves and a narrow spike of tiny orchid flowers. The flowers grow in a dense, one-sided (or spiraling) terminal spike (raceme). Each greenish white flower has a hood formed by the upper petals and sepal. The lower lip is pouch-like and beaked and is covered with glandular hairs.

The specific epithet is Latin, meaning "oblong leaf," referring to the relatively oblong leaves of this species.

Natural History Notes and Uses. The species grows primarily in the western United States. Archibald Menzies, surgeon-naturalist on Captain James Cook's ship *Discovery*, which explored the west coast of North America, presumably was the first European collector of this plant more than two hundred years ago.

The leaves sometimes have scaly markings. To early settlers, this was an indication that the plant would be useful for treating snake bites, hence the name *rattlesnake plantain*. Many native groups made use of the plant. In Okanagan-Colville culture, women wishing to get pregnant would split the leaves open

DOCTRINE OF SIGNATURES

In ancient times it was thought that if a plant part was shaped like or in some other way resembled a human organ or disease characteristic, then the plant was useful for that particular organ or ailment. This concept has no scientific basis, though sometimes uses conceived centuries ago have persisted and may even have been corroborated by scientific evidence of their efficacy. A couple of examples of this doctrine are that kidney beans were good for the kidneys and that the leaves of arrowhead (*Sagittaria* spp.) were useful for wounds caused by arrows.

and blow on them. Women would chew on the plant before and during childbirth. A poultice of the leaves was also useful for treating wounds. For the Thompson Indians, chewing the leaves helped determine the gender of a child and ensure a smooth delivery.

Twayblade (genus *Listera*)

NORTHERN TWAYBLADE (*L. borealis*)

NORTHWESTERN TWAYBLADE (*L. caurina*)

BROAD-LIPPED TWAYBLADE (*L. convallarioides*)

HEARTLEAF TWAYBLADE (*L. cordata*)

Description and Habitat. These small, slender, often rhizomatous perennials have 2 broad, prominently nerved leaves opposite each other in the middle of the stem. Small green or yellowish flowers occur in a short, narrow, usually glandular-hairy inflorescence. The 3 sepals and 2 upper petals are similar and spreading or reflexed. The prominent lip petal projects down or outward. These are small orchids.

FIELD KEY TO THE TWAYBLADES

1. Lip dissected about half the length into two divergent lobes; leaves subcordate . . . . *L. cordata*
1. Lip only shallowly bilobed; leaves not as above 2

2. Lip covered with minute hairs, at least along the margins, oblong in outline and not narrowed toward the base. . . . . *L. borealis*
2. Lip not covered with minute hairs, usually narrowed near the base 3

3. Lip with a fringe of marginal hairs, tip of lip distinctly notched, abruptly narrowed to a short claw . . . . *L. convallarioides*
3. Lip margin glabrous, tip round to only slightly notched, gradually narrowed to base . . . . *L. caurina*

Natural History Notes and Uses. The intricate pollination mechanisms of *Listera* species fascinated Charles Darwin, who studied them intensively. The central furrow holds nectar, which leads insects toward the stigma. The pollen is blown out explosively within a drop of viscous fluid that glues the pollinia to unsuspecting insects (or to your finger if you touch the top of the column). The pollen will then be deposited on the next twayblade the insects visit. The genus *Listera* is named in honor of Dr. Martin Lister, an English naturalist who lived from 1638 to 1711.

SPOTTED ORCHID (*Orchis rotundifolia = Amerorchis r.*)

Description and Habitat. This orchid produces a single, roundish to oval basal leaf up to 3½ inches long and slightly yellow-green in color. The stout, solitary stems are terminated by 10–15 or more small white flowers in a loose raceme. The upper sepals and petals converge to form a pale pink to white hood, above a spurred, flat, 3-lobed lower lip that is typically spotted with dark purple.

Natural History Notes and Uses. Orchid taxonomists now tend to place this species in the genus *Amerorchis*. This is a boreal orchid that ranges from Alaska to southern Greenland, extending south to Wyoming, the northern Great Lakes region, and northern New England.

ALASKA REIN ORCHID (*Piperia unalascensis* = *Habenaria u.*)

Description and Habitat. This species has a leafless stem 8–16 inches tall, with narrow bracts above the base and 1–3 fleshy, egg-shaped roots. The 2–4 basal leaves have a sheathing base and a lance-shaped blade. There are numerous small, green to yellowish flowers in an open, spike-like inflorescence. The upper sepal and lip petal are egg-shaped, while the upper petals and lower sepals are narrower and spreading. The cylindrical spur is curved and longer than the lip. Rein orchids are any of several American species that have a kidney-shaped lip. Alaska rein orchid is found in low- to mid-elevation forests.

Natural History Notes and Uses. The taxonomy of the more than 20,000 members of the Orchidaceae is under continuous change. New discoveries and species reclassifications are just some of the complications involved in describing the evolutionary progression and physical characteristics of orchids. This species was formerly placed in the genus *Habenaria*.

Bog Orchid (genus *Platanthera* = *Habenaria*)

BOREAL BOG ORCHID (*P. dilatata*)

NORTHERN GREEN ORCHID (*P. hyperborea*)

BLUNT BOG ORCHID (*P. obtusata*)

Description and Habitat. Bog orchids are perennials, often with fleshy or tuberous roots. The small white to yellowish green flowers are in spike-like racemes. At the base of the lip is a spur. *Habenaria* is Latin for "reins" or "narrow strap," which refers to the lip of some species. Bog orchids can be found in forest understories or in wet areas and meadows.

FIELD KEY TO THE BOG ORCHIDS

1. Leaves on the lower third of the stem; plant having a leafless stalk arising from a cluster of leaves; flowers greenish........ *P. obtusata*
1. Leaves several to many, all cauline; flowers often white................ 2

2. Flowers white.. *P. dilatata*
2. Flowers greenish...................................... *P. hyperborea*

Natural History Notes and Uses. The tuber-like roots of many species may be eaten raw or cooked (Coffey 1993; Kirk 1975; Weedon 1996). However, Pojar and MacKinnon (1994) recommend a cautious approach until the poisonous nature of the plants is clarified. Some Native Americans used extracts from these plants as poison on baits for coyote and grizzly bear.

HOODED LADIES'-TRESSES (*Spiranthes romanzoffiana*)

Description and Habitat. Hooded ladies'-tresses is a small herb with fleshy roots. The white flowers are in a dense, spirally twisted spike, with the sepals and lateral petals appearing to be fused and forming a hood around the column. Hooded ladies'-tresses can be found in dry to moist areas—meadows, lakeshores, bogs, and marshes—up to timberline. The species was named after Nikolei Rumliantzev, Count Romanzoff (1754–1826), a Russian patron of science.

Natural History Notes and Uses. According to Weedon (1996), *S. romanzoffiana* has strong diuretic properties, making it undesirable for eating. Foster and Duke (1990) indicate that Native Americans used a plant tea of *S. cernua* (nodding ladies'-tresses) of the eastern United States as a diuretic for urinary disorders, for venereal disease, and as a wash to strengthen infants. They also note that other North American, European, and South American species have been used as diuretics and aphrodisiacs.

Caution: *S. diluvialis* (Ute ladies'-tresses) is known to occur in Wyoming, Montana, and Idaho. The plant is listed as threatened by the U.S. Fish and Wildlife Service. Care should be taken not to disturb these plants. Report any findings to the National Park Service, Forest Service, and/or state native plant society.

GRASS FAMILY (POACEAE = GRAMINAE)

With 600 genera and 10,000 species worldwide, the grass family is the most common family of flowering plants, found in practically all habitats and on all continents. There are more than 180 genera and nearly 1,000 species in the United States. Grasses have round, hollow stems with linear sheathing leaves. Because grasses are wind-pollinated, they have no showy flowers to attract insects. The flowers have been reduced to scaly bracts that enclose the male and female parts. The grains form within the papery bracts after pollination. The grass family contains many of the most economically important plants in the world, including cereal crops and forage grasses essential to raising domesticated livestock. Grasses can be found from alpine meadows to sea level.

All grasses in the area have edible grains that are generally small and tedious to collect. The small seeds are tightly enclosed in scales, which are hard to remove. The larger grains of some grass species were a staple among aboriginal peoples. The grains were harvested with a beater and ground for mush and flour. The grains are rich in protein and can be eaten raw but are better if roasted, ground into flour, or boiled into mush. They may also be boiled in the same way as rice and added to soups or stews. The reported toxicity of some grass species may be that of a fungus (*Claviceps purpurea*) associated with the grasses. Any inflorescences containing black grains should be discarded since they may have a harmful fungus infection.

Brown (1985) suggests that all bladed grasses are edible and are rich in vitamins and minerals. Animals often consume grasses to get nutrients they can't get elsewhere. The young shoots are edible raw and are not as fibrous and therefore easier to digest than mature leaves. The green or dried leaves can be steeped to make a tea.

WILD FLOURS (PINOLE)

Pinole was made using small grains that were parched by tossing in a basket with glowing coals or hot pebbles, keeping the grains in constant motion. The grains were then pulverized and eaten. Sometimes they were pressed into cakes held together by the grains' natural oil, with no other liquids added. Grains from several species of plants were often mixed together to enhance the flavor of the pinole.

Caution: Grasses that are infected with *Claviceps purpurea* develop purple sclerotia (ergots) in place of the healthy grain. The sclerotia contain a number of toxic alkaloids and, if eaten, can cause severe illness and sometimes death. One effect of the toxins is constriction of the blood vessels, and the impaired circulation may result in gangrene or loss of limbs. Another effect is on the nervous system, resulting in convulsions and hallucinations (Webster 1980). The sclerotia of *C. purpurea* are used medicinally to hasten uterine contractions during childbirth. Commercial ergot is produced by cultivating the fungus on rye (*Secale cereale*) and other plants.

FIELD KEY TO THE GRASS FAMILY

1. Spikelets mostly replaced by small bulbs enclosed by leaf-like lemmas . . . . *Poa bulbosa*
1. Spikelets not replaced by bulbs. 2

2. Spikelets sessile, forming terminal or lateral spikes . . . . Group A
2. Spikelets stalked in open to contracted and spike-like inflorescence 3

3. Spikelets 1-flowered. . . . . Group B
3. Spikelets 2- to many-flowered. 4

4. At least one glume equal to or greater than the lowest lemma. . . Group C
4. Both glumes shorter than the lowest lemma . . . . Group D

— Group A —

1. Spikes usually more than 1 and borne on the side of the culm, not continuous with the main axis. . . . . *Beckmannia*
1. Spikes solitary and borne at the top of the culm, continuous with the main axis 2

2. Spikelets mostly 1 per node . . . . *Agropyron*
2. Spikelets more than 1 per node. 3

3. Spikelets egg-shaped or elliptical at maturity, awnless with 1 fertile flower. . . . . *Setaria*
3. Spikelets not as above. 4

4. Most spikelets with 1 flower 5
4. Spikelets usually with more than 1 flower. 7

5. Awns of lemma more than ½ inch long. . . . . *Hordeum*
5. Lemmas unawned or with an awn less than ⅛ inch long. 6

6. Lemmas with short awns . . . . *Alopecurus*
6. Lemmas unawned . . . . *Phleum*

7. Florets subtended by numerous awns; axis of spike fragile and easily breaking at the nodes. . . . . *Sitanion*
7. Florets not subtended by awns; spike not easily breaking . . . . *Elymus*

— Group B —

1. Lemmas with terminal 3-parted awn. . . . . *Aristida*
1. Awns of lemmas lacking or not 3-parted 2

2. Lemmas hardened, often with sharp-pointed base. 3
2. Lemmas not hardened, of the same texture as the glumes, base not noticeably hard and sharp 4

3. Awns spirally twisted; lemmas usually more than 6 times as long as wide . . . . *Stipa*
3. Awn not twisted; lemmas usually less than 6 times as long as wide . . . . *Oryzopsis*

4. Spikelets with 2 narrow, sterile lemmas at base
   of fertile lemma .... *Phalaris*
4. Spikelets lacking such sterile lemmas 5

5. Spikelets very short-stalked and congested in dense,
   cylindrical inflorescence usually at least ⅛ inch wide.................. 6
5. Spikelets borne on erect or spreading branches in narrow
   or open but not cylindrical inflorescence 7

6. Lemmas awned; glumes unawned .... *Alopecurus*
6. Glumes awned; lemmas unawned, sometimes with
   small awn tip.. *Phleum*

7. One or both glumes as long as or greater than the lemma 8
7. Glumes shorter than the lemma.................................. 11

8. Plant a rhizomatous perennial, more than 28 inches tall
   with leafy blades about ½ inch wide............................ *Cinna*
8. Plant not as above 9

9. Floret without conspicuous tuft of hairs at the base............ *Agrostis*
9. Floret with conspicuous tuft of hairs at the base.................... 10

10. Awn arising from the tip of the lemma .... *Muhlenbergia*
10. Awn arising from near middle of back of lemma.......... *Calamagrostis*

11. Glumes and/or lemmas awned .... *Muhlenbergia*
11. Glumes and lemmas unawned .... *Catabrosa*

— *Group C* —

1. Two staminate florets below the single perfect
   (staminate and pistillate) floret............................ *Hierochloe*
1. Spikelets not as above.. 2

2. Lemmas without awns 3
2. Lemmas awned 5

3. Leaf tips not shaped like prow of a boat .... *Trisetum*
3. Leaf tips shaped like prow 4

4. Spikelets congested on short branches
   in a spike-like inflorescence................................. *Koeleria*
4. Spikelets not congested on short branches;
   inflorescence sometimes narrow but not congested................ *Poa*

5. Spikelets more than ½ inch long............................ *Trisetum*
5. Spikelets ½ inch long or less 6

6. Lemmas 2-lobed at the tip; awn arising
   from above the middle .... *Trisetum*
6. Lemmas with jagged tips but not distinctly 2-lobed;
   awn arising from near or below the middle .... *Deschampsia*

— *Group D* —

1. Plants stout reeds at least 80 inches tall;
   occurring in wet habitat.................................. *Phragmites*
1. Plants rarely over 60 inches tall; rarely occurring in wet habitat 2

2. Unisexual grasses, staminate and pistillate florets on different plants.... 3
2. Plants with at least some florets with both stamens and pistils 4

3. Leaf sheaths long-hairy at the top .... *Distichlis*
3. Leaf sheaths glabrous or short-hairy .... *Poa*

4. Tips of leaf blades usually shaped like prow of a boat.................. 5
4. Leaf tips not shaped like prow...................................... 6

5. Spikelets congested on short branches in spike-like inflorescence. . . . . *Koeleria*
5. Spikelets not congested on short branches; inflorescence sometimes narrow but not congested . . . . *Poa*

6. Base of floret with tuft of straight hairs; lemmas otherwise hairless. . . . . *Trisetum*
6. Florets not as above. 7

7. Spikelets clustered at the ends of stiff, nearly erect branches of the inflorescence; lemmas with long, stiff hairs on the midvein and margins . . . . *Dactylis*
7. Inflorescence and lemmas not combined as above 8

8. Lemmas with 3 prominent nerves . . . . *Catabrosa*
8. Lemmas with more than 3 prominent nerves or nerves indistinct 9

9. Lemmas obtuse at the top, usually with ragged margins, unawned. 10
9. Lemmas pointed at the tip, often awned. 11

10. Second glume 3-nerved; leaf sheaths with free margins, at least toward the top . . . . *Puccinella*
10. Second glume 1-nerved; leaf sheaths closed to the top. . . . . *Glyceria*

11. Stems usually bulbous at the base just below ground level; uppermost floret usually sterile. . . . . *Melica*
11. Plants not as above 12

12. Lemmas 2-lobed at the tip; awn, if present, arising from just below base of notch . . . . *Bromus*
12. Lemmas not 2-lobed; awn arising from tip. . . . . *Festuca*

Wheatgrass (genus *Agropyron*)

Description and Notes. Species in the genus are commonly referred to as crested-wheatgrasses. All the North American species were introduced. Most species of wheatgrass provide forage for livestock and are among the most valuable species for that purpose in the Rocky Mountains. They have horizontal creeping stems that help hold the soil, especially on slopes or in sandy areas. The flowers are produced in spikes, the several-flowered spikelets placed flatwise, usually single, at each joint of the axis of the spike.

The rhizomes of some species, especially *A. repens* (= *Elytrigia r.*) (creeping quackgrass), may be dried and ground and the meal made into bread. The rhizomes were also scorched as a coffee substitute. A tea made from the plant was a diuretic and was used for kidney stones, to expel intestinal worms, and as a wash for swollen limbs (Foster and Duke 1990).

Bentgrass, Redtop (genus *Agrostis*)

Description and Notes. Most species of bentgrass are important forage plants either in cultivation or in the mountain meadows. Many are extensively used for lawns and golf courses. The spikelets are 1-flowered, the glumes are equal or nearly so, and the lemmas are shorter and thinner than the glumes, awnless or awned from the back, and often minutely hairy at the base. These are medium-sized grasses with very small and numerous spikelets in somewhat feathery panicles that are often dark purple.

Ross' bentgrass (*A. rossiae*) is a rare endemic plant restricted to thermal areas within the geyser basin area of Yellowstone National Park. The species is known from at least four areas, where their population numbers and locations may fluctuate greatly. Other species of bentgrass are commonly used for lawns. This is a desirable grass for golf course tees, fairways, and greens. Bentgrass has numerous advantages for turf applications: it can be mowed to a very short length without damage, it can handle a great amount of foot traffic, it has a shallow root system that is thick and dense, allowing it to be seeded and grow rather easily, and it has a pleasing, deep green appearance.

Foxtail (genus *Alopecurus*)

Description and Notes. Foxtails have spike-like panicles, as in timothy (*Phleum*), but are much softer and with glumes awnless and the lemmas awned. Foxtails occur in northern temperate regions. They can be annual or perennial. A few are considered weeds; others are decorative and are used in dried flower arrangements.

Many perennial grasses have evolved methods of vegetative reproduction that allow them to propagate when conditions are unfavorable for sexual reproduction. One method involves modified stems known as stolons. These creeping "runners" spread over the ground surface and produce new plants by developing roots and shoots along the prostrate stem joints. Another method involves producing underground stems called rhizomes. The rhizomes bear scale-like leaves and send up new plants from underground nodes.

Three-awn (genus *Aristida*)

Description and Notes. This genus is characterized by a 1-flowered spikelet, the lemma of which is tipped by a conspicuous 3-branched awn. The lateral branches of the awn are sometimes much shorter than the central branch. The narrow leaves are wiry.

SLOUGHGRASS (*Beckmannia syzigachne*)

Description and Notes. The spikelets are 1-flowered, in 2 rows along one side of the rachis. The lemma is about as long as the glumes, 5-nerved, awnless but acuminate. Sloughgrass is characteristically found in damp habitats such as wet ground or shallow water and at the margins of ponds, lakes, and streams but may also grow along ditches. This species is readily distinguishable from all other wetland grasses by the strongly flattened spikelets that are round in side view and tightly packed in two rows along the branches of the flowering head.

Beckmannia is named for Johann Beckmann, an eighteenth-century German botanist and author of one of the first botanical dictionaries; *syzigachne* is from the Greek *syzygos* for "joined" and *achne* for "chaff," referring to the prominent glumes that stay joined when the whole spikelet falls from the flowering head.

The cooked seeds have a mild flavor; they can be ground into a flour and used as a cereal. The seed is very small but is easily harvested, although it then must be separated from its husk. Some North American Indian tribes burned the husks of grass seeds. The plant is used for making bedding and pillows.

Brome-grass (genus *Bromus*)

Description and Notes. *Bromus* is a large genus with about 160 species. Estimates in the scientific literature range from 100 to 400. Commonly known as brome-grasses, *Bromus* species occur in many habitats in temperate regions around the world. There is a wide variety between some species, but the morphological differences between others are subtle and difficult to distinguish, making the taxonomy of the genus complicated. The spikelets are several- to many-flowered and in a panicle. The glumes are unequal to subequal. The lemmas are 5–9-nerved, 2-toothed, awned from between the teeth or awnless.

Taxonomists have generated several classification schemes to reflect the morphological variation that is seen in *Bromus.* In North America five sections are generally recognized: Bromus, Genea, Ceratochloa, Neobromus, and Bromopsis. Sections Bromus and Genea are native to the Old World (Eurasia), but many species have been introduced into North America. Sections Bromopsis, Neobromus, and Ceratochloa have several native species in North America.

One *Bromus* species of interest in the area, as well as elsewhere in the West, is cheatgrass (*B. tectorum*). Cheatgrass is a weedy annual grass, growing up to 2 feet tall; it has a branched base and is typically rusty red to purple at maturity. The seeds germinate in the late fall or early spring. Cheatgrass grows rapidly in the spring, and the seeds mature within two months of beginning growth.

Cheatgrass is native to Europe, southwestern Asia, and northern Africa but was introduced into North America and has become an aggressive weed. The seeds are dispersed by wind, small rodents, or attachment to animal fur within a week of maturity. They are also moved as a contaminant in hay, grain, straw, and machinery. It is an abundant seed producer, with a potential in excess of 300 seeds per plant, depending on plant density.

Cheatgrass is widely adapted. It grows on all exposures and all types of topography from desert valley bottoms to the tops of the highest mountain peaks, 2,500 to 13,000 feet in elevation. It quickly invades heavily grazed rangeland, roadsides, waste places, burned areas, and disturbed sites. Because cheatgrass rapidly develops a large root system in the spring, by the time native grass seedlings start to grow in April or May, cheatgrass has stolen most of the water out of the top foot of soil. Although mature native grasses can get water from lower soil regions, seedlings cannot get their roots deep enough into the soil to reach water before drought sets in, and thus they die of thirst. Without this ability to reproduce, native grasses inevitably decline, and so over time, cheatgrass becomes more and more common until eventually it dominates. Cheatgrass often opens the way for secondary invaders such as knapweed and thistle.

Reedgrass (genus *Calamagrostis*)

Description and Notes. The spikelets are 1-flowered, in open to spike-like panicles. The glumes are about equal. The lemma is shorter than the glumes, awned from the back, the callus bearing a tuft of hairs mostly a quarter as long as the lemma to as long as the lemma. The genus contains about 230 species that occur mainly in temperate regions of the Northern Hemisphere. They are commonly adventive. These tufted perennials usually have hairless narrow leaves. The word *Calamagrostis* is derived from the Greek *kalamos* ("reed") and *agrostis* (a kind of grass).

Many species of *Calamagrostis* are morphologically similar, but they generally occur in distinct habitats and have unique geographical distributions. Given the subtle distinctions between many closely related taxa, there are several species complexes that could benefit from additional systematic study. Even the generic boundaries of the genus are controversial. For example, species in the genus *Deyeuxia*, distributed largely in the Southern Hemisphere, are morphologically very similar to species of *Calamagrostis*. It is probably appropriate to recognize all these species in a single genus, but that will require detailed scientific study of DNA of species from around the world.

BROOKGRASS (*Catabrosa aquatica*)

Description and Notes. This aquatic grass has a creeping base, soft leaves, and an open panicle of mostly 2-flowered spikelets. The glumes are unequal in size and shorter than the lower floret. Brookgrass grows around springs and along streams in mountain meadows. *Catabrosa* is Greek, meaning "eating," referring to the eroded glumes, and *aquatica* means "growing in or near water."

DROOPING WOODREED (*Cinna latifolia*)

Description and Notes. This tall plant has 1 or a few stems and loose, drooping panicles of small, 1-flowered spikelets. The florets are not hairy at the base, and the lemma has a very short awn. The glumes are about equal and fall off with the florets. Look for the species in moist woods.

The name *Cinna* was take from the Greek *kinni*, which referred to an unknown grass. The species name is derived from the Latin *latus*, "broad," and *folium*, "leaf," in reference to the wide leaves of this grass. Drooping woodreed increases tremendously on disturbed sites. Fresh woodreed leaves can be burned slowly to produce a mosquito-repelling smoke.

ORCHARD GRASS (*Dactylis glomerata*)

Description and Notes. The stems are 2–4 feet tall and often in large clumps. The few-flowered, flattened spikelets are nearly sessile in dense, 1-sided clusters near the ends of the few, stiff branches of an open panicle. Both glumes and lemmas are hairy on the back along the midribs. *D. glomerata* is usually treated as the sole species in the genus but is commonly divided into several regional subspecies; some botanists treat some of these as distinct species or at the lower rank of variety.

Dactylis is used as a hay grass and for pastures because of its high yields. It is also extensively naturalized in the United States and Australia; in some areas it has become an invasive species. In the United States it is commonly called orchard grass because it tolerates moderate shade.

Oatgrass (genus *Danthonia*)

Description and Notes. These grasses are rather low plants of hillsides and wooded areas. They have medium to large, several-flowered spikelets in short, mostly narrow panicles. The glumes are broad, about equal in size, sharp-pointed, and about as long or longer than the uppermost floret. The lemmas are 2-lobed with a stout, twisted, spreading awn between the lobes.

Hairgrass (genus *Deschampsia*)

Description and Notes. Most of the species in this area are moderately tall grasses with pale or purplish, shining, 2-flowered spikelets in narrow or open panicles. The glumes are about equal and do not fall with the florets. The axes of the spikelets are conspicuously hairy and extend beyond the uppermost floret. The blunt, 2–4-toothed lemmas bear a slender awn from below the middle of the back. The genus is named for the eighteenth-century French naturalist Louis Deschamps.

SALTGRASS (*Distichlis spicata*)

Description and Notes. This grass grows only in alkaline soil. It has extensive, creeping, scaly stems from which it sends up erect stems that are usually 4–16 inches tall and may bear about a dozen large spikelets. The flowers are imperfect, and the staminate and pistillate spikelets are on separate plants. The pistillate spikelets are usually 7–9-flowered, and the staminate are 8–15-flowered. The leaves are numerous, conspicuously in two rows, and generally less than 4 inches long. Saltgrasses have salt glands to excrete excess salts.

Distichlis spicata is the official Latin name for what is commonly referred to as saltgrass, spike grass, or alkali grass. The name has Greek origins, *Distichlis* being the Greek for "two-ranked" (referring to the forked arrangement of the plant's blades) and *spicata* referring to the spiked appearance of the plant's flowers.

Wild Rye (genus *Elymus*)

Description and Notes. The wild ryes are tall, stout, perennial grasses with the spikelets borne at each node of the zigzag axis of a spike. This arrangement looks very much like that of wheat except that the spikelets are usually in pairs at each node instead of one as in wheat. The spikelets are 2–6-flowered. The glumes are nearly side by side in front of each spikelet.

The genus *Elymus* probably consists of around 200 species. However, estimating its size is difficult because its greatest diversity is in Asia, many parts of which are poorly known botanically. An additional complication arises from the differing species interpretations that have been adopted. North American species are thought to have low barriers to interspecific hybridization. The genera *Agropyron* and *Elymus* have recently been combined because some of their members share characteristics such as spikelets per rachis node.

Fescue (genus *Festuca*)

Description and Notes. *Festuca* is a large genus of about 100 species. The fescues are low or medium-sized grasses with medium-sized, several-flowered, often purple spikelets borne in narrow or open panicles. The glumes are unequal, the first sometimes very small. The lemmas are rounded on the back and usually awned from the tip.

Mannagrass (genus *Glyceria*)

Description and Notes. The mannagrasses are tall water or marsh plants with creeping or rooting bases or with creeping underground stems. They have medium-sized or small, few- to many-flowered spikelets in open or contracted panicles. The very small glumes are more or less unequal, usually chaff-like, and 1-nerved. The lemmas are broad, chaff-like at the tip, and with 5–9 prominent nerves.

These are all palatable grasses, but because they usually occur in marshes and wet places, they are not useful as forage. This genus is distinctive in having glumes and lemmas with parallel veins. *Glyceria* is from the Greek *glukeros* ("sweet"), referring to the taste of the seed.

SWEETGRASS (*Hierochloe odorata*)

Description and Notes. Sweetgrass is a slender, erect, sweet-smelling perennial grass with small, open panicles of broad, bronze-colored spikelets about ¼ inch long. Each spikelet has 1 terminal, perfect floret and 2 florets with stamens only. The glumes are equal in length, and the lemmas are about as long as the glumes. Sweetgrass can be found in moist or wet meadows from lower elevations to the lower alpine zone.

Sweetgrass was used by Native Americans as an incense in religious ceremonies. The plant contains a glycoside that when dried produces coumarin, a sweet-smelling crystalline compound important in perfumes. The leaves can also be added to vodka as flavoring. Sweetgrass is in demand for sachet making and basket weaving because it holds its fragrance for years. The oil from the plant was used to flavor candy, soft drinks, and tobacco. The stems were soaked in water and used to treat chapping and windburn and as a tea for coughs and sore throats (Willard 1992; Foster and Duke 1990).

Barley (genus *Hordeum*)

Description and Notes. The barleys differ from the wheats and ryes in having 3 spikelets at each node of the zigzag axis, which readily breaks up, the joints falling with the spikelets attached. The 2 lateral spikelets are imperfect and in some species are reduced to mere bristles. The central spikelet is usually 1-flowered. The glumes are narrow and stand side by side in front of the spikelet. The lemmas are awned from the tip.

Hordeum is a genus of about 20 species of annual and perennial grasses, native throughout the temperate Northern Hemisphere and also temperate South America. One species, *H. vulgare* (barley), is of major commercial importance as a cereal grain. The cultivated barley is a native of southeastern Asia and was used by the Romans to make the heavy bread eaten by soldiers and peasants. Since this species contains little gluten, it cannot be made into light bread. Several of the other species have been used to a small extent, notably *H. distichon*. Some are also nuisance weeds in cereal crops because, being grasses like the crop, they cannot be selectively removed; any herbicide that would kill them would also damage the crop.

JUNEGRASS (*Koeleria macrantha*)

Description and Notes. Junegrass is a slender, medium-sized perennial with erect, dense, pale, spike-like panicles of small, 2–4-flowered spikelets without awns. It is frequently encountered in sagebrush habitats. The seed can be cooked and then ground into a powder and eaten like porridge or used as a flour for making bread. Medicinally, it was used as a styptic in the treatment of cuts.

Oniongrass (genus *Melica*)

Description and Notes. Oniongrasses are rather tall plants with the base of the stem swollen into a bulb-like corm. The spikelets are relatively large, purple-tinged, and borne in mostly simple panicles. The spikelets are 2- to several-flowered, and the axis is prolonged beyond the perfect florets and bears 2 or 3 progressively smaller, empty lemmas, each one enclosing the one above it. The glumes are unequal.

Muhly (genus *Muhlenbergia*)

Description and Notes. *Muhlenbergia* is a large and variable genus. All species have very small, 1-flowered spikelets with the glumes usually shorter than the lemma, the first sometimes minute. The lemma is membranous but firm and 3-nerved. It is usually awned from the tip or awnless. Muhly is found in wet areas from the middle to high elevations.

All species were widely used in basket making. The stalk of the plant was used as the horizontal or fundamental element around which the coils were wrapped. Native Americans also made brooms from some species; others provide shelter for wildlife.

Ricegrass (genus *Oryzopsis*)

Description and Notes. These mostly slender perennial grasses have 1-flowered spikelets in terminal panicles. The glumes and the lemma are all about the same length, with the lemma being awned and more or less hairy. Ricegrass is found in dry rocky or sandy ground, grasslands, and valleys from low to middle elevations.

The relatively large grains of this species have been used by Native Americans as food for centuries. The grains were collected with the stems and were held over a fire to singe off the fine white hairs. They can also be collected in a pan or basket with hot coals or rocks and shaken to burn off the hairs. The grains were then ground into flour and used as mush, made into cakes, or used to thicken soup.

Cereals are an excellent source of protein. However, the proteins they contain are incomplete, as they do not have all the essential amino acids. When cereals (lacking lysine but rich in methionine) and legumes (lacking methionine but rich in lysine) are eaten together, the amino acids found in cereals complement those in legumes, providing all the necessary proteins.

PANIC GRASS (*Panicum capillare*)

Description and Notes. *Panicum* is the largest genus of American grasses in the United States. These grasses have small, 1-flowered spikelets that are borne in open or rather loose panicles. The glumes are very unequal, the first often minute, the second about as long as the sterile lemma. The second glume and the sterile lemma are usually green, thin, and soft, the lemma sometimes subtending a staminate flower. The fertile lemma is hard and awnless, the margins inrolled and enclosing a palea of the same texture. *Panicum* is the Latin name for millet.

Canary Grass (genus *Phalaris*)

Description and Notes. These annual or perennial grasses have numerous flat leaves and narrow, often spike-like panicles. The spikelets are laterally flattened, with 1 perfect terminal floret and 2 very small sterile lemmas below, these remaining attached to the floret. The glumes are about equal and boat-shaped. The sterile lemma is leathery and shorter than the glumes.

Timothy (genus *Phleum*)

Description and Notes. Timothys are perennial grasses with very dense, bristly, spike-like panicles. The spikelets are 1-flowered, and the glumes are equal, awned, and ciliate on the keel. The lemma are shorter than the glumes and awnless. Common timothy (*P. pratense*) is cultivated in many parts of the United States.

COMMON REED (*Phragmites australis*)

Description and Notes. Common reed is native to every continent except Antarctica. It often forms dense thickets along streams, ditches, and marshes at low elevations. The record of human uses for this species is extensive. The reeds have been used for roof insulation since ancient times. Because they are light, they are best known for their use in arrow shafts. The roots can be eaten raw or cooked.

Meal can be obtained from the pulverized stems. In the fall the leaves and stems may become encrusted with grayish exudate. This exudate, actually honeydew (excreta of whitefly and aphids), was obtained from stalks. Stalks were cut and flayed to remove honeydew crystals, which were winnowed and cooked into a stiff dough. The dough was formed into cakes, sun-dried, and stored. The split culms provided fiber. Common reed was used to make flutes and other musical instruments in addition to carrying nets and cordage. The honeydew was given to pneumonia patients to loosen phlegm and soothe pain in the lungs.

Bluegrass (genus *Poa*)

Description and Notes. This large genus contains important forage and lawn grasses. The small or medium-sized spikelets are 2- to several-flowered and are borne in open or contracted panicles. The glumes are unequal, the first usually 1-nerved and the second 3-nerved. The glumes do not fall with the florets. The lemmas are awnless and usually 5-nerved.

Alkali Grass (genus *Puccinella*)

Description and Notes. This rather low, pale, smooth grass has several-flowered spikelets occurring in panicles. The glumes are unequal, the first 1-nerved or sometimes 3-nerved and the second 3-nerved. The lemmas are rounded on the back and 5-nerved.

BRISTLEGRASS (*Setaria viridis*)

Description and Notes. This exotic grass seems to occur all over the United States. It is characterized by greenish spikelets borne in a densely flowered, spike-like panicle with 1–3 bristles below each spikelet. These bristles remain on the stem after the spikelets fall.

Squirrel-tail (genus *Sitanion*)

Description and Notes. These low, tufted perennial grasses have a very bristly spike that is usually 1–3 inches long and breaks up at maturity with the joints attached to the spikelets. The spikelets are 2-flowered, sometimes more, and there are usually 2 at each joint of the axis. The glumes are narrow with 1 or 2 prominent nerves that extend into rough awns. The firm lemmas, which are nearly cylindrical, are also awned. The awns of both glumes and lemmas are wide-spreading and vary from 1 to 4 inches long, making a very bristly spike.

Needlegrass (genus *Stipa*)

Description and Notes. These tufted perennial grasses have 1-flowered spikelets in usually narrow panicles. The floret, when separated from the glumes, is pointed and bearded at the base, and the lemma terminates in a conspicuous awn that is twisted and bent, often twice, above.

Stipa is a genus of around 300 large, perennial, hermaphroditic grasses collectively known as needlegrass. One species, *S. tenacissima* (esparto grass), is used extensively in paper making. Species such as *S. brachytricha, S. arundinacea, S. splendens, S. calamagrostis, S. gigantea,* and *S. pulchra* are used as ornamental plants. Many species are also important livestock forage.

Trisetum (genus *Trisetum*)

Description and Notes. The trisetums are tufted perennial grasses with flat leaf blades and narrow, almost spike-like panicles. The spikelets are mostly 2-flowered, the glumes somewhat unequal, the second usually longer than the first floret. The lemmas are short-bearded at the base and 2-toothed at the apex.

PONDWEED FAMILY (POTAMOGETONACEAE)

Of the two genera and 100 species worldwide, *Potamogeton* is native to the United States. The family is of no direct economic importance to humans but provides a valuable source of food for ducks and other wildlife.

Pondweed (genus *Potamogeton*)

Description and Habitat. Pondweeds are perennial aquatic plants with extensive, slender rhizomes and simple, branched stems that often root at the nodes. The leaves have stipules that clasp the stem. The lower leaves are alternate and submerged, and the upper, often floating, leaves are wider and opposite. The small, greenish flowers are clustered in a spike that arises from the upper leaf axils. Most pondweed species can be identified without flowers and fruits, but the floating or submersed leaves may be necessary. Hybridization among the many species in this genus is fairly common.

Natural History Notes and Uses. The flowers of pondweeds are inconspicuous and often not seen. The tiny spikes extend above the water surface in quiet weather, but the spikes and fruits are drawn below the surface at maturity. Probably all pondweeds have starchy, edible rhizomes, but species with larger rootstalks are preferred for gathering. *P. diversifolius* (waterthread pondweed) was an important source of strong fibers, which were rolled into cordage to make carrying nets, rabbit-trap nets, and other items.

DITCH-GRASS FAMILY (RUPPIACEAE)

The ditch-grass family consists of a single genus (*Ruppia*) of submerged aquatic herbs. Ruppiaceae probably represents a family that has evolved from the Potamogetonaceae. Pollination takes place underwater in some species, whereas in others the pollen floats on the water surface and contacts floating stigmas.

WIDGEON-GRASS (*Ruppia maritima* = *R. cirrhosa*)

Description and Habitat. Widgeon-grasses are bushy, fan-like, underwater plants (occasionally the flowers may extend above the water) with slender, grass-like leaves attached to sheathing bases. They are freshwater species with a high tolerance for salinity and alkalinity. They can be perennials or annuals and are highly variable in form depending on environmental conditions. Some botanists believe that *R. maritima* and *R. cirrhosa* are a singe variable species with most differences in appearance related to habitat.

Natural History Notes and Uses. As with many aquatic plants, the flowers are reduced and the plants are very plastic in form so there is little agreement as to how many species exist. It appears that there are about seven species, but some authorities recognize only one or perhaps two polymorphic species.

The plant provides cover and food for many aquatic species. All the plant parts are eaten by waterfowl (over 5,000 seeds were found in one duck). It is often used for habitat rehabilitation.

BUR-REED FAMILY (SPARGANIACEAE)

One genus with about 20 species occurs in this family. Bur-reeds are found primarily in the cooler regions of the north temperate zone, Australia, and New Zealand. Several species are native to the United States. The plants arise from creeping rootstalks, and the roots are fibrous. The species have no direct economic importance to humans. This family is considered by some authorities to be part of the cattail family (Typhaceae).

Bur-reed (genus *Sparganium*)

NARROW-LEAVED BUR-REED (*S. angustifolium*)

BUR-REED (*S. minimum* = *S. natans*)

Description and Habitat. Bur-reeds are aquatic perennials with erect or floating, unbranched stems. The leaves are linear and sheath the stem. The flowers are borne in dense, round clusters. Bur-reeds can be found in shallow waters of marshes, ponds, and slow-moving steams.

FIELD KEY TO THE BUR-REEDS

1. Male heads 2–5 .... *S. angustifolium*
1. Male heads 1, sometimes 2............................... *S. minimum*

Natural History Notes and Uses. The bulbous bases of the stems and tubers of *S. eurycarpum* (broadfruit bur-reed), *S. simplex* (simplestem bur-reed), and *S. angustifolium* (narrow-leaved bur-reed) can be used as food in much the same way as cattails (*Typha* spp.) and bulrushes (*Scirpus spp.*): dried and pounded into flour. All species provide excellent cover and food for muskrats and waterfowl, as well as for deer.

CATTAIL FAMILY (TYPHACEAE)

The cattail family has one genus (*Typha*) with about 15 species worldwide. Members are marsh or aquatic perennials with creeping rootstalks. Two kinds of flowers are borne in crowded spikes, with the staminate (male) flowers above and the pistillate (female) flowers below. The family is of little economic importance. Typhaceae now incorporates the previous family, Sparganiaceae.

Cattail (genus *Typha*)

NARROW-LEAF CATTAIL (*T. angustifolia*)

BROAD-LEAF CATTAIL (*T. latifolia*)

Description and Habitat. These two cattail species can be found over much of North America. A cattail plant produces a basal cluster of narrow, ribbon-like leaves that are several feet long and stand almost vertically. The upright stem is unbranched, not quite as long as the leaves, and it bears a long, dense, brown spike at the upper end. The spike may vary from 4 to more than 12 inches long. Its upper part bears stamens intermixed with long hairs, each stamen constituting a flower, while its lower part bears pistillate flowers, each flower consisting only of an ovary with an abundance of dark hairs at its base. *Typha* is Greek for "cattail." It is often referred to as "Cossack asparagus." Cattails reproduce rapidly in marshy areas that are usually unsuited for agriculture.

FIGURE 113. Cattail (*Typha latifolia*).

FIELD KEY TO THE CATTAILS

1. Leaves mostly over ½ inch wide	*T. latifolia*
1. Leaves mostly less than ½ inch wide	*T. angustifolia*

Natural History Notes and Uses. In his book *Mountain Man*, Vardis Fisher (1965:76) illustrates how Indians used cattails: "It was a marvel what they did with the common cattail—from the spikes to the root, they ate most of it. The spikes they boiled in salt water, if they had salt; of the pollen they made flour; of the stalk's core they made kind of a pudding; and the bulb sprouts on the ends of the roots they peeled and simmered."

Virtually every part of the plant has a use, from food to fiber. Euell Gibbons considered the cattail the "supermarket of the swamps." Although both cattail species have edible rhizomes, they should never be eaten raw since they may cause vomiting (Jencks 1919). The rhizomes should be boiled or roasted, or dried and then ground into meal or flour. Another way to obtain the starch from the rhizomes is to follow a technique described by Gibbons (1962:50):

> After scrubbing the root [i.e., rhizome] and peeling off the spongy outer layer surrounding the white stiff core, cut the core into small sections, and place the pieces in a bowl of cold water. Work the core with your hands, separating the fibers and scraping out the starch. Slosh the fiber around in the

water until you have removed all the starch. Pour off the water through a coarse sieve to extract the fibers. Allow the water to settle for a little while the starch settles to the bottom of the container. Then carefully pour off the water, leaving the starch in the bowl. For a cleaner starch, pour in some more water and let it settle again. Then pour off the water. After this, you can use the starch almost immediately to make pancakes, breads, and biscuits.

Harrington (1967) reports that one acre of cattails can yield over three tons of nutritious flour, although extraction techniques need to be refined for commercial exploitation.

When pulling up the rhizome, you may notice newly emerging buds. These can be scrubbed, peeled, and eaten raw or boiled. The swollen joint between bud and rhizome is also starchy. Peel it, then roast or boil it for a potato-like vegetable. Like the rhizomes, this part should not be eaten raw. The young green shoots can be peeled of their green outer layer and eaten raw or cooked. It is always good to boil them in a couple of changes of water if there is any bitterness. The peeled core can also be sliced and added to salads.

While the flower spikes are still green, remove the papery sheath and boil the cluster for a few minutes. The flower spikes can then be eaten like corn on the cob, although the core of the cluster is inedible. Cattail pollen is high in protein and can be used in flour for breads or eaten raw. However, if you are allergic to pollen, it should be avoided. The seeds from the female portion of the flower spike can be pulverized to make a nutritious, protein-rich flour. The seeds can be extracted from the fluff by parching them.

Useful fibers can be derived from cattails. Fibers in the stems can be loosened by soaking the plant material in water for several days. The silky fluff on the seeds is buoyant and water repellent and makes a good insulator, especially in boots. The silk can be used for stuffing items from pillows to down vests. It can also be used for tinder. The fuzz will explode into flame with a spark from a flint-and-steel set. The leaves can be woven to make mats, sandals, baskets, and the like. The stems provide a good coil foundation for baskets. Additionally, the stalks have been used as arrows and hand drills. A toothbrush can be fashioned from the fuzzy stem with the flowers removed.

Medicinally, the chopped or pounded rhizome was applied to the skin for minor wounds and burns. Cattail down was used as dressings for wounds. Brown (1985) indicates that a sticky juice derived from between the young leaves can be used as a styptic, antiseptic, and anesthetic. The jelly from between the young leaves can be applied to wounds, sores, external inflammations, and boils to soothe pain. Brown (1985) also notes that the jelly was rubbed on the gums as a novocaine substitute for tooth extraction.

CATTAIL JELLY

Jelly from cattails? The following is a recipe from Jan Phillips, author of *Wild Edibles of Missouri*, who says she makes cattail jelly after the first starchy material has been rubbed out from the rhizome. The jelly is made by boiling the roots (rhizomes) for 10 minutes in enough water to cover them. For every cup of liquid, add an equal amount of sugar, and add a package of pectin for every four cups of juice. The jelly somewhat resembles honey in both taste and color.

HORNED PONDWEED FAMILY (ZANNICHELLIACEAE)

These aquatic herbs have opposite leaves that are entire.

HORNED PONDWEED (*Zannichellia palustris*)

Description and Habitat. Horned pondweed is native and widespread in the United States and has a near cosmopolitan distribution. It is a delicate, underwater, branching perennial that can grow to a length of about 3 feet. It has opposite, threadlike leaves that emerge in such a way as to give the plants a uniform shape. Unlike many look-alike aquatic plants whose flowers emerge from the water on spikes, horned pondweed has inconspicuous underwater flowers and fruits located at the leaf bases. Look for this plant in brackish or alkaline streams, ponds, ditches, and lakes.

Natural History Notes and Uses. The flowers are small, lack sepals and petals, and are solitary or clustered at the leaf bases. Male and female flowers are separate but grow on different parts of the same plant, although often both occur together in the leaf bases. The female flowers are surrounded by a sheathing bract. Because the flowers remain entirely underwater, pollination occurs in the water.

Horned pondweed is not reported to cause major problems in the United States. However, it sometimes grows with other submersed species such as naiads and narrow-leaved pondweeds, clogging waterways. Horned pondweed is an important food for waterfowl in the early growing season (Brooks and Hauser 1978).

Both the seeds and vegetative parts of this native species are consumed by waterfowl (lesser scaup and shovelers). A submerged water plant, horned pondweed is a useful oxygenator for cool-water ponds.

APPENDIX 1

Field Key to Plant Families of Yellowstone and Grand Teton National Parks

As in the text, the dichotomous key presented here consists of a series of contrasting statements called leads. Read both leads of each pair, then select the one that best fits the specimen. The lead either will end at the name of a family or other group or will direct you to the next pair of leads. Continue this procedure until you obtain the name of the plant or group of plants. It is necessary to have flowering material to use this key. In some cases fruit may also be required. When possible, examine numerous specimens.

---

1. Plants reproducing by spores (ferns, fern allies, clubmosses, horsetails) 2
1. Plants reproducing by seeds (conifers and flowering plants) 8

2. Stems jointed and grooved lengthwise, often easily pulled apart at the nodes; leaves reduced to papery scales, plants apparently consisting only of simple or branched stems .... Equisetaceae
2. Stems not as above; leaves present though sometimes small 3

3. Leaves like a 4-leaf clover .... Marsileaceae
3. Leaves not as above 4

4. Plants consisting of a cluster of long narrow leaves united at the base; spores borne in a pocket at the base of each leaf; plants growing in water or mud .... Isoetaceae
4. Plants not as above 5

5. Leaves small and scale-like; stems often resembling small juniper or cedar branches; spores borne at the top in cone-like structures 6
5. Leaves larger and usually lobed or divided (fern-like); spores not borne in cones 7

6. Plants less than 1 inch tall; generally occurring on rocks or in rocky or dry soil .... Selaginellaceae
6. Plants usually more than 1 inch tall; occurring mostly in moist soil of forests or occasionally meadows .... Lycopodiaceae

7. Spores borne on a specialized branch that rises from the base of the vegetative leaf .... Ophioglossaceae
7. Spores borne on leaves that usually are similar to the vegetative leaves, and both kinds of leaves arise directly from the roots ..... Polypodiaceae

8. Spore clusters (sori) borne along the leaf or leaflet margins, elongate and usually partly covered by the inrolled leaflet edge 9
8. Sori borne on veins between midrib and margin of leaf or leaflet, often round in outline .... Aspleniaceae

9. Rhizomes hairy, not scaly; leaves longer than 12 inches, hairy beneath .... Dennstaedtiaceae
9. Rhizomes scaly; leaves often shorter than 12 inches, often not hairy beneath .... Pteridaceae

10. Plants with flowers (sometimes inconspicuous); woody or, more often, herbaceous plants, usually with broad leaves that are mostly not evergreen .... Angiosperms (Flowering Plants)
10. Plants lacking flowers; trees or shrubs with needle- or scale-like, mostly evergreen leaves 11

11. Leaves scale-like and pressed flat to the stem or needle-like and whorled around the branches............. Cupressaceae
11. Leaves needle-like, often borne in clusters.................... Pinaceae

ANGIOSPERMS: THE FLOWERING PLANTS

1. Plants truly aquatic with submerged or floating leaves that become limp when withdrawn from the water (emergent plants with self-supporting stems are not included)............. Group I
1. Plants emergent or nonaquatic..................................... 2

2. Plants herbaceous with undivided leaves that are often narrow and have the main vein parallel to each other; petals and sepals in 3's or multiples thereof .... (Monocots) Group II
2. Plants herbaceous or woody with simple or divided, usually net-veined leaves; petals and sepals usually in 4's or 5's or multiples thereof .... (Dicots) 3

3. Plants trees, shrubs, or woody vines; a large portion of the aboveground stems woody........................... Group III
3. Plants herbaceous or woody only at the base......................... 4

4. Flowers without two distinct whorls of parts differentiated into petals and sepals (i.e., petals or sepals or both lacking, or petals and sepals in a single undifferentiated series) .... Group IV
4. Both petals and sepals present and differentiated from each other...... 5

5. Petals separate from each other all the way to the base .... Group V
5. Petals united, at least toward the base....................... Group VI

— *Group I* —

(Plants truly aquatic with submerged or floating leaves that become limp when withdrawn from the water)

1. Stems lacking; entire plant floating and smaller (usually much smaller than ¾ inch in diameter) .... Lemnaceae
1. Plants larger and usually with obvious stems......................... 2

2. Leaves large (longer than 4 inches) and round to broadly elliptic in outline, borne on the ends of long petioles and floating flat on the surface of the water (water lilies) .... Nymphaeaceae
2. Leaves smaller, narrower, or not always floating 3

3. Plants without leafy stems; all leaves grass-like and attached at the base... 4
3. Plants with leafy stems; leaf shape variable 7

4. Leaves with a sac of white spores at the base .... Isoetaceae
4. Leaves without spore sacs at the base 5

5. Leaves with a well-differentiated blade, broader than the long petiole......................... *Limosella* in Scrophulariaceae
5. Leaves grass-like, without a well-differentiated blade 6

6. Leaves needle-like and round in cross section. . .*Eleocharis* in Cyperaceae
6. Leaves flattened, not needle-like . . .vegetative forms in Alismataceae and/or Sparganiaceae

7. Underwater leaves highly dissected into hair-like segments. 8
7. Underwater leaves linear to elliptic but not dissected 11

8. Leaves with small, egg-shaped bladders among the leaflets . . . . Lentibulariaceae
8. Leaves without bladders. 9

9. Leaves 1 per node; flowers conspicuous, with yellow or white petals . . . . *Ranunculus* in Ranunculaceae
9. Leaves whorled (more than 2 per node); flowers inconspicuous, without petals. 10

10. Ultimate leaflets with toothed margins; stems usually green . . . . Ceratophyllaceae
10. Margins of ultimate leaflets smooth; stems often reddish . . . . Haloragidaceae

11. Leaves whorled (more than 2 per node) 12
11. Leaves alternate or opposite (1 or 2 per node) 14

12. Leaves less than ⅛ inch wide . . . . Najadaceae
12. Leaves more than ⅛ inch wide 13

13. Leaves mostly 6 per node. . . . . Hippuridaceae
13. Leaves mostly 3 or 4 per node. . . . . Hydrocharitaceae

14. Submersed leaves opposite (2 per node) 15
14. Submersed leaves alternate (1 per node) 16

15. Leaves at least ¾ inch long. . . . . Zannichelliaceae
15. Leaves less than ¾ inch long . . . . Callitrichaceae

16. Submersed leaves narrowly elliptic to nearly round in outline (mostly under 5 times as long as wide 17
16. Submersed leaves linear and grass-like (mostly more than 10 times as long as wide). 19

17. Leaves with 4 leaflets, resembling a 4-leaf clover . . . . Marsileaceae
17. Leaves not divided into 4 leaflets 18

18. Leaf veins parallel to each other or nearly so . . . . Potamogetonaceae
18. Leaf veins branching off the midrib . . . . *Polygonum* in Polygonaceae

19. Base of the leaf expanded into a pale membranous appendage (stipule) that surrounds the stem . . . . Potamogetonaceae
19. Base of the leaf without stipules 20

20. Leaves on or near the surface shaped like an arrowhead . . . . Alismataceae
20. Surface leaves linear or grass-like. 21

21. Base of leaves Y-shaped in cross section; flowers and fruits borne in round cluster . . . . Sparganiaceae
21. Leaves nearly flat at the base with a pale, membranous appendage on the inner surface (ligule) where it joins the stem; flowers and fruits not in round clusters . . . . Poaceae

— *Group II: Monocots* —

(Plants herbaceous with undivided leaves that are often narrow and have the main vein parallel to each other; petals and sepals in 3's or multiples thereof)

1. Plants leafless and parasitic on stems and branches of trees . . . . Loranthaceae
1. Plants not parasitic on trees 2

2. Flowers unisexual, borne in dense globose or cylindrical clusters, male above female 3
2. Flowers not as above 4

3. Flowers borne in a cylindrical spike (cattails) . . . . Typhaceae
3. Flowers borne in globose clusters . . . . Sparganiaceae

4. Stamens and ovaries enclosed by bracts; petals and sepals lacking or reduced to bristles or scales (grasses and sedges) 5
4. Flowers with sepals and petals or undifferentiated tepals 6

5. Stems mostly solid and often 3-sided; each flower subtended by 1 (rarely 2) bract(s); leaf sheaths (grasses) . . . . Poaceae
5. Stems hollow and usually round in cross section with conspicuous swollen nodes; each flower subtended by 2 bracts . . . . Cyperaceae

6. Flowers bilaterally symmetrical, at least 1 petal different than the other 2 (orchids) . . . . Orchidaceae
6. Flowers radially symmetrical, all 3 petals or 6 tepals similar in size and shape 7

7. Leaves pinnately divided into leaflets . . . . Limnanthaceae (a dicot with flower parts in 3's)
7. Leaves sometimes lobed at the base but not divided into leaflets 8

8. Leaves more than 3 per node . . Rubiaceae (a dicot with flower parts in 3's)
8. Leaves not more than 3 per node 9

9. Base of petals and sepals (or tepals) attached to the top of the (inferior) ovary . . . . Iridaceae
9. Base of petals and sepals arising from base of the (superior) ovary 10

10. Sessile flowers borne on an unbranched spike at the top of a leafless stem . . . . Juncaginaceae
10. Inflorescence not as above 11

11. Fruit a many-seeded capsule 12
11. Fruit a berry, a globose cluster of 1-seeded achenes, or 1- or 2-seeded capsules (follicles) 13

12. Flower of 6 small undifferentiated petals and sepals (tepals) that are brown or green . . . . Juncaceae
12. Petals or tepals white, yellow, blue, or orange, or if brown or green then more than ¼ inch long . . . . Liliaceae

13. Fruit a berry . . . . Liliaceae
13. Fruit a globose cluster of achenes . . . . Alismataceae

— *Group III: Woody Dicots* —

(Plants trees, shrubs, or woody vines; a large portion of the aboveground stems woody)

1. Plants twining or climbing, woody vines 2
1. Plants trees or shrubs, erect or prostrate but not climbing or twining 6

2. Leaves alternate (1 per node) 3
2. Leaves opposite or whorled (more than 1 per node) 4

3. Small, curling stems or modified leaves that grasp the supporting plant; tendrils present . . . . Vitaceae
3. Tendrils absent . . . . *Solanum* in Solanaceae

4. Leaves divided into leaflets . . . . *Clematis* in Ranunculaceae
4. Leaves sometimes lobed but not divided into leaflets 5

5. Leaves lobed or toothed with long petioles .... Moraceae
5. Leaves with entire margins; petioles lacking
or very short .... *Lonicera* in Caprifoliaceae

6. Leaves and branches opposite or whorls (more than 1 per node)........ 7
6. Leaves and branches alternate (1 per node) or clustered at the base 26

7. Leaves divided into leaflets 8
7. Leaves sometimes lobed or toothed but not divided into leaflets 11

8. Fruit a berry; inflorescence flat-topped or
pyramid-shaped with numerous flowers..... *Sambucus* in Caprifoliaceae
8. Fruit dry and 1-seeded, with or without a wing;
inflorescence not as above 9

9. Plants shrubs usually fewer than
12 inches tall .... *Clematis* in Ranunculaceae
9. Plants trees more than 3 feet tall.................................. 10

10. Lower leaflets with at least 1 shallow lobe as well
as toothed margins .... Aceraceae
10. Lower leaflets finely toothed .... Oleaceae

11. At least some leaves with 3–5 lobes 12
11. Leaves with toothed or entire margins but not lobed................. 13

12. Leaves indented at point where petiole is attached;
fruit dry, 1-seeded, and winged............................. Aceraceae
12. Leaves not indented at the base;
fruit berry-like .... *Viburnum* in Caprifoliaceae

13. Leaves scale-like and 4 per node.................. *Cassiope* in Ericaceae
13. Leaves not as above... 14

14. Leaves covered with brownish or silvery,
mealy scales on 1 or both surfaces .... *Shepherdia* in Elaeagnaceae
14. Leaves not mealy 15

15. Leaves with pointed teeth on the margins 16
15. Leaves with entire or wavy margins, lacking sharp-pointed teeth...... 22

16. Flowers tubular and about 1 inch long.... *Penstemon* in Scrophulariaceae
16. Flowers not as above.. 17

17. Leaves less than 1 inch long....................................... 18
17. Leaves mostly more than 1½ inches long 19

18. Plants creeping; stems hairy .... *Linnaea* in Caprifoliaceae
18. Plants upright or spreading but not creeping;
stems glabrous or nearly so .... Celastraceae

19. Inflorescence with fewer than 12 flowers 20
19. Inflorescence with more than 12 flowers............................ 21

20. Petals ¼ to ¾ inch long; fruit a capsule .... Hydrangeaceae
20. Petals minute or lacking; fruit berry-like ...... *Rhamnus* in Rhamnaceae

21. Leaves with 3 main veins starting at
or near the base .... *Ceanothus* in Rhamnaceae
21. Leaves with only 1 main vein arising
at the base .... *Viburnum* in Caprifoliaceae

22. Flowers tubular and more than
¾ inch long .... *Penstemon* in Scrophulariaceae
22. Flowers not as above.. 23

23. Year-old twigs red . . . . Cornaceae
23. Twigs usually not red 24

24. Flowers with united petals, tubular to urn-shaped . . . . Caprifoliaceae
24. Petals separate 25

25. Plants less than 1 foot tall and found in moist to wet areas;
leaves leathery and less than 1 inch long . . . . *Kalmia* in Ericaceae
25. Plants more than 1 foot tall, occurring in dry to moist areas;
leaves more than 1 inch long . . . . Hydrangeaceae

26. Leaves divided into leaflets 27
26. Leaves sometimes lobed but not divided into leaflets 35

27. Leaves divided into linear segments 28
27. Leaflets lance-shaped or wider 29

28. Leaflets spine-tipped . . . . *Leptodactylon* in Polemoniaceae
28. Leaflets not spine-tipped; foliage aromatic . . . . . . *Artemisia* in Asteraceae

29. Leaflets with spine-tipped teeth on the margins . . . . Berberidaceae
29. Leaflets without spine-tipped teeth on the margins 30

30. Stems with spines or prickles 31
30. Stems lacking spines or prickles 32

31. Leaflets toothed or lobed . . . . Rosaceae
31. Leaflets with entire margins . . . . Fabaceae

32. Leaflets mostly 3–7 33
32. Leaflets mostly 9 or more 34

33. Leaflets mostly 3 . . . . Anacardiaceae
33. Leaflets mostly 5–7 . . . . Rosaceae

34. Flowers with 10 or fewer stamens;
fruit densely reddish-hairy . . . . Anacardiaceae
34. Flowers usually with more than 10 stamens;
fruit not reddish-hairy . . . . Rosaceae

35. Plants with thorns, prickles, or spine-tipped branches 36
35. Plants unarmed 42

36. Leaves with entire margins 37
36. Leaves with toothed or lobed margins 39

37. Leaves silvery with hairs or scales
on the lower surface . . . . *Elaeagnus* in Elaeagnaceae
37. Leaves green on both surfaces 38

38. Petals separate; stems angled
and grooved . . . . *Glossopetalon* in Celastraceae
38. Petals united; stems more or less round
in cross section . . . . *Lycium* in Solanaceae

39. Leaves with 3–9 palmate (like a maple) lobes 40
39. Leaves mostly not lobed 41

40. Blades of many leaves more than 4 inches long . . . . Araliaceae
40. Leaf blades mostly less than 3 inches long . . . . Grossulariaceae

41. Spines mostly branched, teeth of leaves tipped
with a slender bristle . . . . Berberidaceae
41. Spines not branched; teeth of leaves not bristle-tipped . . . . Rosaceae

42. Leaves or their principal lobes less than 1 inch wide 43
42. Leaves or their principal lobes more than 1 inch wide 49

43. Leaves evergreen with 1 or 2 grooves beneath, resembling needles of a fir tree .... *Phyllodoce* in Ericaceae
43. Leaves not as above 44

44. Leaves succulent, nearly round in cross section .... *Suaeda* in Chenopodiaceae
44. Leaves not succulent, mostly flat 45

45. Plants low and cushion-forming with grayish-hairy leaves .... *Eriogonum* in Polygonaceae
45. Plants not as above 46

46. Few to many flowers clustered in heads, each head surrounded by bracts forming a cup or vase shape, each cluster appearing like a single flower .... Asteraceae
46. Flowers not as above 47

47. Male and female flowers borne in catkins on separate plants; shrubs of wet areas .... *Salix* in Salicaceae
47. Flowers bisexual; plants generally of dry habitats 48

48. Plants prostrate or nearly so; flowers with united petals .. Polemoniaceae
48. Plants rarely prostrate; flowers with separate petals .... Rosaceae

49. Leaf margins entire or nearly so 50
49. Leaf margins toothed or lobed, sometimes only slightly so 58

50. Leaves pale beneath with yellow, resinous dots ...... *Ledum* in Ericaceae
50. Leaves not as above 51

51. Male and female flowers borne in dense, cylindrical catkins on separate plants; shrubs of wet or moist habitats ... *Salix* in Salicaceae
51. Flowers bisexual 52

52. Leaves all basal or with a whorl of leaves in the middle of the otherwise naked stem .... *Eriogonum* in Polygonaceae
52. Stem leaves present and alternate 53

53. Leaves covered with silvery scales, especially beneath .... *Elaeagnus* in Elaeagnaceae
53. Leaves not covered with silvery scales 54

54. Few to many flowers clustered in heads, each head surrounded by bracts forming a cup or vase shape, each cluster appearing like a single flower .... Asteraceae
54. Flowers not as above 55

55. Flowers with more than 10 stamens .... Rosaceae
55. Stamens 10 or fewer 56

56. Inflorescence of numerous (more than 20) flowers; leaves paler beneath than above .... *Ceanothus* in Rhamnaceae
56. Inflorescence of fewer than 20 flowers 57

57. Shrubs of disturbed areas or riparian habitats .... Solanaceae
57. Plants mostly in forested areas in the mountains .... Ericaceae

58. Many leaves with 3 lobes at the tip; crushed leaves with a sage odor .... *Artemisia* in Asteraceae
58. Leaves not as above 59

59. Plants alpine, mostly less than 4 inches tall 60
59. Plants not as above 61

60. Flowers bisexual with showy white petals .... Rosaceae
60. Flowers unisexual and borne on separate plants; petals lacking .... *Salix* in Salicaceae

61. Flowers without petals and borne in dense, cylindrical, spike-like inflorescences (catkins) 62
61. Flowers not borne in catkins 63

62. Catkins soft and somewhat fleshy, male and female catkins on separate plants . . . . Salicaceae
62. Female catkins more brittle and cone-like; male and female catkins borne on the same plant. . . . . Betulaceae

63. Trees. 64
63. Shrubs and subshrubs. 65

64. Flowers without petals; fruit a winged seed (samara). . . . . Ulmaceae
64. Flowers with petals; fruit fleshy and berry-like or apple-like . . . . Rosaceae

65. Leaves palmately lobed (like a maple) 66
65. Leaves toothed but not palmately lobed 67

66. Flowers with 5 stamens. . . . . Grossulariaceae
66. Flowers with more than 5 stamens. . . . . Rosaceae

67. Petals united at least at the base 68
67. Petals separate 69

68. Some leaves with 1 or 2 narrow lobes at the base . . . . *Solanum* in Solanaceae
68. Leaves without lobes. . . . . Ericaceae

69. Flowers with 5 stamens. . . . . Rhamnaceae
69. Flowers with 10 or more stamens. . . . . Rosaceae

— *Group IV: Herbaceous Dicots* —

(Flowers without two distinct whorls of parts differentiated into petals and sepals)

1. Few to many flowers clustered in heads, each head surrounded by bracts forming a cup or vase shape, each cluster appearing like a single flower 2
1. Flowers not as above. 3

2. Stamens 4; corolla 4-lobed. . . . . Dipsacaceae
2. Stamens 5; corolla mostly 5-lobed or with only 1 strap-shaped lobe . . . . Asteraceae

3. Plants parasitic on conifers; stems jointed; leaves reduced and scale-like. . . . . Loranthaceae
3. Plants not as above 4

4. Stems and leaves not green; leaves reduced to scales . . . . Ericaceae
4. Plants with green leaves or stems. 5

5. Plants with milky sap; flowers consisting of a cup with 5 conspicuous lobes, stamens, and a 3-lobed, stalked ovary . . . . Euphorbiaceae
5. Plants not as above 6

6. Middle and lower leaves opposite or whorled (2 or more leaves per node). 7
6. Middle and lower leaves alternate or all basal or nearly so 17

7. Plants with whorled leaves and sessile flowers in leaf axils; plants of wet to aquatic habitats . . . . Hippuridaceae
7. Plants not as above 8

8. Plants with sharp stinging hairs on the stems and leaves. . . . . . Urticaceae
8. Plants not as above 9

9. Middle stem leaves divided or lobed 10
9. Middle stem leaves with entire, toothed, or wavy margins 13

10. Flowers small; flower parts green; introduced species of disturbed areas 11
10. Flower parts colored or white; native species.......... 12

11. Leaves deeply, palmately, 3–7-lobed, plants usually twining.... Aceraceae
11. Leaves palmately divided into 5–9 leaflets; coarse annual herb.......... Cannabaceae

12. Pistils and stamens numerous.......... Ranunculaceae
12. Flowers with 1 pistil and 1–4 stamens .... Valerianaceae

13. Stamens and pistils more than 10 each .... Ranunculaceae
13. Stamens and pistils 10 or fewer each 14

14. Leaves whorled (more than 2 per node) .... Rubiaceae
14. Leaves opposite (2 per node).......... 15

15. Petals absent and sepals green .... Caryophyllaceae
15. Colored (not green) petals present or sepals petal-like and non-green 16

16. Plants perennial; flowers radially symmetrical; stamens mostly more than 4; stem leaves with short petioles ...... Nyctaginaceae
16. Plants annual; flowers bilaterally symmetrical; stamens 3; stem leaves sessile .... Valerianaceae

17. Plants vines with thin, curly tendrils for climbing......... Cucurbitaceae
17. Plants not as above 18

18. Flower 3-parted, reddish purple, and lying on the ground; leaves spade-shaped and wider than long or nearly so... Aristolochiaceae
18. Plants not as above 19

19. Leaves deeply lobed (more than halfway to midvein) or divided.......20
19. Leaves with entire, toothed, or wavy margins, not deeply lobed or divided.......... 26

20. Stamens more than 10 per flower.......... 21
20. Stamens 10 or fewer 22

21. Plants with colored or milky sap; flowers with 1 pistil ...... Papaveraceae
21. Plants with clear sap; flowers with numerous pistils; mostly native species.......... Ranunculaceae

22. Flowers green, lacking colored petals; foliage usually fleshy .... *Chenopodium* in Chenopodiaceae
22. Flowers with colored petals; foliage usually not fleshy.......... 23

23. Petals united, at least toward the base; flowers bilaterally symmetrical.......... Fumariaceae
23. Petals separate; flowers radially symmetrical.......... 24

24. Flowers and fruits borne in a dense cylindrical spike .... *Sibbaldia* in Rosaceae
24. Inflorescence not as above.......... 25

25. Petals 4, attached at the base of the (superior) ovary .... *Lepidium* in Brassicaceae
25. Petals 5, attached at the top of the (inferior) ovary .... Apiaceae

26. Stamens more than 10 in each flower .... Ranunculaceae
26. Stamens 10 or fewer 27

27. Flowers bilaterally symmetrical, one of the sepals sac-like .... Balsaminaceae
27. Flowers radially symmetrical, all the petals or sepals similar in shape 28

28. Flowers with 2–4 green sepals, color due entirely
to the strongly exserted purple stamens . . . . . *Besseya* in Scrophulariaceae
28. Flowers not as above 29

29. Fruits disc-shaped and broadly elliptic to nearly round
in outline with a notch at the top; annual plants
often in disturbed soil . . . . . *Lepidium* in Brassicaceae
29. Fruits not as above 30

30. Flower bracts or sepals with a spine tip . . . . . Amaranthaceae
30. Flower bracts and sepals lacking spine tips 31

31. Membranous appendages (stipules) sheathing the stems
above attachment of petiole . . . . . Polygonaceae
31. Plants without sheathing stipules 32

32. Fruit berry-like; leaves fleshy or leathery;
plants rhizomatous . . . . . Santalaceae
32. Plants not as above 33

33. Flowers white, yellow, or pink . . . . . *Eriogonum* in Polygonaceae
33. Flowers green, sometimes tinged purple 34

34. Stamens 10; foliage not covered with silvery scales . . . . . Saxifragaceae
34. Stamens 2–6; foliage sometimes covered
with silvery scales . . . . . Chenopodiaceae

— *Group V: Herbaceous Dicots* —

(Petals separate from each other all the way to the base)

1. Flowers resembling a daisy or dandelion (these are not single flowers
but actually clusters of flowers, each with united petals) . . . . . Asteraceae
1. Flowers not as above 2

2. Flowers bilaterally symmetrical; petals not all the same shape and size . . 3
2. Flowers radially symmetrical; petals all the same shape and size 9

3. Flowers with more than 10 stamens . . . . . Ranunculaceae
3. Flowers with 10 or fewer stamens 4

4. Flowers with 4 petals 5
4. Flowers with 5 petals 6

5. Petals attached on top of the (inferior) ovary . . . . . Onagraceae
5. Petals attached at the base of the (superior) ovary . . . . . Fumariaceae

6. Flowers with 5 stamens (violets) . . . . . Violaceae
6. Flowers with 10 stamens 7

7. Lower 2 petals united to form a canoe-shaped (keel) petal; flowers
strongly bilaterally symmetrical, resembling those of a pea . . . . . Fabaceae
7. Flowers not as above, petals nearly the same size and shape 8

8. Petals attached at the base of the ovary . . . . . Ericaceae
8. A portion of the ovary below the point where
petals are attached . . . . . Saxifragaceae

9. Stems thick, fleshy, green, and spiny;
leaves minute or lacking . . . . . Cactaceae
9. Plants not as above 10

10. Flowers with 3 long sepals and 12 stamens, usually lying on the ground;
leaves spade-shaped, at least as wide as long . . . . . Aristolochiaceae
10. Plants not as above 11

11. Flowers with more than 10 stamens 12
11. Stamens 10 or fewer or absent 21

12. Stem leaves opposite (2 per node) 13
12. Stem leaves alternate (1 per node) or clustered at the base 14

13. Flowers with 4 small, white petals, sessile. . . . . Lythraceae
13. Petals yellow and usually 5; flowers borne on stalks . . . . Hypericaceae

14. Anther filaments fused into a tube
that surrounds the style(s) . . . . Malvaceae
14. Stamens not fused into a tube. 15

15. Plants with white or colored sap. . . . . Papaveraceae
15. Plants with clear sap 16

16. Foliage covered with recurved hairs that cause it
to stick to cloth like Velcro. . . . . Loasaceae
16. Foliage not adhering to cloth like Velcro. 17

17. Leaves divided into leaflets 18
17. Leaves lobed or with entire or toothed margins
but not divided into leaflets. 19

18. Sepals usually separate to the base,
not united to the ovary. . . . . Ranunculaceae
18. Basal part of sepals united to form a cup that is united
to the lower portion of the ovary . . . . Rosaceae

19. Flowers each with 1 pistil and 3–8 stigmas . . . . Portulacaceae
19. Flowers with more than 1 pistil. 20

20. Sepals usually separate to the base,
not united to the ovary. . . . . Ranunculaceae
20. Basal part of sepals united to form a cup that is united
to the lower portion of the ovary . . . . Rosaceae

21. Sepals, petals, and stamens appearing to arise
from the top of the (inferior) ovary 22
21. Sepals, petals, and stamens appearing to arise
from the base of the (superior) ovary. 27

22. Flowers with 2 or 4 petals, sepals, and stamens. 23
22. Flowers with 5 petals and sepals and 5 or 10 stamens. 24

23. Clusters of flowers subtended by 4 white,
petal-like bracts; fruit berry-like. . . . . Cornaceae
23. Flowers not subtended by petal-like bracts;
fruit a nutlet or capsule. . . . . Onagraceae

24. Leaves divided or lobed at least one-third of the way to the midvein . . . 25
24. Leaves with entire or toothed margins, not lobed or divided 27

25. Fruit fleshy and berry-like . . . . Araliaceae
25. Fruit dry, not berry-like 26

26. Inflorescence umbrella-like, flowers borne on stalks
joined at a common point . . . . Apiaceae
26. Inflorescence sometimes head-like but flowers not borne
on stalks joined at a common point. . . . . Saxifragaceae

27. Basal part of sepals united to form a cup that is joined
to the lower portion of the ovary 28
27. Sepals usually separate to the base, not united to the ovary 29

28. Petals yellow . . . . Rosaceae
28. Petals white or purplish . . . . Saxifragaceae

29. Styles more than 5 per flower . . . . Ranunculaceae
29. Styles 1–5 per flower or lacking. 30

30. Flowers with 2 or 3 sepals. 31
30. Flowers with 4 or 5 sepals 33

31. Leaves pinnately divided or lobed . . . . Limnanthaceae
31. Leaves with entire margins or with a pair if lobed at the base 32

32. Petals white or pinkish . . . . Portulacaceae
32. Petals green or brownish . . . . Polygonaceae

33. Lower leaves divided into leaflets or at least halfway to the midvein . . . 34
33. Lower leaves lobed less than halfway to the midvein or
with entire or toothed margins but not divided into leaflets. 38

34. Flowers with 4 petals 35
34. Flowers with 5 petals. 36

35. Leaves with 3 leaflets. . . . . Caprifoliaceae
35. Leaves not as above. . . . . Brassicaceae

36. Leaves with 3 leaflets, each with a rounded notch at the tip . . Oxalidaceae
36. Leaves not as above. 37

37. Sepals separate to the base. . . . . Geraniaceae
37. Sepals united at least below. . . . . Saxifragaceae

38. Flowers with 4 petals 39
38. Flowers with 5 petals. 42

39. Stems and leaves red, pink, white, or yellow but not green . . . . . Ericaceae
39. Stems and leaves green, at least in part 40

40. Leaves alternate (1 per node) or all basal. . . . . Brassicaceae
40. Leaves opposite or whorled (more than 1 per node). 41

41. Leaves fleshy, nearly round in cross section . . . . Crassulaceae
41. Leaves flat, not particularly fleshy . . . . Caryophyllaceae

42. Leaves fleshy; each flower with 4–5 ovaries that are separate
or nearly so . . . . Crassulaceae
42. Plants not as above 43

43. Flowers with only stamens. 44
43. Flowers bisexual. 45

44. Leaves fleshy; each flower with 4–5 separate ovaries
that are separate or nearly . . . . Crassulaceae
44. Leaves flat; each flower with a single ovary . . . . Caryophyllaceae

45. Leaves covered with long, reddish hairs tipped with sticky
secretions that capture insects; plants of *Sphagnum* bogs. . . . Droseraceae
45. Plants not as above 46

46. Leaves opposite or whorled (more than 1 per node). 47
46. Leaves alternate (1 per node) or all basal. 48

47. Leaves finely translucent-dotted (this feature can best
be seen by holding a leaf up to the light and looking for
numerous small "holes" in the blade). Hypericaceae
47. Leaves without translucent dots . . . . Caryophyllaceae

48. Leaves deeply palmately lobed or divided; sepals separate to the base;
fruits with a beak at least twice as long as the sepals . . . . Geraniaceae
48. Plants not as above 49

49. Leaves linear; flowers with 5 stamens . . . . Linaceae
49. Plants not as above 50

50. Flowers with 1 style and a shallowly 5-lobed ovary . . . . Ericaceae
50. Flowers with 2 or more styles; ovary 2-lobed. . . . . Saxifragaceae

— *Group VI: Herbaceous Dicots* —

(Petals united, at least toward the base)

1. Few to many flowers clustered in heads, each head surrounded by bracts forming a cup or vase shape, each cluster appearing like a single flower; petals attached to top of the ovary 2
1. Flowers not as above. 3

2. Flowers with 4 stamens and a 4-lobed corolla . . . . Dipsacaceae
2. Flowers mostly with 5 stamens, usually with united anthers; corolla 1-, 2-or (mostly) 5-lobed, rarely 4-lobed. . . . . Asteraceae

3. Plants with white, yellow, brown, pink, red, or purple stems and leaves; lacking green tissue. 4
3. Plants with green leaves and stems, at least in part. 6

4. Plants with orange, twining stems attached to foliage of other plants . . . . .Cuscutaceae
4. Stems not twining, not attached to foliage of other plants 5

5. Plants less than 5 inches tall; corolla tubular and more than ¼ inch long . . . . Orobanchaceae
5. Plants mostly more than 6 inches tall; corolla urn-shaped and less than ¼ inch long . . . . *Pterospora* in Ericaceae

6. Stems thick, green, and succulent, covered with spines; leaves lacking or minute. . . . . Cactaceae
6. Plants not as above 7

7. Plants with milky sap 8
7. Plants with watery sap 9

8. Corolla reflexed, each flower with 5 horn-like appendages surrounding the style . . . . Asclepiadaceae
8. Corolla bell-shaped to tubular 10

9. Corolla attached on top of the ovary (inferior ovary). . . . .Campanulaceae
9. Corolla attached at the base of the ovary 11

10. Plants erect with opposite leaves . . . . Apocynaceae
10. Plants twining; leaves alternate. . . . . .Convolvulaceae

11. Flowers all unisexual; petals arising from base of ovary; stamens 10; leaves opposite (2 per node) and entire-margined . . . . . . Caryophyllaceae
11. Plants not as above 12

12. Leaves undivided with entire margins; flowers radially symmetrical (petals all alike) with 2 sepals. . . . . Portulacaceae
12. Plants not as above 13

13. Anthers more numerous than petals or lobes of the corolla 14
13. Anthers as numerous or fewer than petals or corolla lobes. 24

14. Flowers radially symmetrical (petals all the same size and shape) 15
14. Flowers bilaterally symmetrical (petals of different size or shape). 21

15. Leaves divided into leaflets 16
15. Leaves lobed or with toothed to entire margins but not divided into leaflets. 17

16. Leaves divided into 3 leaflets like a shamrock . . . . Oxalidaceae
16. Leaflets irregularly shaped, not like a shamrock . . . . Malvaceae

17. Stamens numerous, united to form a tube that surrounds the style(s) .... Malvaceae
17. Stamens 10 or fewer, not united to form a tube 18

18. Leaves fleshy and succulent 19
18. Leaves more or less flat, not fleshy and succulent 20

19. Sepals 2 .... *Portulaca* in Portulacaceae
19. Sepals mostly 4 or 5 .... Crassulaceae

20. Leaves opposite (2 per node) .... Caryophyllaceae
20. Leaves alternate (1 per node) or all basal .... Ericaceae

21. Flowers with more than 10 stamens .... Ranunculaceae
21. Flowers with 10 or fewer stamens 22

22. Flowers with 9–10 anthers .... Fabaceae
22. Flowers with 4–8 anthers 23

23. Flowers with 6 anthers .... Fumariaceae
23. Flowers with 4 (or apparently 8) anthers .... Scrophulariaceae

24. Ovary inferior (corolla attached at top of ovary) 25
24. Ovary superior (corolla attached at base of ovary) 31

25. Leaves opposite or whorled (more than 1 per node) 26
25. Leaves alternate (1 per node) 29

26. Leaves whorled (more than 2 per node), at least in part .... Rubiaceae
26. Leaves opposite (2 per node) 27

27. Ovary and fruit with hooked hairs .... Rubiaceae
27. Ovary and fruit lacking hooked hairs 28

28. Flowers with 3 stamens .... Valerianaceae
28. Flowers with 4 stamens .... *Linnaea* in Caprifoliaceae

29. Leaves divided into 3 leaflets; petals covered with short, hair-like scales .... Menyanthaceae
29. Leaves with lobed to entire margins but not divided into leaflets 30

30. Plants climbing vines with tendrils (curling modified stems or leaves) .... Cucurbitaceae
30. Plants not vines, tendrils lacking .... Campanulaceae

31. Flowers bilaterally symmetrical (petals of different size or shape) 32
31. Flowers radially symmetrical (petals all the same size and shape) 39

32. Flowers with 5 anther-bearing stamens 33
32. Flowers with 2–4 anther-bearing stamens 36

33. Flowers yellow or orange .... *Verbascum* in Scrophulariaceae
33. Flowers not yellow or orange 34

34. Flowers usually bright blue; stamens exserted beyond the mouth of the corolla .... Boraginaceae
34. Flowers pale with purple spots; stamens not protruding beyond the mouth of the corolla .... *Hyoscyamus* in Solanaceae

35. Flowers with 3 stamens .... Valerianaceae
35. Stamens 2–4 36

36. Ovary not 4-lobed; leaves alternate, opposite, or whorled; stems mostly not 4-angled .... Scrophulariaceae
36. Ovary 4-lobed; leaves opposite (2 per node); stems 4-angled 37

37. Anthers as long as or longer than their stalks; corolla nearly radially symmetrical, petals nearly the same size and shape .... Verbenaceae
37. Anthers definitely shorter than their stalks; corolla often 2-lipped with the corolla lobes obviously of different sizes .... Lamiaceae

38. Anther-bearing stamens 2–4, fewer than the lobes of the corolla 40
38. Anther-bearing stamens as many as corolla lobes or at least 5 47

39. Flowers with 2 sepals; leaves with entire margins, often somewhat fleshy .... Portulacaceae
39. Plants not as above 40

40. Flowers with 4 anther-bearing stamens 41
40. Flowers with 2 or 3 anther-bearing stamens......................... 43

41. Ovary not 4-lobed; leaves alternate, opposite, or whorled; stems mostly not 4-angled .... Scrophulariaceae
41. Ovary 4-lobed; leaves opposite (2 per node); stems 4-angled.......... 43

42. Anthers as long as or longer than their stalks .... Verbenaceae
42. Anthers definitely shorter than their stalks .... Lamiaceae

43. All leaves basal; petals thin and papery................. Plantaginaceae
43. Plants mostly with leafy stems; petals not noticeably thin and papery 44

44. Plants climbing vines with tendrils (curling modified stems or leaves) .... Cucurbitaceae
44. Plants not vines, tendrils lacking 45

45. Flowers with 3 stamens................................ Valerianaceae
45. Flowers with 2 stamens.. 46

46. Fruit and ovary 4-lobed; flowers in dense clusters surrounding the stem just above where the leaves join .... *Lycopus* in Lamiaceae
46. Fruit and ovary not 4-lobed; inflorescence not as above .... Scrophulariaceae

47. Flowers without an ovary.................................. Rubiaceae
47. Flowers with an ovary.. 48

48. Ovary 4-lobed or 4-grooved, splitting into 4 nutlets at maturity....... 49
48. Ovary not 4-lobed or 4-grooved, not splitting into 4 at maturity 50

49. Stamens 4, leaves opposite (2 per node) .... *Mentha* in Lamiaceae
49. Stamens 5, leaves alternate (1 per node) at least in part ..... Boraginaceae

50. Plants creeping and twining around adjacent vegetation; leaves arrowhead-shaped............................ Convolvulaceae
50. Plants not as above 51

51. Flowers with 2 sepals .... Portulacaceae
51. Flowers with more than 2 sepals.................................. 52

52. Leaves opposite or whorled (more than 1 per node).................. 53
52. Leaves alternate or all basal 54

53. Flowers with a 3-branched style (3 stigmas)............. Polemoniaceae
53. Flowers with an unbranched style .... Gentianaceae

54. Flowers with 4 stamens; petals thin and papery .... Plantaginaceae

APPENDIX 2

Plants and Their Uses, by Habitat

COMMON NAME *SCIENTIFIC NAME*	FOOD, BEVERAGE	MEDICINE, PERSONAL CARE	DYE, SOAP	GUM, RUBBER, SMOKING	FIBER, FIRE	TOOLS, CRAFTS, ABRASIVES	CONTAINERS, BASKETS	POISONS
FRESH WATER, MARSHES, LAKES, STREAMS								
Narrow-leaf Cattail *Typha angustifolia*	X	X			X	X	X	
Broad-leaf Cattail *Typha latifolia*	X	X			X	X	X	
Narrow-leaf Water Plantain *Alisma gramineum*	X	X						
Northern Water Plantain *Alisma triviale*	X	X						
Arum-leaf Arrowhead *Sagittaria cuneata*	X	X						X
Camas *Camassia quamash*	X	X						
Bog Birch *Betula nana*	X	X		X	X	X		X
Yellow Pond-lily *Nuphar lutea*	X	X						
White Marsh-marigold *Caltha leptosepala*	X							X
Alpine Laurel *Kalmia microphylla*								X
Needlegrass *Stipa* sp.	X	X			X		X	
Rocky Mountain Iris *Iris missouriensis*		X						X
Sweetberry Honeysuckle *Lonicera caerulea*	X	X				X		X
Bunchberry Dogwood *Cornus canadensis*	X	X				X		
Common Cowparsnip *Heracleum maximum*	X	X						
Darkthroat Shooting Star *Dodecatheon pulchellum*	X							
American Bistort *Polygonum bistortoides*	X	X						
GRASSLANDS — Grasses								
Bristlegrass *Setaria viridis*	X	X				X		X
Needlegrass *Stipa comata*	X	X				X		X
Timothy *Phleum pratense*	X	X				X		X
Tufted Hairgrass *Deschampsia caespitosa*	X	X			X		X	X
Spike Trisetum *Trisetum spicatum*	X	X			X		X	X
Junegrass *Koeleria macrantha*	X	X			X		X	X
Oatgrass *Danthonia californica*	X	X			X		X	X
Sandberg Bluegrass *Poa secunda*	X	X			X		X	X
Kentucky Bluegrass *Poa pratensis*	X	X			X		X	X

COMMON NAME *SCIENTIFIC NAME*	FOOD, BEVERAGE	MEDICINE, PERSONAL CARE	DYE, SOAP	GUM, RUBBER, SMOKING	FIBER, FIRE	TOOLS, CRAFTS, ABRASIVES	CONTAINERS, BASKETS	POISONS
Annual Bluegrass *Poa annua*	X	X			X		X	X
Bulbous Bluegrass *Poa bulbosa*	X	X			X		X	X
Tall Mannagrass *Glyceria elata*	X	X			X		X	X
Blue-bunch Fescue *Festuca idahoensis*	X	X			X		X	X
Cheatgrass *Bromus tectorum*	X	X			X		X	X
Wheatgrass *Agropyron* sp.	X	X			X		X	X
Rye Grass *Elymus* sp.	X	X			X		X	X
Squirrel-tail Grass *Sitanion* sp.	X	X			X		X	X
GRASSLANDS — Herbs								
Wild Hyacinth *Triteleia grandiflora*	X							
Sego Lily *Calochortus nuttallii*	X	X						
Idaho Blue-eyed Grass *Sisyrinchium idahoense*		X						X
Nuttall's Larkspur *Delphinium nuttallianum*		X						X
Northern Bedstraw *Galium boreale*	X	X						
Common Yarrow *Achillea millefolium*	X	X	X	X				
Camas *Camassia quamash*	X	X						
Smooth-stem Blazing Star *Mentzelia laevicaulis*	X	X						
Arrowleaf Balsamroot *Balsamorhiza sagittata*	X	X		X	X			
Tufted Evening-primrose *Oenothera ceaspitosa*	X							
Common Gaillardia *Gaillardia aristata*		X						
Spotted Missionbells *Fritillaria atropurpurea*	X							
Wild Bergamot Beebalm *Monarda fistulosa*	X	X						
GRASSLANDS — Shrubs								
Big Sagebrush *Artemisia tridentata*	X	X		X	X	X	X	X
Silver Sagebrush *Artemisia cana*	X	X		X	X	X	X	X
Common Chokecherry *Prunus virginiana*	X	X				X		X
Oregon-grape *Berberis repens*	X	X	X					X
MEADOWS — Trees and Shrubs								
Mountain Alder *Alnus incana*	X	X			X	X		
Rocky Mountain Maple *Acer glabrum*	X	X				X	X	
Quaking Aspen *Populus tremuloides*	X	X			X	X	X	
Narrow-leaf Cottonwood *Populus angustifolia*	X	X			X	X	X	
Bog Birch *Betula nana*	X	X						X
Redosier Dogwood *Cornus sericea*	X	X			X	X	X	X
Golden Currant *Ribes aureum*	X	X						
Scarlet Elderberry *Sambucus racemosa*	X	X			X	X		X
Prickly Currant *Ribes lacustre*	X	X						
MEADOWS — Herbs								
Boreal Bog Orchid *Platanthera dilatata*	X	X						
Columbian Monkshood *Aconitum columbianum*		X						X
Windmill Fringed Gentian *Gentianella detonsa*		X						X

COMMON NAME *SCIENTIFIC NAME*	FOOD, BEVERAGE	MEDICINE, PERSONAL CARE	DYE, SOAP	GUM, RUBBER, SMOKING	FIBER, FIRE	TOOLS, CRAFTS, ABRASIVES	CONTAINERS, BASKETS	POISONS
Camas *Camassia quamash*	X	X						
White Marsh-marigold *Caltha leptosepala*		X						X
Western Columbine *Aquilegia formosa*	X	X						X
Western Meadow-rue *Thalictrum occidentale*								X
Fireweed *Epilobium angustifolium*	X	X	X		X	X		
Rocky Mountain Iris *Iris missouriensis*		X						X
Lance-leaf Springbeauty *Claytonia lanceolata*	X	X						
California False Hellebore *Veratrum californicum*								X
American Bistort *Polygonum bistortoides*	X	X						
Goosefoot or Pine Violet *Viola purpurea*	X	X						
Colorado Blue Columbine *Aquilegia caerulea*		X						X
Nuttall's Larkspur *Delphinium nuttallianum*		X						X
Ballhead Waterleaf *Hydrophyllum capitatum*	X							
Common Cowparsnip *Heracleum maximum*	X	X	X			X		
Streambank Globe Mallow *Iliamna rivularis*		X						
Wyoming Paintbrush *Castilleja linariifolia*	X							
Wavyleaf Paintbrush *Castilleja applegatei*	X							
Lewis's Monkeyflower *Mimulus lewisii*	X	X						
Western Valerian *Valeriana occidentalis*	X	X						
Arrowleaf Groundsel *Senecio triangularis*								X
MEADOWS — Grasses								
Junegrass *Koeleria macrantha*	X	X			X		X	X
Oatgrass *Danthonia californica*	X	X			X		X	X
Sandberg Bluegrass *Poa secunda*	X	X			X		X	X
Kentucky Bluegrass *Poa pratensis*	X	X			X		X	X
Annual Bluegrass *Poa annua*	X	X			X		X	X
Bulbous Bluegrass *Poa bulbosa*	X	X			X		X	X
Tall Mannagrass *Glyceria elata*	X	X			X		X	X
Blue-bunch Fescue *Festuca idahoensis*	X	X			X		X	X
Cheatgrass *Bromus tectorum*	X	X			X		X	X
Wheatgrass *Agropyron* sp.	X	X			X		X	X
Rye Grass *Elymus* sp.	X	X			X		X	X
Squirrel-tail Grass *Sitanion* sp.	X	X			X		X	X
SAGEBRUSH HABITAT — Shrubs								
Antelope Bitterbrush *Purshia tridentata*	X	X	X		X	X		
Big Sagebrush *Artemisia tridentata*	X	X		X	X	X	X	X
Silver Sagebrush *Artemisia cana*	X	X		X	X	X	X	X
Oregon-grape *Berberis repens*	X	X						X
Golden Currant *Ribes aureum*	X	X		X				
SAGEBRUSH HABITAT — Grasses								
Columbia Needlegrass *Stipa columbiana*	X	X			X		X	X

COMMON NAME *SCIENTIFIC NAME*	FOOD, BEVERAGE	MEDICINE, PERSONAL CARE	DYE, SOAP	GUM, RUBBER, SMOKING	FIBER, FIRE	TOOLS, CRAFTS, ABRASIVES	CONTAINERS, BASKETS	POISONS
Blue-bunch Fescue *Festuca idahoensis*	X	X			X		X	X
Cheatgrass *Bromus tectorum*	X	X			X		X	X
Wheatgrass *Agropyron* sp.	X	X			X		X	X
Rye Grass *Elymus* sp.	X	X			X		X	X
Annual Bluegrass *Poa annua*	X	X					X	X
Bulbous Bluegrass *Poa bulbosa*	X	X			X		X	X
SAGEBRUSH HABITAT — Herbs								
Smooth-stem Blazing Star *Mentzelia laevicaulis*	X	X						
Tufted Evening-primrose *Oenothera ceaspitosa*	X	X			X			
Skyrocket Gilia *Ipomopsis aggregata*			X					
Arrowleaf Balsamroot *Balsamorhiza sagittata*	X	X						
Common Gaillardia *Gaillardia aristata*		X						
Nuttall's Larkspur *Delphinium nuttallianum*		X						X
Northern Bedstraw *Galium boreale*	X		X					
Common Yarrow *Achillea millefolium*	X	X						
Streambank Globe Mallow *Iliamna rivularis*		X						
Richardson's Geranium *Geranium richardsonii*	X							
Bluebell Bellflower *Campanula rotundifolia*	X	X						
Aspen Fleabane *Erigeron speciosus*								
Ballhead Waterleaf *Hydrophyllum capitatum*	X							
Sulphur Wild Buckwheat *Eriogonum umbellatum*	X							
American Bistort *Polygonum bistortoides*	X							
JUNIPER–MOUNTAIN MAHOGANY WOODLANDS — Trees and Shrubs								
Rocky Mountain Juniper *Juniperus scopulorum*	X	X		X	X	X		
Mountain-mahogany *Cercocarpus ledifolius*		X				X		X
Ponderosa Pine *Pinus ponderosa*	X	X		X	X	X	X	
Antelope Bitterbrush *Purshia tridentata*		X	X		X	X		
Big Sagebrush *Artemisia tridentata*	X	X			X	X	X	X
Silver Sagebrush *Artemisia cana*	X	X			X	X	X	X
JUNIPER–MOUNTAIN MAHOGANY WOODLANDS — Grasses								
Columbia Needlegrass *Stipa columbiana*	X	X			X		X	X
Blue-bunch Fescue *Festuca idahoensis*	X	X			X		X	X
Cheatgrass *Bromus tectorum*	X	X			X		X	X
Wheatgrass *Agropyron* sp.	X	X			X		X	X
Rye Grass *Elymus* sp.	X	X			X		X	X
Annual Bluegrass *Poa annua*	X	X			X		X	X
JUNIPER–MOUNTAIN MAHOGANY WOODLANDS — Herbs								
Sego Lily *Calochortus nuttallii*	X							
Skyrocket Gilia *Ipomopsis aggregata*		X	X					
ASPEN WOODLAND — Trees and Shrubs								
Quaking Aspen *Populus tremuloides*	X	X			X	X		

COMMON NAME *SCIENTIFIC NAME*	FOOD, BEVERAGE	MEDICINE, PERSONAL CARE	DYE, SOAP	GUM, RUBBER, SMOKING	FIBER, FIRE	TOOLS, CRAFTS, ABRASIVES	CONTAINERS, BASKETS	POISONS
Big Sagebrush *Artemisia tridentata*	X	X		X	X	X	X	X
Silver Sagebrush *Artemisia cana*	X	X		X	X	X	X	X
Sweetberry Honeysuckle *Lonicera caerulea*	X	X						
Whitestem Gooseberry *Ribes inerme*	X							
Oregon-grape *Berberis repens*	X			X				
Western Snowberry *Symphoricarpos occidentalis*								
ASPEN WOODLAND — Herbs								
Camas *Camassia quamash*	X							
Nuttall's Larkspur *Delphinium nuttallianum*		X						X
Northern Bedstraw *Galium boreale*	X							
Prairie Flax *Linum lewisii*	X				X	X	X	
Silvery Lupine *Lupinus argenteus*		X						X
Arrowleaf Balsamroot *Balsamorhiza sagittata*	X	X						
Sticky Geranium *Geranium viscosissimum*	X	X						
Richardson's Geranium *Geranium richardsonii*	X	X						
Common Cowparsnip *Heracleum maximum*	X	X						
Western Meadow-rue *Thalictrum occidentale*		X						X
Ballhead Waterleaf *Hydrophyllum capitatum*	X							
STREAMSIDE WOODLAND — Trees and Shrubs								
Narrow-leaf Cottonwood *Populus angustifolia*	X	X			X	X		
Blue Spruce *Picea pungens*	X	X				X		
Golden Currant *Ribes aureum*	X	X						
Whitestem Gooseberry *Ribes inerme*	X	X						
Woods' Rose *Rosa woodsii*	X	X					X	
Common Chokecherry *Prunus virginiana*	X	X						X
Bitter Cherry *Prunus emarginata*	X	X						X
Black Hawthorn *Crataegus douglasii*	X							
Russet Buffaloberry *Shepherdia canadensis*	X							
Redosier Dogwood *Cornus sericea*	X	X				X		
Scarlet Elderberry *Sambucus racemosa*	X	X				X		X
Twinberry Honeysuckle *Lonicera involucrata*	X							
Quaking Aspen *Populus tremuloides*	X	X			X	X		
Ponderosa Pine *Pinus ponderosa*	X	X			X	X	X	
Alderleaf Buckthorn *Rhamnus alnifolia*		X						
Alpine Laurel *Kalmia microphylla*								X
Bog Birch *Betula nana*	X	X				X	X	X
Rocky Mountain Maple *Acer glabrum*	X	X				X	X	X
Big Sagebrush *Artemisia tridentata*	X	X	X			X	X	X
Silver Sagebrush *Artemisia cana*	X	X	X			X	X	X
Prickly Currant *Ribes lacustre*	X	X						
Thimbleberry *Rubus parviflorus*	X	X						

COMMON NAME *SCIENTIFIC NAME*	FOOD, BEVERAGE	MEDICINE, PERSONAL CARE	DYE, SOAP	GUM, RUBBER, SMOKING	FIBER, FIRE	TOOLS, CRAFTS, ABRASIVES	CONTAINERS, BASKETS	POISONS
Saskatoon Serviceberry *Amelanchier alnifolia*	X	X						
Utah Serviceberry *Amelanchier utahensis*	X	X						
Common Juniper *Juniperus communis*	X	X			X	X	X	
Shrubby Cinquefoil *Pentaphylloides floribunda*		X						
STREAMSIDE WOODLAND – Herbs								
Rocky Mountain Iris *Iris missouriensis*		X						X
Rock Clematis *Clematis columbiana*		X						X
Mountain Yellow Pea *Thermopsis montana*		X						X
Sticky Geranium *Geranium viscosissimum*	X	X						
Richardson's Geranium *Geranium richardsonii*	X	X						
Common Cowparsnip *Heracleum maximum*	X	X			X			
Streambank Globe Mallow *Iliamna rivularis*		X						X
Arrowleaf Groundsel *Senecio triangularis*								X
Camas *Camassia quamash*	X							
Nuttall's Larkspur *Delphinium nuttallianum*		X						X
Northern Bedstraw *Galium boreale*	X	X	X					
Boreal Bog Orchid *Platanthera dilatata*	X	X						
Windmill Fringed Gentian *Gentianella detonsa*		X						
Spotted Missionbells *Fritillaria atropurpurea*	X							
Fairy Slipper Orchid *Calypso bulbosa*	X							
Columbian Monkshood *Aconitum columbianum*		X						X
Fireweed *Epilobium angustifolium*	X	X			X		X	
Western Columbine *Aquilegia formosa*	X	X						
Wyoming Paintbrush *Castilleja linariifolia*	X	X						
Sulphur Wild Buckwheat *Eriogonum umbellatum*	X							
Lewis's Monkeyflower *Mimulus lewisii*	X							
Western Valerian *Valeriana occidentalis*	X	X						
CONIFEROUS FORESTS – Trees								
Subalpine Fir *Abies lasiocarpa*	X	X			X	X	X	
Lodgepole Pine *Pinus contorta*	X	X			X	X	X	
Ponderosa Pine *Pinus ponderosa*	X	X			X	X	X	
Engelmann Spruce *Picea engelmannii*	X	X			X	X	X	
Douglas-fir *Pseudotsuga menziesii*	X	X			X	X	X	
Whitebark Pine *Pinus albicaulis*	X	X			X	X	X	
Quaking Aspen *Populus tremuloides*	X	X			X	X	X	
CONIFEROUS FORESTS — Shrubs								
Oregon-grape *Berberis repens*	X	X	X					
Prickly Currant *Ribes lacustre*	X	X						
Mallow Ninebark *Physocarpus malvaceus*		X						X
White Spiraea *Spiraea betulifolia*		X						
Thimbleberry *Rubus parviflorus*	X	X						

COMMON NAME *SCIENTIFIC NAME*	FOOD, BEVERAGE	MEDICINE, PERSONAL CARE	DYE, SOAP	GUM, RUBBER, SMOKING	FIBER, FIRE	TOOLS, CRAFTS, ABRASIVES	CONTAINERS, BASKETS	POISONS
Woods' Rose *Rosa woodsii*	X	X				X		
Mountain Ash *Sorbus scopulina*	X	X						
Utah Serviceberry *Amelanchier utahensis*	X	X						
Boxleaf Myrtle *Paxistima myrsinites*		X						
Alderleaf Buckthorn *Rhamnus alnifolia*		X						
Snowbrush Ceanothus *Ceanothus velutinus*		X						
Huckleberry *Vaccinium membrabaceum*	X	X						
Rusty Menziesia *Menziesia ferruginea*		X						
Kinnikinnick *Arctostaphylos uva-ursi*	X	X						
Utah Honeysuckle *Lonicera utahensis*	X	X						
Mountain Alder *Alnus incana*	X	X			X	X	X	
Rocky Mountain Maple *Acer glabrum*	X	X			X	X	X	
Mountain-mahogany *Cercocarpus ledifolius*		X				X		X
Common Chokecherry *Prunus virginiana*	X	X						
Bitter Cherry *Prunus emarginata*	X	X						
Russet Buffaloberry *Shepherdia canadensis*	X	X						
Twinberry Honeysuckle *Lonicera involucrata*		X						
Common Juniper *Juniperus communis*	X	X		X	X	X	X	
Shrubby Cinquefoil *Pentaphylloides floribunda*		X						
Scarlet Elderberry *Sambucus racemosa*	X	X						X
Redosier Dogwood *Cornus sericea*	X	X						
CONIFEROUS FORESTS — Grasses and Herbs								
Oniongrass *Melica bulbosa*	X	X			X	X	X	X
Junegrass *Koeleria macrantha*	X	X			X		X	X
Oatgrass *Danthonia californica*	X	X			X		X	X
Sandberg Bluegrass *Poa secunda*	X	X			X		X	X
Kentucky Bluegrass *Poa pratensis*	X	X			X		X	X
Annual Bluegrass *Poa annua*	X	X			X		X	X
Bulbous Bluegrass *Poa bulbosa*	X	X			X		X	X
Tall Mannagrass *Glyceria elata*	X	X			X		X	X
Blue-bunch Fescue *Festuca idahoensis*	X	X			X		X	X
Cheatgrass *Bromus tectorum*	X	X			X		X	X
Wheatgrass *Agropyron* sp.	X	X			X		X	X
Rye Grass *Elymus* sp.	X	X			X		X	X
Squirrel-tail Grass *Sitanion* sp.	X	X			X		X	X
Bunchberry Dogwood *Cornus canadensis*	X	X						
Pipsissewa *Chimaphila umbellata*		X						
Ballhead Waterleaf *Hydrophyllum capitatum*	X							
Scarlet Paintbrush *Castilleja miniata*	X							
Twinflower *Linnaea borealis*		X						
Bluebell Bellflower *Campanula rotundifolia*		X						

COMMON NAME *SCIENTIFIC NAME*	FOOD, BEVERAGE	MEDICINE, PERSONAL CARE	DYE, SOAP	GUM, RUBBER, SMOKING	FIBER, FIRE	TOOLS, CRAFTS, ABRASIVES	CONTAINERS, BASKETS	POISONS
Western Pearly-everlasting *Anaphalis margaritacea*		X		X				
Heartleaf Arnica *Arnica cordifolia*		X						X
SUBALPINE — Trees and Shrubs								
Whitebark Pine *Pinus albicaulis*	X	X			X	X	X	
Common Juniper *Juniperus communis*	X	X			X	X	X	
Subalpine Fir *Abies lasiocarpa*	X	X			X	X	X	
Lodgepole Pine *Pinus contorta*	X	X			X	X	X	
Engelmann Spruce *Picea engelmannii*	X	X			X	X	X	
SUBALPINE — Herbs and Grasses								
Common Beargrass *Xerophyllum tenax*		X			X	X		
Mountain Death Camas *Zigadenus elegans*								X
California False Hellebore *Veratrum californicum*								X
Wild Chives *Allium schoenoprasum*	X	X						
Sulphur Wild Buckwheat *Eriogonum umbellatum*	X	X						
American Bistort *Polygonum bistortoides*	X	X						
American Pasqueflower *Anemone patens*		X						X
St. John's-wort *Hypericum formosum*		X						
Pink Mountain Heath *Phyllodoce empetriformis*		X						X
Lewis' Monkeyflower *Mimulus lewisii*	X	X						
Elephanthead Lousewort *Pedicularis groenlandica*		X						X
Western Valerian *Valeriana occidentalis*		X						
Western Columbine *Aquilegia formosa*	X	X						X
Western Pearly-everlasting *Anaphalis margaritacea*		X		X				
ALPINE								
Alpine Mountain Sorrel *Oxyria digyna*	X	X						
Mountain Avens *Dryas octopetala*		X						X
Sticky Polemonium *Polemonium viscosum*						X		
Western Columbine *Aquilegia formosa*	X	X						X
Western Pearly-everlasting *Anaphalis margaritacea*		X		X				
Pink Mountain Heath *Phyllodoce empetriformis*		X						X
Lewis's Monkeyflower *Mimulus lewisii*	X	X						
Wild Chives *Allium schoenoprasum*	X	X						
Sulphur Wild Buckwheat *Eriogonum umbellatum*	X	X						
American Bistort *Polygonum bistortoides*	X	X						
American Pasqueflower *Anemone patens*		X						X

Glossary

BOTANICAL TERMS

abscise. To shed flowers, leaves, or fruit following formation of scar tissue.
achene. Small, dry, hard, 1-celled, 1-seeded, indehiscent fruit.
acaulescent. Having no stem above ground.
acuminate. Tapering to a point.
acute. Sharp-pointed.
adnate. United with.
adventive. Not native to and not fully established in a new habitat or environment; locally or temporarily naturalized.
allelopathy. The inhibition of the growth of one plant by another through the release of chemical substances.
alpine. Occurring above treeline.
alternate. Arranged with one structure (e.g., leaf, flowers, stem, etc.) per node.
angiosperm. A plant producing flowers and bearing ovules (seeds) in an ovary (fruit).
annual. A plant, usually with a slender taproot, completing its life cycle in a single growing season.
anther. The pollen-bearing portion of the stamen.
appressed. Pressed close to the stem.
aquatic. Growing in water.
areola. On the surface of a cactus, a small defined area that bears the spines.
aromatic. Having a strong, usually agreeable odor.
asexual. Without sex.
auricle. Ear-like appendage.
awl-shaped. Sharp-pointed from a broader base.
awn. A slender, bristle-like organ.
axil. The upper angle formed by a leaf or branch with the stem.
axile. Describing placentae in a compound ovary which are attached to the center axis of the ovary.
axillary. Borne at or pertaining to an axil.
banner. The upper petal of a papilionaceous (butterfly-shaped) flower.
basal. At the base.
beak. A prolonged, slender, tapering projection.
bearded. With long or stiff hairs.
berry. A fleshy or pulpy fruit developed from a single ovary with more than one seed, such as a grape or blueberry.
bi-. Prefix meaning twice.
biennial. A plant completing its life cycle in two growing seasons, usually forming a basal rosette the first season and flowering the second.
bifid. Two-cleft.

borne. Produced or arising from.

bract. A leaf subtending a flower or flower cluster.

bracteole. A small bract-like structure borne singly or in pairs on the pedicel of the calyx of a flower.

bulb. An underground organ constituted mostly of fleshy storage leaves and scale-covered (e.g., onion).

calyx. A flower's sepals considered as a unit.

campanulate. Bell-shaped.

canescent. Grayish white with small hairs.

capillary. Hair- or thread-like.

capitate. Having a head-like cluster.

capsule. A dry, dehiscent fruit composed of more than one carpel.

carpel. A modified leaf forming the ovary.

catkin. In plants such as willows, birches, and alders, the elongated, pendulous, or cone-like flower cluster with minute flowers that lack or almost lack petals and sepals.

caudex. The thickened base of some perennial herbs.

cauline. Growing on the stem.

chaff. Scales or bracts on the receptacles of some members of the Asteraceae (sunflower family).

ciliate. With a fringe of hairs on the margin.

circumscissile. Splitting or opening along a circumference, with the top coming off as a lid.

clasping. Partly surrounding the stem.

clavate. Club-shaped.

claw. The narrow or stalk-like base of some petals.

coma. The tufts of hairs at the ends of some seeds.

compound leaf. A leaf that is divided into two or more distinct leaflets.

cone. A dense cluster of modified, leaf-like organs bearing pollen, spores, or seeds, as in horsetails, clubmosses, and conifers (e.g., pinecone).

conifer. A cone-bearing tree.

coniferous. Having cones or strobili.

cordate. Heart-shaped.

corm. A bulb-like, underground thickening of the stem.

corolla. The petals considered as a unit, usually brightly colored.

corymb. Convex or flat-topped flower cluster of the racemose type, with the pedicels arising from a different point on the axis.

cruciform. Cross-shaped (e.g., the position of petals in the mustard family, Brassicaceae).

culm. The hollow or pithy stem of grasses and sedges.

cyathium. The inflorescence in the genus *Euphorbia* (family Euphorbiaceae), which consists of unisexual flowers crowded within a cup-like involucre.

cyme. A flower cluster, often flat-topped or convex, in which the central or terminal flower blooms the earliest.

deciduous. Falling off once a year, usually at the end of a growing season.

decompound. More than once compound or divided.

decumbent. Reclining or lying on the ground with the tip ascending.

decurrent. Running or extending downward along the stem.

dehiscent. Opening to emit the contents.

dichotomous. Forking regularly by pairs.

diffuse. Widely spreading.

discoid. Having a flowering head without ray flowers.

disk. In the sunflower family (Asteraceae), the central portion of the head, which gives rise to the disk flowers.

disk flowers. In the sunflower family (Asteraceae), the flowers with slender, tubular corollas at the central part of the head.

distinct. Completely separated.
divided. Cut or lobed to the base or to a midrib.
dorsal. Pertaining to the back.
drupe. A fleshy, one-seeded fruit (e.g., cherry).
elliptic. Oval or oblong with the ends regularly rounded.
emergent. With the lower portion in water and the upper portion extending out.
endemic. Found only within a limited geographic area.
entire. With margins not cut, cleft, or otherwise toothed.
ephemeral. Lasting for only a short time.
erose. Eroded as if gnawed.
evergreen. Retaining leaves through the winter.
exotic. Not native; introduced from somewhere else.
exserted. Extending beyond the corolla.
family. A group of related genera.
fascicle. A small cluster of leaves, flowers, etc.
filament. The stalk of a stamen.
flower. The reproductive portion of the plant, consisting of stamens, pistils, or both, and including petals, sepals, or both.
foliage. The leaves of the plant, collectively.
follicle. A fruit consisting of a single carpel, dehiscing by the ventral suture.
forb. A non-grass-like herbaceous plant.
frond. The leaf of a fern.
fruit. The mature ovary, which includes the attached external structures and the enclosed seeds.
galea. The upper lip of a two-lipped corolla.
genus. A group of related species; pl. genera.
glabrate. Becoming glabrous with age.
glabrous. Devoid of hairs.
gland. A secreting cell or group of cells.
glandular. Having glands, usually hairs.
glaucous. Covered or whitened with a bloom.
globose. Spherical or nearly so.
glochid. A barbed hair or bristle (as in the fine hairs of *Opuntia*).
glume. The outer bracts of a grass spikelet.
gymnosperm. A member of the plant group that is characterized as having ovules not enclosed in an ovary (e.g., pines, spruces, firs, junipers).
habit. The general appearance or growth form of a plant.
habitat. The environmental conditions or kind of place in which a plant grows.
head. A type of inflorescence with mostly sessile flowers densely set on a very short axis or disk, thereby having a round outline. The terminal collection of flowers surrounded by an involucre, as in the sunflower family (Asteraceae).
herb. A plant with the aerial portion being nonwoody, dying back to the ground at the end of the growing season.
herbaceous. Not woody, dying back at the end of the growing season.
hispid. Having stiff hairs.
hoary. Grayish white.
host. The plant from which a parasite (e.g., mistletoe) obtains nutrients.
hyaline. Transparent, or nearly so.
imbricate. Overlapping one another like the shingles of a roof.
immersed. Growing under water.
imperfect flowers. Flowers that lack either stamens or pistils.
included. Used to describe a part that does not project beyond another part.
indehiscent. Not splitting open.
indusium. A small growth covering or surrounding the sorus in ferns; pl. indusia.
inferior. Lower or below.
inferior ovary. Ovary positioned below the base of other flower parts.

inflated. Turgid and bladdery.
inflorescence. The flowering part of plants; the arrangement of flowers.
inner bark. The cambium layer.
involucel. A secondary involucre, as at the base of an umbel within a compound umbel.
involucre. A whorl of bracts subtending a flower or flower cluster.
irregular. Used to describe a flower one or more of whose organs are unlike the rest.
keel. Plant part that resembles the keel of a boat.
labellum. The part of an orchid flower that attract insects for pollination.
lanceolate. Lance-shaped.
latex. The milky sap of certain plants.
leaflet. One of the divisions of a compound leaf.
legume. A simple, dry fruit that is dehiscent along both sutures.
lemma. A bract in a grass spikelet just below the pistil and stamens.
ligulate flower. The same as a ray flower in the sunflower family.
linear. Long and narrow.
lip. One of the two divisions of a bilabiate corolla or calyx.
locule. A compartment or cell.
loment. A modified legume that breaks apart at constrictions between segments of seeds.
many. For botanical purposes, numbering more than ten.
meal. Flour.
-merous. Parted, having sections; a 5-merous flower has 5 petals and 5 sepals.
montane. Of or pertaining to the mountains.
mucronate. Tipped with an abrupt, short point.
naturalized. Plants introduced from somewhere else and now established.
node. A joint or point of origin for leaves or branches.
numerous. In botanical terms, more than 10.
nut. A dry, hard-walled, indehiscent fruit, usually with one seed.
nutlet. A small nut.
ob-. A prefix meaning inversely.
oblong. Two to four times as long as broad.
opposite. Nodes having two leaves or branches directly across from each other.
orbicular. Circular in outline.
ovary. The ovule-bearing part of the pistil.
ovate. Egg-shaped.
palea. Upper of the two enclosing bracts in the grass family.
palmate. Leaflets of a compound leaf borne on the apex of the petiole.
panicle. A branching raceme.
papilionaceous. Butterfly-shaped.
pappus. In the sunflower family, the highly modified calyx composed of scales, bristles, awns, or a short crown at the tip of the achene.
parasitic. Growing on and deriving nourishment from another living plant.
parietal. Describing placentae in which the ovules are borne on the wall of the ovary.
parted. Cleft almost to the base.
pedicel. Stalk of a single flower.
peduncle. Flowerstalk of a single flower or of a cluster of flowers.
perennial. Plant with the potential to live more than two years.
perfect flower. A flower with both stamens and pistils.
perianth. The corolla and calyx considered collectively.
perigynium. The sac that encloses the ovary and fruit in *Carex* and *Kobresia*.
petal. A member of the whorl of floral organs, just interior to the sepals and below the stamens.
petaloid. Brightly colored and petal-like.
petiole. Leafstalk.
phyllaries. Bracts under the flowering head.

pilose. Hairy; with soft slender hairs.
pinna. A primary division of a fern frond; pl. pinnea.
pinnate. Having a main central axis with secondary branches or units arranged in two lines on either side of the central axis.
pinnatifid. Pinnately cleft or parted.
pinole. The flour from various seeds and grains mixed together.
pistil. The organ formed from the combination of the stigma, style, and ovary.
pistillate. Having pistils and no stamens.
placenta. Place of attachment of the seeds in the ovary; pl. placentae.
plumose. Feathery and soft.
pod. Any kind of dry, dehiscent fruit, particularly in the pea family (Fabaceae).
pollen. Dust-like microspores produced in the anther.
pollinium. A mass of waxy or coherent pollen grains (notably in the Asclepiadaceae and Orchidaceae).
poly-. A prefix meaning many.
potherb. An herb that is boiled and eaten as a vegetable.
prostrate. Lying flat on the ground.
puberulent. Minutely pubescent.
pubescent. Covered with hairs.
punctate. With translucent or colored dots or depressions.
raceme. Inflorescence with one main axis and subequal primary branches each bearing one flower.
rachis. The axis of a spike.
ray flower. In the sunflower family (Asteraceae), the strap-like flowers attached to the disk.
receptacle. The end of a flowerstalk that bears the floral organs.
reflexed. Bent backward.
regular. Having members of each part alike in size and shape.
rhizomatous. Possessing rhizomes.
rhizome. A creeping underground, usually horizontally oriented stem.
root. The underground part of a plant.
rootstalk. Underground, creeping stem.
rosette. A dense, usually basal cluster of leaves radiating in all directions from the stem.
rosulate. In the form of a rosette.
saccate. Having a sac or the form of a sac.
sagittate. Arrowhead-shaped.
salverform. Having a long slender tube that abruptly flares into a circular limb.
samara. A winged fruit that does not split at maturity.
saprophyte. A plant with little or no chlorophyll that obtains nutrients from dead organic matter by a root association with a fungus.
scabrous. Rough to the touch.
scape. A leafless peduncle arising from the ground.
scarious. Thin, dry, and membranous.
schizocarp. A dry, indehiscent fruit that splits into one-seeded segments at maturity.
seed. A mature ovule that after germination gives rise to a new plant.
seedbank. Reserves of viable seeds present on the surface and in the soil, including new seeds recently shed by a plant and older seeds that have persisted in the soil for several years.
seed cone. The female seed-producing cone of conifers.
sepal. One of a whorl of typically green or greenish, leaf-like, floral organs originating below the petals.
septum. A partition that separates the locules of an ovary.
serrate. Having sharply toothed edges.
sessile. Lacking a stalk.

sheath. The tubular structure surrounding an organ or part.
shoot. A young stem or branch.
shrub. A woody plant, sometimes only at the base, and generally with several stems originating from the base.
silicle. A dry, dehiscent fruit of the mustard family, typically less than twice as long as wide.
silique. A dry, dehiscent fruit of the mustard family, typically more than twice as long as wide.
simple. In one piece.
sinus. An indentation between lobes of a leaf or corolla.
sorus. The fruit dot of a fern; pl. sori.
spathe. A bract or pair of bracts, often large and colored, subtending or enclosing a spadix (fleshy flower spike) or flower cluster.
spatulate. Spatula-shaped; having a long, narrow base and a widened, roundish tip.
spike. Like a raceme except the flowers are sessile.
spikelet. One of the ultimate clusters in the inflorescence of grasses.
sporangium. The spore case of a fern; pl. sporangia.
spore. A single cell or a small group of undifferentiated cells, each capable of producing a plant.
sporocarp. A receptacle containing sporangia or spores.
sporophyll. A spore-bearing leaf.
spur. A hollow, slender, sac-like extension of some part of the flower (e.g., sepal in *Delphinium* or petal in *Viola*).
stamen. The male organ of a flower, which produces pollen. It is composed of an anther and a filament.
staminate. Having stamens and no pistil.
standard. The upper petal of a papilionaceous (butterfly-shaped) flower.
stellate. Star-shaped.
stem. The main axis of a plant.
stigma. The part of the pistil that receives the pollen.
stipe. The stalk of a pistil or other small organ.
stipitate. Furnished with a stipe.
stipules. Appendages on each side of the base of certain leaves.
stolon. Commonly referred to as a runner, an aerial shoot from a plant with the ability to produce adventitious roots and new offshoots of the same plant.
strobilus. An aggregation of sporophylls resembling a cone, as in the spikemosses; pl. strobili.
style. The more or less elongated part of the pistil between the ovary and the stigma.
stylopodium. The enlargement at the base of the style in the Apiaceae.
subalpine. Occurring in the mountains below the alpine zone and above the montane zone.
subcordate. Somewhat heart-shaped.
subtending. Occurring below.
subulate. Awl-shaped.
succulent. Thick, fleshy, and juicy.
superior. Above.
superior ovary. An ovary positioned above the base of the other flower parts.
talus. A slope of rock rubble, usually at a cliff base.
taproot. The primary plant root, considerably larger than any other branches of the root system.
tendril. A thread-shaped organ used for climbing.
tepal. The perianth part when the perianth is not clearly differentiated into calyx or corolla.
terete. Round in cross section.
ternate. In threes.

terrestrial. Growing on ground, not aquatic.
three-ranked. Originating in threes from a common point or level.
throat. Usually the dilated upper part of the tube of the corolla or calyx.
tomentose. Having matted, woolly hairs.
torulose. Having a cylindrical body that is swollen and constricted at intervals.
tree. A large woody plant with a single main stem or trunk.
tricarpellate. Having three carpels.
trifoliate. Three-leaved.
tuber. A swollen underground stem tip (e.g., potato).
tubular. In the form of a tube or cylinder.
umbel. A flower arrangement resembling an umbrella.
urn-shaped. Ovoid and with a small opening at the tip.
valve. One of the pieces into which a dehiscent pod splits.
vascular. Pertaining to specialized tissue that allows food and water to move through the plant.
vascular plant. A plant having vascular tissue.
vegetative. Pertaining to the portion of the plant not producing reproductive structures such as cones or flowers.
vernation. The arrangement of leaves in the bud.
villous. Sticky.
weed. An aggressive plant that colonizes disturbed habitats and cultivated lands; a plant out of place.
whorled. Having three or more similar structures (e.g., leaves, petals, bracts) encircling a node.
woolly. Having soft, curled, or entangled hairs.

MEDICAL TERMS

acrid. Sharp, irritating, or biting to the taste.
alkaloid. A nitrogen-containing, slightly alkaline, often poisonous substance.
alterative. Gradually restoring the normal functions of the body.
analgesic. Relieving pain.
anaphrodisiac. Reducing sexual desire or potency.
anesthetic. Producing anesthesia.
anodyne. A substance that helps quiet or relieve pain.
anthelmintic. Destroying or eliminating parasitic worms.
antibiotic. Destroying pathogenic action of microbes.
anti-inflammatory. Reducing or neutralizing inflammation.
antimicrobial. Inhibiting the growth or multiplication of microorganisms or killing them.
antipyretic. Reducing fever.
antiscorbutic. Effective against scurvy.
antiseptic. Preventing infection.
antispasmodic. Relieving or curing spasms or irregular and painful action of the muscles (as in epilepsy).
antisyphilitic. Relieving or curing venereal disease.
antiviral. Effective against viruses.
astringent. Causing tissue to shrink or bind.
bitters. Sharp, acrid, or biting medicines, prescribed to stimulate an appetite.
cardiac. Having an effect on the heart.
carminative. Dispelling flatulency or griping pains of the stomach and bowels.
cathartic. Stimulating the action of the bowels, purgative.
decoction. The essence of a plant extracted by boiling it down.
demulcent. Having a soothing or emollient effect on inflamed surfaces.
depurative. Blood purifier.

dermatitis. Inflammation of the skin.
diaphoretic. Promoting or increasing perspiration.
diuretic. Increasing the flow of urine by acting on the kidneys.
emetic. An agent that causes vomiting.
emollient. Having a soothing and softening effect on the body tissues.
expectorant. A substance that causes an increase in expectoration, promoting the excretion of mucous from the chest.
febrifuge. A substance that helps reduce or control fevers.
glycoside. Any of the numerous acetal derivatives of sugars that yield a sugar on hydrolysis.
hydrocarbon. An organic compound containing only hydrogen and carbon and often occurring in petroleum, natural gas, and coal.
infusion. A preparation made by soaking a plant in hot water ("tea").
insecticidal. Effective against insects.
irritant. An agent that causes inflammation or abnormal sensitivity in living tissue.
laxative. A substance used to loosen the bowels and relieve constipation.
liniment. A liquid or semiliquid preparation of an herb used to relieve skin irritation and muscle pain.
mucilaginous. Resembling or containing mucilage, slimy.
narcotic. A substance that diminishes the action of the nervous and vascular systems, causing drowsiness, lethargy, stupor, and insensibility.
nervine. A substance that soothes and calms the nerves, restoring them to a natural state.
nitrate. A salt or ester of nitric acid.
nutritive. A substance that nourishes the body, promoting growth or health.
panacea. A cure-all.
parch. To toast or scorch with heat.
photosensitivity. A sensitivity to light.
poultice. A moist, usually warm or hot mass of plant material applied to the skin, or with a cloth between the skin and the plant material, to effect a medicinal action.
purgative. A substance used to evacuate the bowels; more forceful than a laxative.
salve. A healing ointment.
saponin. A glycoside in plants that when shaken with water has a foaming or soapy action.
sedative. A substance used to lessen nervous excitement, irritation, and pain.
selenium. A nonmetallic element that resembles sulfur chemically.
stimulant. A substance that produces energy.
styptic. Controlling bleeding by contracting the tissues or blood vessels.
sudorific. Producing profuse and visible sweating when taken hot.
tincture. A diluted alcohol solution of plant parts.
tonic. A substance that invigorates or stimulates, producing a feeling of well-being or strength.
topical. Applied locally.
vasoconstrictor. An agent that causes the blood vessels to constrict.
vasodilator. An agent that causes the blood vessels to dilate.
vermifuge. A substance used to expel worms or other parasites from the body.
volatile. Readily vaporizes at low temperatures.
vulnerary. Used to promote the healing of wounds.

References

Alexander, R. R. 1985. *Major Habitat Types, Community Types, and Plant Communities in the Rocky Mountains.* U.S. Forest Service General Technical Report RM-123. Fort Collins, Colorado.

Allen, E. B., and D. H. Knight. 1984. The Effects of Introduced Annuals on Secondary Succession in Sagebrush-Grassland. *Wyoming Southwestern Naturalist* 29:407–421.

Altschul, S. 1973. *Drugs and Food from Little-Known Plants.* Harvard University Press, Cambridge, Massachusetts.

Anderson, J. P. 1939. Plants Used by the Eskimo of the Northern Bering Sea and Arctic Regions of Alaska. *American Journal of Botany* 26(9):714–716.

Angier, B. 1966. *Free for the Eating.* Stackpole Books, Harrisburg, Pennsylvania.

———. 1969a. *More Free for the Eating Wild Foods.* Stackpole Books, Harrisburg, Pennsylvania.

———. 1969b. *Feasting Free on Wild Edibles.* Stackpole Books, Harrisburg, Pennsylvania.

———. 1974. *Field Guide to Edible Wild Plants.* Stackpole Books, Harrisburg, Pennsylvania.

———. 1978. *Field Guide to Medicinal Wild Plants.* Stackpole Books, Harrisburg, Pennsylvania.

Arnason, T., R. J. Hebda, and T. Johns. 1981. Use of Plants for Food and Medicine by Native Peoples of Eastern Canada. *Canadian Journal of Botany* 59(11): 2189–2325.

Ashton, R. J., and R. D. Walmsley. 1976. The Aquatic Fern *Azolla* And Its *Anabaena* Symbiont. *Endeavour* 35:39–43.

Avataq Cultural Institute. 1984. *Traditional Medicine Project.* Quebec.

Avery, A. G., S. Satina, and J. Rietsema 1959. *Blakeslee: The Genus* Datura. Ronald Press, New York.

Bacon, A. E. 1903. An Experiment with the Fruit of Red Baneberry. *Rhodora* 5:77–79.

Bailey, F. L. 1940. Navajo Foods and Cooking Methods. *American Anthropologist* 42(2):270–290.

Baker, M. A. 1981. The Ethnobotany of the Yurok, Tolowa, and Karok Indians of Northwest California. Unpublished Master's thesis, Humboldt State University, Arcata, California.

Balls, E. K. 1970. *Early Uses of California Plants.* University of California Press, Berkeley.

Bank, T. P, II. 1951. Botanical and Ethnobotanical Studies in the Aleutian Islands. I. Aleutian Vegetation and Aleut Culture. *Botanical and Ethnobotanical Studies Papers* 37:13–30. Michigan Academy of Science, Arts, and Letters.

Barrett, S. A. 1908. Pomo Indian Basketry. *University of California Publications in American Archaeology and Ethnology* 7:134–308.

——. 1917. The Washoe Indians. *Bulletin of the Public Museum of the City of Milwaukee* 2(1):1–52.

——. 1952. *Material Aspects of Pomo Culture.* Bulletin of the Public Museum of the City of Milwaukee, No. 20. Milwaukee.

Barrett, S. A., and E. W. Gifford. 1933. *Miwok Material Culture.* Yosemite Natural History Association, Yosemite National Park, California.

Barrows, D. P. 1967. *The Ethno-Botany of the Coahuilla Indians of Southern California.* Malki Museum Press, Banning, California. Originally published 1900.

Bartlett, K. 1943. Edible Wild Plants of Northern Arizona. *Plateau* 16(1):11–17. Northern Arizona Society of Science and Art, Museum of Northern Arizona, Flagstaff.

Baumhoff, M. A. 1963. Ecological Determinants of Aboriginal California Populations. *University of California Publications in American Archaeology and Ethnology* 49(2):155–236.

Bean, L. J., and K. S. Saubel. 1972. *Temalpakh-Cahuilla Indian Knowledge and Usage of Plants.* Malki Museum Press, Morongo Indian Reservation, California.

Bomhard, M. L. 1936. Leaf Venation as a Means of Distinguishing *Cicuta* from *Angelica. Journal of the Washington Academy of Sciences* 26(3):102–107.

Boxer, A. 1974. *Nature's Harvest.* Regnery, Chicago.

Boyden, T. C. 1982. The Pollination Biology of *Calypso bulbosa* var. *americana* (Orchidaceae): Initial Deception of Bumblebee Visitors. *Oecologia* 55:178–184.

Brill, S. 1994. *Identifying and Harvesting Edible and Medicinal Plants in Wild (and Not so Wild) Places.* Hearst Books, New York.

Brown, T. 1985. *Tom Brown's Guide to Wild Edible and Medicinal Plants.* Berkeley Books, New York.

Bryan, N. G., and S. Young. 1940. *Navajo Dyes: Their Preparation and Use.* Education Division Publication, U.S. Office of Indian Affairs, Washington, D.C.

Burgess, R. L. 1966. Utilization of Desert Plants by Native People. *Bulletin of the American Association for the Advancement of Science* 8:6–21.

Burt, P. 2000. *Barrenland Beauties: Showy Plants of the Canadian Arctic.* Northern Publishers, Yellowknife, Northwest Territories, Canada.

Callegari, J., and K. Durand. 1977. *Wild Edible and Medicinal Plants of California.* El Cerrito, California.

Camazine, S., and R. A. Bye. 1980. A Study of the Medical Ethnobotany of the Zuni Indians of New Mexico. *Journal of Ethnopharmacology* 2:365–388.

Castetter, E. F. 1935. *Uncultivated Native Plants Used as Sources of Food.* Ethnobiological Studies in the American Southwest. University of New Mexico Bulletin 266, Biological Series, Vol. 4, No. 1. University of New Mexico Press, Albuquerque.

Chamberlain, L. S. 1901. Plants Used by the Indians of Eastern North America. *American Naturalist* 35:1–10.

Chamberlain, Ralph V. 1911. The Ethno-Botany of the Gosiute Indians of Utah. *Memoirs of the American Anthropological Association* 2(5):331–405.

Chatfield, K. 1997. *Medicine from the Mountains: Medicinal Plants of the Sierra Nevada.* Range of Light Press, South Lake Tahoe, California.

Chestnut, V. K. 1902. Plants Used by the Indians of Mendocino County, California. *Contributions to the U.S. National Herbarium* 7:295–408.

Clarke, C. B. 1977. *Edible and Useful Plants of California.* University of California Press, Berkeley.

Classen, P. W. 1919. A Possible New Source of Food Supply (Cat-tail Flour). *Scientific Monthly* 9:179–185.

Coffey, T. 1993. *The History and Folklore of North American Wildflowers.* Facts on File, New York.

Coon, N. 1974. *The Dictionary of Useful Plants.* Rodale Press, Emmaus, Pennsylvania.

Correll, D. S. 1950. *Native Orchids of North America North of Mexico.* Stanford University Press, Stanford, California.

Coulter, J. M. 1885. *Manual of the Botany (Phaenogamia and Pteridophyta) of the Rocky Mountain Region, from New Mexico to the British Boundary.* American Book Co., New York.

Coulter, J. M., and A. Nelson. 1909. *New Manual of Botany of the Central Rocky Mountains (Vascular Plants).* American Book Co., New York.

Coville, F. V. 1897. *Notes on the Plants Used by the Klammath Indians of Oregon.* Contributions to the U.S. National Herbarium 5(2). Washington, D.C.

———. 1904. Desert Plants as a Source of Drinking Water. *Smithsonian Institution Annual Report,* 499–505.

Craighead, J. J., F. C. Craighead, and R. J. Davis. 1963. *A Field Guide to the Rocky Mountain Wildflowers.* Houghton Mifflin, Boston.

Cronquist, A., A. H. Holmgren, N. H. Holmgren, and J. L. Reveal. 1972. *Intermountain Flora: Vascular Plants of the Intermountain West, U.S.A.* Vol. 1. Hafner, New York.

Culley, D. D., and E. A. Epps. 1973. Use of Duckweed for Waste Treatment and Animal Feed. *Journal of Water Pollution Control Federation* 45:337–347.

Culpeper, N. 1972. *English Physician and Complete Herbal.* Arranged for Use as a First Aid Herbal by C. F. Leyel. Wilshire Book Co., North Hollywood, California.

Curtin, L. S. M. 1957. Some Plants Used by the Yuki Indians… I. Historical Review and Medicinal Plants. *Masterkey* 31:40–48.

Dall, W. H. 1868. Useful Indigenous Alaskan Plants. *Report of the Department of Agriculture,* 172–189. Washington, D.C.

Darlington, W. 1859. *American Weeds and Useful Plants.* A. O. Moore, New York.

Davidson, J. 1919. Douglas-fir Sugar. *Canadian Field Naturalist* 33(1):6–9.

Dawson, R. 1985. *Nature Bound.* Omnigraphics, Boise, Idaho.

Densmore, F. 1974. *How Indians Use Wild Plants for Food, Medicine, and Crafts.* Dover, New York.

Despain, D. G. 1975. *Field Key to the Flora of Yellowstone National Park.* Yellowstone Library and Museum Association, Yellowstone National Park, Wyoming.

———. 1999. *Yellowstone Vegetation.* Roberts Rinehart, Niwot, Colorado; Yellowstone Association, Yellowstone National Park, Wyoming.

Dixon, Royal. 1917. *The Human Side of Trees: Wonders of the Tree World.* Frederick A. Stokes, New York.

Doebley, J. F. 1984. "Seeds" of Wild Grasses: A Major Food of Southwestern Indians. *Economic Botany* 38(1):52–64.

Dorn, R. D. 1977. *Manual of the Vascular Plants of Wyoming.* 2 vols. Garland Publishing, New York.

———. 1988. *Vascular Plants of Wyoming.* Mountain West Publishing, Cheyenne, Wyoming.

———. 1992. *Vascular Plants of Wyoming.* 2nd ed. Mountain West Publishing, Cheyenne, Wyoming.

Douglas, J. S. 1978. *Alternative Foods: A World Guide to Lesser-Known Edible Plants.* Pelham Books, London.

Duke, J. A. 1992. *Handbook of Medicinal Plants.* CRC Press, Boca Raton, Florida.

———. 1992. *Handbook of Phytochemical Constituents of Grass, Herbs, and Other Economic Plants.* CRC Press, Boca Raton, Florida.

Dunmire, W. W., and G. D. Tierney. 1997. *Wild Plants and Native Peoples of the Four Corners.* Museum of New Mexico Press, Santa Fe.

Ebeling, W. 1986. *Handbook of Indian Foods and Fibers of Arid America.* University of California Press, Berkeley.

Elias, T. S., and P. A. Dykeman. 1982. *A Field Guide to North American Edible Wild Plants.* Outdoor Life Books, New York.

Elliott, D. B. 1976. *Roots: An Underground Botany and Forager's Guide.* Chatham Press, Old Greenwich, Connecticut.

Ellis, C. 1941. Wild Vegetables of the Desert Indians. *Primitive Man* 3:9–10.

Elmore, F. H. 1944. *Ethnobotany of the Navajo.* School of American Research, Santa Fe, New Mexico.

Erichsen-Brown, C. 1979. *Medicinal and Other Uses of North American Plants.* Dover, New York.

Evert, E. F. 1984a. A New Species of *Antennaria* (Asteraceae) from Montana and Wyoming. *Madrono* 31:109–112.

———. 1984b. *Penstemon absarokensis,* a New Species of Scrophulariaceae from Wyoming. *Madrono* 31:140–143.

Evert, E. F., and L. Constance. 1982. *Shoshonea pulvinata,* a New Genus and Species of Umbelliferae from Wyoming. *Systematic Botany* 7:471–475.

Evert, E. F., and R. L. Hartman. 1984. Additions to the Vascular Flora of Wyoming. *Great Basin Naturalist* 44:482, 483.

Farris, G. 1980. A Re-assessment of the Nutritional Value of *Pinus monophylla. Journal of California and Great Basin Anthropology* 2(1):132–136.

Fernald, M. L., and A. C. Kinsey. 1958. *Edible Wild Plants of Eastern North America.* Revised by R. C. Rollins. Harper and Row, New York.

Fertig, W. 1992a. *Checklist of the Vascular Plant Flora of the West Slope of the Wind River Range and a Status Report on the Sensitive Plant Species of Bridger-Teton National Forest.* Report to Bridger-Teton National Forest. Prepared by the Rocky Mountain Herbarium, University of Wyoming.

———. 1992b. *Checklist of the Vascular Plant Flora of the West Slope of the Wind River Range and Status Report on Sensitive Plant Species Occurring in the Rock Springs District, Bureau of Land Management.* Report to Rock Springs District, BLM. Prepared by the Rocky Mountain Herbarium, University of Wyoming.

———. 1999. A Potpourri of Weeds. *Castilleja* 18(2):3–10.

Fisher, Vardis. 1965. *Mountain Man.* Simon and Schuster, New York.

Flora of North America (FNA) Editorial Committee. 1992. *Flora of North America: Guide for Contributors.* New York Botanical Garden, New York.

———. 1993. *Flora of North America. Vol. 2: Pteridophytes and Gymnosperms.* Oxford University Press, New York.

Foster, S., and J. A. Duke. 1990. *Eastern/Central Medicinal Plants.* Houghton Mifflin, New York.

Fowler, C. S. 1989. *Willard Z. Park's Ethnographic Notes on the Northern Paiute of Western Nevada, 1933–1940.* University of Utah Press, Salt Lake City.

Frankton, C., and G. A. Mulligan. 1987. *Weeds of Canada.* N. C. Press and Agricultural Canada, Ottawa, Ontario.

Frye, T. C. 1934. *Ferns of the Northwest.* Metropolitan Press, Portland, Oregon.

Gail, F. W. 1916. *Some Poisonous Plants of Idaho.* Agricultural Experiment Station Bulletin 86. University of Idaho, Moscow.

Gibbons, E. 1962. *Stalking the Wild Asparagus.* David McKay, New York.

———. 1966. *Stalking the Healthful Herbs.* David McKay, New York.

———. 1971. *Stalking the Good Life.* David McKay, New York.

Gifford, E. W. 1967. Ethnographic Notes on the Southwestern Pomo. *Anthropological Records* 25:10–15.

Goodrich, J., and C. Lawson. 1980. *Kashaya Pomo Plants.* American Indian Studies Center, University of California, Los Angeles.

Gottesfeld, L. M. J. 1992. The Importance of Bark Products in the Aboriginal Economies of Northwestern British Columbia, Canada. *Economic Botany* 46(2): 148–157.

Grant, A. L. 1924. *A Monograph of the Genus* Mimulus. Ph.D. dissertation, Washington University. Publications of Washington University Series V, St. Louis.

Gresswell, R. E. 1984. The Ecological Profile as a Monitoring Tool for Lakes in Yellow-

stone National Park. Report on file, Yellowstone National Park Research Office.
Grillos, S. J. 1966. *Ferns and Fern Allies of California*. University of California Press, Berkeley.
Gunther, E. 1973. *Ethnobotany of Western Washington*. Rev. ed. University of Washington Press, Seattle.
Haines, J. 1988. A Flora of the Wind River Basin and Adjacent Areas, Fremont, Natrona, and Carbon Counties. Unpublished Master's thesis, University of Wyoming, Laramie.
Hall, A. 1976. *The Wild Food Trail Guide*. Holt, Rinehart, and Winston, New York.
Hamel, P. B., and M. U. Chiltoskey. 1975. *Cherokee Plants and Their Uses: A Four Hundred Year History*. Herald Publishing, Sylva, North Carolina.
Hardin, J. W., and J. M. Arena. 1974. *Human Poisoning from Native and Cultivated Plants*. Duke University Press, Durham, North Carolina.
Harrington, H. D. 1967. *Edible Native Plants of the Rocky Mountains*. University of New Mexico Press, Albuquerque.
Hart, J. A. 1981. The Ethnobotany of the Northern Cheyenne Indians of Montana. *Journal of Ethnopharmacology* 4:1–55.
———. 1996. *Montana: Native Plants and Early Peoples*. Montana Historical Society, Helena.
Hartman, H. 1984. Ecology of Gall-Forming Lepidoptera on *Tetradymia*. I. Gall Size and Shape. *Hilgardia* 52:1–16.
Hartman, R. L. 1990. The Flora of the Rocky Mountains Project. *American Journal of Botany* 77(6):134, 135.
Hartman, R. L., and R. W. Lichvar. 1980. Additions to the Vascular Flora of Teton County, Wyoming. *Great Basin Naturalist* 40:408–413.
Harvard, V. 1895. The Food Plants of North American Indians. *Bulletin of the Torrey Botanical Club* 22:98–123.
Haskin, L. L. 1929. A Frontier Food, Ipo or Yampa, Sustained the Pioneers. *Nature Magazine* 14:171–172.
Haughton, C. S. 1978. *Green Immigrants*. Harcourt Brace Jovanovich, New York.
Hedges, K. 1986. *Santa Ysabel Ethnobotany*. San Diego Museum of Man Ethnic Technology Notes No. 20. San Diego.
Hellar, C. A. 1958. *Wild, Edible, and Poisonous Plants of Alaska*. University of Alaska Extension Bulletin No. 40. Fairbanks.
———. 1966. *Wild, Edible, and Poisonous Plants of Alaska*. Division of Statewide Services, Cooperative Extension Service, University of Alaska, Fairbanks.
Hellson, J. C. 1974. *Ethnobotany of the Blackfoot Indians*. National Museums of Canada, Ottawa, Ontario.
Herrick, J. W. 1977. Iroquois Medical Botany. Unpublished Ph.D. dissertation, State University of New York, Albany.
Hill, A. F. 1937. *Economic Botany: A Textbook of Useful Plants and Plant Products*. McGraw-Hill, New York.
Hinton, L. 1975. Notes on La Huerta Diegueno Ethnobotany. *Journal of California Anthropology* 2:214–222.
Hitchcock, C. L., A. Cronquist, M. Ownbey, and J. W. Thompson. 1955. *Vascular Plants of the Pacific Northwest*. Part 5. University of Washington Publications in Biology No. 17. Seattle.
———. 1959. *Vascular Plants of the Pacific Northwest*. Part 4. University of Washington Publications in Biology No. 17. Seattle.
———. 1961. *Vascular Plants of the Pacific Northwest*. Part 3. University of Washington Publications in Biology No. 17. Seattle.
———. 1964. *Vascular Plants of the Pacific Northwest*. Part 2. University of Washington Publications in Biology No. 17. Seattle.
———. 1969. *Vascular Plants of the Pacific Northwest*. Part 1. University of Washington Publications in Biology No. 17. Seattle.

Hocking, G. M. 1949. From Pokeroot to Penicillin. *Rocky Mountain Druggist*, November 1949, pp. 12, 38.

Holloway, P., and G. Alexander. 1990. Ethnobotany of the Fort Yukon Region, Alaska. *Economic Botany* 44:214–225.

Holt, Catharine. 1946. Shasta Ethnography. *Anthropological Records* 3(4):308.

Hough, W. 1897. The Hopi in Relation to Their Plant Environment. *American Anthropologist* 10(2):33–44.

Hussey, P. B. 1939. *A Taxonomic List of Some Plants of Economic Importance*. Science Press, Lancaster, Pennsylvania.

Jacobson, C. A. 1915. *Water Hemlock* (Cicuta). Nevada Agricultural Experiment Station, Technical Bulletin No. 81. Reno.

Jencks, Z. 1919. A Note on the Carbohydrates of the Root of the Cattail (*Typha latifolia*). *Proceedings of the Society for Experimental Biology and Medicine* 17(2):45–46.

Johnson, A. E. 1974. Predisposing Influence of Range Plants on *Tetradymia*-Related Photosensitization in Sheep: Work of Drs. A. B. Clawson and W. T. Hoffman. *American Journal of Veterinary Research* 35(12):1583–1585.

Johnston, A. 1970. Blackfoot Indian Utilization of the Flora of the Northwestern Great Plains. *Economic Botany* 24:301–324.

Kartesz, J. 1994. *A Synonymized Checklist of the Vascular Flora of the United States, Canada, and Greenland*, vols. 1 and 2. Timber Press, Portland, Oregon.

Kavash, B. 1979. *Native Harvests: Recipes and Botanicals of the American Indian*. Random House, New York.

Kelly, I. 1932. Ethnography of the Surprise Valley Paiute. *University of California Publications in American Archaeology and Ethnology* 31(3):67–210.

Kephart, H. 1909. Edible Plants of the Wilderness. In *The Book of Camping and Woodcraft*. Century, New York.

Kindscher, K. 1987. *Edible Wild Plants of the Prairie*. University Press of Kansas, Lawrence.

Kinghorn, D. 1979. *Toxic Plants*. Columbia University Press, New York.

Kingsbury, J. M. 1964. *Poisonous Plants of the United States and Canada*. Prentice-Hall, Englewood Cliffs, New Jersey.

———. 1965. *Deadly Harvest: A Guide to Common Poisonous Plants*. Holt, Rinehart, and Winston, New York.

Kirk, D. R. 1975. *Wild Edible Plants of the Western United States*. Naturegraph, Heraldsburg, California.

Kirkpatrick, R. S. 1987. A Flora of the Southeastern Absarokas, Wyoming. Unpublished Master's thesis, University of Wyoming, Laramie.

Knap, A. H. 1975. *Wild Harvest: An Outdoorsman's Guide to Edible Wild Plants in North America*. Pagurian Press, Toronto.

Knight, D. H. 1994. *Mountains and Plains: Ecology of Wyoming Landscapes*. Yale University Press, New Haven.

Krochmal, A., and C. Krochmal. 1973. *A Guide to the Medicinal Plants of the United States*. Quadrangle, New York.

Krochmal, A., S. Paur, and P. Duisberg. 1951. Useful Native Plants in the American Deserts. *Economic Botany* 8(1):3–20.

Lackschewitz, K. 1991. *Vascular Plants of West-Central Montana—Identification Guidebook*. General Technical Report INT-277. U.S. Department of Agriculture, Forest Service, Intermountain Research Station, Ogden, Utah.

Lands, M. 1959. Folk Medicine and Hygiene. *Anthropological Papers of the University of Alaska* 8:1–75.

Lee, D. 1989. *Exploring Nature's Uncultivated Garden*. Havelin Communications, Takoma Park, Maryland.

Le Strange, R. 1977. *A History of Herbal Plants*. Arco, New York.

Lewis, W. H., and M. Elvin-Lewis. 1977. *Medical Botany*. John Wiley and Sons, New York.

Lichvar, R. W. 1979a. The Flora of the Gros Ventre Mountains. Unpublished Master's thesis, University of Wyoming, Laramie.

Life-Support Technology, Inc. 1963. *Foods in the Wilderness.* Boise, Idaho.

Linn, J. G., E. J. Sraba, R. D. Goodrich, J. C. Meiske, and D. E. Otterby. 1975. Nutritive Value of Dried and Ensiled Aquatic Plants. I. Chemical Composition. *Journal of Animal Science* 41:601–609.

Lust, J. B. 1987. *The Herb Book.* 20th ed. Bantam Books, New York.

Mabey, R. 1977. *Plantcraft: A Guide to Everyday Use of Wild Plants.* Universe Books, New York.

Mahar, J. M. 1953. Ethnobotany of the Oregon Paiutes of the Warm Springs Indian Reservation. Unpublished Bachelor's thesis, Reed College, Portland, Oregon.

Markow, S. 1992. Preliminary Report on a General Floristic Survey of Vascular Plants of the Targhee National Forest. Report to Teton National Forest. University of Wyoming, Laramie.

Marriott, H. 1985. Flora of the Northwestern Black Hills, Crook and Weston Counties, Wyoming. Master's thesis, University of Wyoming, Laramie.

Martin, L. C. 1984. *Wildflower Folklore.* Globe Pequot Press, Chester, Connecticut.

McArthur, E. D., A. C. Blauer, and S. C. Sanderson. 1988. Mule Deer–Induced Mortality of Mountain Big Sagebrush. *Journal of Range Management* 41(2):114–117.

McHarg, I. L. 1969. *Design with Nature.* Doubleday, Garden City, New York.

Medsger, O. P. 1974. *Edible Wild Plants.* Collier-Macmillan, New York.

Merriam, C. H. 1918. The Acorn, a Possibly Neglected Source of Food. *National Geographic Magazine* 34(2):129–137.

——. 1966. *Ethnographic Notes on California Indian Tribes.* University of California Archaeological Research Facility, Berkeley.

Merrill, R. E. 1923. Plants Used in Basketry by the California Indians. *UC-PAAE* 20: 215–242.

Meuninck, J. 1988. *The Basic Essentials of Edible Wild Plants and Useful Herbs.* ICS Books, Merrillville, Indiana.

Miller, J. A. 1973. Naturally Occurring Substances That Can Induce Tumors. In *Toxicants Occurring Naturally in Foods.* National Academy of Sciences, Washington, D.C.

Millspaugh, C. F. 1974. *American Medicinal Plants: An Illustrated and Descriptive Guide to Plants Indigenous to and Naturalized in the United States Which Are Used in Medicine.* Dover, New York.

Mitchell, R. S., and J. K. Dean. 1982. *Ranunculaceae (Crowfoot Family) of New York State.* Bulletin No. 446. New York State Museum, New York.

Moerman, D. E. 1977. *American Medical Ethnobotany: A Reference Dictionary.* Garland, New York.

——. 1986. *Medicinal Plants of the Native Americans.* University of Michigan Museum of Anthropology Technical Report No. 19. Ann Arbor.

Moore, M. 1979. *Medicinal Plants of the Mountain West.* Museum of New Mexico Press, Santa Fe.

——. 1989. *Medicinal Plants of the Desert and Canyon West.* Museum of New Mexico Press, Santa Fe.

Morton, J. 1963. Principal Wild Food Plants of the United States, Excluding Alaska and Hawaii. *Economic Botany* 17:319–330.

——. 1975. Cattails (*Typha* spp.)—Weed Problem or Potential Crop? *Economic Botany* 29(1):7–29.

Mosquin, T. 1970. The Reproductive Biology of *Calypso bulbosa* (Orchidaceae). *Canadian Field Naturalist* 84:291–296.

Muenscher, W. C. 1962. *Poisonous Plants of the United States.* Macmillan, New York.

Muir, John. 1988. *My First Summer in the Sierra.* Sierra Club Books, San Francisco. Originally published 1911.

Murphey, E. V. A. 1990. *Indian Uses of Native Plants.* Meyerbooks, Glenwood, Illinois.

National Park Service. 1986. *Exotic Vegetation Management Plan*. Yellowstone National Park, Wyoming.

National Research Council. 1985. *Amaranth: Modern Prospects for an Ancient Crop*. Rodale Press, Emmaus, Pennsylvania.

Nelson, R. A. 1992. *Handbook of Rocky Mountain Plants*. 4th ed. Roberts Rinehart, Niwot, Colorado.

Nequakewa, E. 1943. Some Hopi Recipes for the Preparation of Wild Plant Foods. *Plateau* 18:18–20.

Newberry, J. S. 1887. Food and Fiber Plants of the North American Indians. *Popular Scientific Monthly* 32:31–46.

Norton, C. 1942. Would You Starve? *Nature Magazine* 35(6):295–297.

Norton, H. 1981. The Association Between Anthropogenic Prairies and Important Food Plants in Western Washington. *Northwest Anthropological Research Notes* 13:175–200.

Olsen, L. D. 1990. *Outdoor Survival Skills*. Brigham University Press, Provo, Utah.

Oswalt, W. H. 1957. A Western Eskimo Ethnobotany. *Anthropological Papers of the University of Alaska* 6:17–36.

Palmer, E. 1878. Plants Used by the Indians of the United States. *American Naturalist* 12:593–606 (Sept.) and 646–655 (Oct.).

Perry, E. 1952. Ethno-Botany of the Indians in the Interior of British Columbia. *Museum and Art Notes* 2(2):36–43.

Peterson, L. A. 1978. *A Field Guide to Edible Wild Plants of Eastern and Central North America*. Houghton Mifflin, New York.

Pfeiffer, N. E. 1922. Monograph of the Isoetaceae. *Annals of the Missouri Botanical Gardens* 9:79–232.

Pojar, J., and A. MacKinnon, eds. 1994. *Plants of the Pacific Northwest Coast*. Lone Pine Publishing, Auburn, Washington.

Porsild, A. E. Edible Plants of the Arctic. *Arctic* 6(1):15–34.

Powers, S. 1874. Aboriginal Botany. *Proceedings of the California Academy of Science* 5:373–379.

Price, L. W. 1981. *Mountains and Man*. University of California Press, Berkeley.

Reagan, A. B. 1929. Plants Used by the White Mountain Apache Indians of Arizona. *Wisconsin Archeologist* 8:143–161.

———. 1934. Various Uses of Plants by West Coast Indians. *Washington Historical Quarterly* 25:133–137.

Risk, P. 1983. *Outdoor Safety and Survival*. John Wiley, New York.

Ritchie, G. A., ed. 1979. *New Agricultural Crops*. Westview Press, Boulder, Colorado.

Rogers, D. J. 1980. *Lakota Names and Traditional Uses of Native Plants by Sicangu (Brule) People in the Rosebud Area, South Dakota*. Rosebud Educational Society, St. Francis, South Dakota.

Romero, J. B. 1954. *The Botanical Lore of the California Indians*. Vantage Press, New York.

Rydberg, P. A. 1900. *Catalogue of the Flora of Montana and the Yellowstone National Park*. Memoirs of the New York Botanical Garden, Vol. 1. New York.

———. 1917. *Flora of the Rocky Mountains and Adjacent Plains: Colorado, Utah, Wyoming, Idaho, Montana, Saskatchewan, Alberta, and Neighboring Parts of Nebraska, South Dakota, North Dakota, and British Columbia*. Published by the author.

Saunders, C. F. 1976. *Edible and Useful Wild Plants of the United States and Canada*. Dover, New York.

Sawyer, J. O., and T. Keeler-Wolf. 1995. *A Manual of California Vegetation*. California Native Plant Society, Sacramento.

Schery, R. W. 1972. *Plants for Man*. 2nd ed. Prentice Hall, Englewood Cliffs, New Jersey.

Schofield, J. J. 1989. *Discovering Wild Plants: Alaska, Western Canada, the Northwest*. Alaska Northwest Books, Seattle.

Scott, R. W. 1966. The Alpine Flora of Northwestern Wyoming. Unpublished Master's thesis, University of Wyoming, Laramie.

Scully, V. 1970. *A Treasury of American Indian Herbs: Their Lore and Their Use for Food, Drugs, and Medicine.* Crown, New York.

Shaw, R. J. 1976. *Field Guide to the Vascular Plants of Grand Teton National Park and Teton County, Wyoming.* Utah State University Press, Logan.

Smith, C. E. 1973. *Man and His Foods: Studies in Ethnobotany and Nutrition—Contemporary, Primitive, and Prehistoric Non-European Diets.* University of Alabama Press, Tuscaloosa.

Smith, H. H. 1923. *Ethnobotany of the Menomini Indians.* Bulletin of the Public Museum of the City of Milwaukee, vol. 4, no. 1. Milwaukee. Reprinted 1978.

Snow, C. R. 1935. Vegetables of the Alaska Wilderness. *Alaska Sportsman* 1(4):6–8.

Snow, N. 1989. Floristics of the Headwaters Region of the Yellowstone River (Wyoming). Unpublished Master's thesis, University of Wyoming, Laramie.

———. 1990. Phytogeographical Affinities of the Absaroka Range (Wyoming-Montana). *Proceedings of the Southwestern and Rocky Mountain Division,* American Association for the Advancement of Science, Colorado Springs, Colorado, *Program and Abstracts,* p. 30.

Snow, N., B. E. Nelson, and R. L. Hartman. 1990. Additions to the Vascular Flora of Yellowstone National Park, Wyoming. *Madrono* 37:214–216.

Sparkman, P. S. 1908. The Culture of the Luiseño Indians. *University of California Publications in American Archaeology and Ethnology* 8(4):187–234.

Spellenberg, Richard 1979. *The Audubon Society Field Guide to North American Wildflowers.* Alfred A. Knopf, New York.

Spier, L. 1930. Klamath Ethnography. *University of California Publications in American Archaeology and Ethnology* 30:1–338.

Steward, J. H. 1933. Ethnography of the Owens Valley Paiute. *University of California Publications in American Archaeology and Ethnology* 33(3):233–350.

Stewart, Kenneth M. 1965. Mohave Indian Gathering of Wild Plants. *Kiva* 31(l):46–53.

Stone, D. E. 1959. A Unique Balanced Breeding System in the Vernal Pool Mouse-Tails. *Evolution* 13:151–174.

Strike, S. S. 1994. *Ethnobotany of the California Indians, vol. 2. Aboriginal Uses of California's Indigenous Plants.* Koeltz Scientific Books, Champaign, Illinois.

Stuart, J. D., and J. O. Sawyer. 2001. *Trees and Shrubs of California.* California Natural History Guides No. 62. University of California Press, Berkeley.

Swartz, B. K., Jr. 1958. A Study of Material Aspects of Northeastern Maidu Basketry. *Kroeber Anthropological Society Publications* 19:67–84.

Sweet, M. 1976. *Common Edible and Useful Plants of the West.* Naturegraph, Healdsburg, California.

Taylor, S. J. Elk. 1994. *Eat the Weeds at Your Feet: An Edible Plant Guide of Sonoma County.* Rose of Sharon Press, Citrus Heights, California.

Teit, J. A. 1930. *Ethnobotany of the Thompson Indians of British Columbia.* U.S. Government Printing Office, Washington, D.C.

Terrell, E. E. 1977. *A Checklist of Names for 3,000 Vascular Plants of Economic Importance.* U.S. Department of Agriculture Handbook No. 505. Washington, D.C.

Thompson, S., and M. Thompson. 1972. *Wild Plant Foods of the Sierra.* Dragtooth Press, Berkeley, California.

———. 1977. *Huckleberry Country: Wild Food Plants of the Pacific Northwest.* Wilderness Press, Berkeley, California.

Tilford, G. L. 1993. *The EcoHerbalist's Fieldbook: Wildcrafting in the Mountain West.* Mountain Weed Publishing, Conner, Montana.

———. 1997. *Edible and Medicinal Plants of the West.* Mountain Press Publishing, Missoula, Montana.

Train, P., J. R. Henriches, and W. A. Archer. 1957. *Medicinal Uses of Plants by Indian Tribes of Nevada.* Quarterman Publications, Lawrence, Massachusetts.

Truax, R. E., D. D. Culley, M. Griffith, W. A. Johnson, and J. P. Wood. 1972. Duckweed for Chick Feed. *Louisiana Agriculture* 16(1):8–9.

Tull, D. 1987. *A Practical Guide to Edible and Useful Plants.* Texas Monthly Press, Austin, Texas.

Turner, N., And H. V. Kuhnlein. 1991. *Traditional Plant Foods of Canadian Indigenous Peoples.* Gordon and Breach Science, Philadelphia.

Turner, N. J., and A. F. Szczawinski. 1991. *Common Poisonous Plants and Mushrooms of North America.* Timber Press, Portland, Oregon.

Turney-High, H. 1933. Cooking Camass and Bitterroot. *Scientific Monthly* 36:262–263.

Tweedy, F. 1886. *Flora of the Yellowstone National Park.* Washington, D.C.

Tyler, V. E. 1987. *The Honest Herbal.* George F. Stickley, Philadelphia.

Underhill, J. E. 1974. *Wild Berries of the Pacific Northwest.* Superior Publishing, Seattle.

Uphof, J. C. T. 1959. *Dictionary of Economic Plants.* H. R. Engleman, New York.

Usher, G. 1976. *A Dictionary of Plants Used by Man.* Constable, London.

Van Etten, C. H., R. W. Miller, I. A. Wolff, and Q. Jones. 1963. Amino Acid Composition of Seeds from Two Hundred Angiosperm Plant Species. *Journal of Agriculture and Food Chemistry* 11(5):399–410.

Vestal, P.A. 1952. *Ethnobotany of the Ramah Navaho.* Papers of the Peabody Museum of American Archaeology and Ethnology, Vol. 40, No. 4. Harvard University, Cambridge.

Vizgirdas, R. 1999a. The Fallacy of Plant Edibility Tests. *Wilderness Way Magazine* 5(2).

———. 1999b. Fireweed. *Wilderness Way Magazine* 4(4).

———. 1999c. Courting the Conifers: The Pines. *Wilderness Way Magazine* 4(1).

———. 2000a. The Mustard Family. *Wilderness Way Magazine* 6(1).

———. 2000b. Butterflies and Edible Plants. *Wilderness Way Magazine* 5(3).

———. 2003a. *Useful Plants of Idaho.* Idaho State University Press, Pocatello.

———. 2003b. Useful Plants of the Southern California Mountains. *San Bernardino County Museum Association Quarterly* 50(2). Redlands, California.

Vizgirdas, R. S., and E. M. Rey-Vizgirdas. 2006a. *Wild Plants of the Sierra Nevada.* University of Nevada Press, Reno.

———. 2006b. *Discovering Sawtooth's Butterflies.* Idaho State University Press, Pocatello.

Walker, M. 1984. *Harvesting the Northern Wild: A Guide to Traditional and Contemporary Uses of Edible Forest Plants of the Northwest Territories.* Yellowknife, Northwest Territories, Canada.

Weber, W. A. 1987. *Colorado Flora, Western Slope.* University of Colorado Press, Niwot.

———. 1990. *Colorado Flora, Eastern Slope.* University of Colorado Press, Niwot.

———. 2001. *Colorado Flora, Western Slope.* 3rd ed. University of Colorado Press, Niwot.

Webster, J. 1980. *Fungi.* 2nd ed. Cambridge University Press, Cambridge.

Weedon, N. F. 1996. *A Sierra Nevada Flora.* 4th ed. Wilderness Press, Berkeley, California.

Weiner, M. A. 1972. *Earth Medicine—Earth Food: Plant Remedies, Drugs, and Natural Foods of the North American Indians.* Macmillan, New York.

Welsh, S. L., N. D. Atwood, S. Goodrich, and L. C. Higgins. 1993. *A Utah flora.* 2nd ed., revised. Print Service, Brigham Young University, Provo, Utah.

Wherry, E. T. 1942. Go Slow on Eating Fern Fiddleheads. *American Fern Journal* 32(3):108–109.

Whipple, J. J. 1999. The Yellowstone Sand Verbena. *Castilleja* 18(4):1–3.

———. 2001. Annotated Checklist of Exotic Vascular Plants in Yellowstone National Park. *Western North American Naturalist* 61(3):336–346.

Whiting, A. F. 1939. *Ethnobotany of the Hopi.* Museum of Northern Arizona, Bulletin No. 15. Northern Arizona Society of Science and Art, Flagstaff.

Whittlesey, R. 1985. *Familiar Friends: Northwest Plants.* Rose Press, Portland, Oregon.

Wilford, W. R., J. P. Harrington, and B. Freire-Marreco. 1916. *Ethnobotany of the Tewa Indians.* Smithsonian Institution, Bureau of American Ethnology, Bulletin No. 55. U.S. Government Printing Office, Washington, D.C.

Willard, T. 1992. *Edible and Medicinal Plants of the Rocky Mountains and Neighboring Territories.* Wild Rose College of Natural Healing, Calgary, Alberta, Canada.

Williams, R. L. 1984. *Aven Nelson of Wyoming.* Colorado Associated University Press, Boulder.

Wyman, L. C., and S. K. Harris. 1941. *Navajo Indian Medical Ethnobotany.* University of New Mexico Press, Albuquerque.

Zigmond, M. L. 1981. *Kawaiisu Ethnobotany.* University of Utah Press, Salt Lake City.

Zwinger, A. H., and B. E. Willard. 1972. *Land above the Trees: A Guide to American Alpine Tundra.* Harper and Row, New York.

Index

Page numbers in *italics* refer to figures, maps, and tables.